TECHNICAL COMMUNICATION

Principles and Practice

Fourth Edition

Meenakshi Raman
Former Professor and Head
Department of Humanities and Social Sciences
BITS, Pilani (K.K. Birla Goa Campus)

Sangeeta Sharma
Professor
Department of Humanities and Social Sciences
BITS, Pilani

OXFORD
UNIVERSITY PRESS

OXFORD
UNIVERSITY PRESS

Oxford University Press is a department of the University of Oxford.
It furthers the University's objective of excellence in research, scholarship,
and education by publishing worldwide. Oxford is a registered trade mark of
Oxford University Press in the UK and in certain other countries.

Published in India by
Oxford University Press
22 Workspace, 2nd Floor, 1/22 Asaf Ali Road, New Delhi 110002

First Edition published 2004
Fourth Edition published 2022
Third impression 2024

ISBN-13: 978-93-5497-225-6
ISBN-10: 93-5497-225-X

eISBN-13 (eBook): 978-93-5497-487-8
eISBN-10 (eBook): 93-5497-487-2

Typeset in Adobe Garamond Pro and Benguiat Gothic Std

by B2K-BYTES 2 KNOWLEDGE, Tamil Nadu

Printed in India by Gopsons Papers Pvt. Ltd, Noida -201301

Cover image: © fizkes/Shutterstock

For product information and current price, please visit www.india.oup.com

Preface to the Fourth Edition

'You can have brilliant ideas, but if you can't get them across, your ideas won't get you anywhere.'

– Lee Iacocca

Effective communication is the key to your personal and professional success. You, as students or professionals, are expected to get across your concepts, thoughts, and ideas—whether they are technical or non-technical—clearly, concisely, and effectively to your peers or superiors to justify/assert your views on various matters. However, please remember that acquiring, developing, and sustaining effective communication skills cannot be achieved in one day! In fact, it not only requires considerable knowledge and awareness of various forms and documents of technical communication but also needs adequate practice in preparing and presenting them effectively.

Though today's technology-driven world has enabled you to communicate across countries and cultures effortlessly and quickly, you may be aware that the need for the four basic language skills, namely listening, speaking, reading, and writing (LSRW) remain unchanged. While you may use the Internet acronyms such as LOL, IMY, ASAP, etc., in your personal emails and messages, you cannot use them in your academic or professional documents and presentations. For instance, you may have to write formal letters, technical reports, proposals, research papers, thesis, etc., and deliver professional presentations. Hence, it's important for you to be well-versed with the principles and practices of oral and written technical communication which may help you in your seminars, group discussions, job searches, and interviews. Such an understanding coupled with your technical knowledge would also enable you to perform various assignments such as participating in discussions, meetings, developing profiles and websites, etc., with ease and elan.

About the Book

The fourth edition of *Technical Communication: Principles and Practice* has been brought out keeping in mind the ever-increasing demand for effective speaking and writing in the academia and workplaces and the expectations of the employers from graduating students. All the four language skills (LSRW) have been given adequate coverage to make your communication more effective and appropriate to the contexts. Extensive discussions on the most used technical documents such as reports, proposals, letters, research papers, thesis, etc. would make you more confident in preparing and presenting them. In short, you may find this book a useful manual to your communication requirements.

New to This Edition

- Important and relevant topics as per current scenario: Technology in Communication along with Ethics and Etiquettes in Communication
- Some other new topics to match syllabi coverage: Paralinguistic Features, Mother Tongue Influence, Negotiation Skills, Journal Writing, Overcoming Barriers to Communication and more
- Additional sample interview questions (HR, Skills testing, Behavioural types) and types of and sample résumés for readers to prepare for job interviews

- Additional exercises on vocabulary, pronunciation, and grammar for practice
- Discussions supplemented with videos of professional learnings—Time Management, Negotiation Skills, Online Meetings to enhance the employability skills of readers
- New topics: Communication Network, Mother Tongue Influence, Paralinguistic Features, Public Speaking, PQRST Technique

Online Resources

The new edition includes augmented Online Resources for both students and faculty.

For Faculty

- PowerPoint slides
- Instructor manual
- Assessment exercises: Multiple-choice questions (Grammar related and more)/Quizzes
- Web links

For Students

- Additional readings: Effective Listening, Effective Presentation Strategies, Listening & Speaking, Paragraphs & Essays, Phonetics, Prepositions, Referencing & Styling
- Sample résumés, reports, letters, etc.
- Flashcard glossaries
- Author audio podcasts on tips for better etiquettes, presentation, tone, pronunciation
- Videos and PowerPoint presentations on group discussions and interviews
- Listening and speaking practice exercises on telephonic conversation, phonetics, and more
- Web links for interview tips, technical writing tips, online technical writing, copyright basics, style manuals, and ethics
- Modules on Communication, Time Management, Negotiation Skills, Online Meetings

Meenakshi Raman
Sangeeta Sharma

The publisher and the authors would like to thank the following reviewers for their valuable feedback:
1. Dr Itishri Sarangi, Associate Professor of English, KIIT-DU, Bhubaneshwar
2. Prof. Banani Das, Asst. Professor, Dept. of English, The Assam Royal Global University, Guwahati
3. Dr Bonala Kondal, Assistant Professor, SR University, Warangal
4. Dr Mahesh B. Shinde, Assistant Professor, KIT's College of Engineering (Autonomous), Kolhapur
5. Prof. Manashi Devi, Assistant Professor, Girijananda Chowdhury Institute of Management Technology, Assam Science and Technology University, Guwahati
6. Dr Nira Konar, HOD Humanities and Associate Prof. of English College of Engineering and Management, Kolaghat
7. Prof. Sathyendra Bhat J., Assistant Professor, St Joseph Engineering College, Vamanjoor, Mangaluru
8. Dr Sunanda Mahesh Shinde, Assistant Professor, Sanjay Ghodawat University

Acknowledgements

We sincerely acknowledge the support extended to us by Professor Souvik Bhattacharya, Vice-Chancellor, Birla Institute of Technology and Science (BITS) Pilani and Professor Sudhir Kumar Barai, Director, BITS Pilani Campus. They had encouraged us throughout the journey of writing this fourth edition of the book. Having served as an Adjunct Professor at the Department of HSS at IISER Bhopal, Meenakshi Raman wishes to thank the Director of the institution and her colleagues there for providing the rich experience of interacting with Master's and PhD students.

We are immensely thankful to the editorial and production team at Oxford University Press India, for meticulously collating the material and pursuing us for timely publication of the book. We also appreciate the reviewers' constructive feedback. Their comments enabled us to revise the third edition, which would cater to the latest syllabi of various universities.

We are highly thankful to the end-users of this book, the students who had shown tremendous enthusiasm in providing feedback every time they used the book. This helped us incorporate the right kind of examples and illustrations.

We wish to thank our spouses and family members profusely for their unstinted support and cooperation extended during the course of this project.

We genuinely hope that this revised edition will be more beneficial as it is contemporary, updated, and aligned with students' requirements. We would be happy to receive any feedback for further improvement.

Meenakshi Raman

Sangeeta Sharma

Preface to the First Edition

Technical communication is one of those courses that are essential for all students and professionals. With the advent of new technologies, the world is shrinking into a global village. Therefore, the need for effective communication is becoming all the more essential. While students feel the need for communicating successfully through oral and written media in their academic tasks everyday, professionals face the challenges of communicating effectively and efficiently in their workplace.

The main aim of this book is to enable the reader to face the challenges in communication primarily in a technical milieu as communicating formal and technical messages is essential for both students and professionals. This has been accomplished by underscoring the relevance of both oral and written communication in myriad situations. A number of strategies have been provided to enhance the efficiency and effectiveness of communication in varied situations. We have made a sincere attempt to introduce certain techniques to improve confidence and effectiveness both for making presentations and for working in a team. The key strategies have been substantiated with examples from workplace situations. By the amalgamation of theory and practice in this book, we have tried to bring forward the real communication situations in our discussion. The contents, writing style, text organization, and format for various types of technical documents have been elaborately discussed. These discussions, we hope, would enable the reader to take up all writing tasks with ease and confidence.

The main strength of the book is its audience-centred approach, which we hope will make for an enjoyable reading. It directly addresses the students, giving real-life examples and orienting the strategies to practical application.

The contents of this book have been divided into four major parts. *Part I* explains to the readers the *basics of communication*, including the *barriers in communication* and the *use of technology*. *Part II* elaborates on *effective presentation strategies* as also the nuances of *interview* and *group discussion*. One has to have a firm grasp of the fundamentals of effective writing to achieve a commendable style in writing. Realizing this need, we have devoted *Part III* entirely to the *constituents of effective writing*. Finally, a variety of technical documents, namely letters, reports, proposals, research papers, and manuals, which are widely used in academic and professional environments, have been discussed at length in *Part IV* of this book.

The preponderance of communication in the academic and professional arena motivated us to take up this assignment of writing a book on technical communication. We hope that this book, with comprehensive coverage of all aspects and types of technical communication, will prove to be relevant and useful for the reader. Especially, the chapters on interviews, group discussions, effective presentation strategies, business letters, and reports will be of immense help to both students and professionals. This book can serve not only as a textbook for the technical communication course offered by universities, colleges, and polytechnic institutions but also as a reference book for training programmes offered by various businesses and industries.

We would like to extend our sincere gratitude to Prof. S. Venkateswaran, Vice Chancellor, BITS, Pilani, under whose patronage we were able to write this book. We are also indebted to our Director Prof. L.K. Maheshwari and our Deputy Directors, Prof. K.E. Raman, and Prof. V.S. Rao for their constant support and encouragement. Our sincere thanks to Prof. Ashok Sarkar, Dean, Faculty Division I and Instruction Division, and Prof. R.N. Saha, Dean, Educational Development Division and Faculty Division II, for constantly motivating us. With great pleasure, we acknowledge the compatible environment shared by our colleagues and faculty members of the Languages Group.

We are indebted to Prof. Rajiv Gupta, Dean, Educational Hardware Division, and Mr Anirudh Gautam, Deputy Chief Project Manager, Indian Railways, for providing us the samples of technical proposals. Our students Karkare Vaibhav Pradeep, S. Sarmistha, Jimit Arora, Vikram Sampath, J. Arun, Amit Goyal, and Gaurav Kapoor deserve our special thanks for providing the sample speech and sample reports.

At the home front, from our respective sides, we acknowledge the great support from our loved ones. Meenakshi Raman would like to thank the encouragement received from her husband Prof. K.E. Raman and children Prabhu and Priya, without which the book would not have been possible. Sangeeta Sharma acknowledges the constant motivation received from her husband Suresh and her children Betu and Reshu, who cheerfully bore with her absence.

We would like to thank Mr P.P. Mehta for word-processing the manuscript with speed and efficiency.

Meenakshi Raman

Sangeeta Sharma

CHAPTER 2

Barriers to Communication

OBJECTIVES

You should study the chapter to know
- ○ what communication barriers are and what causes them
- ○ how to avoid communication failures caused by noise
- ○ how to classify communication barriers

Introduction

Having studied the basic principles of technical communication in Chapter 1, we now move on to a very important factor related to communication. Imagine you are a junior executive who has just joined as a trainee in an automobile company. You go to meet your boss, to seek permission for visiting the automobile exhibition being held in Delhi. While you are talking, two of your colleagues also arrive to get your boss' signature on some bills. You fall silent while he signs the bills. After they leave, you resume talking but you find that your boss has not really grasped what you have said earlier. You later analyse this situation, and realize that your colleagues' intervention led to your boss' lack of concentration. As the sender, you had patiently tried to express your wish. But because of the interruption by your colleagues, the receiver, your boss, could not decode your message fully. Hence the communication process failed.

This discussion brings us to the introduction of the term 'barrier' in communication. A barrier is defined as something that prevents or controls progress or movement. This definition implies that a barrier is something that comes in the way of the desired outcome. In the example given above, notice that the arrival of your colleagues was an event that prevented your boss from concentrating completely on what you were saying. So, we say that this incident was a barrier to the communication between you and your boss. All of us have come across such situations while communicating with parents, friends, or colleagues. Let us now learn how to avoid such communication failures and make our interactions more effective.

Chapter Outline
All chapters in the book begin with a chapter outline that gives an overview of the contents covered in the chapter.

- Conversational flavour along with vocal inflection takes a back seat here, which is a great asset for a speaker.

Impromptu mode

The impromptu mode, as the word suggests, is what we use when we have to deliver an informal speech without preparation. For example, at a formal dinner party you may be invited to deliver a vote of thanks. Do not panic and babble something in an unmethodical way. Instead, calmly state your topic and then preview the points you are to make. Support your points with whatever examples, quotes, and anecdotes

Mr Naidu was called upon after dinner to give an impromptu speech.

Illustrations
Illustrations, interspersed with the text in the chapters, make the book a more lively and interesting read.

the Book

Exercises

Exercises

A series of practice exercises highlight the major topics covered in the chapter. The questions enhance learning and can be used for review and classroom discussion.

EXERCISES

1. Identify the communication barrier that describes each of the following situations:
   (a) 'Every time I have a meeting with Mr Gupta, I end up disagreeing with him about a particular issue.'
   (b) *Manager*: 'Reeta, where is the report that I asked you to submit on the financial matters of the company?'
   Reeta: 'I do not remember you asking me to submit a report.'
   (c) *Teacher*: 'Students, why have you not submitted the report within the fortnight?'
   Students: 'But you asked us to submit it bimonthly!'
   (d) 'This room is horrible to work in. I am able to hear everyone around, and there is no scope for privacy.'
   (e) 'It is quite tedious to manually work on the students' records, but I fear using the computer as it might corrupt all our data.'
   (f) 'Why every time I get a meeting with John, I usually end up showing my disinterest with him about a particular topic?'
   (g) 'If you want some more information from me, ask only the specific questions and do not waste my precious time!'

2. Identify and explain a communication barrier which may hinder each of the process components given in the schematic representation of the human communication process.

3. 'A barrier acts like a sieve, allowing only a part of the message to filter through; as a result, the desired response is not achieved.'
   Keeping in mind the above statement mention the various types of interpersonal barriers which hinder the communication process. Substantiate your answer with suitable examples.

4. Imagine you are the Sales Manager in Ramanath Paper and Pulp Company. Some of your co-employees are spreading rumours that you are involved in fraudulent activities. In order to defend yourself, you are asked to address the same co-employees. You are angry due to these rumours and you find it difficult to put your views before them. Identify the communication barrier that hinders this communication situation.

5. Explain the following terms with reference to communication barriers and give two examples for each term.

   (a) Organizational barrier
   (b) Emotional outburst
   (c) Cultural differences
   (d) Information overload
   (e) 'Know-it-all' attitude

6. You have been assigned to host a group of American university students who are visiting your institute for the next two weeks. What can you tell them that will help them fit into the culture on your campus? Make a list of the important behavioural rules they should understand in order to communicate effectively with students and faculty on your campus. Also point out some problems that might occur if the American students disregard these rules.

7. Identify the barriers that lead to miscommunication in the following scene. How can the manager overcome this barrier?

'I called for a meeting with the supervisors. But none turned up . . . nobody pays attention to me in this place . . . it would be better to quit'.

8. Write the possible solutions to overcome these barriers:
   (a) Dealing with people working in isolated office or environment.
   (b) Dealing with a customer who is very talkative
   (c) Working in a noisy surrounding
   (d) Dealing with a frustrated and angry client
   (e) Dealing with visual distractions at workplace

9. Give at least two situations for the barriers mentioned below:
   (a) Physical barrier
   (b) Negative tendency
   (c) Wrong inferences
   (d) Transfer station
   (e) Difference in background and language

Detailed Contents

Preface to the Fourth Edition   iii
Acknowledgements   v
Preface to the First Edition   vi
Features of the Book   viii

PART I: INTRODUCTION TO TECHNICAL COMMUNICATION

1. Basics of Technical Communication   2
Introduction   2
Importance of Technical Communication   3
General and Technical Communication   3
Objectives and Characteristics of Technical
  Communication   4
Process of Communication   4
  Communication Cycle   4
Communication Styles   6
  Style Matrix   7
Levels of Communication   7
  Extrapersonal Communication   7
  Intrapersonal Communication   7
  Interpersonal Communication   8
  Organizational Communication   8
  Mass Communication   8
Flow of Communication   9
  Vertical Communication   10
  Horizontal Communication   10
  Diagonal Communication   11
Communication Networks   11
  Formal Network Models   11
  Informal Network Models   12
Perspectives in Communication   14
Technology in Communication   15
Positive Impact of Technology-enabled
  Communication   16

Negative Impact of Technology-enabled
  Communication   16
  Overcoming Negative Impact   17
Selection of Appropriate Technology   17
Effectiveness in Technology-based
  Communication   19

2. Barriers to Communication   22
Introduction   22
  Noise   23
Classification of Barriers   24
  Intrapersonal Barriers   24
  Interpersonal Barriers   27
  Organizational Barriers   31
Overcoming Barriers   33

3. Non-verbal Communication   35
Introduction   35
Kinesics   36
Proxemics   40
Chronemics   41
Haptics   41
Correlating Verbal and Non-verbal
  Communication   42
Cross-cultural Variations   43
  Significance of Understanding Culture   44

PART II: LISTENING AND SPEAKING

4. **Active Listening**   48

Introduction 48
    Meaning and Art of Listening 48
    Importance of Listening and Empathy in
      Communication 49
Reasons for Poor Listening 49
    Listening versus Hearing 50
    Poor Listening Habits 50
Traits of a Good Listener 53
    Being Non-evaluative 53
    Paraphrasing 53
    Reflecting Implications 54
    Reflecting Hidden Feelings 54
    Inviting Further Contributions 55
    Responding Non-verbally 55
Listening Modes 55
    Active versus Passive Listening 56
    Global versus Local Listening 58
Types of Listening 58
Barriers to Effective Listening 60
Listening for General Content and Specific
  Information 63
    Listening and Note-taking 63
    Intensive Listening 65

5. **Effective Speaking**   68

Introduction 68
Basic Sounds of English 68
    Vowels and Consonants 69
    Phonetic Transcription 73
    Rules of Pronunciation 74
    Problem Sounds 75
Word Stress 76
    Weak Forms 77
Sentence Stress 78
    Sentence Rhythm 79
Intonation 80
    General Uses of Tone I and Tone II 81
Achieving Confidence, Clarity, and Fluency 83
    Confidence 84
    Clarity 85
    Fluency 86
    Developing Voice Quality 87
    Mother Tongue Influence (MTI) 87
    Sociolinguistic Competence 88
Paralinguistic Features 89
    Quality 90
    Volume 90

    Pace/Rate 90
    Pitch 91
    Articulation 91
    Pronunciation 91
    Voice Modulation 92
    Pauses 92
    Linguistic Features of Voice Control 93
    Public Speaking 94
    Drafting the Speech 95
    Practising Public Speaking 96
Speaking and Negotiation Skills 96

6. **Conversations and Dialogues**   100

Introduction 100
Conversations 101
    Types of Conversation 101
    Strategies for Effectiveness 102
    Conversation Practice 106
Argumentative Skills 107
Dialogue Writing 108
Types of Dialogues 109
    Situational Dialogues 110

7. **Formal Presentations**   114

Introduction 114
Planning 115
    Occasion 115
    Audience 116
    Purpose 117
    Thesis Statement 118
    Material 118
Outlining and Structuring 118
    Introduction 119
    Main Body 121
    Conclusion 124
Nuances of Delivery 125
    Modes of Delivery 125
    Guidelines for Effective Delivery 128
Controlling Nervousness and Stage Fright 132
    Strategies for Reducing Stage Fright 133
    Visualization Strategies 133
    On-camera Techniques 134
Visual Aids in Presentations 134
Expert Technical Lecture 136

8. **Interviews**   139

Introduction 139
Objectives of Interviews 139
Types of Interviews 140

Job Interviews   142
  Stages of Interview   143
  Face-to-face Interviews: Campus and On
    Site   143
  Telephonic Interviews   155
  Sample Interview Questions   158
Résumés   158
  Résumé, Biodata, and Curriculum Vitae   159
  Résumé Design and Structure   159
Media Interviews   161
Press Conferences   162
  Preparation   163
  Process   163

9. **Group Communication**   **166**
Introduction   166
Forms of Group Communication   167
Use of Body Language in Group
  Communication   169
Group Discussions   170
  Speaking in Group Discussions   171
  Discussing Problems and Solutions   171
  Creating a Cordial and Cooperative
    Atmosphere   172
  Using Persuasive Strategies   173
  Being Polite and Firm   173

Turn-taking Strategies   173
Effective Intervention   175
Reaching a Decision   175
Organizational GD   175
  Brainstorming   176
  Nominal Group Technique   176
  Delphi Technique   176
GD as Part of Selection Process   177
  Characteristics   177
  Evaluation and Analysis   178
  Approach to Topics and Case Studies   181
Meetings   183
  Purposes   184
  Preparation   184
  Procedure—Conducting Effective
    Meetings   189
Conferences   194
  Significance   195
  Planning and Preparation   196
  Procedure   197
Symposia and Seminars   197
Brochure   198
Bulletin   200
Newsletter   200

PART III: READING AND WRITING

10. **Reading Comprehension**   **206**
Introduction   206
Improving Comprehension Skills   206
Techniques for Good Comprehension   207
  Skimming and Scanning   207
  Non-verbal Signals   207
  Structure of the Text   209
  Structure of Paragraphs   209
  Punctuation   209
  Author's Viewpoint (Inference)   209
  Reader Anticipation: Determining the Meaning
    of Words   210
  Summarizing   210
  Typical Reading Comprehension
    Questions   211
Predicting the Content   213
Understanding the Gist   214
PQRST Technique   215
Study Skills   218
  Note-making   218

Critical Reading, Critical Thinking and
  Effective Googling   221
Understanding Discourse Coherence   224
Sequencing of Sentences   226

11. **Elements of Effective Writing**   **238**
Introduction   238
Right Words and Phrases   238
  Use Familiar Words   238
  Use Concrete and Specific Words   239
  Use Acronyms and Abbreviations
    Sparingly   240
  Avoid Clichés   242
  Avoid Excessive Use of Jargons   242
  Avoid Foreign Words and Phrases   243
  Avoid Redundancy and Circumlocution   243
  Avoid Discriminatory Writing   244
Sentences   244
  Sentence Patterns   245
  Salient Points of Sentence Construction   246

Journal Writing 253
Writing for the Web 254
Theme Clarity 255
Audience Awareness 255
Information Design and Development 255
Effective Style of Writing 256
Formatting 256
Proofreading 256
Collaborative Writing 256
Guidelines for Success 257
Creating Indexes 258

12. The Art of Condensation **264**
Introduction 264
Steps to Effective Précis Writing 265
Guidelines 265
Samples 267
Original Text 1 267
Original Text 2 268
Original Text 3 269

13. Technical Reports **273**
Introduction 273
Importance of Reports 274
Objectives of Reports 275
Characteristics of a Report 275
Categories of Reports 276
Informative Reports 277
Analytical Reports 277
Periodic and Special Reports 277
Event Report 277
Oral and Written Reports 281
Long and Short Reports 281
Formal and Informal Reports 282
Individual and Group Reports 282
Formats 282
Prewriting 284
Purpose and Scope 284
Audience 284
Sources of Information 288
Organizing the Material 289
Interpreting Information 289
Making an Outline 289
Structure of Reports (Manuscript Format) 292
Prefatory Parts 293
Main Text 300
Supplementary Parts 303
Types of Reports 304
Writing the Report 305
First Draft 305

Revising, Editing, and Proofreading 306

14. Technical Proposals **313**
Introduction 313
Definition and Purpose 313
Types 314
Characteristics 315
Structure of Proposals 316
Prefatory Parts 316
Body of the Proposal 317
Supplementary Parts 319
Style and Appearance 319
Evaluation of Proposals 320

15. Formal Letters, Memos, and Email **340**
Introduction 340
Formats of Written Correspondence 340
Types of Messages 341
Letter Writing 342
The Seven Cs of Letter Writing 343
Significance 343
Purpose 343
Structure 344
Layout 349
Principles 353
Planning a Letter 358
Business Letters 359
Credit Letters 359
Collection Letters 360
Letters of Enquiry 366
Order Placement Letters 368
Claim Letters 368
Sales Letters 372
Instruction Letters 377
Cover Letters 378
Writing the Cover Letter 379
Academic and Business Cover Letters 379
Cover Letters Accompanying Résumés 379
Memos 382
Classification and Purpose 384
Structure and Layout 386
Style 390
Emails 393
Advantages and Limitations 393
Style, Structure, and Content 394
Email Etiquette 396
Effectiveness and Security 400

16. Research Papers, Thesis, and Technical Descriptions 408
Introduction 408

Research Paper 409
  Characteristics 409
  Components 409
Thesis 415
  Outline 416
  Organization 416
  Timetable 417
  Iteration 417

Style 418
Presentation 418
Structure 419
Technical Description 423
  *Guidelines for Writing Good
    Descriptions 423*
  Checklist 424
  Writing Technical Descriptions 425

PART IV: REVIEW OF GRAMMAR

**17. Grammar and Vocabulary
    Development** **430**
Introduction 430
A Brief History of Words 430
Using the Dictionary and Thesaurus 434
Changing Words from One Form to
  Another 435
Word Formation: Prefixes and Suffixes 435
Synonyms and Antonyms 435
  Synonyms 436
  Antonyms 437
Idioms 437
Confusables 438
One-word Substitutes 438
Homonyms 442
Homophones 442
Eponyms 442
Phrasal Verbs 443
  Phrasal Verb Patterns 444
Nouns 446
  Compound Nouns 446
  Noun Phrases 447
Gerunds 449
  Uses of Gerunds 450
Infinitives 450
  Uses of Infinitives 451

Subject–verb Agreement 452
Tenses 454
  Present Tense 455
  Past Tense 457
  Future Tense 458
Active and Passive Voice 460
Conditional Sentences 462
Adjectives and Degrees of Comparison 465
  Types of Adjectives 466
  Comparison of Adjectives 467
Adverbs 469
  *Adverbs and Adjectives with the Same
    Form 469*
  Adverbs with Two Forms 469
  Adverb Forms 470
  Adverbs as Intensifiers 470
  Kinds of Adverb 470
Conjuctions 471
  Conjunction Classes 472
  *Conjunctions Used in Adverbial Phrases
    and Clauses 473*
Prepositions 474
Articles 475
  Indefinite Articles 475
  Definite Article 476
  Omission of Articles 476

Index 484
About the Authors 488

PART I

Introduction to Technical Communication

Chapter 1: Basics of Technical Communication

Chapter 2: Barriers to Communication

Chapter 3: Non-verbal Communication

CHAPTER
1

Basics of Technical Communication

OBJECTIVES

You should study the chapter to know

- the importance of technical communication
- how general-purpose communication is different from technical communication
- the objectives and characteristics of technical communication
- the constituents of the communication process
- the different levels of communication
- how communication flows in an organization

Introduction

In the academic environment, we encounter various situations involving speech or writing: conversation with friends, professors, or colleagues to achieve various purposes; seminars, group discussions, written tests, and examinations; and laboratory or project report submissions on diverse topics. Likewise, at the workplace, we interact with superiors and subordinates, converse with them face-to-face or over the telephone, and read and write emails, letters, reports, and proposals.

All these activities have a common denominator—*the sharing of information*. For example, when you request your professor to explain a concept you could not understand very well in class, you transmit the information to him/her that you need some clarification. Now, the professor receives this information, understands it, and responds by giving an explanation which clears your doubt. If you are satisfied with this explanation, you thank the professor and the communication comes to an end. If you are still in doubt, you once again request clarification, and the process continues. This process involving the transmission and interchange of ideas, facts, feelings, or courses of action is known as the *process of communication*. We give, get, or share information with others during this process. Whether the communication is oral or written, this process essentially remains the same.

When one becomes a part of any organization, one needs to communicate, and communicate effectively. No organization can survive without communication. All the activities an organization undertakes have communication at their hub. The better our communication skills, the greater are our chances of quick progress. However skilful one may be in other aspects such as work, knowledge, thoughts, and organization, without proper communication, those are of little use. For instance, though you may have an excellent academic record, you may not be successful in an interview if you are not able to express your ideas clearly to those on the other side of the table. It has been observed that people who are successful in their careers generally have excellent communication skills, which is one of the very reasons for their success.

Most of this is technical *communication*, so let us find out more about what role technical communication plays in an organization.

Importance of Technical Communication

Technical communication plays a pivotal role in any set-up, whether it is a business enterprise, an industry as a whole, or an academic institution. All managerial or administrative activities involve communication, be it planning, organizing, recruiting, coordinating, or decision-making. When you write reports, give instructions, or read brochures and manuals, you are involved in the process of communication. Communication serves as an instrument to measure the success or growth of an organization. For example, papers published by R&D organizations bring to light their progress. When the chief executive officer (CEO) of an organization presents his/her company's achievements in a meeting, each of the participants comes to know of these milestones. The higher one's position is, the greater is their need to communicate. A labourer, for example, may not be as involved in formal communication as a top-level executive. The various types of communication not only help an organization to grow, but also enable the communicators to develop the required skills.

However, though most professionals are well aware of the importance of communication, they do not develop their skills to good effect in their sphere of work. The more people participate in the communication process, the better they develop their skills in collecting and organizing information, analysing and evaluating facts, appreciating the difference between facts and inferences, and communicating effectively. To become an effective communicator, one needs to communicate, communicate, and communicate. There is no other way out.

General and Technical Communication

Communication is important not only in an organization but also in one's daily life. It is an integral part of daily activity. When an alarm clock goes off, it is communication through sound, urging one to get out of bed. When one feels loyal towards a particular brand of toothpaste, it is possible that the television (TV) commercials for that brand have been successful in communicating the message. Watching news on TV, saying goodbye to one's family, or calling a cab and giving directions are all different types of communication. At the workplace, all activities revolve around oral or written communication. Interacting with one's boss, reading the newspaper at home, or even dreaming in one's sleep are all examples of communication.

Messages that are non-technical or informal in nature are categorized as general-purpose communication, whereas messages pertaining to technical, industrial, or business matters belong to the category of technical or business communication. Table 1.1 shows the differences between the two categories.

Communication in everyday life

Table 1.1 Differences between general and technical communication

General communication	Technical communication
• Contains a general message	• Contains a technical message
• Informal in style and approach	• Mostly formal
• No set pattern of communication	• Follows a set pattern
• Mostly oral	• Both oral and written
• Not always for a specific audience	• Always for a specific audience
• Does not involve the use of technical vocabulary or graphics, etc.	• Frequently involves jargon, graphics, etc.

Objectives and Characteristics of Technical Communication

Technical communication takes place when professionals discuss a topic with a specific purpose with a well-defined audience. Technical communication usually has the following objectives:
- To provide organized information that aids in quick decision-making
- To invite corporate joint ventures
- To disseminate knowledge in oral or written form

Let us take an example of a customer who has bought a washing machine and does not know how to use it. The customer reads the instructions in the user manual and gradually learns to operate the washing machine without any assistance. This is an example of successful technical communication. When you are confused about which camera to buy, the salesperson explains all the technical features of each model to you. If that helps in your buying decision, it is successful technical communication again.

Technical communication has to be correct, accurate, clear, appropriate, and to the point. Correct information is objective information. The language should be clear and easy to understand. If the communication is through a user manual for a phone, remember that people will usually never use it unless they are stuck. And if they are stuck, they will look for instant information to solve their problem. The information must be brief and arranged sequentially so that it is easy for a user to find relevant information. It is also vital that the technical information provided in the manual be accurate.

Process of Communication

For sharing information, two parties are required—the sender and the receiver—without whom communication, which is an interactive process, cannot take place. At any given time, one is active and the other is passive. However, this is not sufficient; there should also be cooperation and understanding between them. Through what they have to communicate, the sender and receiver mutually influence each other. They should have a mutually accepted code of signals making up a common language. So, communication can be defined as the exchange of information, ideas,

> Communication can be defined as the exchange of information, ideas, and knowledge between a sender and a receiver through an accepted code of symbols.

and knowledge between a sender and a receiver through an accepted code of symbols. It is termed effective only when the receiver receives the message intended by the sender in the same perspective. Otherwise it becomes miscommunication.

Communication Cycle

Consider the communication process shown in Figure 1.1. The communication cycle involves various elements, as discussed in the following paragraphs.

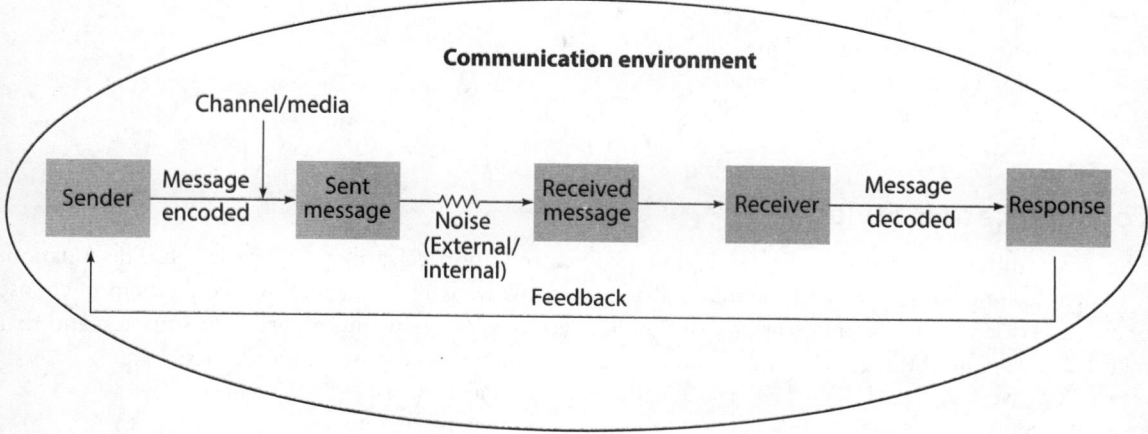

Figure 1.1 The process of communication

The first step is *formulation*, wherein the sender forms the content of the message to be sent. This formulation depends on the level of experience, intelligence, knowledge, and purpose of the sender. The content, once formed, is called the *message*. The sender *encodes* the message using a basic *tool*. This tool is nothing but the language used—words, actions, signs, objects, or a combination of these. Once encoded using proper language, the message is ready to be delivered. This delivery happens through *channels* or media of communication. It can be face-to-face, on paper, or through electronic or digital media such as the Internet. The *receiver* receives the message, *decodes* it, and acts on it. If the message received is the same as the message sent, there will be an appropriate response; if not, there will still be a response, but probably an inappropriate or unexpected one, as there has been a breakdown or interference in the communication. This may happen because of *noise*, which has been discussed in detail in Chapter 2. *Noise affects the decoding part of the communication process.*

> 'How well we communicate is determined not by how well we say things, but how well we are understood.'
> –Andrew Grove

The transmission of the *receiver's response* to the sender is called *feedback*. Feedback is essential, as it measures the effectiveness of communication. When a message is sent, the communication cycle is complete only when there is a response from the recipient of the message. Otherwise, the message needs to be re-sent. When a response is received, the message has been successfully delivered to the other party. For example, you put up a notice asking the members of your student council to attend a meeting on a specified date, at a particular time, at the location mentioned. On the day of the meeting, you find that some of them have come while the others have not turned up. In this case you have obtained both positive and negative responses to your message. However, since you have got some feedback, the communication process is complete. You also know that it has been effective, at least from your side. Hence, to know whether the communication has been successful or not, there must be some feedback, which is nothing but an observation of the recipient's response. *The communication is fully effective only when there is a desired response from the receiver.*

Effective communication takes place in a well-defined set-up. This is called the *communication environment*. A classroom is the communication environment when a teacher delivers lectures to students. If such a communication is attempted without a proper environment, it will not have the desired effect. Similarly, a teacher's cubicle becomes the communication environment when a student privately approaches the teacher. Thus, the essentials of *effective communication* are as follows:

- A well-defined communication environment
- Cooperation between the sender and the receiver
- Selection of an appropriate channel
- Correct encoding and decoding of the message
- Feedback

Communication Styles

There exist different communication styles and people need not necessarily possess just one. The situation, who people are talking with, and the number of people involved are the factors that guide them to choose one style. For effective communication, we should adapt the style that suits a particular situation and that may vary throughout the day. The most common are the following communication styles:

- Assertive
- Aggressive
- Passive
- Passive-aggressive

These styles are dependent on the type of message, that is, analytical, intuitive, functional, and personal.

Assertive communication style   This is the most efficient communication style where you can share your thoughts considering others' opinion as well. Assertive speakers tend to be more confident and positive. Listening and involving others' opinion is the key feature of this communication style. The speaker using this communication style never tries to ridicule the listener, rather maintains his/her dignity. Assertive communicators tend to resort to personal pronoun rather than directly using incisive remarks. For example, rather than saying 'You have to postpone the meeting', assertive communicators will say, 'I have potential conflict at the suggested time, therefore can we meet at another time?'

By adopting assertive style, one can manage any situation amicably.

Aggressive communication style   This is quite opposite to the previous style as an aggressive communicator thrusts his/her opinion on others as there is a strong desire to be heard by others. An aggressive communicator considers himself/herself to be the focal point of conversation; however, the content of the conversation is neglected because of the imposing nature. With this communication style, listeners might feel demeaned and frightened.

Team members often react negatively to aggressive communication style without absorbing the content of the message. They are put off by the jarring message content and they naturally go against it.

Passive communication style   This communication style is adopted by people pleasers. Such communicators tend to be submissive and avoid conflict. They want to keep the communication environment light and pleasant.

The drawback of this communication style is that in long durations, a passive communicator fails to steer the team and the members lack clarity in absence of direction. That is why with passage of time, the assertive or aggressive communicator will take the leadership role. Passive communicators never want to take up confrontation. That is why they don't put their ideas across strongly. Sometimes highly valuable ideas get lost in their politeness.

Passive-aggressive style   Passive-aggressive style entails the aspects of both passive and aggressive styles of communication. Such communicators appear easy-going but they have a strong desire to shift decisions towards their thought processes. Superficially the communicator appears to be relaxed and soft but he/she is a strong decision maker inside.

People around remain confused as they cannot anticipate the next move. The restive behaviour become known to all through grapevines. People are not comfortable to work with this style of communicators. This style is quite suffocative in the work environment leading to a lot of dissatisfaction.

Style Matrix

People adopt different styles as per the requirement of the situation. One should be flexible to adjust with a change in environment. See Figure 1.2.

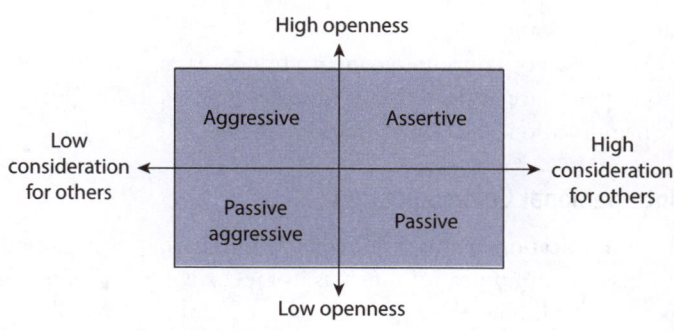

Levels of Communication

Having understood the communication process, let us now study the various levels at which human communication takes place:

- Extrapersonal
- Intrapersonal
- Interpersonal

Figure 1.2 Style matrix

- Organizational
- Mass

Extrapersonal Communication

Communication between human beings and non-human entities is extrapersonal. For example, when your pet dog comes to you wagging its tail as soon as you return home from work, it is an example of extrapersonal communication. A parrot responding to your greeting is another example. More than any other form, this form of communication requires perfect coordination and understanding between the sender and the receiver because at least one of them transmits information or responds in sign language only.

Intrapersonal Communication

Intrapersonal communication takes place within an individual. We know that the brain is linked to all parts of the body by an electrochemical system. For example, when you begin to 'feel hot', this information is sent to the brain and you may decide to 'turn on the cooler', responding to instructions sent from the brain to the hand. In this case, the relevant organ is the sender, the

electrochemical impulse is the message, and the brain is the receiver. Next, the brain assumes the role of sender and sends the feedback that you should switch on the cooler. This completes the communication process. This kind of communication pertains to thinking, which is the basis of information processing. Without such internal dialogue, one cannot proceed to the further levels of communication—interpersonal and organizational. In fact, while we are communicating with another party, our internal dialogue with ourselves continues concurrently—planning, weighing, considering, and processing information. You might have noticed that at times you motivate yourself or consciously resolve to complete a certain task. Self-motivation, self-determination, and the like take place at the intrapersonal level.

Interpersonal Communication

Communication at this level refers to the sharing of information among people. To compare it with other forms of communication, such as intrapersonal, organizational, etc., we need to examine how many people are involved, how close they are to one another physically, how many sensory channels are used, and the feedback provided.

Interpersonal communication differs from other forms of communication in that there are few participants involved, they are in close physical proximity to each other, many sensory channels are used, and feedback is immediate. Also, the roles of the sender and receiver keep alternating. This form of communication is advantageous because direct and immediate feedback is possible. If a doubt occurs, it can be instantly clarified. Note that non-verbal communication plays a major role in the interpretation of a message in this form of communication due to the proximity of the people involved.

Interpersonal communication can be *formal* or *informal*. For example, your interaction with a sales clerk in a store is different from that with your friends and family members; the interaction between the panel members and the candidate appearing at an interview is different from the conversation between two candidates waiting outside. Hence, depending upon the formality of the situation, interpersonal communication takes on different styles.

Moreover, most interpersonal communication situations depend on a variety of factors, such as the psychology of the two parties involved, the relationship between them, the circumstance in which the communication takes place, the surrounding environment, and finally the cultural context.

Organizational Communication

Communication in an organization takes place at different hierarchical levels. As we have learnt, it is extremely necessary for the sustenance of any organization. Since a large number of employees are involved in several different activities, the need to communicate effectively becomes greater in an organization. With a proper networking system, communication in an organization is possible even without direct contact between employees. Organizational communication can be further divided into the following.

Internal-operational All communication that occurs in the process of operations within an organization is classified as internal-operational.

External-operational The work-related communication that an organization has with people outside the organization is called external-operational communication.

Personal All communication in an organization other than that for business or official purposes is called personal communication.

We will learn more about communication in organizations later in this chapter.

Mass Communication

Mass communication is meant for large audiences and requires a medium to transmit information. There are several mass media such as journals, books, television, and newspapers. The audience is heterogeneous and anonymous, and thus the approach is impersonal. Press interviews given by the chairman of a large firm, advertisements for a particular product or service, and the like take place through mass media. This type of communication is more persuasive in nature than any other form, and requires utmost care on the part of the sender while encoding the message. Oral communication through mass media requires equipment such as microphones, amplifiers, etc., and the written form needs print or visual media. The characteristics of mass communication are as follows:

Large reach Mass communication has the capacity to reach audience scattered over a wide geographical area.

Impersonality Mass communication is largely impersonal, as the participants are unknown to each other.

Presence of a gatekeeper Mass communication needs additional persons, institutions, or organizations to convey the message from a sender to a receiver. This 'gatekeeper' or mediator could be a person or an organized group of persons active in transferring or sending information from the source to the target audience through a mass medium. For example, in a newspaper, the editor decides which news makes it to the hands of the reader. The editor is therefore the gatekeeper in this mass communication process.

Flow of Communication

Information flows in an organization both formally and informally. *Formal communication refers to communication that follows the official hierarchy and is required to do one's job.* In other words, it flows through formal channels—the main lines of organizational communication. Internal-operational and external-operational communication is formal. In fact, the bulk of communication that a business needs for its operations flows through formal channels. For example, when a manager instructs a subordinate on some matter or when an employee brings a problem to a supervisor's attention, the communication is formal. Similarly, when two employees interact to discuss a customer's order, the communication is formal. Information of various kinds flowing through formal channels, such as policy or procedural changes, orders, instructions, and confidential reports, is formal communication. Formal communication can flow in various directions—vertical, lateral, or diagonal—as shown in Figure 1.3.

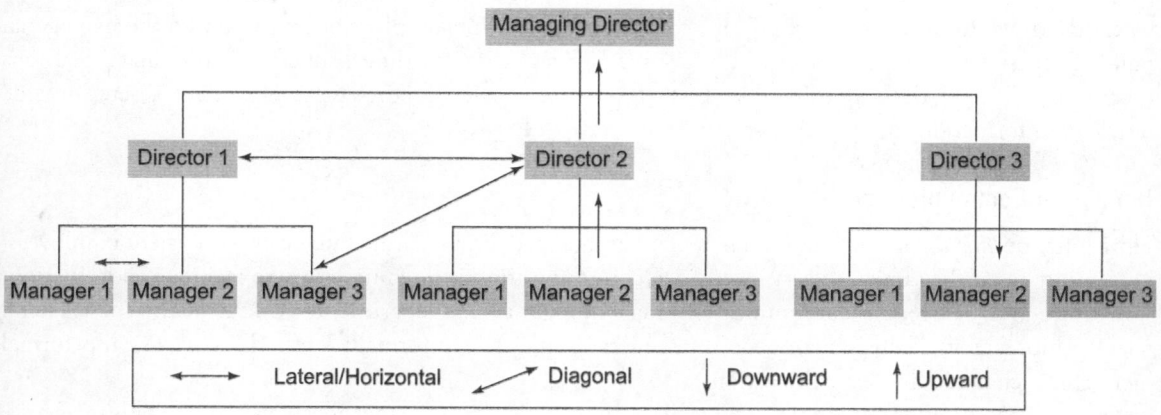

Figure 1.3 Flow of communication in an organization

Vertical Communication

Figure 1.3 shows that communication can flow in any direction in an organization. Vertical communication consists of communication up and down the organization's chain of command. Vertical communication can be classified as downward communication and upward communication according to the direction of its flow.

Downward communication

Downward communication flows from a manager down the chain of command. When managers inform, instruct, advise, or request their subordinates, the communication flows in a downward pattern. This is generally used to convey routine information, new policies or procedures, seek clarification, ask for an analysis, etc. People also send feedback to their subordinates on their actions through this channel. Downward communication can take any form—emails, memos, notices, face-to-face interactions, or telephone conversations. However, it should be adequately balanced by an upward flow of communication.

Upward communication

When subordinates send reports to inform their superiors, or to present their findings and recommendations to their superiors, communication flows upward. Upward communication keeps managers aware of the business operations as well as of how employees feel about their jobs, colleagues, and the organization in general. Managers also rely on upward communication for making certain decisions or solving problems concerning the organization.

The extent of upward communication, especially that initiated at the lowest level, depends on the organizational culture. In an open culture without too many hierarchical levels, i.e., in a flat structure, managers are able to create a climate of trust and respect, and implement participative decision-making or empowerment. In such an environment, there will be a considerable amount of upward communication. This happens mainly because the employees provide the input for managerial decisions. In a highly authoritative environment, where downward flow dominates, upward communication still takes place but is limited to the managerial ranks. Suggestion boxes, employee attitude surveys, grievance procedures, superior–subordinate decisions (decisions taken for the subordinate by his/her superior), review reports, statistical analyses, etc. provide restricted information to top management.

Horizontal Communication

Horizontal or *lateral* communication takes place among peer groups or hierarchically equivalent employees, i.e., employees at the same seniority level. Such communication is often necessary to facilitate coordination, save time, and bridge the communication gap among various departments. Occasionally, these lateral relationships are formally sanctioned. But generally, they are informally created to bypass the formal hierarchical channels and expedite action.

From the organization's point of view, lateral communication can be either advantageous or disadvantageous. As compared to vertical (downward or upward) communication, which can at times hold up and delay timely and accurate transfer of information, lateral communication can be beneficial. Nevertheless, they can also create conflicts when formal vertical channels are bypassed by employees in order to accomplish their goals, or when superiors find out that they had not been consulted before certain decisions were taken.

Lateral communication enables the sharing of information with a view to apprise the peer group of the activities of a department. The Vice President (Marketing) sending some survey results in the form of a memo to the Vice President (Production) for further action is an example of lateral communication. This type of communication is vital for the growth of an organization as it builds cooperation among the various branches. It plays a greater role in organizations where work is decentralized, because there is a higher probability of communication gaps in such set-ups.

Diagonal Communication

Diagonal or cross-wise communication flows in all directions and cuts across the various functions and levels in an organization. For example, when a sales manager communicates directly with the Vice President (Production), who is not only in a different division, but also at a higher level in the organization, they are engaged in diagonal communication. Though this form of communication deviates from the normal chain of command, there is no doubt that it is quick and efficient.

> 'A coordinated flock of birds or a shoal of fish maintain their relative positions, or alter direction simultaneously due to lateral communication amongst members; this is achieved due to tiny pressure variations.'
> –Wikipedia

In some situations, ignoring vertical and horizontal channels expedites action and prevents other employees from being used merely as messengers between the actual senders and receivers.

The increased use of email also encourages cross-wise communication. Any employee can communicate via email with another employee, regardless of the receiver's function or status. Since there is no specific line of command, diagonal communication is also referred to as *cross-wise*, *radial*, or *circular* communication, depending upon the structure of the organization. For instance, a managing director could directly call a supervisor and give instructions.

Communication Networks

A variety of patterns emerge when communication through vertical, horizontal, and diagonal channels is combined. These patterns are termed as *communication networks*.

Formal Network Models

Five common communication networks exist in formal communication in an organization—chain, Y, wheel, circle, and all-channel.

The *chain network* represents a vertical hierarchy in which communication can flow only upward or downward. This network is used in direct line of authority communications, with no deviations (Figure 1.4).

The *Y-network* is in effect a multi-level hierarchy and a combination of horizontal and vertical flow of communication. If we turn the Y upside down, we see two subordinates reporting to one senior, with two levels of authority above the latter (Figure 1.5).

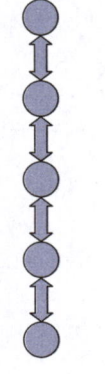

Figure 1.4 Chain network

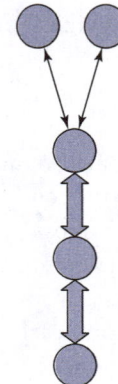

Figure 1.5 Y-network

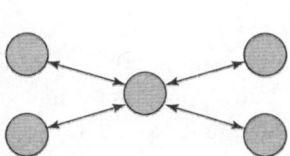

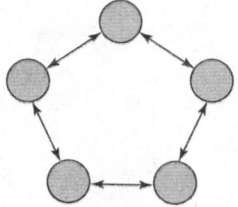

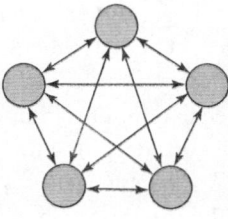

Figure 1.6 Wheel network   **Figure 1.7** Circle network   **Figure 1.8** All-channel network

The *wheel network* refers to several subordinates reporting to a superior. This is a combination of horizontal and diagonal flow of communication. But here, though the subordinates are of equal rank, all of them report to one superior, and without any interaction between themselves (Figure 1.6).

The *circle network* allows employees to interact with adjacent members but no further than that. Look, for example, at this three-level hierarchy. There is vertical communication between superiors and subordinates, and horizontal communication only at the lowest level (Figure 1.7).

The *all-channel network*, which is the least structured, enables each employee to communicate freely with all the others. There are no restrictions on who can communicate with whom. All are equal, as no one employee formally or informally assumes a leading role. Hence, everybody's views are equally and openly shared (Figure 1.8).

Finally, for effective use of communication networks, we need to remember the following:
- No single network is suitable for all occasions.
- The wheel and all-channel networks are preferred if speed of communication is a priority.
- The chain, Y, and wheel networks serve best when accuracy is crucial.

Informal Network Models

Besides flowing through formal networks, communication in an organization also travels along an informal network—*the grapevine.* This informal network is very active in almost every organization. In fact, some organizations consciously use it to great effect, to develop their human resources, and in turn, their productivity. For example, Michael Eisner, the former Chairman of Walt Disney, adopts management by walking around (MBWA). He goes around the company, talks to employees, observes them talking among themselves, and uses this knowledge effectively to adopt or change certain strategies.

There appear to be patterns emerging for this type of informal communication as well (Figures 1.9–1.12). Among the four patterns shown in these figures, *single strand* is the way in which most people view the grapevine. Here, the message is passed from one person to another along a single strand. In the *gossip* network, one person passes information to all the others. In the *probability* type of network, each person tells others at random. The cluster pattern, the most popular pattern of grapevine communication, refers to the flow of information in which some people tell something to a select few. Which individuals are active on the grapevine often depends on the message. For example, a message that sparks the interest of an employee may stimulate him or her to tell someone else, whereas another message that is perceived to be of lesser interest may never be transmitted further.

'The term "grapevine" can be traced to Civil War days when vine-like telegraph wires were strung from tree to tree across battlefields and used by army intelligence. The messages that came over these lines were often so confusing or inaccurate that soon any rumor was said to come from the grapevine.'
                                    –*Jitendra Mishra*

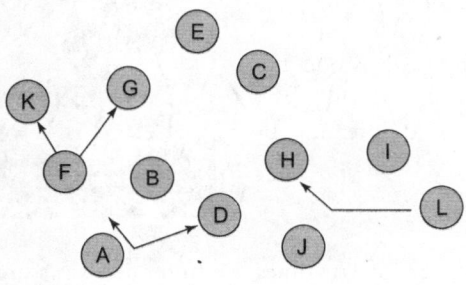

Figure 1.9 Single strand

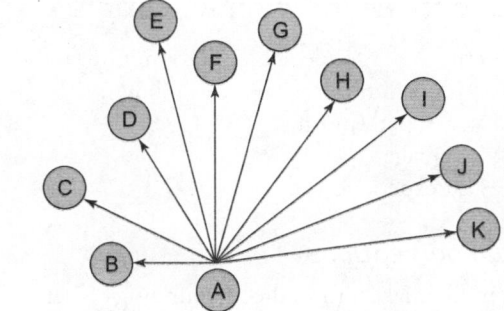

Figure 1.10 Gossip

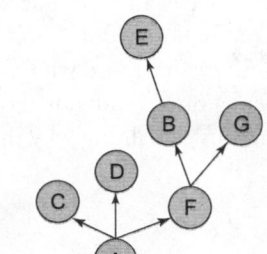

Figure 1.11 Probability

Figure 1.12 Cluster

Since only a small number of employees typically pass on information to more than one other person, managers can analyse grapevine information and predict its flow. Both advantages and disadvantages arise from the grapevine (Table 1.2). However, the advantages to an organization of using the grapevine seem to outweigh the drawbacks. To effectively use this channel, organizations should:

- Not ignore information received through the grapevine
- Use this channel to supplement the formal channel
- Identify but not threaten the main sources of information
- Try to understand the human relationships involved in grapevine communication.

Table 1.2 Characteristics of the grapevine

Advantages	Disadvantages
In general, the grapevine is	If information from the grapevine is blindly accepted, it may
• not expensive	• reveal some degree of error
• rapid	• be harmful in case it is baseless or imaginary
• multidirectional	• lead to misunderstandings because of incomplete information
• if carefully cultivated, capable of resolving conflicts	• unreliable, as nobody takes the responsibility for it
• an outlet for anxieties, worries, frustration	• cause damage to the organization because of its excessive swiftness at times (e.g., a baseless rumour that is not clarified in time will be accepted as the truth, causing a negative impact)
• a promoter of unity and solidarity	
• voluntary and unforced	

Perspectives in Communication

Perspectives refer to ideologies that limit our understanding of the communication. The various factors that can restrict communication are as follows:

- Visual perception
- Language
- Past experiences
- Prejudices and feelings
- Environment
- Ethics and etiquette*

Visual perceptions

Each individual captures the surrounding objects differently because of variation in one's experiences and aesthetic senses. The difference in visual perception may affect communication.

When the audience is asked to guess this lady's age (lady in the image), the responses received are both 'young' and 'old'. The young respondents probably interpret it as young, and those who are old call it old. The truth is one, but a lot more depends on an individual's perception. Perception directly affects communication, thus explaining why the same thing is perceived differently by different people.

Language

Inadequate vocabulary can be a major hindrance in communication. At times, we find ourselves searching for the exact word or phrase that would be appropriate for what we are trying to express. For example, during a speech, if you are at a loss for words, your communication will be very ineffective, and you will leave a poor impression on the audience. On the other hand, if you have a varied and substantial vocabulary, you can create a favourable impression on your listeners.

Merely having a wide vocabulary is of no use unless the communicator knows how to use it. In communication, the denotative (literal or primary) and connotative (implied or suggested) meanings of the words used should be absolutely clear to the receiver. Therefore, one should make constant efforts to increase one's vocabulary by regularly reading a variety of books and listening to native speakers of the language. Thereafter, using a wide vocabulary regularly will also help to make its usage comfortable and natural. Chapter 24 discusses vocabulary development in detail.

Past experiences

The outlook and attitude of people are formulated based on past experiences. Throughout life, experiences leave an individual with positive or negative impressions, making them form a future opinion about things. There are common experiences amongst human beings like love, kindness, hatred, guilt, anger, etc., but everyone goes through different situations, and therefore, their experiences are different for each attribute. If someone has undergone a bad experience, it is very difficult to change his/her opinion. If someone ridicules you for a trivial mistake during a presentation, you will always be apprehensive of giving good presentations.

Prejudices and feelings

Communication is governed adversely by prejudices and stereotypes. Unconsciously an individual formulates an opinion about others based on religion, caste, body type, skin colour, etc. The expression of prejudice is displayed through the choice of words and the tone of the conversation. To be honest in communication, one should have minimal prejudices against others. For example, if a speaker is not able to even stand correctly, opinions are formulated and just assumed that the presentation will not be very effective. But that might be a wrong presumption.

Emotions and feelings also play a major role in communication. If an individual is upset about a failure, he/she cannot concentrate and might even unintentionally hurt others. Similarly, if a person is too excited or happy, he/she might not find appropriate words to express his/her thoughts.

Environment

Surroundings set the mood of a communicator. Environment is crucial for effective communication as it can impact both positively and negatively. In cordial settings, you are enthusiastic about communicating, and there will be a better flow of ideas. Imagine sitting in a dimly lit room and being asked to make a persuasive speech—the task is a little complicated. Similarly, if the room has too many people creating suffocation and chaos, it will be difficult to put across your thoughts. Some people get creative ideas in a secluded place, while others might get brilliant ideas amidst a crowd. Room settings, lighting, arrangements, noise, etc. are part of communication environment.

Ethics and etiquette

Ethics is a general notion of perception of what is right. Society is governed by ethics for acceptable behaviour. In the absence of ethics, there will be no discipline. Ethical communication is based on the premise that communication should be honest, clear, and open. It is unethical to use a language that hurts the sentiment of a particular community, caste, or race. One should be sensitive to others and not offend anyone. The key aspect in communication is to be open minded and value others' opinion rather than pushing your thoughts on others. The major concerns for ethical communication are truthfulness, honesty, accuracy, and reason. If one abides by these concerns, the environment of the organization will be conducive to work. Breach in ethical issues in communication may have legal implication, therefore one has to be cautious in the word choice, tone, and content of communication.

Etiquette is how you behave in a given setting. Etiquette is usually learnt from society and surroundings. There might be subtle differences in behavioural patterns of people according to place and region but majorly they remain the same. One must be aware of the right practices; else embarrassment might follow. Etiquette may also change from one organization to other. For example, you may call your senior by first name, while it may not be allowed in another organization. Following the right etiquette can help you in progressing while not following may pull you back.

Technology in Communication

Technology plays a crucial role in enhancing the quality of communication. In today's scenario, it is almost impossible to communicate without technological aid. Technology helps us to connect with people at different geographical locations. It makes communication faster and easier. People can communicate with others immediately and provides a communication channel 24 × 7. There are various facilities available on technological gadgets, such as text, pictures, emojis, and others, that help the communicators use them according to their convenience. Technology has brought down the cost of stationery and tedious exercise of posting documents. The deadlines of submission can be met with the facilities available on machines. The information is in the hands of the user all the time.

Positive Impact of Technology-enabled Communication

We live in an increasingly interconnected world. We are more mobile than the previous generation—and that mobility is extending to our data. Process data are routed from the factory floor to the boardroom. A technician can access wireless sensors in remote locations to see what is happening, and an engineer in Bengaluru can troubleshoot an automation problem in Hyderabad while sitting at his/her PC.

Technology has tremendously changed the way companies, professionals, and institutes operate. Phenomenal advancements in computer technology have enabled small as well as huge organizations to communicate more closely and frequently within and outside. In today's technology-driven world, organizations that do not embrace technology may not be able to survive.

Of course, there are many organizations in our nation that still follow the traditional way of communicating, not because they do not realize the importance of technology but because they may not have adequate resources to establish technology-oriented communication networks. Given the resources, every organization would wish to own the new technologies as they have the following advantages:

- Distance is no longer a major barrier. The importance of personal contact between people has been replaced with face-to-face communication. Communication across the country, or even around the world, has become as easy as communication with the office down the hall.
- The organizational structure has become more streamlined as managers have increased direct contact with subordinates. Since this also means fewer intermediaries (people who pass messages on), the organization itself is more flattened.
- More people in an organization have access to more information. This reduces the 'information float'—the rate of information flow—and tends to change the traditional role of managers as primary information sources.
- The time required to make decisions has decreased because managers have access to increased information resources. The time taken to consider decisions, though, has also decreased because of increased pressure to act quickly.
- The timelines and quality of information are increasingly important as more people have access to more sources of information. The difficulty is that more information does not necessarily mean better information.
- The implementation of projects, particularly those depending on communication or involving strict time schedules, has been enhanced.
- Teamwork in organizations has increased. More people, with a broader range of skills, can provide inputs on projects. In fact, many newer organizational charts are designed around computer links.
- Finally, as technology increases in scope, managers are required to learn more about the communication process at all levels.

> 'Once a new technology rolls over you, if you're not part of the steamroller, you're part of the road.'
> —Stewart Brand

Negative Impact of Technology-enabled Communication

Despite the advantages of the electronic office, technology at work and in communication also creates its own set of barriers—those specifically related to the interactions between people and machines. One such barrier that we can observe in formal communication is technophobia, or the fear of new technology itself. This can be resolved through familiarization and education. Besides technophobia, there are other, more subtle, barriers that can create a negative impact on the parties involved in organizational communication:

Information overload The acceleration of change is accompanied by an increase in the information needed to keep up with all these developments. This also leads to psychological, physical, and social problems. A worldwide survey found that two-thirds of managers suffer from increased tension and one-third from ill health because of information overload. Other effects of excessive information include anxiety, poor decision-making, difficulties in memorizing and remembering, and reduced attention span. These effects merely add to the stress caused by the need to constantly adapt to a changing situation.

Less time for organizational activities Because of the time spent on checking emails and Internet browsing, the time required for decision-making or problem-solving is curtailed; this trend may lead to increased problems in future.

Blurring lines between professional and personal lives Technology-enabled communication has blurred the lines between the professional and personal lives of many people. For example, with the increased use of mobile phones, just as executives can be disturbed by their family members while they are at an important meeting, they can also be disturbed by their superiors or subordinates on a vacation with their family.

People isolation Technology's isolating effects are easy to observe. Prior to the widespread use of the printing press, people had to congregate and verbally interact with each other in person in order to communicate. With the advent of media technologies, such as record players, televisions, and VCRs, people no longer have to gather together in one place for entertainment. Innovations in telephone services such as voicemail and caller ID make it possible for people to deliver messages without a personal interface at all. However, several groups of people who do not have access to recent technologies feel isolated from the privileged lot. For example, there are many promising men and women in India who may not be able to apply for a job advertised on the Internet.

Overcoming Negative Impact

As the advantages override the limitations, it is appropriate to find some solutions to overcome the negative effects of technology-enabled communication. One strategy to consider is instituting information 'filters'. To cope with volumes of data, email filters can be used to screen out less-than-critical messages and prevent an overwhelming amount of data from being thrown at a person. Deleting our name from list servers may be another way to limit the influx of email. Likewise, properly managing the usage of mobile phones can reduce the blurring of lines between personal and professional lives.

Simply realizing the fact that computer technology is not the only means of communication would surely enable one to resort to other means of communication, thereby achieving one's predetermined purpose of communication.

Selection of Appropriate Technology

As already discussed, new technologies have given people a wider range of choices for communication than ever before, and each channel has its uses. Hence, it is important not only to decide as to which communication channel to use but also when to use each one most effectively.

Each channel has its own advantages as well as drawbacks. It is, therefore, important to identify some parameters to decide the technology that would be most efficient in accomplishing a specific communication goal. We suggest that the following criteria may be considered for selecting an appropriate communication channel: purpose, audience, nature of message, time, cost, and presentation needs.

Purpose

Generally, the purposes for which we communicate in an organization may fall into three major categories: to inform, to persuade, and to collaborate. For example, as the vice-president of a company, you may like to inform your managers about the company's new set of policies, you may like to motivate the managers for better sales, or you may have discussions with your counterparts in other branches to implement a new marketing strategy.

Audience

Adapting oneself or choosing a communication technology to suit the audience is vital for successful communication. Knowing the number, status (designation), composition, possible reactions, level of understanding, relationship, and needs of the audience before selecting a technology for communication purposes is essential. Imagine the impatience and anxiety that can be caused by the delayed response or no response for an email message sent by us to somebody who does not have easy access to email.

Nature

Besides the purpose of a message and the type of recipients, there is yet another parameter to be considered for choosing the communication technology—the nature of the message. This can be ordinary, confidential, or even strictly confidential. For example, it would be inappropriate to convey a strictly confidential and sensitive message such as suspending the services of an employee through email.

Time

The urgency of a message is also an important factor in selecting a method of communication. For instance, there might be important decisions that have to be quickly communicated to the individual or group concerned. Employees would be happy to receive an email announcing a 10 per cent increase in their DA quickly rather than transmitting the message through the formal channels of communication, such as memos and circulars.

Cost

The cost factor has to be given due importance while deciding on a communication technology. For instance, using conventional phone lines is not only expensive but also does not provide several benefits offered by Internet phone (voice calls made over the Internet). There are many web servers, such as MSN, Yahoo, and Skype, which are efficient and have made communication very affordable. Hence, it is important to consider the cost of conveying messages.

Presentation needs

If a presentation—oral or written—contains complex data, graphs, maps, diagrams, or photographs, one has to carefully select an appropriate technology to present them effectively. For example, visual presentations can be made more effective through animation or colour contrasts, etc. by using any suitable presentation software.

We may not always have the luxury of choosing a technology. However, when we do, it would serve well to consider all the factors and select the most appropriate means. Table 1.3 presents in a nutshell the various factors that have to be considered before selecting a suitable technology for communication.

Table 1.3 Factors affecting the choice of technology

Technology	Nature of message	Speed of establishing connection	Type of information conveyed	Time to receive feedback	Control over composing and delivering	Receivers' attention	Permanency
Email	Formal/ informal, routine, non-routine	Fast	Brief, non-complex; text only, no formatting	Delayed/ not always immediate	High	Low	Yes
Fax	More formal	Fast	Words, numbers, and images	Delayed	High	Low	Yes
Telephone	Formal/ informal	Varied	Vocal, not visual	Immediate	Moderate	High	Usually none
Voicemail	Mostly personal	Fast	Vocal, not visual	Delayed	High	Low	Possible
Teleconferencing	Formal/ informal	Slow	Vocal, not visual	Immediate	Moderate	High	Usually none
Videoconferencing	Formal	Slow	Both verbal and non-verbal messages	Immediate	Moderate	High	Possible
Computer conferencing	Formal/ informal	Moderate	Text and sometimes visual images	Immediate	Moderate	Low	Possible

Effectiveness in Technology-based Communication

Like any other activity, technology-based business communication also requires planning before its implementation.

Before starting on a venture, plan for the technology that will be required. Get those extra telephone lines, purchase and learn how to use a computer, decide how you will and will not use the Internet. Keep in mind that monitoring employees' use of the Internet while at work is not a violation of their rights to privacy. Any policy should be in writing and signed by the employees. The appropriate use of technology can make work more efficient and easy from the very beginning. Table 1.4 lists some dos and don'ts regarding the use of technology.

Table 1.4 Top ten do's and don'ts when using technology-based communication

Do's	Don'ts
1. Learn to work on word processing, spreadsheets, presentations, and email.	1. Do not begin work without a basic knowledge about and possession of computer tools.
2. Consider using an accounting software program suitable to the work.	2. Do not overlook making regular external backups of computer files.
3. Consider a laptop computer if the work demands mobility.	3. Do not overlook the Internet as an important communication tool.

(Contd)

Do's	Don'ts
4. Learn the use of digital technology including that of digital pictures.	4. Do not purchase more equipment than that required for the next two years.
5. Consider using a headset for cordless and cell phones.	5. Do not spend on a top-of-the-line computer unless it is really required.
6. Plan ongoing internal communications including awards, newsletters, and discussions.	6. Do not sign up for extended time periods on any service including fixed-line and mobile phones.
7. When leaving messages, clearly and slowly repeat your name and number.	7. When tariff plans change, do not overlook to request information on communication plans that more closely meet the new requirements.
8. Use a remote voicemail answering system rather than an answering machine.	8. Do not fail to exercise rights on return policies within allowed time limits.
9. Use a separate dedicated phone line for business and fax lines.	9. Do not think that a toll-free telephone number is important unless orders are received by phone.
10. Develop a logo to represent oneself on stationery, signs, cards, and websites.	10. Do not sign up for long-term plans with Internet service providers.

SUMMARY

Technical communication is process of sharing information through various modes with a specific audience for a specific purpose. The process involves the transmission and interchange of ideas, facts, feelings, or courses of action. Technical communication is different from general communication. The objective of technical communication is to present correct, accurate, concise, clear, and appropriate information.

The communication process includes six main elements—sender, message, channel, receiver, response, and feedback. The success of communication lies in positive feedback. Sometimes the message received is not the same as the message intended by the sender; this is because of the presence of noise.

There exist different communication styles, including Assertive, Aggressive, Passive, and Passive-Aggressive. According to the circumstances and types of message, an individual adopts one of them.

Communication takes place at different levels: extrapersonal, intrapersonal, interpersonal, organizational, and mass communication. In an organization the flow of communication can be vertical, horizontal, or diagonal.

Communication is divided into formal and informal communication. Formal networks include the flow in horizontal, vertical, and diagonal directions. Informal communication is called grapevine and it has both positive and negative impact on organizations.

The perspectives in communication refer to the understandings that limit communication. These perspectives include visual perception, language, past experiences, prejudices and feelings, environment, ethics, and etiquette.

Technology plays a crucial role in enhancing the quality and speed of communication. The world today is interconnected through technology. There are both positive and negative sides to technology, therefore one should be cautious while communicating through gadgets. While selecting the technology keep in mind the purpose, audience, nature of messages, time, cost, and presentation needs.

EXERCISES

1. Answer the following questions in about 200 words each:
   (a) How is general-purpose communication different from technical communication?
   (b) Communication is the process of sending and receiving information. Explain the communication process in the light of this statement. Draw the communication cycle to support your answer.
   (c) How is feedback important in communication? Give two examples of delayed feedback.
   (d) Explain 'flow of communication'. Illustrate it with examples from the existing communication patterns in your college/institute.

2. What do you understand by the term technical communication? Explain its importance with examples.

3. Human communication takes place at different levels. How can you distinguish between intrapersonal and interpersonal communication?

4. What are the characteristics of mass communication? Explain the term gatekeeper.

5. What are the various modes of communication flow in an organization? What is upward flow and what is the purpose of this mode in an organization?

6. Write the origin of the word 'grapevine'. How is grapevine useful for organizations?

7. Project: Visit a few organizations (academic institutions/business enterprises/industries) and determine the communication patterns existing there. Classify them into oral and written categories. Also figure out the direction in which these flow. Prepare a two-page report on each of your visits.

8. There is an important project to be undertaken on 'Women and Society' in your organization. You as the head of the workforce have to communicate this information to the subordinates. Which communication style will you adopt? Elaborate.

9. Is technology a boon or bane for communication effectiveness? Give instances from your life experiences to substantiate your standpoint.

CHAPTER
2

Barriers to Communication

OBJECTIVES

You should study the chapter to know

- ○ what communication barriers are and what causes them
- ○ how to avoid communication failures caused by noise
- ○ how to classify communication barriers

Introduction

Having studied the basic principles of technical communication in Chapter 1, we now move on to a very important factor related to communication. Imagine you are a junior executive who has just joined as a trainee in an automobile company. You go to meet your boss, to seek permission for visiting the automobile exhibition being held in Delhi. While you are talking, two of your colleagues also arrive to get your boss' signature on some bills. You fall silent while he signs the bills. After they leave, you resume talking but you find that your boss has not really grasped what you have said earlier. You later analyse this situation, and realize that your colleagues' intervention led to your boss' lack of concentration. As the sender, you had patiently tried to express your wish. But because of the interruption by your colleagues, the receiver, your boss, could not decode your message fully. Hence the communication process failed.

This discussion brings us to the introduction of the term 'barrier' in communication. A barrier is defined as something that prevents or controls progress or movement. This definition implies that a barrier is something that comes in the way of the desired outcome. In the example given above, notice that the arrival of your colleagues was an event that prevented your boss from concentrating completely on what you were saying. So, we say that this incident was a barrier to the communication between you and your boss. All of us have come across such situations while communicating with parents, friends, or colleagues. Let us now learn how to avoid such communication failures and make our interactions more effective.

We all know that effective communication is the nerve of all the business activities in an organization. Even a slight break in the communication flow can lead to misunderstandings. Communication is effective only if it creates the desired impact on the receiver. Often, managers get frustrated in their efforts, and end up saying that nobody in the organization understands them. Many employees fail to listen attentively during meetings, or send incomprehensible business letters. Such situations arise due to the presence of barriers in communication, which can take many forms such as inadequate communication skills.

> Communication is effective only if it creates the desired impact on the receiver.

There are numerous such barriers associated with communication. These need to be addressed in order to ensure that no gap occurs in the communication cycle. A common barrier is the wrong assumptions made about the person to whom the message is being sent and *sometimes about the message itself*.

For example, if the sender of the message is talking about a technical proposal, he/she would be wrong if he/she makes assumptions about the receiver's level of technical knowledge. The problem can be resolved to a great extent if the sender of the message analyses his/her message thoroughly and anticipates the likely response before sending it.

If a particular communication fails to evoke the desired response, the following five steps can help solve the problem:

- Identify the problem
- Find the cause/barrier
- Work on alternative solutions
- Opt for the best solution
- Follow up rigorously

The first step—identifying the problem—is the most difficult. We first realize that there is a problem when we do not receive the desired feedback. To identify the problem correctly, it is mandatory that the feedback be analysed carefully. For example, you have asked your subordinate to write a bimonthly report, and until the next month, he has not done so. When you ask him about the delay, he replies that he was asked to produce the report bimonthly. The problem here is that to you the term *bimonthly* meant *twice in a month*, whereas to your subordinate it meant *once in two months*. Later, you look up the dictionary and find that *bimonthly* means *twice a month* as well as *once in two months*!

Having identified the problem, the next step is to find out what caused it. In this situation, we could say that it was the choice of words. The third step is to explore possible solutions. In this case, a way out would be to choose words that are more specific in their meaning, i.e., words that could mean only one thing—the intended meaning. After thinking through the alternatives, apply the best solution that not only solves the problem, but also does not create any new difficulties. Hence, instead of using the troublesome term *bimonthly*, either *twice a month* or *once a fortnight* could be used. After successfully completing all the four steps, the last step requires that we implement the best solution properly. Having once come across a particular communication barrier, there should be a conscious effort to never let it crop up again.

In this chapter, we will discuss the various types of communication barriers, and how they can be identified and overcome. Before going on to consider the different barriers to communication, however, let us first understand the related term 'noise'.

Noise

Any interference in the message sent and the message received leads to the production of 'noise' (see Figure. 2.1).

The term communication barrier, or that which inhibits or distorts the message, is an expansion of the concept of noise. Noise here does

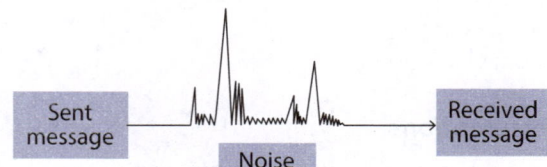

| Sent message | Noise | Received message |

Figure 2.1 Noise interferes in the correct transmission of messages

not mean sound, but a break or disturbance in the communication process. If noise occurs because of technological factors, it is not too much of a problem as it can be removed by correcting the technological faults. However, if the noise is due to human error, the parties involved in the communication process need to take corrective measures.

Noise is defined as any unplanned interference in the communication environment, which affects the transmission of the message. Noise can be classified as *channel* and *semantic*. *Channel noise is any interference in the mechanics of the medium used to send a message.* Familiar examples of channel noise are distortion due to faulty background, noise in telephone

> Noise is defined as any unplanned interference in the communication environment, which affects the transmission of the message.

lines, or too high a volume or pitch from loudspeakers. In written communication, illegible handwriting can be termed as channel noise. Whereas channel noise develops externally, *semantic noise is generated internally, resulting from errors in the message itself.* It may be because of the connotative (implied) meaning of a word that is interpreted differently by the sender and the receiver. For example, the word 'condescend' may have been used in a positive manner, implying *grace* or *dignity of manner*, but the receiver might interpret it in a negative manner, related to a (*baseless*) *assumption of superiority.* Other examples of semantic noise are ambiguous sentence structure, faulty grammar, misspellings, and incorrect punctuation.

Classification of Barriers

A barrier acts like a sieve, allowing only a part of the message to filter through; as a result, the desired response is not achieved. To communicate smoothly and effectively in an organization, irrespective of your position, you need to know how barriers operate, why they cause misunderstandings, and how to minimize their negative impact. How often have you said, 'I meant to say this and not that'? Even with the best intentions, communication barriers crop up and our written and spoken messages are misunderstood. If we classify barriers according to the processes of message formation and delivery, we can identify three types:

- Intrapersonal
- Interpersonal
- Organizational

Intrapersonal Barriers

Individuals are unique because of differences in perceptions, experiences, education, culture, personality, etc. Each of us interprets the same information in different ways, as our thinking varies. These differences lead to certain inbuilt or intrapersonal barriers. Let us explore all the common causes that lead to these intrapersonal barriers:

- Wrong assumptions
- Varied perceptions
- Differing background
- Wrong inferences
- Blocked categories
- Categorical thinking

Wrong assumptions

Many barriers stem from wrong assumptions. For example, when a doctor tells her patient that he has to take some medicine only 'SOS' (i.e., during an emergency), without knowing whether the patient understands the term 'SOS', she is creating a barrier in their communication. Here the doctor has made a wrong assumption about her patient's level of knowledge. Wrong assumptions are generally made because the sender or the receiver does not have adequate knowledge about the other's background or entertains certain false concepts, which are fixed in his/her mind. To strengthen your skills as a

'Take this medicine only SOS.'

communicator, try to put yourself in the shoes of the listener. This exercise will prevent making wrong assumptions about the receiver.

Varied perceptions

We all know the story of the six blind men and their description of an elephant. The elephant was perceived by each man as a fan, a rope, a wall, a sword, a snake, and a tree. None of the blind men were wrong, as the part of the elephant body touched by each man compared well with the various objects they named. This is how different individuals hold different viewpoints about the same situation.

Similarly, individuals in an organization also perceive the same situation in different ways. Let us take the case of disagreement between two individuals. If you are close to one of them, you are likely to be biased. You may perceive your friend's arguments as correct, and hence, may not be able to appreciate his/her opponent's point of view. It is all a matter of perception. The best way to overcome this barrier is to step back and take a wider, unbiased perspective of the issue.

Differing backgrounds

No two persons have the same background. Backgrounds can be different due to different education, culture, language, environment, financial status, etc. Our background plays a significant role in how we interpret a message. At times, something not experienced earlier is difficult to interpret or appreciate. Think of a class where the professor talks about his rock-climbing adventure. Students who have experienced rock climbing may be able to appreciate the professor's talk, while others who have never been into adventure sports may not find it interesting at all. The representative of a computer company would not make much sense to a group of doctors if in his/her presentation he/she goes into details about the hardware aspects of the computer that he/she plans to install in a hospital.

> To enhance communication skills, it is necessary to know the background of the audience. This information can accordingly be used to construct your message.

To enhance communication skills, it is necessary to know the background of the audience. This information can accordingly be used to construct the message. *Empathy or identification with another person is the solution to this barrier.* We must make an effort to understand what the listener can find difficult to comprehend in our message because of the difference between our background and that of the listener. The language understood by the receiver should be used to avoid ambiguity and reduce multiple meanings.

The exact meaning of a word resides in the mind of the speaker; therefore, one ought to be cautious while using words. The multiple

'... and if you clip a carabiner to each end of the sling, you have a quickdraw...'

meanings of a word can astronomically increase the problem of communication barriers. A word can have several connotations (implied meanings) and definitions. The more a word is prone to multiple meanings, the greater are the chances of it being misunderstood. A conscientious speaker is careful to explain her message in context by rephrasing and repeating words that can be confusing.

Confusables

Common groups of words are sometimes confused or ignored by users.
- Ability is a skill that you have mastered through study or practice. Capacity refers to innate talent.
- If something affects you, it has an effect on you. To effect something, however, means to make something happen: 'The new management wanted to effect drastic changes, but the unions felt these would adversely affect workers.' Refer to Chapter 16 for more on this.

Wrong inferences

Suppose you have returned from a business trip and you find that two of your colleagues are absent. They do not turn up for several days. Since there is a recession on, you draw an inference that they have been laid off. The fact is that they have been promoted and sent to another department. This is an example of *fact-inference confusion*. It has happened because you failed to distinguish between what actually exists and what you had assumed to exist.

Inferences are more dramatic than facts, and for this reason they can provide more scope for gossip and rumour to spread. When professionals analyse material, solve problems, and plan procedures, it is essential that inferences be supported by facts. Systems analysts, marketing specialists, advertisers, architects, engineers, designers, and others must work on various premises and draw inferences after collecting factual data. When presenting any inference in the course of your work, you could use qualifiers such as 'evidence suggests' or 'in my opinion' to remind yourself and the receiver that this is not yet an established fact.

Blocked categories

In general, we react positively to information only if it is in consonance with our own views and attitudes. Conversely, when we receive information that does not conform to our personal views, habits, and attitudes, or appears unfavourable to us, we tend to react negatively or even disbelieve it. Rejection, distortion, and avoidance are three common, undesirable, and negative reactions to unfavourable information.

Communication and other technologies are advancing so rapidly today that many people find it difficult to quickly adapt themselves to these developments. Instead of taking advantage of these developments, which help expedite the communication process, such people tend to resist and criticize them. This is a result of having a closed mind. Such people are called *misoneists*. They tend to ignore variations and differences, which leads to unreliable conclusions. Some people have certain prejudices so deeply embedded in their mind that these cannot be challenged.

Similarly, people who are very rigid in their opinions may face problems in communicating effectively. For example, one of your fellow students may think that only students of science are good in reasoning; another might be of the opinion that young executives are more efficient than older ones. Such people fall into blocked categories, because they may not be able to accept any deviation from their points of view.

When Writing or Speaking

1. Think of your audience. How well do they understand the language? How much do they know about your logic?
2. Select your words with care, especially technical terms. Words must be suited to your audience's language skills.
3. If you have to use slightly difficult or unfamiliar words, try to explain these as part of your communication.
4. Do not fall into the trap, however, of oversimplifying your language; your audience could get put off if you use words that sound unprofessional or non-technical.
5. Feedback serves as an effective barometer to find out if the intended message has been put across. Ask the receiver to paraphrase the message and also ask questions on what was said.
6. Even if you have an extensive vocabulary, never use words merely to impress. Rather use them to express your ideas as simply and clearly as possible.

Categorical thinking

People who feel that they 'know it all' are called *pansophists*. This type of thinking exists in people who feel that they know everything about a particular subject, and therefore refuse to accept any further information on that topic. For example, in a general body meeting of your organization, you are to be briefed about the annual budget. However, you do not pay attention because you feel you have already been briefed about it by your secretary the previous day. Later you propose that new vehicles have to be bought. Imagine your embarrassment when you realize that the topic was discussed and a decision has already been taken in the general body meeting. This type of thinking can pose a major barrier, leading to a failure in communication. In such instances, the receivers refuse information because of their 'know-it-all' attitude.

The clue to detecting this barrier in ourselves and in others is the use of words like *all, always, everybody, everything, every time* and their opposites like *none, never, nobody,* and *nothing*. If a message contains too many of these words, then there is a fair chance of the communication getting distorted. To avoid this barrier, substitute these words with phrases like 'in most situations' or 'most likely'. Label your opinions with phrases like, 'it appears to me' or 'the evidence indicates'. If your data is insufficient, it is better to admit that you are unaware of the rest of the information rather than being indirect. To sum up, good communicators should:

- Be non-judgemental
- Be empathetic
- Not assume anything
- Stick to the subject
- Listen, and above all, paraphrase
- Remember that generalizations do not always hold good in all situations

Interpersonal Barriers

Intrapersonal barriers stem from an individual's attitudes or habits, whereas interpersonal barriers occur due to the inappropriate transaction of words between two or more people. The two broad categories into which these barriers can be classified are:

- Inefficient communication skills
- Negative aspect nurturing in the climate

The second point refers to a situation when negative tendencies nurtured by some people affect others around them. This leads to a barrier as individuals start thinking only negative.

Interpersonal barriers creep in as a result of the limitations in the communication skills of the encoder or the decoder, or of both. In addition, they may also occur because of some disturbance in the channel or medium of communication. If two people are involved in communication, the traits that distinguish them as individuals can be the root cause of a communication problem.

In a business environment, we neither attempt to change these traits, nor can we do it; however, we can try and understand the role of differences among individuals that lead to communication breakdowns. The most common reasons for interpersonal barriers are:

- Limited vocabulary
- Incompatibility (clash) of verbal and non-verbal messages
- Emotional outburst

- Communication selectivity
- Cultural variations
- Poor listening skills
- Noise in the channel

Limited vocabulary Inadequate vocabulary can be a major hindrance in communication. At times, we find ourselves searching for the exact word or phrase that would be appropriate for what we are trying to express. For example, during a speech, if you are at a loss for words, your communication will be very ineffective, and you will leave a poor impression on the audience. On the other hand, if you have a varied and substantial vocabulary, you can create a favourable impression on your listeners.

Merely having a wide vocabulary is of no use unless the communicator knows how to use it. In communication, the denotative (literal or primary) and connotative (implied or suggested) meanings of the words used should be absolutely clear to the receiver. Therefore, one should make constant efforts to increase one's vocabulary by regularly reading a variety of books and listening to native speakers of the language. Thereafter, using a wide vocabulary regularly will also help to make its usage comfortable and natural. Chapter 16 discusses vocabulary development in detail.

Incompatibility of verbal and non-verbal messages

Imagine a situation where your CEO introduces the newly recruited middle-level manager to the other employees. In a small speech, he conveys the message that he is very delighted to have the new manager appointed in his office. However, the expression on his face shows just the opposite of what he is saying. The stark difference between the verbal and non-verbal aspects of his communication leaves his listeners feeling confused and puzzled.

A communicator should acclimatize himself to the communication environment, think from the angle of the listener, and then communicate. Misinterpreted non-verbal communication acts as another barrier to effective information flow instead of enhancing and enlivening verbal communication. Non-verbal cues provide a deeper insight into the sender's message. Ignoring non-verbal cues or misinterpreting them can result in the message being completely misunderstood. Thus, one should not only try to accurately gauge others' non-verbal cues but also be aware of one's own body language.

Generalizations, based on assumptions about physical appearance or dress, can also lead to severe communication barriers. Physical appearance often serves as one of the most important non-verbal cues. For instance, many people initially formed negative impressions of Einstein because of his worn-out appearance.

'Did whatever I said in the last half hour make any sense to you?'

Guidelines to improve your appearance:

- Dress according to the occasion.
- Wear neat and clean clothes.
- Choose an appropriate hairstyle.
- Wear clean and polished shoes.

> The first impression about people is most often made on the basis of their physical appearance, which significantly affects the quality of communication.

While interpretation of non-verbal cues requires keen observation, there are also pitfalls to guard against. For instance, there is great disparity in the use and interpretation of non-verbal messages across countries and cultures. For example, in Kenya, a mother-in-law and a son-in-law avoid eye contact. In fact they turn their backs to each other. In America, this would be a sign of disrespect.

In brief, your non-verbal cues should consistently match your verbal messages, adding to their effectiveness and enhancing your image as a competent and interesting communicator.

Emotional outburst

Imagine that you are the President of a well-established company. There are rumours floating amongst your employees that you have indulged in fraudulent activities. You are fully aware that these rumours are baseless. However, when you are asked to address the same employees, you are unable to put your point across, as you are flushed with anger. Despite the fact that you are a confident public speaker, your communication failed as you were overwhelmed by your emotions.

In most cases, a moderate level of emotional involvement intensifies communication, making it more personal. However, excessive emotional involvement can be an obstacle in communication. For example, extreme anger can create such an emotionally charged environment that a rational discussion becomes impossible. Likewise, prejudice, stereotyping, and boredom all hinder effective communication. Positive emotions such as, happiness and excitement, also interfere in communication, but to a much lesser extent than negative feelings.

Emotions are an integral part of our being, whether in business or in personal encounters. By sharpening self-awareness, intuition, and empathy, emotions can help in developing an environment that is highly conducive to good communication. Yet, situations often arise where people react negatively. Depending on their nature and the situation, this negative reaction may be classified as hostile or defensive. Hostility can be considered as a move to counter-attack the threat, whereas defensiveness is resistance to it. Both reactions occur when the receiver of the message perceives some kind of threat. Both these responses have an extremely negative impact on the communication. Messages are misinterpreted, ignored, or overreacted to by people displaying such behaviour. Those who witness such behaviour are most likely to lower their opinion of such people.

It is important to maintain one's composure in all kinds of communication. Viewing issues from different perspectives helps develop objectivity and rational thinking, which in turn can eliminate many of the causes of hostility or defensiveness. When confronted with such negative behaviour, it is essential to avoid reacting. The person displaying these emotions should be calmed down. They should preferably be taken to a quiet place to try and sort out the problem that caused their emotions to spin out of control.

Communication selectivity

When the receiver in a communication process pays attention only to a part of the message, he/she is imposing a barrier known as communication selectivity. This happens because he/she is interested only in that part of the message which may be of use to him/her. In such a situation, the sender is not at fault. It is the receiver who breaks the flow of communication.

Take for example, a meeting held by the CEO of a company. She has called all her senior executives from various divisions—production, marketing, finance, human resource (HR), etc. During the meeting she discusses diverse topics. However, she may not be able to get the entire message across to each one of the participants, unless she gets their undivided attention. If the production manager and the marketing manager pay attention only to matters related to their respective areas, they may not be able to get the total perspective of what the CEO is conveying.

Communication selectivity may act as a barrier in written forms of communication as well. While reading any document, if you read only the parts you consider useful, you are posing this barrier.

Cultural variations

This is one of the predominant interpersonal factors contributing to communication failure. As businesses are crossing national boundaries to compete on a global scale, the outlook of the global and domestic workforce has changed drastically. European, Asian, and American firms have expanded their businesses worldwide to create international ties through partnership, collaborations, and affiliations. The management and employees of such companies need to closely observe the laws, customs, and business practices of their host countries, while dealing with their multinational workforce. To compete successfully in such a business environment, one must overcome the communication inadequacy arising due to different languages and cultures.

This factor holds good in the area of education as well. You will prove to be a successful communicator abroad, during the course of your higher studies, if you take pains to understand the culture of the educational campus in which you would be studying. Success, whether as a student or as a professional, lies in knowing the business practices, social customs, and etiquette of the particular country one is dealing with.

Poor listening skills

A common obstacle to communication is poor listening habits. We should remember that listening and hearing are not the same. Hearing is a passive exercise while listening requires careful attention and accurate decoding of the signals received from the speaker. Misunderstandings and conflicts can be avoided if people listen to the message with attention. The various distractions that hinder listening can be emotional disturbances, indifference, aggression, and wandering attention.

Sometimes, an individual is so engrossed in his/her own thoughts and worries that he/she is unable to concentrate on listening. If a superior goes on shifting the papers on his/her desk while listening to his subordinate, without making eye contact with the latter, he/she pays divided attention to the speaker's message. This divided attention adversely affects the superior–subordinate relationship, besides distorting the communication. Chapter 4 discusses listening skills in detail.

Noise in the channel

As discussed earlier, noise interferes in the transmission of signals. Noise is any unwanted signal that acts as a hindrance in the flow of communication. It is not necessarily limited to audio disturbances, but can also occur in visual, audio-visual, written, physical, or psychological forms. All these forms of noise communicate extraneous matter which may distract the receiver from the message, and even irritate him/her.

Technical or physical noise refers to the din of machines, the blare of music from a stereo system, or other such sounds which make the task of the listener difficult. Human noise can be experienced when, for instance, employees gather for a meeting and a member arrives late distracting everybody's attention. Disturbances in telephone lines, poorly designed acoustics of a room, dim typescripts, and illegible writing are some more examples of technical noise.

Organizational Barriers

Communication barriers are not only limited to an individual or two people but exist in entire organizations. Every organization, irrespective of its size, has its own communication techniques, and each nurtures its own communication climate.

In large organizations where the flow of information is downward, feedback is not guaranteed. Organizations with a flat structure usually tend to have an intricately-knit communication network. Irrespective of size, all organizations have communication policies which describe the protocol to be followed. It is the structure and complexity of this protocol that usually causes communication barriers.

Most large companies are realizing that a rigid, hierarchical structure usually restricts the flow of communication. This is because there are numerous transfer points for communication to flow in these hierarchical systems, and each of these points has the potential to distort, delay, or lose the message. To obviate this, there should be direct contact between the sender and the receiver with minimum transfer stations. If the message is presented orally, this further reduces the dependence on transfer stations.

The main organizational barriers are as follows:

- Too many transfer stations
- Fear of superiors
- Negative tendencies
- Use of inappropriate media
- Information overload

Too many transfer stations

The more links there are in a communication chain, the greater are the chances of miscommunication. Imagine, for instance, that your professor asks you to convey a message to X. You, because of some inconvenience or sheer laziness, ask your friend Y to do this job. Now, there are four people involved in this communication channel. Let us see how the message gets distorted as a result of the increased number of transfer stations:

> Professor: X was supposed to meet me today regarding the submission of an assignment on Magnetic Theory. But I want X to meet me on Friday, as I am going out of station tomorrow.
>
> You (to your friend Y):   Ask X to meet the professor tomorrow, regarding the assignment, as the professor is going out of station today.
>
> Your friend Y (to X):   X, you have to meet the professor today as he will not be available tomorrow.

This is an example of how messages get distorted in huge organizations with several layers of communication channels. The message gets distorted at each level not only because of poor listening or lack of concentration, but also because of several other reasons. Some employees may filter out the parts of the message they consider unimportant. Whatever the reasons for filtering or distorting the message, having too many transfer stations is always an obstacle to effective communication and should be avoided. Transfer stations do serve a purpose, but having too many of them is counter-productive.

Fear of superiors

In rigidly structured organizations, fear or awe of superiors prevents subordinates from speaking frankly. An employee may not be pleased with the way his/her boss extracts work from him/her but is unable to put his/her point across because of fear of losing the boss's goodwill. As a supervisor, it is essential to create an environment which enables people to speak freely. An open environment is conducive to increasing the confidence and goodwill of a communicator.

To avoid speaking directly to their boss, some employees may shun all communication with their superiors. At the other extreme, they may present all the information they have. This is because they feel that they will be viewed in an unfavourable light by leaving out some vital information. In written communication, this results in bulky reports, where essential information is clubbed with unimportant details. Such unfocussed messages result in a lot of wasted time. Such practices need to be eliminated by superiors to ensure that communication flows effectively in their organizations. Moreover, by encouraging active participation from their subordinates, senior officers pave the way for more ideas, resources, or solutions to come forth from their subordinates.

Negative tendencies

Many organizations create work groups. While some groups are formed according to the requirements of the task at hand, such as accomplishing a particular project, many other small groups are also formed for recreational, social, or community purposes. These groups may be formal or informal, and generally consist of people who share similar values, attitudes, opinions, beliefs, and behaviours. Nevertheless, on some occasions, a communication barrier can exist due to a conflict of ideas between the members and non-members of a group.

> In organizations with many levels of communication, messages have a greater chance of being distorted. This occurs due to poor listening, lack of concentration, or a person's tendency to leave out part of the message.

For example, the student members of the sports club of an educational institution may be annoyed with non-members who oppose the club's demand for allocating more funds to purchase sports equipment. This type of opposition gives rise to insider–outsider equations, which in turn pave the way for negative tendencies in the organization. Once these negative tendencies develop, they create noise in interpersonal communication.

Use of inappropriate media

Some of the common media used in organizations are graphs and charts, telephones, facsimile machines, boards, email, telephones, films and slides, computer presentations, teleconferencing, and videoconferencing. While choosing the medium for a particular occasion, the advantages, disadvantages, and potential barriers to communication must be considered. While deciding upon the medium, the following factors should be considered:

- Time
- Cost
- Type of message
- Intended audience

The telephone, for instance, would not be an ideal medium for conveying confidential information. Such messages are best conveyed in person or, if the receiver is located in another office, by private chat messenger. Printed letters, which provide permanence, are preferable for information which requires to be stored for future reference. Usually, a mix of media is best for effective communication. For example, after booking an order online, a follow-up call can be made to verify whether the order has been placed.

Information overload

One of the major problems faced by organizations today is the decrease in efficiency resulting from manual handling of huge amount of data. This is known as *information overload*. The usual results of information overload are fatigue, disinterest, and boredom. Under these circumstances, further communication is simply not possible. Very often, vital, relevant information gets mixed up with too many irrelevant details, and is therefore ignored by the receiver. Thus, the quality of information is much more important than the *quantity*.

To reduce information overload in an organization, screening of information is mandatory. Messages should be directed only to those people who are likely to benefit from the information. Major points should be highlighted, leaving out all irrelevant details.

Bearing in mind all these possibilities and reasons for communication failure, one can take pre-emptive measures to avoid these barriers.

Overcoming Barriers

Constant practice and rigorous implementation of these ideas will help you become an excellent communicator.

- Always keep the receiver in mind.
- Create an open communication environment.
- Avoid having too many transfer stations.
- Do not communicate when you are emotionally disturbed.
- Be aware of diversity in culture, language, etc.
- Use appropriate non-verbal cues.
- Select the most suitable medium.
- Analyse the feedback.

SUMMARY

Communication fails because often the message sent is not always the message received. Various interruptions or barriers prevent the proper passage of information from sender to receiver. This failure can be attributed to various types of 'noise', which could exist either at the source, in the channel, or at the receiver.

If a speaker does not see the desired response from the audience, he/she must identify the problem, find the cause or barrier, work on alternative solutions, select the best solution, and follow up rigorously to ensure that this barrier does not come up again.

Barriers to communication are classified as intrapersonal, interpersonal, and organizational. Intrapersonal barriers occur because of individual attributes, such as wrong assumptions, varied perceptions, differing backgrounds, wrong inferences, blocked categories, and categorical thinking. Interpersonal barriers occur due to inappropriate transactions of verbal and non-verbal messages between two or more people. The different barriers are limited vocabulary, incompatibility, or clash of verbal and non-verbal messages, emotional outburst, communication selectivity, cultural variations, poor listening skills, and noise in the channel.

Organizational barriers stem from organizational attributes such as too many transfer stations, fear of superiors, negative tendencies, use of inappropriate media, media, and information overload. Once we know the reasons for failure of communication, we should take pre-emptive measures to overcome it.

EXERCISES

1. Identify the communication barrier that describes each of the following situations:
   (a) 'Every time I have a meeting with Mr Gupta, I end up disagreeing with him about a particular issue.'
   (b) *Manager*: 'Reeta, where is the report that I asked you to submit on the financial matters of the company?'
   Reeta: 'I do not remember you asking me to submit a report.'
   (c) *Teacher*: 'Students, why have you not submitted the report within the fortnight?'
   Students: 'But you asked us to submit it bimonthly!'
   (d) 'This room is horrible to work in. I am able to hear everyone around, and there is no scope for privacy.'
   (e) 'It is quite tedious to manually work on the students' records, but I fear using the computer as it might corrupt all our data.'
   (f) 'Why every time I get a meeting with John, I usually end up showing my disinterest with him about a particular topic?'
   (g) 'If you want some more information from me, ask only the specific questions and do not waste my precious time!'

2. Identify and explain a communication barrier which may hinder each of the process components given in the schematic representation of the human communication process.

3. 'A barrier acts like a sieve, allowing only a part of the message to filter through; as a result, the desired response is not achieved.'
   Keeping in mind the above statement mention the various types of interpersonal barriers which hinder the communication process. Substantiate your answer with suitable examples.

4. Imagine you are the Sales Manager in Ramanath Paper and Pulp Company. Some of your co-employees are spreading rumours that you are involved in fraudulent activities. In order to defend yourself, you are asked to address the same co-employees. You are angry due to these rumours and you find it difficult to put your views before them. Identify the communication barrier that hinders this communication situation.

5. Explain the following terms with reference to communication barriers and give two examples for each term.

   (a) Organizational barrier
   (b) Emotional outburst
   (c) Cultural differences
   (d) Information overload
   (e) 'Know-it-all' attitude

6. You have been assigned to host a group of American university students who are visiting your institute for the next two weeks. What can you tell them that will help them fit into the culture on your campus? Make a list of the important behavioural rules they should understand in order to communicate effectively with students and faculty on your campus. Also point out some problems that might occur if the American students disregard these rules.

7. Identify the barriers that lead to miscommunication in the following scene. How can the manager overcome this barrier?

'I called for a meeting with the supervisors. But none turned up . . . nobody pays attention to me in this place . . . it would be better to quit'.

8. Write the possible solutions to overcome these barriers:
   (a) Dealing with people working in isolated office or environment.
   (b) Dealing with a customer who is very talkative
   (c) Working in a noisy surrounding
   (d) Dealing with a frustrated and angry client
   (e) Dealing with visual distractions at workplace

9. Give at least two situations for the barriers mentioned below:
   (a) Physical barrier
   (b) Negative tendency
   (c) Wrong inferences
   (d) Transfer station
   (e) Difference in background and language

CHAPTER 3
Non-verbal Communication

OBJECTIVES

You should study the chapter to know
- ○ the meaning of non-verbal communication
- ○ the different aspects of non-verbal communication such as kinesics, proxemics, chronemics, and haptics
- ○ cross-cultural communication differences

Introduction

Effective communication takes into account both the verbal and non-verbal aspects of communication. While verbal communication is organized by language, non-verbal communication is not. This chapter discusses non-verbal communication. Non-verbal communication refers to all communication that occurs without the use of words, spoken or written. It is concerned with body movements (kinesics), space (proxemics), and haptics (touch) features. It includes all unwritten and unspoken messages, both intentional and unintentional. Non-verbal cues, however, speak louder than words, as even though speech can be made up, bodily expressions can rarely be masked well enough to hide one's true feelings and emotions.

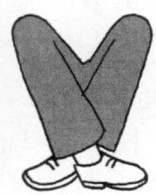

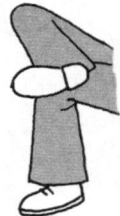

Crossed at the ankle　*Crossed at the knees*　*Open crossed with one ankle on the other thigh*　*Uncrossed and straight closed together*　*Uncrossed and straight far apart*

Personal appearances, facial expressions, postures, gestures, eye contact, voice, proximity, and touch are all non-verbal signals that influence the way in which a message is interpreted and understood. Though they have a profound impact on the receivers, it is difficult to analyse them accurately. This is because the interpretation of non-verbal cues is a very subjective concept, varying based on people's varied backgrounds (refer to cross-cultural communication). Nevertheless, they must not be ignored, but recognized and understood as correctly as possible. See the PowerPoint presentation on body language in the Online Resource Centre. See the GD and interview video to know more about the right use of body language.

Kinesics

> 'You cannot not communicate. You communicate just by being. Nodding your head, blinking your eyes, shrugging shoulders, waving hands, and other such physical activities are all forms of communication.'
> –Watzlawick and associates

> 'He that has eyes to see and ears to hear may convince himself that no mortal can keep a secret. If his lips are silent he chats with his fingertips; betrayal oozes out of him at every pore.'
> –Sigmund Freud

Kinesics is the study of the body's physical movements. It is the way the body communicates without words, i.e., through the various movements of its parts.

Some kinesic behaviours are deliberate. For example, you nod your head to indicate acceptance. While speaking, listening, reading, or writing, we consciously use words to receive or send ideas. Why do we use words? Because they are the primary symbolic forms that convey our thoughts. On paper, words remain static; however, punctuation marks are used to convey pauses, expressions, emotions, etc. But in face-to-face communication, the message is conveyed on two levels simultaneously. One is verbal and the other is non-verbal. For example, suppose you are congratulating two of your friends on their successful interviews. If you extend your hand to them with a big smile on your face along with the utterance, 'Congrats', your appreciation has more impact on them than the word in isolation. Your smile and the handshake are kinesics, which enhance the impact of your verbal communication.

The non-verbal part of any communication is not as deliberate and conscious as the verbal part. Rather, it is subtle and instinctive, and often involuntary. It is important to study body language because it is estimated that the verbal component of oral communication carries less than 35 per cent of the social meaning of the situation, while more than 65 per cent is attributed to body language. People react strongly to what they see.

Body Language

When a speaker presents himself/herself, we see him/her before we start hearing him/her. Immediately, we begin developing impressions of his/her abilities and attitudes based on the non-verbal signals he/she sends. This is why body language is so critical in oral communication.

Body language includes every aspect of our appearance, from what we wear, how we stand, look, and move, to our facial expressions and physical habits, such as nodding the head, jingling change in the pocket, or fiddling with a necktie. Our use of space and gestures are other key indicators.

Personal appearance Personal appearance plays an important role; people see before they hear. Just like we adapt our language to the audience, we should also dress appropriately. Appearance includes clothes, hair, accessories, cosmetics, and so on. Today, the purpose of clothing has altered from fulfilling a basic need to expressing oneself. Clothes also accentuate the body's movements, and the choice of clothes reveals a lot about the wearer's personality and attitude.

Personal appearance must be so planned that it communicates effectively to others. Even before a speaker utters his/her first syllable the audience begins to form an opinion about him/her and visualizes the way he/she is going to talk. One's appearance may put the audience into a resistant or hostile attitude or induce in them a receptive mood. To be clean and well groomed, conforming to the need of the occasion, is of utmost importance. Appearances communicate how we feel about ourselves and how we want to be viewed.

Posture Posture generally refers to the way we hold ourselves when we stand, sit, or walk. One's posture changes according to the situation. If nervous, one would normally be seen pacing, bobbing the shoulders, fidgeting with notes, jingling coins, moving constantly, or staying glued to the ground.

When we are with friends we are probably spontaneous. We are not conscious about our posture and our physical movement is natural. But when we encounter an unfamiliar situation, we become more conscious of our posture. For instance, during an oral presentation, stiff positions, such as *standing akimbo* (with hands on hips and elbows pointing away from the body), send the message of defiance or aggression. It is always better to lower the hands to one's sides in a natural, relaxed, and resting posture. Standing, sitting, or walking in a relaxed way is a positive posture, which will encourage questions and discussion. Also being comfortably upright, squarely facing an audience, and evenly distributing one's weight are aspects of posture that communicate professionalism, confidence, attention to detail, and organization. The way one sits, stands, or walks reveals a lot:

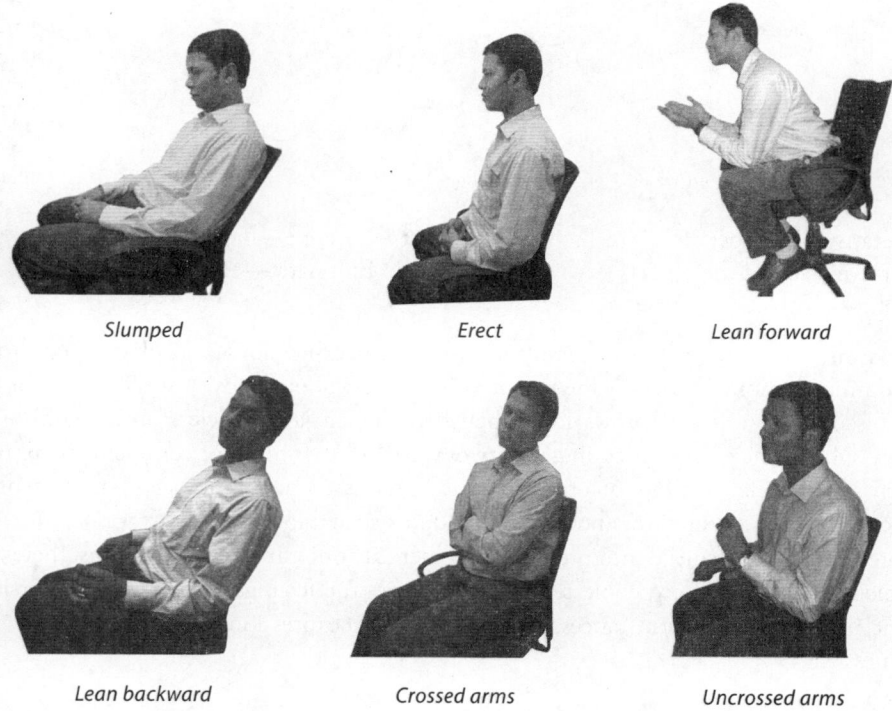

Slumped   Erect   Lean forward

Lean backward   Crossed arms   Uncrossed arms

- Slumped posture—low spirits
- Erect posture—high spirits, energy, and confidence
- Lean forward—open, honest, and interested
- Lean backward—defensive or disinterested
- Crossed arms—defensive and not ready to listen
- Uncrossed arms—willingness to listen

Gesture Gesture is the movement made by hands, head, or face. Skillful and appropriate gestures can add to the impact of verbal communication. A well-timed gesture not only drives a point home but also enhances the value of what is being said. Similarly, an awkward gesture (like playing with a key chain or button) can mar the effectiveness of the message.

Gestures clarify our ideas or reinforce them and should be well suited to the audience and occasion. Gestures are more numerous than any other form of non-verbal communication, and the meanings attached to them are diverse. It has been observed that there are as many as 700,000 varied hand gestures alone (Birdwhistell 1952), and the meanings derived from them may vary from individual to individual. Some hand gestures are shown in a PowerPoint presentation on body language in the Online Resource Centre.

Gestures should not divert the attention of the listener from one's message. They should be quite natural and spontaneous. Be aware of and avoid irritating gestures such as playing with a ring, twisting a key chain, clasping hands tightly, or cracking knuckles. Gestures can roughly be divided into the following types:

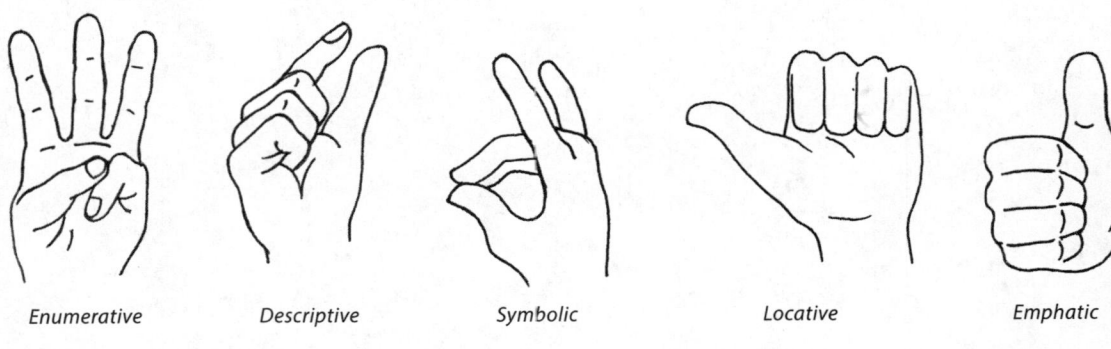

| Enumerative | Descriptive | Symbolic | Locative | Emphatic |

- Enumerative—numbers
- Descriptive—size of the objects
- Symbolic—abstract concepts
- Locative—location of an object
- Emphatic—emphasis

Facial expression Along with postures and gestures, facial expressions also play an important part in non-verbal communication. The face is the most expressive part of our body. A smile stands for friendliness, a frown for discontent, raised eyebrows for disbelief, tightened jaw muscles for antagonism, etc. Facial expressions are subtle. They can be used in a variety of ways to aid, inhibit, or complement communication. The face rarely sends a single message at a time. Instead, it sends a series of messages—facial expressions may show anxiety, recognition, hesitation, and pleasure in quick succession.

Facial expressions are difficult to interpret. Though there are only six basic expressions, there can be many shades and blends of these. Also, people tend to hide their true feelings, and project expressions that are appropriate according to the circumstances. The six basic facial expressions are:

- Happiness
- Disgust
- Anger
- Surprise
- Fear
- Sadness

Eye contact Eyes are considered to be the windows of the soul. We look at the eyes of a speaker to find out the truthfulness of his/her words, intelligence, attitudes, and feelings. Eye contact is a direct and powerful form of non-verbal communication. We use our eyes to cull information. Eyes are also a rich source of feedback.

Looking directly at listeners builds rapport. Prolonging the eye contact for three to five seconds (without, however, giving the impression of staring) tells the audience that the presenter is sincere in what he/she says and that he/she wants us to pay attention. Eye contact is especially important when we start a conversation.

Our upper eyelids and eyebrows help us convey an intricate array of non-verbal messages. Arabs, Latin Americans, and South Europeans look directly into the eyes. Asians and Africans maintain far less eye contact. In the professional world one should make personal and pleasant eye contact with the listeners. Eye contact shows one's intensity and elicits a feeling of trust. A direct look conveys candour and openness. This direct and powerful form is a signal of confidence or sincerity; therefore, experienced speakers maintain longer eye contact. The eyes should convey the message, 'I am pleased to talk to you, do believe in what I am saying?'

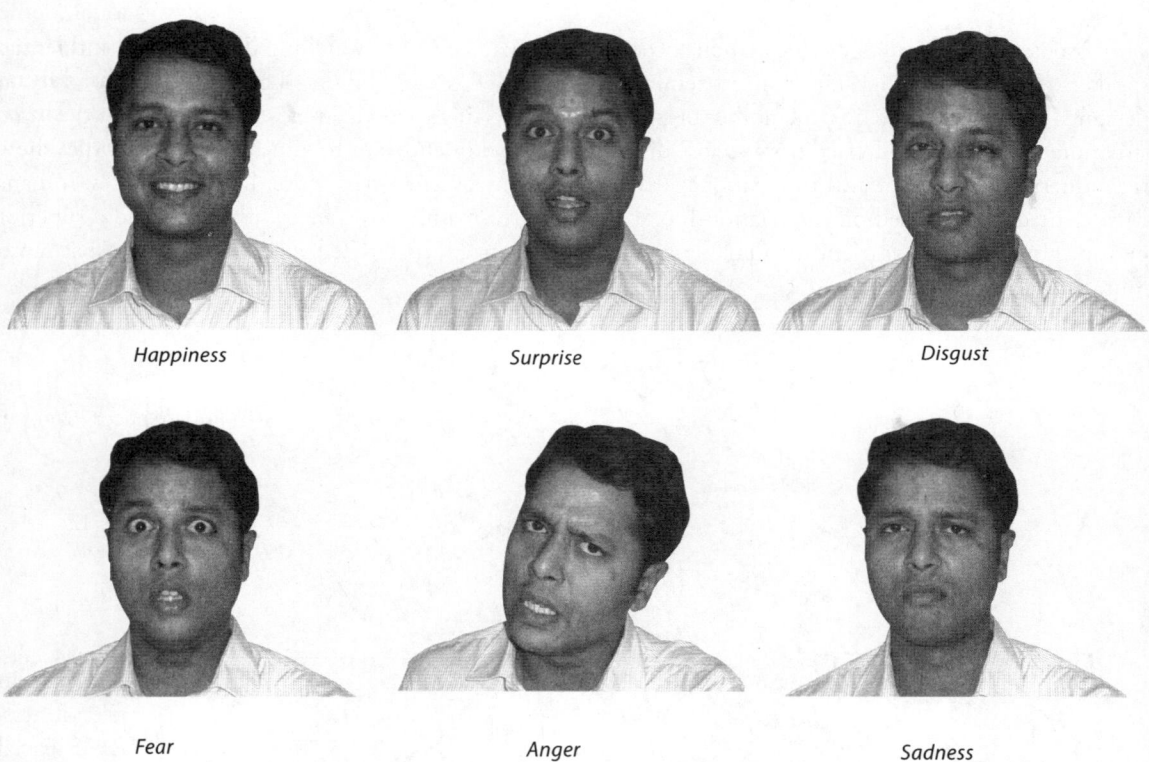

Happiness   *Surprise*   *Disgust*

Fear   *Anger*   *Sadness*

Oculesics is the systematic study of eye contact, including eye movement, gaze, and pupil dilation. The eyes are the most expressive organ on the human face. You can make out the feelings of an individual by looking at the eyes. Eyes play a crucial role in exuding emotions. Eyes can express happiness, boredom, sorrow, envy, excitement, ecstasy, etc.

Eye contact and gaze   When two people involved in communication look at each other reflecting interest in communication, they make eye contact. However, when a person simply stares at another person, it is gaze. Eye contact justifies the genuineness in conversation. Contrary to this, if you indulge in communication without eye contact, you will be labelled untruthful, dishonest, and unstable. The practice of eye contact varies culture to culture. In some cultures, eye contact is preferred, while in others, if you make eye contact, you are considered shameless. In the USA, looking into the eyes of the other person while communicating suggests truthfulness and openness in communication. However, in Korea, Japan, and South Asian countries, including India, you are not supposed to look into the eyes of other people, especially elders. Culture has direct implications on eye contact, as direct eye contact in South Asian culture is considered a sign of disrespect and has sexual connotations are attached to it. Sometimes if a person is shy, direct eye contact becomes difficult for him/her. People are often labelled rude for this particular reason.

Proxemics

Proxemics is the study of physical space in interpersonal relations. Space is related to behavioural norms. The way people use space says a lot about them. In a professional setting, space is used to signal power and status. For instance, the head of a company has a larger office than junior employees.

Gestures should be in accordance with the space available. When there is plenty of space to manoeuvre, one should move more boldly and expand one's gestures. When seated at a table, one should use milder gestures. One can even subtly reach out over the table to extend one's space. This expresses control and authority.

It is possible to learn a great deal about how to manipulate space by watching dynamic and influential speakers. Interestingly, like kinesics, proxemics also has cultural variations. A Latin American or French person is likely to stand closer to another person when conversing than an Anglo-Saxon would in the same situation. Americans, addressed from a close distance, may feel offended or become aggressive. Studies show that Americans, unlike many other nationals, avoid close contact with one another in public places. Indians decide the distance based on the relationship. They prefer to maintain distance with elders and a superior person. However, with a friend they may maintain less distance. Edward T. Hall (1966) divides space into four distinct zones (see Figure 3.1).

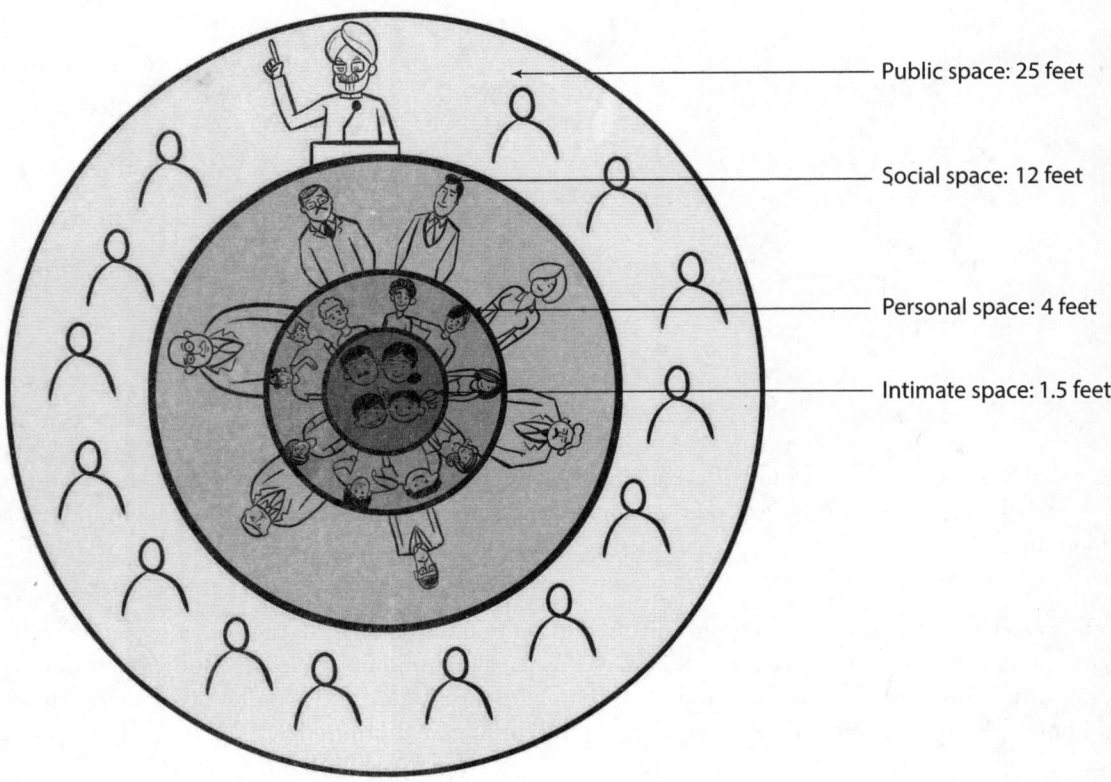

Public space: 25 feet

Social space: 12 feet

Personal space: 4 feet

Intimate space: 1.5 feet

Figure 3.1   The four distinct space zones

Intimate   This zone starts with personal touch and extends just to 18 inches (one and a half feet). Members of the family, lovers, spouses, relatives, and parents fall under this zone. The best relationship that describes it is the mother–child relationship. This zone does not need active conversation. One can whisper or make unintelligible sounds but still be able to communicate. Other individuals come close for a very brief period and only under special circumstances—when they want to congratulate, sympathize, or console. A handshake, a pat on the back, or a hug, all come into this zone.

Personal   This zone stretches from 18 inches (one and a half feet) to 4 feet. Close friends, colleagues, peers, etc. fall in this zone. Instead of whispering sounds or utter silence, there can be normal conversation in this zone. Though this zone is personal, it is quite a relaxed and casual place. It permits spontaneous and unplanned communication. Sitting or standing so close brings one closer to the listener and gives the impression of friendliness and warmth.

Social   Social events take place in the radius of 4 feet to 12 feet. In this zone, relationships are more formal and official. People are more cautious in their movements. These situations involve less emotion and more planning. The number of people decides whether it should be a sitting–sitting or sitting–standing position. It is through experience that one decides which position to take. If the number of people is less and eye contact can be maintained, a sitting–sitting position can be used. To be authoritative with a large audience, a sitting–standing position is used.

Public   This zone starts from 12 feet and may extend to 25 feet or to the range of eyesight and hearing. Events that take place in this zone are formal. Here the audience views what is happening as an impartial observer. The degree of detachment is very high. The audience is free to do whatever it feels like. Here the speaker has to raise his/her voice to communicate to others or use a microphone. Public figures like the prime minister of a country, for example, have to maintain this distance for security reasons.

Chronemics

Chronemics is the study of how human beings communicate through their use of time. How do we communicate with others? What does time mean to us? In order to use time as an effective communication tool, we should understand its impact on the various aspects of our lives and act accordingly. We must attempt to use time as effectively as possible.

In the professional world, time is a valuable resource. When we are late for an appointment, people react negatively. If we arrive early, we are considered either over-eager or aggressive. So, we should always be *on time*. By valuing someone else's time, we communicate our professionalism or seriousness both subtly and explicitly.

People have their own *time language*. To one person who wakes up at 8:00 a.m., 6:30 a.m. may be early; to another, 8:00 a.m. may be late if he/she wakes up at 5:00 a.m. every day!

Time language also varies from culture to culture. In Latin countries, meetings usually begin well after their appointed time. Everyone knows this. It is customary, and no one is offended by the delay. In Scandinavia or Germany, on the other hand, strict punctuality is the rule, and tardiness is frowned upon. In India, time language varies according to the occasion. Punctuality is expected for a professional meeting, but it is not insisted upon for a party. People in India are generally liberal with time.

Haptics

Haptics is a systematic study of touch in communication. It has derived its meaning from the Greek word 'hapticos', which means touch. Haptics is part of non-verbal communication where people use touch to communicate. Sense of touch is crucial as it is an indispensable component of communication and it provides meaningful transmission of feelings during vulnerable times. Touch is an integral part of interpersonal relationship signifying intimacy with the other person. When an individual makes an impactful presentation and gets a pat from the senior, it has lasting impact. Similarly, if one had experienced an abysmally sad event, touch can have a healing effect.

Touch can be classified into different categories, like positive touch, serious touch, control touch, negative touch, intimate touch, etc. Each one has a specific communication message attached to it.

Correlating Verbal and Non-verbal Communication

Imagine you are giving a presentation in front of your professor and friends. They all seem to be listening to you, but their non-verbal behaviour indicates boredom and restlessness. Somewhat puzzled and unsure of yourself, you seek their agreement and several of them concur verbally through verbal expressions such as 'Great!', 'Wow!', 'Perfect!', etc. Nevertheless, their non-verbal language conveys the impression that they are far from confident about the presentation. What would you do in such a situation? You should try to understand the non-verbal cues and pause the presentation to ask a few questions. Questions usually get people involved and make a presentation more interactive and meaningful.

'Words conceal but actions reveal'. This is true because when we speak to somebody, we are constantly sending some non-verbal cues as well. The way we use our voice, our body language including our facial expressions, posture, gestures, eye contact, and the distance we maintain add meaning to the words spoken, or modify the verbal message being conveyed. The tone of our voice can change the meaning perceived from positive to sarcastic, and the stress points of our sentences can highlight the specific points and subtly change the meaning of our utterances.

We should learn to recognize patterns of non-verbal language, beginning with our own. We should always be aware of how non-verbal language operates throughout the organization. According to psychologists, people use non-verbal behaviour to express their emotional attitudes: the degree of like and dislike towards others, the degree of dominance and submissiveness, and the degree of responsiveness, i.e., the intensity of positive and negative feelings aroused in them by others.

A good body posture is usually an indicator of confidence and uprightness. However, without mental and emotional confidence, our words will sound hollow to the audience. For instance, a used-car salesman from a dubious franchise may have a great body posture, and greet you with a warm smile and a firm handshake. However, if in his heart he sees you as just another customer he can take for a ride, then sooner or later, his internal conflict between what he says and what he really thinks will cause him to trip up. His movements and gestures will start giving away his real intentions. You will start feeling uncomfortable around him, even though you may be unable to pinpoint why. However, if the same used-car salesman is genuinely interested in helping you find the right car and puts your needs before his own, then his words and actions will be in harmony with his underlying intentions and you will instinctively trust him, even though you might not be able to identify the reason for such trust.

Non-verbal communication can be divided into two categories—conscious and unintentional. When speaking of the former category, one can think of the silent pauses a speaker takes to emphasize on some point. Also when someone does not intend to continue a conversation in the direction it is taking, they play with their key rings or some other accessory or they avoid making eye contact and look somewhere else. In unintentional non-verbal communication, one is not trying to express certain feelings and thoughts but displays them involuntarily. For instance, you might have observed that when people tell lies, they touch their face unintentionally. However, somebody good in interpreting body language may be able to catch this.

Eyes play a very important part in both intentional and unintentional non-verbal cues. When someone gives you a cold stare, you feel unwelcome. When a speaker makes a point and looks at you for longer than usual, he/she might be trying to say that the point is especially relevant to you. Eyes betray feelings that people try to hide. Hand movements are also very effective non-verbal communication. Usually, hands are used more for conscious communication and do not give unintentional cues. However, when people are tensed or worried, their hand movements give away their real feelings. Even if they utter some confident words, their eye contact, facial expressions, or gestures will reveal the truth. Hence, it is generally said that when your verbal and non-verbal messages do not match, your listeners will rely more on your non-verbal cues.

There are times when we recognize non-verbal cues without putting conscious thought into it. For example, when someone puts their palm up, it is understood that they want to end the conversation. Or when you are speaking, and the listener suddenly jerks their head towards you; you know you have said something of interest.

Cross-cultural Variations

This age of globalization and information technology has entirely changed the face of governments, businesses, and organizations. People are not confined to the geographical walls of their own nations anymore, but have become part of an international network. Communication being the backbone of inter- and intra-organizational coordination, it is essential for people to comprehend the linguistic and cultural differences among organizations to get the desired results at the workplace.

> 'Preservation of one's own culture does not require contempt or disrespect for other cultures.'
> –Cesar Chavez

It is not simple to define culture in concrete terms. Culture is a complex concept, with a variety of definitions. The dictionary meaning of the word 'culture' is a group or community with which we share common experiences that shape the way we understand the world. It consists of groups that we are born into, such as gender, race, or national origin, etc. It also includes groups we join or become part of, or the new habits we acquire as we interact with different people throughout our lives. Culture consists of various elements such as language, religion, politics, etc. Let us now discuss each element individually.

Language

Language forms the basis of all communication. It includes spoken, written, and body language. As mentioned, we are no longer restricted to one state or country during the course of our profession, relations, etc. As such, we often have to deal with people who speak different languages.

Religion and beliefs

Another important element of culture is religion. An individual's religious beliefs and norms, sacred objects, philosophical systems, prayers, and rituals are all parts of culture. Religion and belief affect the communication process. For example in India, some religious rites have become integral to our culture and have been adopted by people from other religions. Before starting any important project, it is a norm in India to follow certain religious rites and offer prayers. In conferences, it is observed that before starting the deliberation, prayers are offered and lamps are lit.

Values and attitudes

Different values and attitudes of individuals towards time management, decision-making, achievement, work, change, etc. are also important parts of culture. These attributes affect communication between people with differing values to a great extent. It becomes important for people of different cultures to respect and appreciate each other's values to be able to communicate effectively without adversely affecting their sensitivities. In the Indian culture, e.g., little importance is attached to the personal space and privacy of an individual. For example, when you meet someone and think of striking up a conversation, you can ask questions about that person's marital status or where he or she works or lives. Some South American and Mediterranean countries also allow this liberty. On the other hand, in Western countries, such unnecessary questions will be treated with a lot of apprehension and be considered highly inappropriate. In India, people do not place much importance on punctuality. It is fashionable to walk into a party late, or acceptable to dial into a teleconference five minutes later. Our Western counterparts may look at these as signs of disrespect. Gradually, in the global workplace, people of various cultures are learning to accept and respect each other better than they could a decade earlier. Knowing these basic differences and shaping our reactions accordingly are essential for overall successful communication.

Politics and law

The political system of a nation consists of national intents, power, ideologies, political risks, sovereignty, law of the country in which the organization works, rules and regulations imposed by the government, etc.

Technology

Technology includes scientific make-up, invention, communication media, urbanization, etc. and these are all essential parts of culture. For example, with the growth of information and communication technology, we have seen a tremendous change in the social and cultural framework of urban India.

Social organization

Social organization is an important element of culture. It consists of social institutions, the authority structure, interest groups, and status systems, etc. All these elements constitute the culture of a nation. For instance, maintaining eye contact with a senior during conversation was considered disrespectful in our culture until a few years ago. However, all this is changing in the professional world. Most organizations are adopting a corporate culture that might take some time to get oneself aquainted with. In most multinational companies, it is okay to address one's senior with his/her first name. It is an American tradition, and does not show disrespect or over-familiarity like we think it does. So, one must be cognizant of the various hidden conventions that are prevalent in the professional world.

Significance of Understanding Culture

There are different cultural groups in the world with different patterns of behaviour, religions, languages, politics, values and norms, etc. Thus, the same action is interpreted differently in different nations. For example, 'thumbs up' in America means approval, but is considered vulgar in Iran and Ghana (see the PowerPoint presentation on body language in the Online Resource Centre). This makes clear the importance of understanding different cultures. However, when we cross cultural boundaries, we carry our own culture with us. We must understand that our own cultural context cannot be used to judge the standards of another. It is imperative to give importance to another culture and not to judge others' behaviour according to our own cultural values.

Conducting international business requires a good understanding of the concepts of business negotiations and ethics. Understanding varying business cultures and different values of management and behaviour is important for success in the international market. Familiarity with the different business practices of different nations aids in sustaining successful business relations.

SUMMARY

Non-verbal communication plays a very significant role in effectiveness of interaction. It is important to project oneself as positive and professional, not just by words, but through actions as well. Non-verbal communication includes kinesics (body movements), proxemics (space), chronemics (time), and haptics (touch).

It is also essential to respect the differences in thinking and culture while dealing with an audience from a different cultural and ethnical background. Language, religion and beliefs, values and attitudes, politics and law, technology, and social organization are the various elements of culture which should be considered to communicate effectively.

EXERCISES

1. What is non-verbal communication? Do you think you can manage any communication situation just with non-verbal cues? Give situational examples.
2. How do kinesics enhance the impact of your verbal communication. Explain with examples.
3. What factors will you bear in mind while communicating with people from a different cultural or ethnic background?
4. Is it more likely to have communication gaps with people from different cultural or ethnic backgrounds than with someone from your own background? How would you overcome these gaps?
5. Answer the following questions:
   (a) Explain the role of eye contact in communication.
   (b) What are the four different space zones according to Hall?
   (c) Proxemics play a prominent role in communication. Justify.
6. How can you reduce the disparity between words and non-verbal communication? Provide one example to illustrate your answer.

PART II

Listening and Speaking

Chapter 4: Active Listening

Chapter 5: Effective Speaking

Chapter 6: Conversations and Dialogues

Chapter 7: Formal Presentations

Chapter 8: Interviews

Chapter 9: Group Communication

CHAPTER 4

Active Listening

OBJECTIVES

You should study the chapter to know
- ◯ the importance of listening
- ◯ why some people are poor listeners
- ◯ the common myths about listening and guard against them
- ◯ the traits of a good listener
- ◯ the different modes and types of listening
- ◯ the barriers to effective listening
- ◯ how to take notes and listen intensively

Introduction

Imagine that a member of your team, in a meeting with potential clients, involuntarily keeps yawning. This behaviour would obviously put the clients off. Their professional evaluation of your organization would be negative and they might not want to go ahead with their business proposal. We would definitely not want someone dozing off when we are speaking to them. Likewise, we should attentively listen when someone else is addressing us.

Listening is a very important skill. Most people are oblivious of the time they spend in purposeful listening. Listening is quite similar to reading, as it involves the reception and decoding of verbal messages from another person. It is unwise to rely solely on receiving the message; meticulous efforts should be made not only to receive these messages but also to interpret them correctly. This is illustrated by the following example: In a biology class, the teacher asked the students to refer to *Origin of Species* by Darwin. One student sincerely rummaged through the library bookshelves looking for *Oranges and Peaches*, but to his utter dismay, he just could not find it. When narrated as a story, this sounds funny, but in reality, we regularly come across such situations, where confusion arises because of poor listening skills.

No *communication process is complete without listening*. Several studies have indicated that business people spend almost 45 per cent of their working time in listening. According to management *guru* Tom Peters, listening is an essential management and leadership skill. Similarly, effective listening is extremely important for students, as they spend most of their time listening to lectures. While we may not necessarily be born good listeners, active listening skills can be learnt and developed. Sharp learners may be poor listeners and, unbelievable as it may sound, those with imperfect hearing may still be excellent listeners. We can define listening as follows: *Listening is a process of receiving, interpreting, and reacting to a message received from the speaker.*

Meaning and Art of Listening

Listening is very different from hearing. We hear numerous little sounds and noises during the day without registering most of them. Birds chirping, cars honking, kids playing, and people talking are examples of such sounds. We do not pay attention to these sounds because we are not interested in them or because these sounds are not meaningful to us. Listening is as important as talking. Good listening involves encouraging the other person (the speaker in this case) by being sensitive to the thoughts and emotions hidden in their expressions. We also need to be patient while listening because otherwise we will lose track of what is being said.

Listening takes a lot of energy. It is becoming a lost art, as people have stopped realizing its importance. Most children have a very short span of attention generally as well as in classrooms, making the job of the teacher very difficult. Like any other art, such as dancing or singing, listening also should be practised with full dedication and concentration. One can practise by listening to chat shows on TV and making mental notes.

Importance of Listening and Empathy in Communication

Listening is very important in the communications process. People believe that a good speaker is essentially a good communicator, but this may not be always true. We must gauge the person's listening skills during a conversation. We realize the importance of listening more when a very close person does not listen to what we say and comes up with 'Oh! I did not get you. Can you say that again?' She may have reacted this way because of divided attention. In our professional life we cannot afford to do that. If you observe a salesperson carefully, you will realize that he/she is successful not because of his/her persuasive power in speaking but because he/she has listened to the customer's need carefully, which has enabled him/her to sell his/her point well.

There is a close relationship between speaking and listening. Empathy plays a very important role in listening. We must put ourselves in the speakers' place and then listen. If we understand the speakers' viewpoint, apart from understanding them better, we will also be able to develop a good rapport with them. It is very important to understand the emotions and feelings of a person to better empathize with his/her viewpoint. We must listen without any biases and prejudices, and be open to the views of other people.

Reasons for Poor Listening

There are several reasons for poor listening. We will examine these reasons in this section.

Listening training is unavailable

Most people are formally trained in the major communicative skills of writing, reading, and speaking. While listening is a skill all of us use most frequently, it is also a skill in which we have least formal training. While workshops and conferences provide opportunities to improve writing and speaking skills, it is difficult to find training to sharpen listening skills.

Speed of thought is more than speed of speech

Another reason for poor listening skills is that people can think faster than they can speak. Most of us speak at the rate of about 125 words per minute. However, we have the mental capacity to understand someone speaking at 400–500 words per minute (if that were possible). This difference between speed of speaking and thought means that when we listen to the average speaker, we are using only 25 per cent of our mental capacity. We still have 75 per cent of unutilized mental capacity. So, our mind starts wandering. This means that we need to make a real effort to listen carefully, and concentrate more of our mental capacity on the act of listening.

How Much are You Paid for Listening?

A manager asked her secretary to keep track of the time she spent listening on the telephone. She was shocked to discover that her company was paying her 35 per cent of her salary, or $18,000, for this function. Amazingly, on the average, people are only about 25 per cent efficient as listeners. With this efficiency rate, she was being paid about $13,500 for the time she spent listening inefficiently!

We are inefficient listeners

Numerous tests confirm that humans are inefficient listeners. Studies conducted by Gail Miller at Washington State University have shown that immediately after listening to a 10-minute oral presentation, the average listener has heard, understood, and retained 50 per cent of what was said. Within 48 hours, that drops off by another 50 per cent, to a final level of 25 per cent efficiency. In other words, we often comprehend and retain only one fourth of what we hear. We all want to be more than 25 per cent efficient. It is not difficult to see the many problems inefficient listeners can create for themselves and others. Poor listening is the cause of many personal and professional problems.

Listening versus Hearing

When we think about listening, we tend to assume that it is basically the same as hearing. This is a misconception because it leads us to believe that effective listening is instinctive. As a result, we make little effort to learn, or develop listening skills, and unknowingly neglect a vital communication function. Consequently, we create unnecessary problems for others and ourselves: misunderstandings, hurt feelings, confused instructions, loss of important information, embarrassment, frustration, and lost opportunities. Listening involves a more sophisticated mental process than hearing. It demands energy and discipline. Effective listening is most often a learned skill. The first step is to realize that effective listening is an active, not a passive, process. The belief that the power of the talker plays a major role in communication is why many managers are poor listeners. In our society, talking is mistakenly viewed as more important, with listening categorized as only a supportive function.

Poor Listening Habits

We often do a lot of things without thinking too much about them. These are our habits, which have formed over many years of doing the same thing and getting by with it. The following are some of these ingrained habits which result in poor listening.

Selective and distractive listening   Sometimes a person listens only to the supporting facts or details or to the way they are presented, and misses the real meaning.

Example: When your professor explains the importance of 'voice' in professional presentations, and you listen to the various features of voice such as quality, pitch, rate, and volume but do not understand the importance of these features in making your presentations effective—then you have missed the real meaning of the topic. Likewise, instead of focusing on the crux of topic 'voice' if you keep thinking about the visual aids, your professor's vocabulary, appearance, etc., you may not get the real meaning of the topic that is presented.

Remedy   Listen to the complete talk/lecture making notes wherever needed. Don't assume that some parts of the talk are more important than the other parts. Do not get distracted by extraneous features.

Rehearsing   Some people listen only until they decide that they want to say something; then they quit listening, start rehearsing what they want to say, and wait for the opportunity to jump in and talk.

Example: Most of us are interested in speaking rather than listening. However, such a habit affects effective listening. When you are attending a conference, if you keep thinking and rehearsing about what kind of comments or questions you can add so that you can impress upon the speaker and audience, you may certainly lose the thread of presentation that is going on.

Remedy   Make notes of what you are listening. You can even record your appreciations. Based on your notes, you can ask a question or give a comment after the speaker finishes the talk.

Interrupting   The listener does not wait to determine the complete meaning, but interrupts so forcefully that the speaker stops in mid-sentence.

Example: Many a time, when you do not understand the point being presented, you immediately interrupt and raise your doubt. Assume that you are attending a lecture on 'Listening' where the speaker says, 'Listening involves a more sophisticated mental process than hearing', and you don't understand the meaning of 'sophisticated mental process'. You immediately interrupt the speaker even before the sentence is completed. This interruption makes the speaker stop abruptly. If the speaker is a novice, she gets confused and goes blank at times. In fact, speakers should always keep in mind to inform the audience whether the latter can ask questions then and there or they can ask only at the end of the presentation. If they don't mention, you as a listener should ask them about the question-answer session. If it is an online lecture, you can post your question in the chat box without interrupting the speaker.

Remedy   Follow the speaker's instructions as to whether you can interrupt or not. Be patient and do not try to display your curiosity.

Hearing what is expected   People frequently hear what they expect to hear; alternatively, they refuse to hear what they do not want to hear.

Example: Perceptions differ from person to person. You may like a particular field of education because of peer-association and hence you expect the speaker also to support your views. Assume that you like to pursue engineering education after your secondary school and when your parents wish you to take up pure science, you do not want to listen to their views, however beneficial they may be.

Remedy   Believe that you may be benefitted by listening to and understanding others from their perspectives. Don't be rigid in holding on to your own beliefs and perspectives.

Feeling defensive   The listener assumes that they know the speaker's intention, or why something was said; or for various reasons, they expect to be attacked and react defensively.

Example: Baseless assumptions always act as barriers to good listening. When your neighbours explain the difficulties they have with your loud music at night, you don't agree with them immediately because you assume that they are old, and they don't like the loud music. So, you argue with them vehemently and refuse to accept their views.

Remedy   Avoid baseless assumptions on the topics or speakers. Keep aside such assumptions while listening to others.

'The sales are low! We need to strategize. Call for a brainstorming session in Goa.'

Listening for a point of disagreement   Some listeners seem to wait for the chance to attack someone. They listen intently for points on which they can disagree and then attack or confront.

Example: If you don't like some speakers or you think that they are showing off, you may wait for a chance to pull them down. Assume that one of your acquaintances John is delivering a talk on 'Effective Public Speaking' and talks about the significance of quotations in public speaking. As you don't like him and feel that he is showing off, you wait for a point on which you can disagree. During his talk, when John changes a few words from a quote by Wordsworth, you are glad that you have got a point to attack John as you know that quotations are to be given verbatim. In short, you have listened to disagree.

Remedy   Don't try to prove that you know better than the speaker. Listen to the entire speech and not only to the points of disagreement.

Labelling the subject matter uninteresting   A poor listener when faced with a known topic might find the discussion dull. They are convinced that the topic is uninteresting, and turn to the many other thoughts and concerns stored up in their minds for just such an occasion—to start using that unoccupied 75 per cent of their mental capacity. A good listener, on the other hand, might start at the same point, but would arrive at a different conclusion. He/she would say, 'That sounds like a dull subject and I do not see how it could help me in my work. But I am here, so I guess I will pay attention and see what this person has to say. Maybe there will be something I can use.'

Example: You may have listened to many talks on 'English Phonetics' and hence think that you are very familiar with the topic. So, when you attend another lecture on the same topic, you may not listen to the talk with complete concentration and your mind may wander about your ongoing project, problems at home, etc. On the contrary if you are a good listener, you may still listen to the talk very carefully so that you can use some examples given by the speaker.

Remedy   Believe that every talk has some 'take home' points which may be useful for you.

How to Solve Problems—by Listening

Employees frequently have excellent ideas about improving the productivity of the work environment. Managers who listen to these ideas solve more problems than those who do not. These managers create a sense of concern for their staff while receiving better-quality information. Ranjan Das, the foreman of a large manufacturing plant, called in Krishnan, a supervisor of a production line, into his office to explain the plans for a new way to assemble machinery. Ranjan described how he thought the procedure should be changed. Krishnan's only response was silence and a frown. The foreman, reading Krishnan's non-verbal communication accurately, realized that something was wrong and sensed that Krishnan might have something to say. So he said, 'Krishnan, you have been in the department longer than me. What is your reaction to my suggestion? I am listening.' Krishnan paused and then began to speak. He realized his manager had opened the door to communication and felt comfortable offering suggestions from his years of experience. As the two employees exchanged ideas, a mutual respect and trust developed, along with a solution to the technical problems. While listening, the manager remained in complete control of the situation. He was an active, not a passive, listener.

Common Myths about Listening

To better understand what it takes to be a good listener, we must discard the following common myths associated with listening.

Myth 1 'I do not have to concentrate: listening comes naturally.'

Truth Being a good listener requires a conscious effort. You must keep your mind constantly engaged and in gear.

Myth 2 'I am a good listener because I always get the facts and figures straight.'

Truth You may be a selective listener. You listen to the facts and figures, but do not hear or comprehend the rest of the information, such as questions raised, emotions of the speaker, context, opinions, and ideas discussed.

Myth 3 'You should not interrupt when someone is speaking.'

Truth A good listener does not hesitate to interrupt if the speaker's information is unclear. You must be an inquisitive listener to be an effective listener.

Myth 4 'A good listener paraphrases everything a speaker says.'

Truth If you constantly parrot people's statements back to them, they will think you are slow or—even worse—proud. It is good to repeat key information and ask the speaker to verify it, but do not rephrase all the comments.

Traits of a Good Listener

Although acquiring the skill of active listening needs much effort, it is not impossible. A few guidelines are given in the following paragraphs. Some actions might seem unnatural and forced but can be inculcated with practice. For instance, it is very difficult to be patient with a speaker if he/she appears to be irrational and illogical. However, we must be tolerant in such circumstances. The guidelines suggested here will definitely help develop an attitude of tolerance and understanding.

Being Non-evaluative

The verbal and non-verbal behaviour of an active listener will suggest to the speaker that he/she is being properly heard and understood. It should not, however, indicate what one thinks about the person. The purpose is to communicate, overlooking the qualities of the ideas, attitudes, and values of the speaker. In other words, the speaker is not to be evaluated for his/her personal qualities. Our behaviour should convey the impression that we accept the speaker without making any judgement of right or wrong, good or bad, suitable or unsuitable.

A good conversationalist is popular, a good listener even more so. Talk only if you have something to say.

Paraphrasing

To clarify a point, one can simply paraphrase what the speaker has said and enquire from the speaker whether one has heard it accurately. Use phrases like the following ones to ensure that the information has been paraphrased correctly:

As I gather, you want to tell …

So you mean to say that …

Oh! Your feeling towards …

Do you mean that …

Reflecting Implications

To reflect on the implications of what has been said, it is necessary to go a bit beyond the direct contents of the speaker, indicating appreciation of his/her ideas and what he/she is leading to. This may encourage the speaker to further extend the ideas. The listener's aim here is to reflect eagerness and zest by verbal or non-verbal means, thereby giving positive feedback. Phrases such as the following can be used:

> I am sure if you did that, you would be in a position to ...
>
> So this might lead to a result which ...
>
> So you are suggesting that we might ...
>
> Will that help us to alleviate the problem of ...

If this technique is used to change the direction of the speaker's thinking, by showing one's cleverness in suggesting ideas that the speaker has not thought of before, it creates distrust between the two parties. However, if the technique is used with the genuine intention of understanding more, one can certainly help the speaker by boosting his/her confidence and making him/her believe that the listener has grasped the content well.

Reflecting Hidden Feelings

Sometimes, one has to go beyond the explicit feelings and content of what is being said to unravel the underlying feelings, intentions, beliefs, or values that may be influencing the speaker's words. The listeners must try and empathize or identify themselves with the speaker, to experience what they feel. One can express one's sentiments by using phrases such as:

> If I were in your place, I would not have handled the situation so well.
>
> That must have been a satisfying experience.
>
> If I were in such a situation, I would have sought the help of seniors.
>
> If that had happened to me, I would have been very upset.

In reflecting the speaker's covert feelings, one must be careful not to overexpose the speaker, by coaxing them to admit more than they desire. We should also avoid suggesting to the speaker that the feelings we reflect are what ought to be felt by the speaker in such a situation. This would make the speaker feel evaluated. Acceptance is conveyed more by one's manner and tone than by words. We have already studied non-verbal communication in Chapter 3.

Attending Responses
Tone of voice can quickly indicate if you are interested. Use a warm, informal, and friendly tone. Address the speaker by his first name, whenever appropriate.
Jargon
Unless you know speakers well enough to know their familiarity with technical language, avoid using language that could be perceived as techno-babble.

As a Good Listener

Use affirmative prompting through verbal or non-verbal means, where appropriate, to encourage the speaker.

Be careful with your own body language. Crossing and re-crossing your legs, slumping, or raising your brows can be misinterpreted by the speaker and interrupt the flow of thought.

Inviting Further Contributions

In a situation where we have not heard or understood enough yet to respond with empathy and understanding, we must prompt the speaker to give more information. Phrases such as the following can be used:

Can you throw more light on …

It would be great if you can expand more on this.

What happened after that?

How did you react when … ?

While it is useful to ask questions, one should guide against asking too many. This may lead speakers to suspect that rather than seeking information, one is interrogating or challenging them.

When we want a speaker to expand on his/her subject, we must ask *open-ended questions*, which require more than a straight *yes* or *no* answer. *Open-ended questions* encourage a speaker and help the listener gain more information. For instance, 'What solutions have you thought of?' is open-ended, whereas 'Have you thought of this solution?' is not. Avoid probing and pointed questions fired at the speaker in a dogmatic way. Open-ended questions create a more supportive, trusting climate, which helps the communication move smoothly.

The listener must keep statements and questions short and easy to understand. They should use conversational English (say *talk* instead of *communicate* and *write* instead of *correspond*). Above all, remember that if we are talking most of the time, we are probably not listening very well.

Responding Non-verbally

Listeners can show active interest by adopting certain postures and sending non-verbal signals that communicate their interest in what the speaker is saying. These include regular eye contact (without staring), body leaning slightly towards the speaker, head nods, and a slightly tilted head (refer to the photographs of appropriate and inappropriate body postures given in Chapter 3). Occasionally, certain receptive utterances such as '*yes*' and '*ur-hum*' can also be used to indicate that one is following what the speaker is saying, while being careful not to overdo it. Attentive silence is more effective than too many receptive utterances.

Non-verbal Techniques to Indicate Your Attentiveness

- Suitable facial expressions and natural smiles (without excessive smiling)
- Open posture (arms not crossed) and open palms (rather than clenched fists or fidgeting)
- Appropriate distance (usually an arm's length away from the speaker)

By giving such non-verbal signals, the listener helps the speaker feel more confident and reflects interest and understanding. This also helps generate more trust between the listener and the speaker.

Listening Modes

People use different kinds of listening modes in different situations. The mode depends on factors such as mood, mindset, topic, time of day, relevance, and importance. It is sometimes not enough to just listen; we should also send positive signals to the speaker that we are with him or her. Let us discuss the various types of listening one by one.

Active versus Passive Listening

Paying attention

We often listen to various forms of communication in an unconscious manner. At dinner, we glance at the television while busy eating or talking to someone else. This is not active listening—either to the TV or to the person we are talking to. How can we develop the ability to listen to others patiently and carefully? Improper listening is very harmful, as such communication is prone to distortion. It is also a waste of time. The responsibility of the listener is to *show keenness* in the speaker's talk through *expressions, alertness,* and by *asking questions* about the speech, if required. By doing this, the listener will encourage the speaker to express his/her ideas clearly and enthusiastically. If the listener looks bored, it will discourage the speaker.

Skills in this area can be improved by listening to commentaries on TV or radio. Concentrate on the theme, supporting ideas, and also the digressions, if any, in the speech. Further, it is helpful to write down the gist of what we have heard. This exercise can certainly help improve listening skills.

So far, we have emphasized on the importance of paying attention to a person's speech. It is equally important not to neglect the physical aspects of the person. Appearance, expressions, bodily movements, and posture are all as significant as words in conveying a meaning. A person's body language, or non-verbal communication is involuntary and, therefore, more truthful. Hence, a listener should pay considerable attention to the physical messages conveyed by a speaker in order to assess the accuracy and sincerity of his/her speech.

> Focus on the speaker. Ignore all distractions so that you can concentrate on the speaker's flow of thought. Try to ignore feelings of hunger, weariness, or discomfort. Alternatively, you can confide your discomfort to the speaker, so that some remedy can be provided (e.g., improved ventilation).

If, as the listener, we feel that the speaker is being critical about us, we tend to become inattentive, as nobody likes criticism. *As a wise listener, one must look for a valid reason for the criticism* and determine the reason for the speaker's dissatisfaction. Adopting a receptive and constructive attitude to criticism can lead to self-improvement. We tend to listen carefully only to those messages that are advantageous or pleasing to us, but in order to have a fruitful conversation, we should also learn to take equal interest in speeches that contain messages of interest to the speaker.

As a Sincere Listener

Accept your role as a listener by listening actively, engaging positively, participating fully, and encouraging the speaker consciously.

Do not pretend to show interest. Your involuntary non-verbal behaviour, such as glazed eye-contact or strained expression, will give you away.

Dealing with distractions

In the process of developing active listening skills, one should train oneself to avoid physical distractions and concentrate completely on the message. An attractive face in the room and sometimes even the fragrance of perfume can act as distractions. However, a careful listener has to exercise a great deal of mental discipline to remain focused on the message conveyed by the speaker.

Often, after a period of continuous listening, people get tired and start losing interest in the message. They have to force themselves hard to stay with the speech and the contents. This usually happens because of 'brain time'. As discussed earlier, our minds have the capacity

to understand more than what can be said by an average speaker in a minute. This mismatch coupled with general disinterest leads to a wandering mind. To overcome this difficulty, the listener should try to stay alert by anticipating what the speaker will say next. If the listener's guess turns out correct, his/her interest in the speech will revive. This will enable the listener to grasp and recall the speech better.

> 'The most important thing in communication is to listen to what isn't being said.'
> –Peter F. Drucker

Since a listener's capacity to absorb information is much more than a speaker's ability to talk, a lot of time is available for the listener to evaluate the message. Sometimes a listener focuses on the message in fragments and is, therefore, unable to take in the entire message. Further, listeners also interrupt the speaker without listening to the complete message. Careful listeners, however, never jump to conclusions about the message unless the speaker has finished. *Effective listening is possible only if the listener patiently listens to the complete speech.*

A sincere listener always puts in a conscious effort by listening with a positive attitude. A pretentious listener will show his/her attentiveness by awkward postures like resting his/her chin on his/her hand, or bending forward too much to show that he/she is paying a lot of attention to the message while his/her mind is actually far away. He/she has no clue about what the speaker is saying. Effective listening does not come easily; it requires great effort to pay complete attention. Listeners require mental preparedness and energy to concentrate on the speaker's words as well as his/her non-verbal cues of communication like posture, gesture, eye contact, facial expressions, etc.

Sometimes noise distracts the attention of the listener. This should be ignored or sidetracked. If you go to your boss to discuss something and he continuously shuffles papers and talks over the telephone, his listening will be distracted. Superiors should take care to provide an ambience conducive to sympathetically hearing their subordinates. Proper listening will enable the speaker to release emotional tension, which will improve the working environment of the organization. Table 4.1 lists a few tips for effective listening.

Table 4.1   Tips for effective listening

Dos	Don'ts
• Be mentally prepared to listen.	• Pay undue emphasis to the vocabulary, as you can use the context to understand the meaning.
• Evaluate the speech, not the speaker.	• Pay too much attention to the accessories and clothing of the speaker.

(Contd)

Dos	Don'ts
• Be unbiased towards the speaker by depersonalizing your feelings.	• Prepare your responses while the speaker is speaking.
• Fight distractions by blocking off sound sources.	• Get distracted by outside influences.
• Be open-minded.	• Hold preconceptions and prejudices.
• Ask questions to clarify and confirm thoughts.	• Concentrate too hard.
• Paraphrase from time to time.	• Interrupt too often.
• Send appropriate non-verbal signals from time to time	• Show boredom even to an uninteresting speaker.

Global versus Local Listening

To become an effective listener in multicultural settings, one must be aware of the cultural variations involved. In the case of a foreign speaker, one must pay attention to the diction and choice of language. One should also be aware of the possible idiomatic expressions that are specific to any particular region. For example, 'cover all the bases' is an idiom originated from baseball, a sport not known in India. So, if an American speaker uses such a phrase, the listener shouldn't be taken by surprise. The non-verbal conversational behaviour of people from different countries also varies. For example, Japanese nod their head when they are in conversation to reflect that they are just hearing; however, Americans will nod while in conversation to indicate that they are listening attentively. Therefore, in global listening, one must make an effort to pick up the differences in culture, politics, demography, non-verbal cues, etc. It is much different from when you are in a local setting, listening to people you are comfortable with.

Types of Listening

While certain skills are basic and necessary for all types of listening (receiving, attending, and understanding), each type requires some special skills. Before we can fully appreciate the skills and apply the guidelines, we must understand the different types of listening.

Appreciative listening

Appreciative listening is listening for deriving aesthetic pleasure, as we do when we listen to a comedian, musician, or entertainer.

Empathetic listening

As we have learnt, empathy is very important in communication, particularly in listening. A good example of empathetic listening is that practiced by nurses. So much so that it gives a healing touch to the patient. For effective empathetic listening, one has to feel what the speaker is feeling. One has to appreciate the speaker's emotions, circumstance, mindset, and perspective, and be able to provide emotional and moral support. When a psychiatrist listens

to her subject, she employs empathetic listening. We must feel the person's nerves. This can be done through phrases like 'I can understand what you have gone through,' 'It must be difficult to face such a situation,' etc. Sometimes the situation is very sensitive and must be handled with caution.

> 'You do not listen with just your ears: you listen with your eyes and with your sense of touch, you listen by becoming aware of the feelings and emotions that arise within yourself because of this contact with others. You listen with your mind, your heart, your imagination.'
> *–Egan Gerard (1988)*

Comprehensive listening

This type of listening is needed in the classroom when students have to listen to the lecturer to understand and comprehend the message. Similarly, when someone is giving you directions to find the location of a place, comprehensive listening is required to receive and interpret the message.

Critical listening

Also known as *evaluative, judgemental,* or *interpretive* listening, critical listening involves analysing, evaluating, and judging what is being said. Just as we formulate opinions about people before they speak based on their physical attributes, we also tend to get judgemental about the contents of their speech. We try to see if the person has said something based on facts or is simply beating about the bush. This type of listening is applicable when the other person is trying to persuade. In such cases, we try to evaluate the tone, the non-verbal signals, and the underlying meaning of the words. We judge the argument based on our knowledge and experience. For example, listening to a salesperson before making a purchase or listening to politicians making their election campaign speech involves critical listening.

Practice 1

Audio Clip Group 1: *Listen to the two World News items and answer the following questions by choosing the most appropriate option for each:*

 Clip 1: Polish President Lech Kaczynski Dies in Air Crash in Russia

1. What is the primary source of information about the air crash?
   - A. ABC live
   - B. Russian Prime Minister
   - C. Russia's Emergencies Ministry
   - D. Polish Foreign Ministry
2. What is the cause of the accident?
   - A. Incompetent pilot
   - B. Dense fog
   - C. High speed of the aircraft
   - D. Unexpected rains
3. Where was the Polish president going along with his delegation?
   - A. To take part in the Katyn memorial ceremonies
   - B. To the site at which Russian officers were executed
   - C. To attend a funeral
   - D. To pay homage to Russian leaders
4. The ill-fated plane was a _____ air craft.
   - A. PI-156   B. PR-155   C. Tu-154   D. Ts-150

 Clip 2: India Aims to Connect 100 Million People with Broadband Internet in the near future

1. What was the news about?
   - A. Improving health
   - B. Improving education
   - C. Power to Panchayats
   - D. Broadband Internet

2. Sam Pitroda is the _____ advisor to Prime Minister.
   A. Education information   B. Information infrastructure
   C. Health information   D. Pension scheme
3. The broadband Internet connection will pave way for_____
   A. Better delivery system for various government schemes
   B. Better food for poor people
   C. Better administration
   D. Better national knowledge
4. The forecasted broadband connectivity is expected to be completed by_____
   A. 2020   B. 2015   C. 2012   D. 2016

Superficial listening

Superficial listening can be compared with hearing or passive listening. One pretends to be listening by giving fake expressions to avoid offending the speaker. However, not much is fed into the brain. This happens, for example, when one is forced to attend a guest lecture on an area that is not of one's interest.

'You' viewpoint

Covey says 'Seek first to understand, and then to be understood'. This is a subtle way of saying that we should put the other person before us. If we give the other person more importance and avoid forcing our opinion, we can listen better. Using this 'you' viewpoint, we boost the speaker's confidence, making him/her more open to communication. Another way of achieving this is to address the speaker by his/her name, which gives the impression that we are listening with full attention.

Barriers to Effective Listening

In this section we will examine some common barriers to listening, which, with some practice, can be overcome. An average person spends 70 per cent waking hours communicating, and 45 per cent of those listening. The following could be some factors that create barriers to listening.

Content

Listeners knowing too much   They feel that their knowledge is so extensive that there is little left to learn.

Listeners knowing too little   They tune out when faced with difficult intellectual or emotional content. They only listen to information that conforms to their beliefs.

Remedies

- Do not sit back passively and allow sound to enter ears.
- Develop a positive attitude towards the message.
- Anticipate the importance of the message content.
- Seek areas of interest in the message.
- Remind yourself that something of value can be learned.

Speaker

Delivery The speaker's accent, organization, clarity, speed, volume, tone, inflections, emotions, and appearance affect the interpretation of the message. (Refer to the telephone conversation in the QR code.)

Attitudes toward speaker Listeners are influenced more by their attitude towards the speaker than the information presented. If the listener likes the speaker, he/she is more likely to empathize and, therefore, comprehend the message.

'When people talk, listen completely. Most people never listen.'

–Ernest Hemingway

Remedy Concentrate on the 'what' of the message, not the 'who' or 'how'.

Medium

Distance and circumstances Listening requires least effort when the speaker is not visible. More effort is needed when the speaker is visible, but not physically present (e.g., a speaker on TV). Maximum effort is needed in face-to-face interactions. This happens because the amount and variety of both verbal and non-verbal stimuli increase. These stimuli can either help or hinder communication.

Remedy Realize the potential for better understanding, and increase listening effort.

Distractions

Extraneous stimuli Sounds, lights, odours, mannerisms, voice inflections, and moving objects can easily distract listeners. Psychological studies indicate that a listener's attention span is sometimes not more than two or three seconds. These stimuli can be categorized as environmental or physical, but most often psychological.

Remedies

- Identify and eradicate distractions.
- If distractions cannot be eliminated, increase concentration.
- Free yourself from preconceptions, prejudices, and negative emotions.

Mindset

Attitudes Attitudes are structured by a listener's unique physical, mental, and emotional characteristics. An individual's mindset can either magnify or diminish stimuli, distorting the message.

Remedies

- Strive to not let personal biases interfere with comprehension.
- Respect others' freedom of values and beliefs.
- Accept that attempting to understand another's viewpoint is not necessarily agreeing with it.
- Realize that there may be more than one acceptable point of view.

Language

Ambiguity   Listeners rarely hear every word spoken and may attach different meanings to words than intended by the speaker.

Misinterpretation   This can occur when the words used are imprecise, emotional, technical, or overly intellectual. It occurs most often when listeners interpret words based on personal definitions, established by background, education, and experience.

Remedies

- Realize that different words may have different meanings for different people.
- Evaluate the context in which the word is used.
- Remember that the meaning is in the mind, not in the word.

Listening speed

Rate   The average speaking rate is 125–150 words per minute. The average listening capacity is 400–500 words a minute, leaving a lot of excess thinking time.

Think time   Poor listeners use the excess time to daydream, often missing part of the message.

Remedies

- Use the excess time to outline messages.
- Identify the purpose and how it is supported.
- Evaluate the soundness of logic; verify and integrate it with existing knowledge.
- Maintain eye contact to observe and interpret non-verbal signals.
- Formulate questions to enhance and verify understanding and provide feedback.

Feedback

Inappropriate   Often, the listener ends up giving premature comments or evaluations without a full understanding of the speaker's viewpoint. Such comments, which may be coloured with emotions of resentment, defensiveness, or suspicion, can hinder the speaker by confusing them or diverting them into tangents.

Remedy   Supportive feedback can demonstrate interest through appropriate eye contact, smiling and animation, nodding, leaning forward, verbal reinforcements such as 'I see' or 'yes', and phrasing interpretations of the comments for verification. These must be timed so as to assist rather than hinder the speaker.

Cultural barriers

Listening is tough, and more so if someone is from another culture or subculture. The problem crops up because of the different choice of words, accents, pronunciation, and many other intangible reasons. One must be extra careful while listening to a person from another region or culture. In fact listening can be improved by talking to more people of different nations and trying to understand them. For example,

At a hotel lobby in Tokyo:

American: I would like a room for two nights.
Japanese speaker of English: For tonight?
American: No, not 'tonight'. Two nights.

Listening for General Content and Specific Information

When listening to a speaker for general content, one's focus should be on an overview of the topic. Do not pay attention to the minor details and the examples, but observe the gist of it. If one is clear with the purpose of listening, one's task will be easier. However, one needs to concentrate even when listening for general content.

Specific information is a lot simpler to collect. The method remains similar to that just described, just that the scope is limited and efforts reduced. We can ignore the peripheral details and look only for the keywords.

Practice 2

Audio Clip Group 2: *Listen to the three short audio clips based on a discussion on diabetes by a professor with his students. Choose the best option for the following questions:*

  (i) What does the professor imply about many people?
      A. Some people should know that sweets are unhealthy.

 **Clip 1**
      B. Some people believe that sweets with organically grown sugar do not cause diabetes.
      C. Some people incorrectly think that sugary foods are the cause of diabetes.
      D. Some people already know that eating sweets can cause diabetes.

 (ii) What does the professor mean when he says: 'And this can be tricky!'?
      A. Controlling blood sugar levels is like performing magic.

 **Clip 2**
      B. Maintaining healthy insulin levels throughout the day is not easy.
      C. Diabetics cannot give themselves insulin without a doctor's assistance.
      D. Determining the correct amount of insulin for an injection is not difficult.

(iii) Referring to question (ii), Why does the professor say this?
      A. To point out to students that they may have experienced low blood sugar.

 Clip 3
      B. To determine how the students manage their blood sugar levels.
      C. To determine whether there were any diabetics in the class.
      D. To point out an error that a student had made earlier.

Listening and Note-taking

You might have noticed that when you present a topic to your class, your teacher takes notes, and after you complete, she refers to her notes and gives her comments on your performance.

Note-taking and note-making are two processes of summarizing the information from the spoken and written material respectively. While you have enough time in your hand for note-making when reading a text, you may have to be very quick and alert when taking notes from a spoken information because you can't go back to the information if you lose some points. In other words, taking notes is a spontaneous process while making notes is a planned one. Hence, you may have to be very brief in taking notes by using acronyms, abbreviations, short words, etc., which you can expand later. This skill requires adequate practice. As a student, you do this exercise daily when you attend your classes. Most probably for every subject you have a notebook in which you write the main points of a lecture. First, one has to listen to what is being said with full attention. Next, one has to recognize quickly the main points that the speaker makes and note them down immediately. As we know, the speed of speech is faster than the speed of writing. So some extra effort is required to keep pace with the speaker. Unlike written material, where one has the advantage of going back and referring to it, speeches, unless recorded, are not available afterwards, and hence require one's complete attention.

You might have found it easy to take notes during certain lectures. In such lectures, the speaker would have given a signal when he/she was about to discuss a main point and/or a summary of the main points.

So you could quickly check the points you had noted. Some speakers write the full text of their speech and read it before an audience. In such a case one might procure a copy·and *make* notes.

While taking notes, one may also take down the important examples or quotations that the speaker may use for support, illustration, or explanation. If visual material is used for this purpose, the note-taker's job is easier, as there is the advantage of being able to see the matter being presented. Usually, the speakers present only the very important points visually. One need not, however, copy everything that is presented visually. One must exercise one's judgement and note only the main points.

The following methods are used in note-making:
- Outline/linear method
- Sentence/categorical method
- Schematic/mapping method

However, as we are dealing with 'listening skills' we will now discuss the following method of note-taking which is most commonly used:

Keyword outline method of note-taking

To use this method, one needs to note the speaker's main points and supporting evidence in a rough outline form. Consider the following speech:

> Reports from a range of scientific disciplines are telling us with certainty that we are making a mess of the earth, we are fouling our nest, and we have to act decisively and against our immediate inclinations for we tend to be superstitious, hierarchical, and self-interested just when the moment requires us to be rational, even-handed, and altruistic. We are shaped by our history and biology to frame our plans within the short term, within the scale of a single lifetime; and in democracies, governments and electorates collude in an even tighter cycle of promise and gratification. Now we are asked to address the well-being of unborn individuals we will never meet and who, contrary to the usual terms of human interaction, will not be returning the favour.

> To concentrate our minds, we have historical examples of civilizations that have collapsed through environmental degradation—the Sumerian, the Indus Valley, Easter Island. They extravagantly feasted on vital natural resources and died. Those were test-tube cases, locally confined; now, increasingly, we are one and we are informed, reliably or not, that it is the whole laboratory, the whole glorious human experiment that is at risk.

Using the keyword outline method of note-taking, you would take the following notes:

- Reasons for making mess of the earth
  - superstitions, selfishness, hierarchies, inadequate planning, clash between governments and electorates
- Examples of environmental degradation
  - the Sumerian, the Indus Valley, Easter Island
- Situation facing us
  - ancient civilizations feasting on natural resources—localized
  - now it is globalized

Notice how brief the notes are. There are only 45 words as compared to the speaker's 193 words, but they accurately summarize the speaker's main points. Also notice how clear the notes are. By separating the main points from the sub-points, the outline format shows the relationship among the speaker's ideas.

When taking notes, the name of the speaker, the topic, and the date of the speech should also be noted. Since notes are taken under pressure of time, it is necessary to read them soon after the lecture is over. This would enable one to insert any omission or delete any irrelevant matter that may have been included. Whenever one's own short forms are used for certain words (e.g., 'sth' for something, 'sb' for somebody, 'exp' for experiment), these must be expanded soon after. It is advisable to make it a habit to use fixed short forms, so that one does not falter in deciphering (e.g., 'exp' should always refer to experiment and not experience).

Intensive Listening

Intensive listening is required when the answer to a specific problem is required. You must have observed that when you are listening to some popular speaker to get the answer to a particular query, your focus is on the answer to that query, while the rest of the details are unimportant. To test your intensive listening skills, you could read the test questions first and then listen to the passage for answers to those questions. The following practice exercise will help you practise intensive listening.

Practice 3

Audio Clip 3: *Listen to this brief lecture on the influence of climate on architecture. Take notes if necessary and then fill in the following blanks with necessary details:*

1. Buildings in Greece have _____ roofs and few _____ openings as there is a moderate rainfall and strong light whereas in Sweden, which has a colder climate, the buildings have _____ roofs.
2. Egypt has a _____ climate with _____ light and hence people constructed buildings with _____ roofs and _____ windows.
3. In places such as Vietnam, which have river deltas, houses are often built on _____ to keep them clear of the water.
4. Brick, wood, stone, and _____ have been the _____ building materials.
5. _____ has generally been chosen for important structures because of its _____ and _____.
6. For constructing a lintel, it is important that the two columns are _____ to one another and that the _____ between the columns is not more than _____ the width of the two columns.

Practice 4

Audio Clip 4: *Ashwin Iyer, Manager of Citibank, Bangaluru, visits credit card manufacturer, Vikram Kumar, in Hyderabad. Over lunch they find out a little more about each other. Listen to their conversation and fill in the following blanks:*

1. Ashwin joined Citibank, Bengaluru, _____ years ago.
2. Vikram hails from _____.
3. Ashwin was with the _____ section when he first joined Citibank.
4. Ashwin's home town is _____, known for its _____.
5. Vikram has _____ children, studying in _____ .
6. Ashwin's children are staying in _____.
7. Vikram's business deals with _____.
8. Vikram moved to Hyderabad at the age of _____.
9. Ashwin was shifted to the investment section because of his knowledge on _____.
10. Vikram is visiting Bengaluru in _____.

SUMMARY

Listening is an often ignored but yet a very important communication skill. Active listening requires the listener to understand, interpret, and evaluate what is being said. The ability to do this can improve personal and professional relationships remarkably by eliminating misunderstandings, strengthening cooperation, and improving understanding. Professionals spend most of their time in attentive listening.

Listening is different from hearing. One has to make a conscious effort to become an effective listener. Empathy plays a very important role in listening. We must listen without any biases and prejudices, and be open to the views of other people.

The reasons for poor listening include unavailability of listening training, mismatch between the speeds of thought and speech, and inefficient listening. Listening skills can be improved by removing the common myths about listening from our minds. The traits of a good listener are being non-evaluative, paraphrasing, reflecting implications, reflecting hidden feelings, inviting further contributions, and responding non-verbally. We employ different types of listening in different situations: appreciative, empathetic, comprehensive, critical, or superficial listening. There are several barriers to listening. Being aware of them, we can take corrective measures. Note-taking and intensive listening are important tools that can help improve active listening abilities.

EXERCISES

1. Define listening. How is it different from hearing?
2. Give two instances from your personal experience where communication failed because of poor listening skills.
3. What do you understand by the following terms?
   (a) Appreciative listening
   (b) Empathetic listening
   (c) Comprehensive listening
   (d) Critical listening
4. Do you agree that listening is more important than speaking? Why or why not?
5. What are the traits of a good listener? How can you make out if the other person is listening to you properly or not?
6. Discuss the barriers to listening in detail and give suggestions to overcome each one of them.
7. If the speaker's verbal and non-verbal cues do not match, how will it affect your listening? Which one will you believe then? Why?
8. (a) Distinguish between the two processes: Note-taking and Note-making
   (b) What is the correlation between listening and taking notes? Give a few tips for effective note-taking.

9. **Listening Activity 1**
   Form a group of 8 to 10 members with your friends. Write down the four questions given below on a sheet of paper and circulate among the group members. Give them one minute to go through the questions. Now read out the passage twice and ask them to answer these questions briefly in their own words.

   Passage
   Why is strategic management considered so important? It is involved in many of the decisions that managers make. Most of the significant current business events reported in the various business publications involve strategic management. For example, on a recent day the reports of business events such as the proposed merger of Lockheed Corporation and Martin Marietta, the departure of one of Walt Disney Company's key executives, and the announcement of the merger of Babbage's and Software, are examples of managers making strategic decisions. Also, one survey of business owners found that 69 per cent had strategic plans, and among those owners, 89 per cent responded that they had found their plans to be effective. They stated, for example, that strategic planning gave them specific goals and provided

their staffs with a unified vision. Other studies of the effectiveness of strategic planning and management have found that, generally speaking, companies with formal strategic management systems had higher financial returns.

Today strategic management has moved beyond the private sector to include government agencies, hospitals, and educational institutions. For example, the skyrocketing costs of college education, cutbacks in federal aid for students and research, and the decline in the absolute number of high school graduates have forced many university administrators to assess their organizations' aspirations and identify a market niche in which they can survive and prosper.

If an organization produced a single product or service, managers could develop a single strategic plan that covered everything it did. But many organizations are in diverse lines of business. For example, General Electric is in a variety of businesses—everything from manufacturing airplane engines and light bulbs to owning the NBC television network. The Gillette Company includes a diverse array of products ranging from blades and razors and toiletry items to writing instruments, stationery products, and small household and personal-care appliances. Each of these different businesses typically demands a separate strategy. Moreover, these multibusiness companies also have diverse functional departments such as finance and marketing that support each of their businesses. As a result, we need to differentiate between corporate-level, business-level, and functional-level strategies.

Questions

(a) Why is strategic management considered important?

(b) According to the studies which companies had higher financial returns?

(c) What are the other organizations apart from the private sectors where strategic management has permeated?

(d) Why is it necessary to differentiate among the three levels of strategy?

10. **Listening Activity 2**

Ask two people, one boy and one girl, to speak out the dialogue given below and record the same. Then play the recording to a group consisting of you and several others. Listen and discuss the answers to the questions given at the end of the conversation:

Rahul: Hi Rima, I saw you at registration yesterday. I sailed right through, but you were standing in a long line.

Rima: Yeah, I waited an hour to sign up for a distance-learning course.

Rahul: Distance learning? Never heard of it.

Rima: Well, it is new this semester—it is only open to psychology. All I have got to do is watch a twelve-week series of televised lessons. The department shows them several different times a day and in several different locations.

Rahul: Do you not ever have to meet your professor?

Rima: Yeah, after each part of the series I have to talk to her and the other students on the phone, you know, about our ideas. Then we will meet on campus three times for reviews and exams.

Rahul: It sounds pretty nontraditional to me, but I guess it makes sense, considering how many students have jobs. It must really help with their schedules—not to mention how it will cut down on traffic.

Rima: You know, last year my department did a survey and they found out that 80 per cent of all psychology majors were employed. That is why they came up with the programme. Look, I will be working three days a week next semester, and it was either cut back on my classes or try this out.

Rahul: The only thing is … does not it seem impersonal, though? I mean, I would miss having class discussions and hearing what other people think.

Rima: Well, I guess that is why phone contact is important. Anyway, it is an experiment. Maybe I will end up hating it.

Rahul: Maybe, but … I will be curious to see how it works out.

Questions

(a) Where did the man see the woman the previous day?

(b) How is the distance learning course different from traditional courses?

(c) What do the speakers agree is the major advantage of the distance learning course?

(d) Why did the woman decide to enrol in the distance learning course?

(e) What does the man think is a disadvantage of distance learning?

CHAPTER 5

Effective Speaking

OBJECTIVES

You should study the chapter to know
- ◯ how to enhance the effectiveness of your speaking
- ◯ the basic sounds of English
- ◯ how to use phonetic transcription
- ◯ the rules of pronunciation and the problem sounds
- ◯ how to employ appropriate word stress, sentence stress, and intonation patterns for effective oral communication
- ◯ how to overcome nervousness, if you have any, while speaking
- ◯ how to be clear and fluent in your speaking
- ◯ paralinguistic features, mother tongue influence, sociolinguistic competence, public speaking, speaking and negotiation

Introduction

Now that we have learnt the importance of listening, let us move on to the next important communication skill of speaking. Speaking effectively and powerfully is a skill that is really worth learning. Regardless of the expertise or responsibility at work, everyone will eventually be expected to give a presentation or make a speech. We may be asked to talk to colleagues, clients, suppliers, or the general public.

For effective oral communication, a basic understanding of the sound system of the English language is very important. The ability to produce individual sounds, both in isolation and in combination with other sounds, also plays a significant role. Pronunciation is far more important than the accurate production of individual sounds. Pronunciation is the way a word or a language is usually spoken or the manner in which someone utters the words of a language.

This chapter gives an insight into the sound system of English as well as facilitates our understanding of effective speaking.

The discussions on the essential components, namely correct stress, right intonation, paralinguistic features and confidence, the impact of Mother Tongue Influence and Sociolinguistic Competence,etc., would enable one to make his/her speaking more effective. In addition, the chapter also adds a discussion on public speaking and also the role of effective speaking in the context of negotiation.

Basic Sounds of English

A phoneme is any one of the set of smallest units of speech in a language that distinguishes one word from another. Changing one phoneme in a word can produce another word, or make the word unintelligible. For example, changing the first phoneme in the word 'cat' can produce a word with a very different meaning, such as 'rat'. A phoneme is represented between slashes (/b/, /k/, /s/, etc.) by convention. The number of phonemes varies widely according to the language. Languages can contain two to thirty vowels and five to more than 100 consonants.

In most written languages, a one-to-one correspondence between letters and phonemes does not exist. That is, there are (i) some letters that can represent more than one phoneme (but only one at a time) and/or (ii) some phonemes that can be represented by alternative individual letters and/or some combination(s) of letters.

The English alphabet has twenty six letters and these represent forty four distinct sounds of English. For example, 'c' in 'car', 'k' in 'kite', 'cc' in 'occasion', and 'ch' in 'chemist' all represent the same sound /k/. Similarly, different sounds are represented by 'ch' in words such as 'chemistry',

'C' mon boy! Put the cheese down and pull the keys out.'

'machine', 'attach', etc. Further, certain letters do not represent any sound. For example, 'e' in 'mine', 't' in 'listen,' 'p' in 'psychology', and 'h' in 'hour' do not represent any sound—they are silent. There are also letters that represent sounds not even hinted by the graphic sign. For example, 'gh' in 'enough' represents the /f/ sound, 'x' in 'examine' stands for /gz/, and 'y' in 'city' is for the /i/ sound. This, however, does not mean that there is no system of classification of English sounds. In fact, a very comprehensive and scientific description of these sounds has been made. A symbol from the internationally accepted system of signs is used to represent each sound. These symbols belong to the International Phonetic Alphabet (IPA). IPA is a system of phonetic notation devised by linguists to provide a standardized, accurate, and unique way of representing sounds of any spoken language. It is also used in some dictionaries and textbooks to indicate pronunciation.

There are various ways in which sounds can be classified.

Oral and nasal   Sounds are produced by the speech organs by forcing the air stream out of the lungs either through the mouth or the nose. All sounds are either oral or nasal. In the production of the former, the air is released through the mouth, whereas, in the latter, it is released either fully or partially through the nose.

Voiced and unvoiced   Sounds are either voiced or unvoiced. All sounds produced with the vibration of the vocal cords are called *voiced sounds*. For example, vowels, diphthongs (combinations of two vowel sounds or vowel letters), and consonants such as /b/, /d/, and /m/ are voiced sounds. In the production of unvoiced sounds such as /p/, /t/, /k/, /s/, etc., on the other hand, the vocal cords do not vibrate and the air passes through the wide-open glottis. The vibration can be felt by putting the palm of the hand on the Adam's apple.

Vowels and consonants   Sounds are also classified as vowels and consonants on the basis of manner and place of production. In English, all the vowels are oral and voiced. Among the consonants, only three are nasal: /m/, /n/, and /ŋ/ (as in 'sing'). Several oral consonants are unvoiced. The list of sounds of English is given in Tables 5.1 and 5.2. Now, let us study vowels and consonants in some detail.

Vowels and Consonants

During the production of vowel sounds, the air from the lungs comes out in an unrestricted manner in a rather continuous stream. There is no closure of the air passage or friction between any speech organs. There are twenty distinct vowel sounds in English. These twenty vowels are further classified as *pure vowels* and *diphthongs*. A pure vowel is a single sound marked by its steady quality. During the production of a pure vowel, its

> Consonants are distinguished from vowels by their manner of production.

quality does not change. In the production of a diphthong, on the other hand, one sound position glides to another, as a result of which the quality of the vowel changes. For example, /i:/ is a pure vowel as in 'feet', whereas /aɪ/ is a diphthong as in 'fight'.

While pronouncing consonants, the air passage is either completely or partially closed and the air passes through the speech organs with an audible friction. There are twenty four distinct consonants in English. Tables 5.1 and 5.2 present a list of vowel sounds and consonant sounds with examples.

Table 5.1   Vowel sounds

Sl. No.	IPA symbols	Initial	Words middle	Final
1.	/i:/	easy	field	see
2.	/ɪ/	it	hill	duty
3.	/e/	enemy	step	—
4.	/æ/	apple	man	—
5.	/ɑ:/	art	mask	car
6.	/ɒ/	office	cross	—
7.	/ɔ:/	all	born	saw
8.	/ʊ/	—	book	to
9.	/u:/	ooze	fool	true
10.	/ʌ/	under	sun	—
11.	/ɜ:/	earth	bird	stir
12.	/ə/	about	police	maker
13.	/eɪ/	eight	snail	say
14.	/əʊ/	old	hope	so
15.	/aɪ/	idea	bite	buy
16.	/aʊ/	out	sound	cow
17.	/ɔɪ/	oil	noise	boy
18.	/ɪə/	ear	hear	mere
19.	/eə/	air	hairy	share
20.	/ʊə/	—	cured	poor

The first twelve sounds are called pure vowels and the remaining eight diphthongs.

Table 5.2   Consonant sounds

Sl. No.	IPA symbols	Description	Initial	Words middle	Final
1.	/p/	unvoiced, oral	pen	speak	leap
2.	/b/	voiced, oral	bet	about	nib
3.	/t/	unvoiced, oral	time	better	beat
4.	/d/	voiced, oral	day	adorn	said
5.	/k/	unvoiced, oral	call	echo	ask
6.	/g/	voiced, oral	gate	ago	league
7.	/tʃ/	unvoiced, oral	church	butcher	attach

(Contd)

Sl. No.	IPA symbols	Description	Initial	Words middle	Final
8.	/dʒ/	voiced, oral	judge	adjust	age
9.	/f/	unvoiced, oral	five	after	enough
10.	/v/	voiced, oral	vine	averse	nerve
11.	/θ/	unvoiced, oral	thin,	atheist	oath
12.	/ð/	voiced, oral	then	other	clothe
13.	/s/	unvoiced, oral	some	biscuit	cots
14.	/z/	voiced, oral	zoo	dozen	buzz
15.	/ʃ/	unvoiced, oral	shape	ashamed	bush
16.	/ʒ/	voiced, oral	genre	pleasure	mirage
17.	/h/	unvoiced, oral	hat	behave	ah
18.	/l/	voiced, oral	lottery	along	bottle
19.	/m/	voiced, nasal	mat	among	balm
20.	/n/	voiced, nasal	nest	sound	bin
21.	/ŋ/	voiced, nasal	—	hanging	sing
22.	/r/	voiced, oral	rest	screech	ever
23.	/w/	voiced, oral	want	question	—
24.	/j/	voiced, oral	university	student	—

The sound represented by the letter 'r' is pronounced only when it is followed by a vowel sound in the same word or when the word immediately following it begins with a vowel sound. For example, in 'heart', it will not be pronounced, whereas in 'real', it will be. Similarly, in expressions such as 'the mother of Ram', 'remember it', and 'later on', it will be pronounced because it is followed by words beginning with vowel sounds /ə/, /i/, and /ɒ/.

You can listen to the vowel and consonant sounds in the QR code.

The following material in the form of minimal pairs, i.e., a set of two words distinguished by only one sound, will further help one recognize and pronounce words.

(i) In producing both the sounds in the following pairs, the lips are rounded but the first one is shorter.

/ɒ/	/ɔ:/
don	dawn
cot	caught
shot	short
pot	port
cod	cord

(ii) In the following pairs, both the sounds are long but in the production of the first, the lips are kept in a neutral position, whereas rounding them is essential for the second.

/ɑ:/	/ɔ:/
cart	court
barn	born
card	cord
car	core
bar	bore
hard	hoard
darn	dawn

(iii) The first sound is shorter and in producing it the lips are in a neutral-spread position. The second is produced from the centre of the mouth with the lips in a neutral position.

/ʌ/	/ə:/
bust	burst
bud	bird
hut	hurt
cut	curt
thud	third
such	search

(iv) In articulating the first sound the lips are rounded and then spread. In the production of the second, the mouth is widely opened with the lips moving from neutral position to loosely spread position.

/ɔi/	/ai/
boy	buy
toy	tie
voice	vice
toil	tile
point	pint

(v) In the production of the first sound, the air is blocked in the mouth and then suddenly released by a slight separation of the lips. In the second, the air is released by pressing the lower lip against the upper front teeth.

/p/	/f/
pat	fat
pen	fen
pale	fail
past	fast
pour	four
pearl	furl

(vi) In producing both the sounds, the air is released partially through the nose and partially through the mouth.

The first is produced from the front of the mouth with the tip of the tongue touching the upper teeth ridge, whereas the second is produced from the back of the mouth.

/n/	/ŋ/
sin	sing
kin	king
ran	rang
ban	bang
ton	tongue
sun	sung

(vii) In producing both the sounds, the air is released without vibrating vocal cords and by keeping the lips in neutral position. However, the first is produced from the front of the mouth, whereas the second is produced from the centre.

/s/	/ʃ/
sake	shape
same	shame
sign	shine
save	shave
mess	mesh

(viii) The sounds /z/ and /ʒ/ are produced like /s/ and /ʃ/ respectively, but in their production the vocal cords vibrate. It is not possible to give minimal pairs for contrasting them as there are no words to distinguish them by the opposition of these sounds alone.

(ix) The first sound is articulated like /f/ but it is voiced. In producing the second, the lips are first sounded and then spread.

/v/	/w/
vet	wet
verse	worse
veil	wail
vine	wine
vent	went

Practice 1

Speak out the following 175 words as correctly as possible. Identify and encircle the words containing any of the long vowels (/iː/, /ɑː/, /ɜː/, /uː/, /ɔː/). Try to find out the meaning of those words that you have not come across so far:

acid, arch, achieve, allergy, ache, afternoon, aunt, agree, blue, burn, bend, boom, bath, brought, bathe, beat, caught, catch, cut, cure, come, calm, cute, curfew, czar, dance, dull, doom, deprive, dearth, diesel, dense, door, eight, ever, eel, English, fight, fill, fool, first, forty, fever, girl, gate, grass, gum, groom, geese, horse, hill, hoot, hurt, heaven, heal, ink, item, irk, immune, ice cream, idea, implant, jar, jelly, jeep, joy, jute, knell, kneel, loop, life, lean, love, last, license, look, lord, loom, loyal, learn, make, mark, mourn, manage, many, mute, meat, merchant, mean, magazine, normal, neat, nagging, news, nark, nation, oodles, ointment, oil, orbit, overt, organic, pen, peel, pan, porch, polite, parole, pearl, pending, pool, palm, root, rain, rack, ruin, reach, rancid, redeem, reserve, rinse, return, sit, same, seal, serve, sober, scourge, seamless, skew, sore, soothe, tea, tempt, torn, token, tool, tight, turn, total, toll, task, ultra, ugly, unclean, unbound, umpire, umbrage, uproot, vaccine, vain, vault, vast, venal, verbose, vibrant, victorious, volume, verdict, vase, video, wooden, work, wallet, wall, welcome, watch, water, worth, woo, wolf, wobble, who

Phonetic Transcription

As discussed earlier, English has twenty vowel sounds and twenty four consonant sounds that can be represented through IPA symbols. We have also learnt that there is no one-to-one correspondence between many letters and the sounds they represent. For instance, in the word 'chemistry', the letters *c* and *h* are together represented by the consonant /k/. Similarly, the sound /ʃ/ represents the letter *s* in the word 'sugar' and /θ/ represents the letters *th* in the word 'thief'. Such sounds may pose problems in pronouncing the words correctly, and hence it is necessary to refer to the dictionary to learn the correct pronunciation. Most dictionaries include the phonetic transcription or the representation of a word in terms of its phonemes or sounds. Such a transcription alone can tell us how to pronounce a word because knowing just the spelling of a word does not ensure the right pronunciation, as some letters and sounds may not match. Generally, dictionaries use IPA symbols for phonetic transcriptions, as shown in Table 5.3.

Transcribing the words phonetically may be quite interesting, provided we are familiar with all the phonemes represented by the IPA. After we transcribe a word according to the way we pronounce it, it can be verified with the dictionary. In this manner we can learn the correct pronunciation faster. *Orthography* refers to the spelling of a word.

Table 5.3 Phonetic transcriptions of some words

Word	Phonetic transcription
find	/faɪnd/
low	/ləʊ/
speak	/spiːk/
image	/imeɪdʒ/
icon	/aɪkon/
English	/ɪŋglɪʒ/
language	/læŋgwɪdʒ/
aim	/eɪm/
bags	/bægz/
passed	/pɑːst/
excellent	/eksələnt/
doctor	/dɒktər/
epic	/epɪk/
chalk	/tʃɔːk/
stadium	/steɪdiəm/

Practice 2

Convert the following phonetic transcriptions into orthography.
ʌndərstænd, ɑːskt, luːs, maɪt, rɒndɪvuː, kləʊz, kənsiːd, dɔːnt, fətɒɡrəfər, ðən

Rules of Pronunciation

The production of individual sounds does not follow a set pattern. It has to be learnt through intensive practice. However, the inflectional endings -s or -es and the past (tense) markers -d or -ed follow a regular pattern. Their pronunciation is guided by the following rules:

(a) The inflectional endings -s or -es are pronounced in three different ways:

   (i) /iz/ after the consonants /s/, /z/, /ʃ/, /ʒʒ/, /tʃ/, and /dʒ/
       Examples: classes, raises, brushes, mirages, catches, edges

   (ii) /s/ after the consonants /p/, /t/, /k/, /f/, and /θ/
       Examples: maps, boats, locks, roofs, oaths

   (iii) /z/ after all other consonants and all vowels
       Examples: bombs, cubs, pads, leaves, buys, feathers, clues, keys, cures, ties, etc.

(b) The past markers -d or -ed are pronounced in three different ways:

   (i) /id/ after the consonants /t/ and /d/
       Examples: quoted, bounded

   (ii) /t/ after the consonants /p/, /k/, /tʃ/, /f/, /q/, /s/, and /ʃ/
       Examples: hopped, poked, matched, coughed, earthed, blessed, blushed

   (iii) /d/ after all other consonants and all vowels
       Examples: clubbed, hugged, paved, erased, edged, called, named, cried, favoured, cured, etc.

(c) There are a few word-endings that are sometimes mispronounced. The correct pronunciations of these endings have been provided herewith:

   (i) -age → /idʒ/ and not /eɪdʒ/
       Examples: adage, cabbage, savage

   (ii) -ate → in adjectives → /ət/ and not /eɪt/
       Examples: intimate, penultimate, delicate
       -ate → in verbs → /eit/
       Examples: cultivate, punctuate, differentiate

   (iii) -ance/-ence, -ant/-ent → vowel /ə/ and not /e/ or /æ/
       Examples: disturbance, preference, arrogant, agreement

   (iv) -cian → /ʃn/ and not /ʃian/
       Examples: magician, politician, technician

   (v) -est/-et → vowel /i/ and not /e/
       Examples: finest, boldest, socket, pocket

   (vi) -cial, -sial, and -tial → /ʃl/ and not /ʃiəl/
       Examples: social, controversial, preferential

(d) 'ng' is pronounced in the following ways:

   (i) In final position 'ng' is always pronounced as /ŋ/ and never /ŋg/
       Examples: bang, ring, lung, speaking, reading, writing

   (ii) In medial position 'ng' is pronounced in two different ways:
       /ŋ/ only in the words formed from verbs

Examples: singer, bringer, ringer
Also, /ŋ/ only when the plural marker -s is suffixed to nouns ending in /ŋ/
Examples: flings, things, songs

(iii) /ŋg/ in all other cases
Examples: linger, finger, hunger

Practice 3

Speak out the following words keeping in mind the rules of correct pronunciation for past tense words and plurals. Then group them under the appropriate phonemes given thereafter.

acted, added, affected, aided, amused, arms, asked, belts, bended, birds, bloated, books, boxes, branded, buses, cars, classes, clothed, combs, cooked, coughed, counted, cried, crossed, danced, days, doomed, drinks, earned, ended, eyes, faces, faded, fetched, flouted, friends, fused, guessed, hands, hated, hedges, helped, hooted, hoped, hops, hushed, laughed, legs, lifted, lived, loaded, loads, loaves, looked, marked, mashed, masked, missed, mixed, needed, noses, pads, painted, parked, passed, phones, pieces, planes, planned, played, plays, poked, prizes, pubs, queued, rebelled, rested, rolled, scored, senses, shops, shouted, smiled, soaked, songs, sounded, started, subjects, switches, talked, topped, touched, uses, walked, wanted, weighed, winds, wished

/d/   /id/   /t/   /s/   /z/   /iz/

Problem Sounds

Go through the following dialogue:

Speaker 1: Did you like the test of our copy?

Speaker 2: What do you mean? Which test and copy are you talking about?

Speaker 1: The copy you have just now finished!

Speaker 2: Oh! Your coffee! It tastes so well. Thank you.

Such confusions can arise when, e.g., Indians speak English with the various accents used in their mother tongues, because many sounds of their language may differ from the sounds of English. Some of the sounds in English are similar to the sounds in the Indian languages. For instance, the sounds /p/, /b/, /t/, /d/, /k/, /g/, /m/, /n/, and /r/ may not pose much of a problem for Indian speakers of English. However, they face problems with the other sounds that they do not use in their regional languages.

Among English consonant sounds, fricatives (sounds made by narrowing the air passage at some place in the mouth so that the air, while escaping, causes audible friction) pose problems for Indian speakers. Some of the fricatives that pose problems are /f/, /v/, /θ/, /ð/, /z/, /ʃ/, and /ʒ/.

In some Indian languages, many words end with a vowel sound, and hence speakers of these languages tend to add a vowel sound even at the end of English words. Similarly, in some Indian languages, consonant clusters (explained in the next section) are rare in the beginning of words, and hence these speakers insert a vowel sound at the beginning of consonant clusters in English words too.

'Thees is e lame daak gorment.'

Examples

Root: /ruːt/ becomes /ruːtu/ or /ruːtə/

Scheme: /skiːm/ becomes /iskiːm/ or /skiːmiː/

As against the twenty vowel sounds in English, Indian English has only seventeen, out of which eleven are pure vowels and six are diphthongs. Hence, problems arise for Indian speakers in producing certain vowels too (see Table 5.4).

Table 5.4   Problem vowels

Word	English usage	Indian usage
Date	/deɪt/	/det/
Post	/pəʊst/	/post/
Court	/kɔːt/	/kort/
Bird	/bɜːd/	/bəd/
Bud	/bʌd/	/baːd/
Tour	/tʊə/	/tuːr/
Hurt	/hɜːt/	/hut/
There	/ðeər/	/ðer/
Ball	/bɔːl/	/baːl/ or /bɔl/

Practice 4

Speak out the following sentences and check in the dictionary whether you have uttered the words set in italics correctly.
1. She is very *tall*.
2. I have great *pleasure* in welcoming you.
3. Please bring a *dozen* bananas.
4. I *wish* to keep a *bird* as my pet.
5. Birds of the same feather flock together.
6. All his *kith* and kin were *present* at the party.
7. Do you want me *to go now*?
8. Have you *taught* me this lesson?
9. Let him take all these chairs.
10. *Rain*, rain, go *away*.

You can strengthen your knowledge of the sounds of English by referring to the following websites:

http://www.antimoon.com

http://fonetiks.org

http://www.learnenglish.de/

http://www.englishmedialab.com/pronunciation.html

http://cambridgeenglishonline.com/Phonetics_Focus/

Word Stress

Accent or stress is an important aspect of spoken English. Developing effective word accent and sentence accent is very important to speak intelligible English. The following discussion would enable one to understand the concepts pertaining to syllables, word stress, and sentence stress, which are necessary to speak English effectively with the right accent.

When we speak, the air from our lungs does not come out in a continuous stream but in small puffs, each puff of air producing a syllable. Such puffs require greater muscular strength when an *accented* or *stressed* syllable is produced. Thus, an accent is the prominence or the relatively greater emphasis given to a particular syllable in the word. In a multisyllablic word, one particular syllable has generally greater stress than the others.

Accent is of two types: primary accent and secondary accent. The *primary accent* is shown by a vertical bar above and in front of the accented syllable, e.g., aˈbroad, examination, etc., and the *secondary accent* is shown by a vertical bar below and in front of the accented syllable, e.g., ˈcalcuˌlate, eˌxamiˈnation, etc.

It is difficult to give a comprehensive list of rules for accenting words. However, the following rules can help one get the proper accent.

(a) If a word ends with -ion or -logy, the syllable preceding these endings is accented.
Examples: ex-hi-'bi-tion, ter-mi-'na-tion, an-thro-'po-logy, bi-'o-lo-gy, etc.
(b) If an adjective ends with -ic, the main accent is normally on the second syllable from the end.
Examples: a-'to-mic, sym-'bo-lic, etc.
(c) If a verb ends with -fy, -ate, -ize, -ise, or -yse, the main accent is on the third syllable from the end.
Examples: ex-'em-pli-fy, 'i-mi-tate, 'fac-to-rize, e-'co-no-mize, 'a-na-lyse, etc.
(d) If an adjective ends with -ical, the main accent is normally on the third syllable from the end.
Examples: psy-cho-'lo-gi-cal, 'gra-phi-cal, etc.
(e) If a word consisting of three or more syllables ends with -ity, the main accent is on the third syllable from the end. If it has only two syllables, the main accent is on the first syllable.
Examples: 'en-mi-ty, ac-'ti-vi-ty, 'ci-ty, 'pi-ty, etc.
(f) If an adverb ends with -ically, the main accent normally falls on the fourth syllable from the end.
Examples: geo-'gra-phi-ca-lly, 'lo-gi-ca-lly, etc.
(g) If a word ends with -ee, -eer, or -ette, the main accent normally falls on the last syllable.
Examples: em-plo-'yee, auc-tion-'eer, ci-ga-'rette, etc.

There are some words in which a mere shift in accent changes their class. A few examples are given below:

Noun	Verb
'Conduct	Con'duct
'Object	Ob'ject
'Project	Pro'ject
'Progress	Pro'gress
'Contact	Con'tact

Weak Forms

The weak form of a word is used when the word has no stress and is phonetically distinct from its strong/full form. A weak form is an unstressed syllable. A word may have multiple weak forms or none.

There are about 45 words in English with two or more pronunciations—one strong pronunciation and one weak pronunciation or weak form. Words of this kind may be called *weak form words*. Since almost all native speakers of RP use weak forms in their pronunciation, it becomes difficult for a learner of RP to understand without learning those words properly. The main words with weak forms in RP are:

a, am, an, and, are, as, at, be, been, but, can, could, do, does, for, from, had, has, have, he, her, him, his, just, me, must, of, shall, she, should, some, than, that, the, them, there, to, us, was, we, were, who, would, you

These words are also known as *function words*—words that do not have a dictionary meaning in the way we normally expect nouns, adjectives, verbs, and adverbs to have. Words such as auxiliary verbs, conjunctions, pronouns, prepositions, articles, etc., fall into this category, all of which are pronounced in their strong forms in certain circumstances but are more frequently pronounced in their weak forms. There are certain contexts where only the strong form is acceptable and others where the weak form is generally used.

Examples:
And: full form: /ænd/; weak form: /nd/, /n/, /nd/, /n/.
bread *and* butter (weak form)
You should not put '*and*' at the end of a sentence. (Full form)
Can: full form: /kæn/; weak form: /kən/
Malini *can* speak English better than I *can*. (The first can is the weak form while the second is the full form.)

Of:	full form: /ɔːv/; weak form: /əv/
	I am fond *of* fruits. (Weak form)
	Fruits are what I am fond *of*. (Full form)
Have:	full form: /hæv/; weak form: /əv/
	Have you finished? (Weak form)
	Yes, I *have*. (Full form)
Should:	full form: /ʃud/; weak form: /ʃd/
	I *should* forget it. (Weak form)
	So you *should*. (Full form)
To:	full form [tuː], [tu] before vowels; weak form: [tə] before consonants
	It is time *to* act. (Full form)
	Try *to* stop. (Weak form)
His:	full form: /his/; weak form: /is/
	His name was mentioned. (Full form)
	Give me *his* books. (Weak form)
Some:	full form: /sʌm/; weak form: /səm/
	Have *some* more coffee. (Weak form)
	I have got *some*. (Full form)

Practice 5

Mark the primary stress in the following words:
tunnel, deafening, teacher, yesterday, physiology, examination, interview, laboratory, economic, effectively

Sentence Stress

Look at the following sentence.

This acid is very harmful.

Try and speak this sentence without stressing any word. Your utterance may not send the intended meaning to your listener. But when you give emphasis on 'this' the importance is on the kind of acid. When you stress the word 'very', the importance shifts to the nature of the harm.

In English sentences, each word does not have the same prominence and some words stand out from the rest due to stress or greater breath force. Just as in a single polysyllabic word, one syllable gets more stress than the others, in a sentence one or two words get more emphasis than the others. The following are some sentences in which stressing on different words may intend different meanings.

This is the latest book I have read.
The company produces ten thousand cars every day.
I want you to look into this matter.
I must start in ten minutes.

Practice 6

Underline the word or words you would like to give emphasis on in the following sentences:
1. He is a charming fellow.
2. He was not by himself when I saw him.

3. Give yourselves plenty of time.
4. You will not hurt me, you'll only hurt yourself.
5. That is what I told you.
6. These are the best.
7. Do you like these flowers?
8. Is the book on the table a good one?
9. That is the thing I wanted.
10. I did not mean the channel, I meant just a channel.
11. Has she brought any sugar?
12. I want some more.
13. Any newspaper will do.
14. Have you done anything yet?
15. It is a disappointing and unconvincing play.
16. They have got a lovely little house in the country.
17. Did the boys see the teacher after the class?
18. He is a fellow I do not like.
19. Somebody must have lost it.
20. Has anything been done for it?
21. There is not much time.
22. He must be either stupid or careless.
23. He made a long, boring, and depressing speech.
24. Do not forget your homework.
25. We are not going.

Sentence Rhythm

English is a rhythmic language, rhythm being a feature of the phonological structure of English. Rhythm generally refers to the timing pattern. There are two factors that influence the rhythm of English: sentence stress and connected speech. Maintaining a rhythm in a speech makes the speech sound natural and fluent.

Words are of two types: *content* or *lexical* words and *function* or *grammatical* words. The content or lexical words carry meaning by themselves, whereas function or grammatical words do not have any significant meaning in themselves and depend on the content words to derive meaning. Sentences contain both content and function words. Content words that include nouns, verbs, adjectives, and adverbs are strong words and carry stress in a sentence, while function words that include determiners, pronouns, prepositions, conjunctions, auxiliary verbs, and interrogatives are weak and unstressed words.

The rhythm produced by the stressed and unstressed words in a sentence is a major characteristic of spoken English. Using only the strong forms (i.e., stressing all the words) in sentences may make a speech sound dull and artificial. The listener also may not understand the intended emphasis or meaning in the speech.

Just as sentence stress, speed is another very important factor in the fluency of English. When we speak, we do not speak words in isolation but group them and speak without any pauses between them. Unstressed words always sound different when used in a sentence as against when used in isolation. The most common feature of connected speech is the weak forms of function words such as *of, have, was, were, to, has,* etc., and contractions such as *can't, haven't, shouldn't, won't, didn't, they've,* etc. However, we ignore other features, namely elision (losing sounds as in the word listen), linking (adding or joining sounds between words (as in 'far away'), and assimilation (changing sounds when two words are combined as in 'good girl'), that help preserve rhythm in our utterances. In addition to these features, there is *schwa*—the most common vowel sound spoken in unstressed words in English (as the sound /ə/ in 'ago', 'about', etc.).

Read aloud the following excerpt from the famous Gettysburg address by Abraham Lincoln to understand how rhythm works in English language and makes a speech look natural and fluent.

> Fourscore and seven years ago, our fathers brought forth on this continent a new nation, conceived in liberty and dedicated to the proposition that all men are created equal. Now we are engaged in a great civil war, testing whether that nation or any nation so conceived and so dedicated can long endure. We are met on a great battlefield of that war.

We have come to dedicate a portion of that field as a final resting place for those who here gave their lives that that nation might live. It is altogether fitting and proper that we should do this.

Parallel grammatical structure, placing important ideas in the beginning or at the end, using subordinating conjunctions for less important ideas, using appositive and absolute phrases to add details to the main idea—all these bring in rhythm in a speech.

It is important to know which words are to be stressed in a sentence and which syllables of polysyllabic words are to receive primary accent depending on the rhythmic balance of the sentence and the relative importance of its different parts. Consider the following sentences:

1. You 'ought to' know the 'way to 'use a 'pen.
2. I 'wish I 'could.
3. 'What an 'interesting 'story!
4. You could 'make a 'point of 'telling him.

In the first and second sentences, one unstressed syllable is followed by a stressed syllable. Hence, a speaker finds it very easy to utter the sentence in such a manner that the stressed syllables *ought, know, way, use, pen* (S1), *wish,* and *could* (S2) occur at regular intervals of time. The third sentence has eight syllables, three of which are stressed. There is only one unstressed syllable between the first two stressed syllables, three unstressed syllables between the second and third stressed syllables, and one unstressed syllable at the end of the sentence. Similarly, the fourth sentence has nine syllables, three of which are stressed. There are six unstressed syllables, with only one unstressed syllable between the two stressed syllables—the first two and also the next two stressed syllables. There are two unstressed syllables at the beginning as well as at the end of the sentence. The time interval between *what, m-, st-* (S3), *make, point,* and *te-* (S4) will be approximately the same in spite of the variation in the number of unstressed syllables between the stressed ones.

The above examples (1–4) explain the importance of stressed and unstressed syllables in creating characteristic rhythm in a sentence. Unstressed syllables are generally crowded together and pronounced rapidly whereas stressed syllables are pronounced more clearly. Thus, accented/stressed syllables with no stressed syllables in between them are prolonged. The speed of delivery, therefore, depends on the number of unstressed syllables between the two stressed ones.

 Refer to the Online Resource Centre to understand the strong and weak forms of English words.

Intonation

The intonation of a language refers to the patterns of pitch variation or the tones it uses in its utterances. In normal speech, the pitch of our voice goes on changing constantly—going up, going down, and sometimes remaining steady. Different pitches of the voice combine to form patterns of pitch variation or tones, which together constitute intonation.

Intonation is closely linked to stress because important changes in pitch occur with stressed syllables. These changes generally take place on the last stressed syllable in an utterance, and hence this syllable is called the nucleus. The following are the main functions of intonation:

- Distinguishing different types of utterances such as statements, commands, requests, and questions
- Differentiating the speaker's emotional attitude such as curiosity, apprehension, friendliness, and politeness
- Drawing the listeners' attention to those segments of an utterance that one considers important

A segment of speech carrying one intonation pattern is called a tone. Sequences of English speech fall into two well-defined tones or intonation patterns:

1. Falling tone (tone I)   2. Rising tone (tone II)

A tone group may, thus, consist of one or more syllables. A short utterance quite often forms a single tone group, while a longer one is made up of two or more. While speaking, we divide long utterances into small and manageable groups of words, between which we pause. Hence, a tone group may also be defined as a stretch of speech between two pauses. A few examples are given below to show the concept of tone groups wherein the tone group boundary is indicated by an oblique (/) sign:

> A stretch of speech over which one pattern of pitch variations, or contour of pitch, extends is called a tone group.

1. Yes.
2. She will meet you tomorrow.
3. Why do not you help me?
4. I am not afraid of them.
5. The door has been opened.
6. I want to complete this work/before my teacher comes.
7. Please do it now/if you have the resources.
8. We can go for a picnic/if we receive permission from them.
9. Do you know/that they are planning to put up a show/for collecting funds for Tsunami victims?
10. You have plenty of time, but we do not, despite the three forthcoming holidays.

General Uses of Tone I and Tone II

Falling tone (tone I)

In utterances with falling tone, the pitch falls from high to low. The falling tone (which is marked by [`]) usually occurs in the following types of sentences:

(i) Statements that are complete and definite
 Examples:

I am well aware of my limi`tations.
They might have already `left.
I `cannot a'fford to `do it.

I 'go for a 'walk `daily.
You must 'know 'how to `do it.

(ii) Question-word questions that are matters of fact and are intended to be neither polite nor impolite
 Examples:

When are you `leaving?
How long will it take you to come `here?
Who is knocking at the `door?

Where do you come `from?
Why does he want to leave the 'company?

(iii) Commands
 Examples:

Open the `door.
Give me that `pen.
'Tell him to be`have.

'Look 'up the `dictionary.
'Turn to page `forty.

(iv) Choice questions
 Examples:

Will you agree or `disagree?

Is he reading or `sleeping?
Do you want tea or `coffee?

Would you like to see him today or `tomorrow?
Has she accepted or `rejected?

(v) Exclamations
Examples:

`Fantastic!                    How `beautiful!
What fine `weather!            `Bravo!
By `Jove!

(vi) Tag questions with a negative tag
Examples:

You agree to this, `don't you?        We have faced it several times, `haven't we?
He can help you, `can't he?           She has worked hard, `hasn't she?
They should come forward, `shouldn't they?

Rising tone (tone II)

In utterances with rising tone the pitch rises from low to high. The rising tone, marked by [,], generally occurs in the following types of sentences:

(i) Yes/No type question
Examples:

Do you 'agree?   Shall we go ,now?
Was he ill ,yesterday?   Have they ,finished?
Did they explain the reasons?

(ii) Statements intended to be a question
Examples:

You will not ,do that?   You do not want to lend me ,your book?
He does not have ,money?   She cannot hear ,you?
They are not going?

(iii) Non-terminal tone group (the first group is spoken with a rising tone while the second with a falling tone)
Examples:

Unless you decide to ,succeed ... (you may not   Before leaving for ,vacation ... (I must
    succeed)   complete this job)
While going ,home ... (I would return this box)   Since this ,morning ... (he has not reported)
If you do not turn up ,tomorrow ... (I'll send you out)

* (iv) Polite requests/commands intended to sound like a request
Examples:

Get me that pen ,please.   Call him ,in.
Listen to me for a ,minute.   Do not mention it to ,him.
Buy me a ticket, ,please.

(v) Tag questions with a positive tag
Examples:

She is not honest, ,is she?   He cannot complete the project, ,can he?
They are not coming today, ,are they?   It was not enough for you, was it?
We should not let them go, ,should we?

Falling-rising tone

Besides using the falling tone and the rising tone discussed above, we can also use the combination of tones, namely the falling-rising tone and the rising-falling tone for our utterances in English. In utterances with falling-rising tone, the pitch falls from high to low and then rises again to high. On the contrary, in utterances with rising-falling tone, the pitch rises from low to high and then falls again to low. These are detailed as follows.

The falling-rising tone [marked as (ˇ)] may be used in expressing special implications not verbally expressed (apology, happiness, doubt, insinuation, indirect insult, etc.) and in incomplete statements leading to one of the following tone groups:

> Do you watch movies? ˇSometimes. (Not frequently)
> I ˋmet her at the ˇclub. (I expected her somewhere else)
> Is he at home? ˇNow? (Doubtful)
> ˋShe cannot? (I'm almost sure she can)
> He ˇcan? (I am almost sure he cannot)
> I am glad to see you doing ˇwell. (Encouraging, sympathetic)
> ˋWhen will you ˏbring me some more books? (A polite rise on *bring* and a mild insistence on when)
> Mind your ˇlanguage. (Strong but sympathetic warning)
> When I come to know of it, /(I will inform you). (Incomplete utterance)
> If you do not behave, /(I will punish you severely). (Incomplete utterance)

The rising-falling tone is generally used to reinforce the meaning expressed by the following falling tone. The initial rising tone may also exhibit warmth, friendliness, anger, or sarcasm.

> Do you believe it? Yes. (Enthusiastic agreement)
> Of course. (Enthusiastic agreement)
> It was awful. (Enthusiastic agreement)
> But is her child so naughty? (Suspicious interest)
> Are you sure he can play? (Suspicious, mocking)
> How wonderful. (Sarcastic)
> Oh, really. (Sarcasm)
> Go and break your head. (Haughty)
> Come and face the challenge. (Irritable)

Table C5.3 in the Online Resource Centre provides a complete list of phonetic symbols and signs used in English. Table C5.4 lists the variations in the sounds of Indian, American, and British English.

Achieving Confidence, Clarity, and Fluency

To grow in our academic or professional career, we must have the ability to stand in front of an audience and deliver a lecture or presentation on a certain topic. It should be convincing, supported with facts and examples, and be able to create an impact. We may have brilliant ideas, but it is very important to get them across to others for them to be appreciated. Similarly, in our personal or community life, we may have to meet people, talk to them about some matter, or persuade them to do something. We may have to speak in various situations, be it with colleagues at the workplace, friends and teachers in college, or others in our neighbourhood. The effectiveness with which we do it shapes the perception of us in others' minds. You probably remember times when you persuaded your parents to agree to something, given a motivational talk to your siblings or friends, or informed your teacher about something. On the other hand, you might have failed in convincing or persuading others at some other occasions. Thus, different experiences might have made you realize that being confident, clear, and fluent is the key to effective oral communication. Besides being confident, clear, and fluent, you need to understand the impact of your mother tongue and sociolinguistic competence on your English usage. Hence, let us discuss these aspects in some detail to make you more effective in your speeches and presentations.

Confidence

Most people tend to become anxious or nervous before doing something important in public. For instance, athletes are nervous before a big game, politicians are nervous to stand in front of a huge gathering during their election campaign, and actors are nervous to face the camera and deliver dialogues. However, we must know that only those people who know how to transform their nervousness into excitement or enthusiasm become successful. The same approach is true for successful speakers. Some speakers may even feel that they would perform well only if they become nervous before a speech and that they would fail if they are cool and self-assured. Hence, becoming nervous before a speech or presentation is natural, normal, and sometimes even desirable. However, we should know how to overcome this *nervousness* or *stage fright* so that we not only feel confident but also appear so in front of our audience. The following are a few tips to overcome nervousness before a speech:

- Believe that fear is your friend; it makes your reflexes sharper, heightens your energy as more adrenaline is pumped into your blood when you are nervous or stressed.
- Given a choice, choose a topic you truly like. If you are given a specific topic, develop genuine interest in it.
- Prepare, prepare, and prepare until you become comfortable with delivering the topic. Thorough preparation is the antidote for nervousness.
- If you are a novice speaker, then prepare more than the required material so that you do not go blank during your speech on any main or subtopics.
- Give a mock speech in front of your friends and find out your strengths and weaknesses.
- Try the positive visualization technique that requires you to concentrate on how good you are and to think *all is well*. You should practice this a day or a few hours before your speech: assume that you are just chatting with your friends; close your eyes and imagine that your audience is intently listening to you, smiling, and applauding; at the end of your speech, many people come to you, shake hands, and appreciate your speech; and then you enjoy your favourite food with your family and friends in a nice restaurant.
- Anticipate easy and difficult questions and prepare answers for the same.
- Work hard on your introduction material. Practise till it is smooth. Generally the speaker's anxiety level begins to drop significantly after the first thirty seconds. Hence, once you are through with the initial part of your introduction you may feel better.
- Take a couple of slow, deep breaths before you start your speech.
- Do not start immediately after reaching the stage. Take a few seconds to look into the eyes of your audience.
- Check the venue and other arrangements. For instance, if the mike or projector does not work, your nervousness may get aggravated. Hence, it is better to check the equipment.
- While waiting for the audience you can take a quick walk in the room.
- Look at the friendliest faces in the audience.
- Do not explicitly show your nervousness; if your hands are trembling keep them close to your body. If your legs are shaking, lean on the lectern or table on the stage.
- Do not comment on your nervousness.
- Remember that nervousness does not show even one-tenth as much as it feels.
- If your speech is being recorded, forget that you are standing in front of a camera. Speak only to your audience and not to the camera.

Clarity

Listeners, unlike readers, cannot look up a dictionary or re-read the words to comprehend the meaning of the words they listen to. A speaker's meaning must be immediately understandable; it must be so clear that there is practically no chance of misunderstanding. Many speakers despite having a complete control of what they were speaking find it very difficult to speak clearly. Even if they spend hours in preparing for their speeches or presentations and deliver it confidently, an audience may not understand most part of the speech if there is no clarity in their speech. A speech may lack clarity, if the speaker:

- Speaks either very fast or very slow
- Does not articulate the words properly
- Pronounces incorrectly or does not follow the standard pronunciation
- Gives wrong emphasis on words
- Does not have a well-organized material
- Uses too many unfamiliar words

The following are some of the ways in which the clarity of speech can be improved:
- The average number of words that can be spoken per minute is about 120–140. Inexperienced speakers generally tend to speak faster because of their nervous energy. Hence, a conscious effort to slow down can reduce the speed significantly. Likewise, if the speed is very slow, an effort to speed up should be made. However, in both the cases the speaker should always look natural.
- The following exercises that involve speech organs such as jaw, tongue, etc., should be practised:
  (a) Open your mouth wide and then close. Repeat it several times.
  (b) Touch the inner parts of your mouth with your tongue or rotate your tongue, thereby touching all parts of your mouth.
  (c) Make wide chewing motion while humming gently.
  (d) Stretch your jaw muscles (as though yawning) while moving your jaw sideways and in circles.
  (e) Puff air into your mouth, keep for a few seconds, and then release it. Repeat it several times.
- One should learn the correct pronunciation of words while preparing for a speech. The preceding chapter on Phonetics would facilitate in this process. Listening to English news in television channels may also help as the news readers speak Standard English. One can also seek the help of friends who speak good English.
- While speaking, the volume should be adjusted keeping in mind the number of audience, the size of the room, etc. If using a mike, one needs to adjust the volume before one begins the speech or presentation and has to get it checked thoroughly.
- All the words in each sentence of one's speech is not equally important. Take for example a statement 'We all want the best solution'. As a speaker, one should know which word should be stressed— *all*, *best*, or *solution*. One's misplacing of emphasis may make the audience confused. Hence, while rehearsing, it is always better to highlight the words that are to be emphasized and to speak them with the right accent. One should believe in the saying 'Practice makes one perfect'.
- Whether be it a discussion with one's professors, a conversation with an official, or a speech to a group of audience, one needs to think and organize the contents of the message before speaking. Assume that you wish to enquire the Passport officer about the status of your passport. If you ask him or her 'What happened to my Passport?', and then provide the details about application number, date, etc., chances are that you may not get the desired response. On the other hand, if you provide a context and then ask your query, the officer may understand and respond. Similarly, in a speech one needs to give a preview to the topic, discuss all the points in a logical manner, and then give the highlights of the discussion at the end. Otherwise, the speech will lack clarity.

- Simple and unambiguous words and expressions should be used in a speech. One should always avoid too many technical terms, acronyms, complicated words and phrases that the audience may not be familiar with. Rather than impressing the audience it may just end up putting them off. Knowing whom one is addressing is as important as what one is going to speak. If one really has to use technical terms, then those should be explained to the audience.
- Recording one's presentations may help know how one sounds and where the problems lie. A video recording would be an even better idea. One may request some friends to take a look at the recordings and provide a frank assessment of the clarity of the presentation. Then, based on the feedback and on one's observations, one can work on improving the clarity of the speech.

Fluency

We can ask these following questions to ourselves and try to answer them:

- Do I say 'umm…' 'ah…' because I do not get the right words to continue my statements?
- Do I pause a lot when I speak?
- Do I use certain expressions such as 'you know', 'I mean', 'actually', etc., during my speaking assignments?
- Do I speak very slowly and carefully because I feel that I may commit mistakes?
- Do I mumble some words because I am not very sure about my pronunciation?
- Do I feel irritated when a member of the audience interrupts me during my speech?

If the answer is 'yes', then we have a problem in speaking English fluently. Fluency is nothing but delivering continuous flow of message at an appropriate rate with necessary pauses. The fluency in speaking English depends mostly on our proficiency in English. We must realize that gaining good command over a language will pave the way for fluency, which in turn may enable us to face the audience confidently. The following guidelines may help enhance our fluency in English:

- Aim for clear oral communication devoid of speech errors.
- Write personal diary in English.
- Listen to good speakers and read material written in good English. Listen to news in English. Watch English documentaries. Read the newspaper both for content and language. Read fiction/non-fiction to understand the use of figures of speech such as metaphor, simile, alliteration, etc.
- Expand vocabulary by learning at least five words per day, along with their meanings and usages. Refer to Chapter 17 for some guidelines on how to improve vocabulary.
- Practise correct pronunciation, accent, and tone to make the speech impressive.
- Learn from your mistakes. For instance, during your initial attempts on speaking, you may go wrong in grammar or vocabulary but make sincere efforts to correct them in the subsequent attempts.
- Concentrate on your ideas rather than your appearance, the impact you are creating, etc.
- Read aloud the passages from books or magazines that you enjoy reading.
- Watch English movies to understand the manner of speaking conversational English.
- Think in English what you have done or what you are going to do.
- Practise your speaking skills in small talks—on weather, game, hobbies, current affairs, etc.—with your friends or family members. Ask them to correct you if you make some mistakes in grammar, vocabulary, or the use of appropriate pauses.
- Believe in what you're speaking.
- Spare at least an hour every day for developing your English fluency.
To enhance fluency, one needs to keep speaking English and continue to learn from one's mistakes.

Developing Voice Quality

Our voice reflects our personality, and hence it is important to cultivate a good voice. Although our voice quality depends on various factors such as our vocal habits developed since our childhood and the structure and condition of our speech organs, which cannot be changed much, we can improve on our voice quality by manipulating other aspects of voice such as pitch, rate, and volume.

To develop voice quality, we need to first identify the strengths and weaknesses of our voice. Just like effective body language enables us to enhance the impact of our speech, a proper use of our voice makes our speech more lively and dynamic. The features associated with our voice are known as paralinguistic features. The following section discusses these features in detail.

Mother Tongue Influence (MTI)

Pronunciation plays a crucial role which is as important as grammar and vocabulary in acquiring English proficiency. However, not much attention is given to the way in which each English sound is uttered while pronouncing a word. Even if adequate attention and practice is given for pronouncing individual sounds, the Mother Tongue Influence (MTI) is difficult to overcome in most of you as 'habits die hard'. You have been speaking in your mother tongue or native language since your childhood and in schools.

Whenever you face something new, say new music, food, place, etc., you tend to compare it with the one you are familiar with. Similarly, when you learn the speech system of a new language (here, English), you try to find similarities between your own language and English. In fact, you try to learn the new system through the known system and in the process, you may find problems with some sounds which are completely different in the two systems. For instance, the sound /ʒ/ which is represented in words like 'pleasure', 'measure', etc. is specific to English and hence speakers of Hindi tend to pronounce it as the sound /ʃ/ or /dʒ/ mainly because of their MTI.

Your 'mother tongue' can be defined as the language which you have grown up speaking from your early childhood or the first language you've learnt. Alternatively, it can also be called your 'native language'. 'Second language' is any language that you speak other than your first/native language. For instance, if your mother tongue is Hindi, it is your first language (L1); when you learn English at your school, it is your second language (L2).

Your mother tongue is closely associated with your culture as it is not only the language you learn from your mother but also your dominant and home language. It is the language community of your mother and the language spoken in your region. It enables your growing in a particular system of linguistic perception of the world. Hence, it is difficult to avoid the influence of your L1 on the acquisition of L2.

It is redundant to talk about the importance of English in today's globalized, digital society. Hence, regardless of the learning environment, your goal may be to master this language so that you can perform well in your academic, professional, and social assignments. As already mentioned, the sounds, structure, and usage of your L1 and L2 may be different. When you depend on your L1 to pronounce, form words, and frame sentences in L2, you are sure to encounter a lot of errors in the latter because of the interference of your L1 on L2. For instance, the Hindi sentence 'Mai kal Chennai ja raha hoon' would be literally translated into 'I tomorrow to Chennai going' by speakers of Hindi if they literally borrow the syntax from their L1without getting into the English (L2) sentence structure. Similarly, there are many words that are pronounced incorrectly by speakers of Hindi and other regional languages such as Tamil, Telugu, Bengali, Punjabi, or Marathi because of their MTI (e.g., *post, go, ball, talk, pleasure, bird, west, etc.*).

It's interesting to know that your mother tongue (L1) has both positive and negative effects on your learning of English. In fact, your L1 is a good resource which you use consciously or subconsciously to arrange or rearrange your L2 data in the input and to perform as best as you can. L1 can be a contributing factor to your L2 acquisition as it is a development process. You would have heard teachers or others say, 'If you wish to speak good English, think in English.' It is impossible for us to think in any language other than our native language unless we have gained a mastery over the other language because our native language is the one which we have been culturally attached to.

Though it is difficult to eliminate MTI on your L2 acquisition, it is very much possible to restrict it from interfering with your L1 by adapting certain tips given below:

- Don't feel shy/Don't hesitate to admit that you have a problem.
- Ask your friends/colleagues to correct you.
- Listen to English news bulletins in TV and good speakers of English.
- Read aloud so that you can hear your own pronunciation of words.
- Keep trying to correct your mistakes; practice makes you perfect.
- Try to identify the specific sounds which you are unable to pronounce correctly.
- Learn the 44 sounds in English with their correct symbols and pronunciation so that you can look up the dictionary to understand their correct pronunciation.

Remember that language is a tool to communicate and exchange your ideas, thoughts, feelings, and actions to others. Hence, please do not try to develop British or American accent while learning and speaking English. You can do it only when you master over English by learning and practising its sounds and structure thoroughly, thereby acquiring the required proficiency. But you can certainly minimize the influence of your mother tongue on your English learning by making conscious efforts and rigorous practice of certain problem sounds and using the tips mentioned earlier.

Sociolinguistic Competence

You may perform well in your text-based examinations with your good grammatical knowledge, fluency, and ability to frame correct sentences but you may not be able to communicate properly with others in different cultural contexts without the sociolinguistic competence. As you know, you can't be satisfied with your textual knowledge of English as you would like to communicate with others in different social contexts. Hence it is important to understand 'Sociolinguistic Competence', a valuable tool which helps you enhance your speaking skills. Let's see what we mean by the term 'Sociolinguistic Competence' and how it enables you to speak meaningfully in various social situations.

Look at the following utterances:

Sheela passed out of Patel Arts College.

He is from my batch.

Please tick the correct box.

May I use this tissue?

My dad is out of station.

You may understand the meaning of these utterances which are used in Indian contexts. However, when you use such expressions in other countries or cultures, say the USA, Australia, etc., you may find that the listeners not only find it difficult to understand the underlined words but also misunderstand the utterances. In fact, it is necessary to use the words as they are used in their countries. Table 5.5 given below may explain the point in question.

Now, elucidating the above such differences between Indian and American English usage will bring our focus to the term 'Sociolinguistic Competence'. It refers to the ability to use language appropriately in the respective social contexts in which it is spoken. As you may be aware, social contexts also refer to the culture specific contexts which are formed by the habits, practices, customs, etc., of that society. In addition, sociolinguistic competence also involves selecting appropriate topics according to the situation.

Table 5.5 Different meanings in Indian and American contexts

Indian usage	Indian meaning	USA usage/Meaning	USA usage in a sentence
passed out	graduated	fainted	Hey! She has passed out. Call the doctor.
batch	Group	quantity/consignment	This batch of medicines is the best seller.
tick	mark with a √	check	Check the correct box.
tissue	maper used to clean your hands	napkin	The host set the dinner table with plates, spoons, forks, etc., along with neatly folded napkins.
station	your town	town	Peter is out of town today.

For example, it would be inappropriate to discuss a cricket match with your friend while waiting for a condolence meeting or to boast about your children's achievements in a temple. You may be embarrassed if you do not acquire adequate sociolinguistic competence. Do realize this competence is not only needed in communicating with people of other cultures but also in communicating with people of your own culture because it is required in speaking (using the right words, right expressions) according to the communication context.

Using the right body language as per the communication and cultural context is also an important ingredient of sociolinguistic competence because your speaking will be effective only when your verbal signals match your body signals. For instance, when you seek apology for your mistake, you need to use the right words, right tone, and right body language. Expressions of greeting, apology, request, gratitude, compliment, invitation, and the like are different in different cultures. Though globalization has familiarized with different cultural practices, it is our sociolinguistic competence that makes us effective communicators across cultures.

Paralinguistic Features

Paralinguistic communication refers to the study of human voice and how words are spoken. Paralinguistic features are non-verbal vocal cues that help us to give urgency to our voice. Our voice is our trademark; it is that part of us that adds human touch to words. Writing does not have that immediacy because the words are static on a page. Voice gives extra life to our delivery. Therefore, it is useful to understand the characteristic nuances of voice, namely *quality, volume, rate, pitch, articulation, pronunciation*, and *pauses*.

Projection

Loudness and projection are two different aspects of our voices. It is possible to project our voice without being loud. Stage actors often do this when they speak in a low voice, and yet are heard from the back rows of a theatre. If we begin speaking in a loud voice, we will get the full attention of the audience. Then shifting to a lower, softer voice when appropriate will still hold that attention. Generally, we need to use a louder voice at the beginning of our talk and at the start of each new section.

To make our voice travel through the room, we need to breathe deeply as we need air in our lungs to project. Often, speakers run out of air and let their voices fade just when they are delivering critical information. At the end of a section, our voice need not be loud, but it must project.

Quality

Quality is a characteristic that distinguishes one voice from another. Each one of us has a unique voice and its quality depends on its resonating mechanism. While the quality of one's voice cannot be changed, it can be trained for optimum impact. It may be rich and resonant, soft and alluring, thin and nasal, hoarse and husky, or harsh and irritating. Very few people are naturally blessed with deep and resonant quality; everybody can improve on the quality of the voice and develop it to its fullest potential. Abraham Lincoln and Winston Churchill, for example, adapted the quality of their voices to become speakers par excellence.

Volume

Volume is the loudness or the softness of the voice. Our voice should always project but need not always be loud. If the place we are speaking in is large and open, the volume should be high, and if the place is small and enclosed, the volume should be low. If our volume is too high we may sound boorish and insensitive, whereas if it is too low we may convey an impression of timidity, which has no place in the business world. It may also give the impression that we are not well prepared and lack the confidence to express ourselves. Thus, we should vary our volume so as to make our voice audible and clear.

The volume should be adjusted to the acoustic arrangements of the room. If we are not using a mike, we may have to adjust according to the number of people we are addressing. The level of background noise should also be taken care. Naturally, our voice sounds louder to us than to our audience. Hence, we should keep an eye on the reaction of the audience at the farthest end of the room. If they crane their neck to listen carefully or look puzzled, then we should get the hint and try to adjust our volume.

One way to improve our voice and speaking style is through reading aloud. Reading children's stories, giving each character a unique way of speaking, may develop vocal variety. Reciting tongue twisters, such as *she sells sea shells on the seashore*, may also help improve diction.

Pace/Rate

Rate is the number of words that one speaks per minute. It varies from person to person and from 80 to 250 words per minute. The normal rate is from 120 to 150 words per minute. We should cultivate our pace so as to fit in this reasonable limit. If a person speaks too slowly and monotonously, he/she is most likely to be considered a dull speaker even though the contents of the speech may be highly interesting. Similarly, a fast speaker also causes discomfort because the listeners do not get enough time to grasp the thoughts and switch from one thought to another. Under these circumstances, listeners may just stop listening and their attention may go astray.

Although the average rate at which a person speaks is 120 to 150 words per minute, this rate is not universal. In fact, there is no uniform rate for effective speaking. Martin Luther King began his famous speech 'I have a Dream' at ninety two words per minute and ended it at 145. The former Prime Minister of India, Atal Bihari Vajpayee, reputed for his powerful speeches, speaks ninety words per minute probably because the pauses are more. Hence, the best rate depends on the vocal attributes of the speaker, the mood he/she tries to create, the nature of the occasion, etc. For instance, if we want to give a running commentary of a football match, we need to speak very fast. On the other hand, if we are describing the scenic beauty of a hill station or explaining a complex idea, we may speak slowly. With practice, the rate of our speech can be controlled.

It is best, therefore, to vary the speaking pace. Appropriate pauses should be used to create emphasis. A well-paced, varied message suggests enthusiasm, self-assurance, and awareness of audience.

Pitch

Pitch refers to the number of vibrations of our voice per second. The rise and fall of the voice conveys various emotions. 'Thank you' is such a phrase. We can make out the difference when it is uttered indifferently and when with sincerity. *Inflections* give warmth, lustre, vitality, and exuberance to our speech. Lowness of pitch can indicate sadness, shock, dullness, guilt, etc. When we are excited, joyous, triumphant, and even angry, our pitch automatically becomes high. A well-balanced pitch results in a clear and effective tone. It helps us avoid being monotonous. Intonation refers to the rising and falling pitch of the voice when somebody says a word or a syllable. By learning and adopting an appropriate intonation pattern, we will be able to express our intention very clearly.

'Hi dude, whaddaya gonna do today?'

Pitch is also influenced by the air supply in our body; if we run out of air, we cannot control the pitch of our voice. Like the strings of a guitar, if we tense the vocal chords, a higher pitch results, and vice versa. Tilting our chin up or down reduces our ability to control pitch.

A variety of pitches should be used to hold the listeners' attention. We should always avoid raising the pitch of our voice as we end a sentence. This vocal pattern, called *pitching up,* makes our remarks sound tentative or unfinished.

The best way to cultivate our pitch is to get feedback from our friends or from the recorded version of our speech and to practise speaking with the right inflections.

Articulation

Speakers should be careful not to slop, slur, chop, truncate, or omit sounds between words or sentences. If all the sounds are not uttered properly, the flow of understanding gets interrupted and deters the listener from grasping the meaning of the message. The result is similar to the negative impression that written errors leave with a reader. Lazy articulation, slurred sounds, or skipping over words will lower the

> 'Speak clearly, if you speak at all; carve every word before you let it fall.'
> –Oliver Wendell Holmes

credibility of the speaker. Develop in yourself the ability to speak distinctly; produce the sounds in a crisp and lucid manner without causing any confusion. The audience will better understand '*I do not know*' and '*I want to go*' than '*I dunno*' and '*I wanna go*'.

Pronunciation

Pronunciation requires us to speak out sounds in way that is generally accepted. The best way is to follow British Received Pronunciation. Received Pronunciation RP, also called the Queen's (or King's) English, Oxford English, or BBC English, is the accent of Standard English in England.

One should be careful enough to pronounce individual sounds along with word stress according to the set norms. Do not be taken in by the fancy that you know the correct pronunciation of all the words. Whenever there is confusion, always consult a good dictionary and try to pronounce it accordingly. Given below are a few commonly mispronounced words along with their correct pronunciations:

Word	Common error	Correct pronunciation
arctic	a:tik	a:ktik
gesture	gestʃə(r)	dʒestʃə(r)
3 tier	θrɪ:taɪə(r)	θrɪ:tɪə(r)
Gigantic	dʒaɪdʒæntɪk	dʒaɪgæntɪk

Voice Modulation

While intonation refers to the tonal variations, modulation pertains to the way we regulate, vary, or adjust the tone, pitch, and volume of the sound or speaking voice. Modulation of voice brings flexibility and vitality to our voice, and we can express emotions, sentiments such as impatience, careful planning, despondency, suspicion, etc., in the best possible way. If we do not pay special attention to the modulation of our voice, then our voice becomes flat and we emerge as a languid speaker with no command over our voice. Word stress and sentence stress also play an important role in voice modulation. For example, by accentuating one or two words in a sentence (e.g., in the sentence '*This company produces fifty cars everyday*', one can stress 'this' and 'fifty cars'), we can effectively bring in modulation in our voice. Thus, a novice speaker should better underline the words that he/she may like to stress during the presentation. This helps one avoid sounding dull and monotonous.

Practice 7

 Try the audio exercises given in the Online Resource Centre.

Pauses

A pause is a short silence flanked by words. A pause in speaking helps the listener reflect on the message and digest it accordingly. It also helps the speaker glide from one thought to another. It embellishes the speech as it is a natural process to give a break. However, it should be spontaneous. Being too self-conscious may make the process look artificial.

> 'Better to remain silent and be thought a fool than to speak and remove all doubt.'
> –Abraham Lincoln

Vocalized pauses or vocal segregates such as *uh*, *ah*, *hm*, *ahem*, *a*, *aah* should be substituted by silent pauses. Vocalized pauses make the speech sound evasive and untruthful; they dilute the conviction of the message. Moreover, using repetitive phrases such as '*I mean*', '*well*', '*like*', '*ok*', '*got it*', '*actually*', etc., may sweep away the good impression we have created. Thoughtful use of pauses at definite intervals exhibits assurance, confidence, and self-control. Pauses should be used at the end of certain thought units to let the audience fully absorb the information.

Managing and using the right pauses may be a challenge in the beginning. But with practice, we can realize how useful the correct pauses can be during our speeches. In order to use the right pauses, we will need to listen to seasoned speakers and see how they use pauses to modulate the rate and rhythm of their messages. In case of a novice speaker, it may be a good idea to highlight in the text of the speech where pauses are needed.

Practice 8

 Audio Clip Group 5: *Listen carefully to the following sentences in the Online Resource Centre and practise speaking them in the same way.*

1. I enjoy seeing a project through to completion.
2. In the end, we managed to improve the efficiency of the engine.
3. If the polar ice caps melt, sea levels would rise and low-lying areas of the world would be flooded.
4. Aviation engineering is concerned with the design and production of aircraft.
5. Increasingly, glass fibre is being used for long-distance telephone links.
6. We arrived early at the rendezvous.
7. The industrial revolution marked a new epoch in the history of mankind.
8. I met a singer near the monument this morning and he sang a song for me.
9. It was a pleasure to work with Mohan to work in this garage.
10. Engineers need to continue their education throughout their careers.

 Audio Clip Group 6: *Listen carefully to the following short passages in the Online Resource Centre and practise speaking them in the same way, with the correct paralinguistic features.*

1. Silence is solitude; it is company. We find books in running brooks and sermons in stones. Silence is the caravan of ideas. The quiet aspects of nature are not silent in the real sense, but they have an association of ideas.
2. A gentleman is an ornament, a delight of society. He is always conscious of his social responsibilities. He harbours malice towards none and does not allow winged jealousies to hover over his head. His thoughts are limpid like crystal, his judgements balanced and free from prejudices.
3. I enjoy living downtown. Of course, it is very noisy; the traffic is loud, and the young people often shout when they come out of the clubs. But there are lots of good points too. There is a wide range of shops and it is easy to get around.
4. There was nothing else to do, so he leaned back in the chair and went to sleep. When he woke up, he noticed that the others had also gone off to sleep. He turned to the window and looked out. The sky was clear now and in the afternoon light he saw a sight whose beauty left him breathless.

Linguistic Features of Voice Control

Vocal cues such as quality, volume, rate, pitch, articulation, pronunciation, modulation, pauses, etc., are commonly classified under Paralinguistic Features as all of them are associated with the language that we speak. However, out of the above-mentioned aspects, we can group articulation, pronunciation, modulation, pauses, etc. into Linguistic Features because they involve production of individual sounds, word and sentence accent, and punctuation which are linguistic aspects. Though your voice quality may be very good, your manner of articulation and pronunciation of words may be inaccurate. Likewise, your intonation or the rise and fall of your utterances may be incorrect and may mislead the listeners about your intentions and feelings. In addition, the pauses in your speech act as punctuation marks you use in written communication. So, if your speech-punctuation/pause is incorrect, you may end up in miscommunication though the pitch, rate, volume, etc. of your voice are perfect.

Assume that you are quoting the following text:

'In grammar, the notion of *potential pause* is sometimes used as a technique for establishing word units in a language—pauses being more likely at word boundaries than within words.' (David Crystal, *Dictionary of Linguistics and Phonetics*, 6th ed. Blackwell, 2008)

Here, the text has a few punctuation marks which convey the author's intentions to us. When you speak the author's words, you should also try to bring forth the intended meaning. For instance, the long dash '—'sets off the phrase, 'word units in a language' from the word 'pauses'. So, while quoting the author, you need to give a pause between the phrase and the word mentioned here. Similarly, the quotation marks in the beginning and end tell us that these words were used by the author David Crystal. When you speak, you have to verbalize this by saying, 'I quote' and then begin the text.

You can refer to the topic 'Basic Sounds of English' discussed in the beginning of this chapter to know more about pronunciation, accent/stress, intonation, etc., so that you can use your voice more effectively by appropriately using these linguistic features.

Tips for Effective Communication

- Create an open communication environment.
- Always keep the receiver in mind.
- Do not communicate when you are emotionally disturbed.
- Be aware of diversity in culture, language, etc.
- Use appropriate non-verbal cues.
- Speak with confidence, clarity and fluency.
- Focus on vocal cues like volume, pitch, rate, etc. to enhance the effectiveness of speaking.

Public Speaking

Public speaking need not necessarily mean the speech we deliver in front of a large number of people. Look at the following situations:

- As an intern in a company, you have been involved in two projects. Towards the end of your internship, you have the opportunity of getting absorbed by that industry. For this chance, you need to speak to your project manager and the vice president about the tasks you have handled in the projects.
- During your graduation party, you need to speak to a gathering of your batchmates and your faculty members about your experience at the college.
- You are participating in a group discussion and you have to convince your listeners about your views.

All these situations involve public speaking. To perform well in these situations, one needs to know the characteristics of good public speaking. All these situations demand you to plan, visualize, organize, draft, and organize your speaking. Given below are the essentials of good public speaking:

(a) Clarity of purpose: Know whether you want to inform, persuade, or entertain the audience.
(b) Audience awareness: Find about who will be your audience—their background, age, gender, education, status, interest.
(c) Familiarity with the location: Know where you need to deliver the speech—the audio equipment, the position of lectern, lighting, etc.
(d) Collection and selection of content: Collect as much material as possible for your topic and select according to the focus of the topic and the time given. For instance, when you need to speak for 20 minutes on 'Cloning', look for content in books, journals, newspapers, TV, etc., but then edit and compile as per your purpose and time.
(e) Outlining: Frame topics and subtopics for your speech.
(f) Organization of content: Introduction, main body, conclusion.
(g) Selection of suitable mode of delivery: Extempore, reading from text (to be carried out with adequate practice), speaking from memory (to be avoided).

(h) Effective use of body language, voice, and visual aids.

(i) Starting on time.

(j) Objectivity: Be unprejudiced and present a balanced point of view.

(k) Planning and preparation: After drafting the speech, go for one or two rehearsals in front of your friends and get their comments.

(l) Answering questions: Say clearly at the beginning whether you would answer the questions as and when they are posed or answer at the end. Listen patiently and answer.

> 'If you would impress an audience be impressed yourself. Your spirit, shining through your eyes, radiating through your voice, and proclaiming through your manner, will communicate itself to your audience.'
>
> –Dale Carnegie

Drafting the Speech

A beginner in public speaking needs to write down the entire speech. Preparing the first draft of a speech may be the most challenging task. But the first draft should not be considered as the final draft. Hence, the first draft needs to be edited according to our purpose, audience, and time. You can follow the tips given below to prepare the first and final drafts of your speech:

Preparing the first draft

- Set a deadline and complete your first draft at least three days before the date of delivery so that you have time for editing, revising, and preparing the final draft.
- Arrange your points in bullet form in words, phrases, or sentences.
- Do not bother about the sequence. If you cannot get ideas for introduction, you can start with the main text.
- Do not worry about the coherence or flow of your speech now. Transitions can be inserted later.
- Do not mind if your speech is longer or shorter than that is required for the given time. This can be adjusted later.
- Do not bother about vocabulary at this point. You can correct or change words during the preparation of the final draft.

Preparing the final draft

Experienced speakers say that the best speeches are not written but are rewritten. However, many speakers either do not write their speech at all or stop with the first draft. If you wish to be an effective speaker, revise and edit your first draft with the help of the following tips.

- Check the contents for focus: Every point, statistics, example, visual aid, should support the core idea.
- Edit for clarity: Check whether the ideas have been arranged logically so that they do not confuse the audience.
- Check the conciseness: Ask yourself the question 'Is this essential?' If the answer is 'no', cut that part of the first draft. Replace long and difficult words with short and easy words. Eliminate sentences from paragraphs if they do not support the main idea.
- Edit for smooth flow.
- Check for impact: Insert analogies, vivid images, quotes, similes, metaphors, etc., for creating an impact on the audience. Check whether the introduction and conclusion are impressive.
- Highlight the words or phrases you need to emphasize during your speech.

> 'Brood over your topic until it's mellow and expansive... then put all the ideas down in writing, just a few words enough to fix the idea... put them down on scraps of paper... you will find it easier to arrange and organize these loose bits when you come to set your material in order.'
>
> –Charles Reynold Brown

Practising Public Speaking

Even if you rehearse once, you may be able to improve the effectiveness of your public speaking. Given below are a few guidelines for practising your speech:

- Check whether your speech is delivered within the allotted time.
- Check whether you are able to retain your energy and enthusiasm.
- Try to control your nervousness.
- Check whether you are using appropriate words and phrases.

Speaking and Negotiation Skills

Negotiation is the process in which two or more individuals or groups, having both common and conflicting goals, state and discuss proposals for specific terms of a possible agreement. It normally occurs between companies, groups, or individuals because one has something the other wants and is willing to bargain to get it. We all need good negotiation skills not only in the business environment but also in our personal lives. Whether it is with an employer, family member, or business, we all negotiate for things or issues big or small each day at our place of work, home, or any other relevant locale. Whenever we attempt to influence another person through an exchange of ideas, or something of material value, we are negotiating. Consider the following situations:

- Deciding on a date for the next meeting of the project group
- Deciding with the family which car to purchase
- Agreeing on realistic project deadlines
- Selecting a contractor to build a new home
- Agreeing on a change of election procedure with the students' union
- Choosing a new location for the office

The ventures may be different in each case, but all of these need negotiation skills before arriving at the final decision. The aim of a negotiation is to explore the situation to find a solution that is acceptable to both parties.

Given below are the six steps of negotiation:

1. Beginning the negotiation
2. Stating your purpose
3. Starting the process
4. Expressing disagreement and conflict
5. Reassessing and compromising
6. Agreeing in principle or settling

Whether you are negotiating for your personal purchase or for your company, all these 6 steps are necessary, and your speaking skills play a significant role in making your negotiations more effective and possibly successful.

Assume that you are planning to buy a car and you are at the car company. Look at some sample English expressions which you can use for each of these steps:

1. Beginning the negotiation

Hi, how are you?

How was your day so far?

You seem to be tired; looks like you've had a lot of customers!

I've heard from a friend of mine about your ABC model car and I really loved it.

Recently, I came across your company ad about the ABC model car which caught my attention and thought of going for that model.

2. Stating your purpose

Shall I tell you what exactly I am looking for?

Though I liked three of your car models, I would like to see ABC model and see if I can really go for it.

I am looking for the recently introduced ABC model of your company and understand its features in detail.

As already mentioned over phone, we are here to discuss the ABC model and see if we can go ahead with our plan of buying it.

3. Starting the process

Good. Thanks for explaining the features. I am sure you offer free one-year maintenance warranty. Don't you?

In fact, it would be impossible for me to go for this car unless a free one-year warranty is included.

Could you be more specific on the interior accessories?

I feel that the leg space could have been better in this model.

Will it be difficult for senior citizens to get in easily into the car?

Ok. Let's get into business. I like the car but not happy with the price!

4. Expressing disagreement and conflict

I'm afraid, I may have to disagree with you on this point as this accessory doesn't match what I had expected from your ad.

I can't exactly agree with you on this point, and you may have to offer some concession.

I'm indeed shocked at the price you've just mentioned. I my opinion it should not be more than … if I've understood your ad correctly.

You know, this is not exactly what I had thought of!

As already mentioned, I'm not very happy with the boot space and the rear view mirror.

5. Reassessing and compromising

I'm sorry, I can't pay that much. Let me look elsewhere.

Rs…..? You see, I can't go beyond Rs….

Ok, agreed. Will you investigate this accessory once again?

Your price is still very high. Could you speak to your manager?

Would you be able to compare your ABC model to the same model by PQR company and justify your offer price?

6. Agreeing in principle or settling

Well, OK. If that includes delivery and installation you can take it as 'confirmed' from my side.

Anyway, there are still some loose ends to tie up. Hope you will be able to do that.

I think we can agree on that.

If you feel we have discussed everything, shall we agree and seal the deal?

OK. That sounds reasonable and we can agree on that.

As your speaking skills are extremely important both in your professional and personal communication as discussed in this chapter, you need to focus on all the necessary ingredients of effective speaking to hone your skills.

Refer to the samples for a few dialogues on negotiation.

SUMMARY

Effective speaking is an important requirement not only in our academic and professional career, but also in our personal life. To develop oral communication skills, it is required to understand the sound system of English, including various aspects such as individual sounds—consonants and vowels, production of speech, word stress, and sentence stress—and intonation or the variation of pitch. It is also very useful to understand and practise the rules of pronunciation, to improve the way we articulate speech. One should also be aware of the common problem sounds, and practise to eliminate these.

Confidence, clarity, and fluency are also considered to be vital for the effectiveness in our speaking. Paralinguistic features, such as our voice quality, pitch, rate and volume, pauses, modulation, etc., play a key role in creating a desired impact in speaking.

You need to understand and minimize your Mother Tongue Influence (MTI) on your English learning. Another factor that enhances your speaking skills is your Sociolinguistic Competence.

EXERCISES

Part A

1. What is a phoneme? Classify the basic sounds of English.
2. How is the production of vowel sounds different from that of consonant sounds? How many distinct vowel and consonant sounds does English have?
3. Give examples of words with pronunciations that use the following sounds:
   (a) /i/
   (b) /ɑ:/
   (c) /au/
   (d) /ʌ/
4. List the rules of pronunciation.
5. What is meant by accent?
6. What is intonation? What are the general uses of tones I and II?
7. 'Human voice is an extremely valuable resource and contributes significantly to the effectiveness of speaking.' Discuss this statement in about 500 words.
8. Define each of the following terms and bring out the difference between them: Articulation and Pronunciation
9. What is MTI and how can you restrict it from interfering with your L2 learning?
10. What do you mean by 'Sociolinguistic Competence'? Discuss with three examples.

Part B

1. Make ten words each with the sounds mentioned in the columns of the table given below.

Words with /ə/	Words with /ð/	Words with /aɪ/	Words with /dʒ/	Words with /ɑ:/

2. Add the consonant sound to the word to make another word:
   Example: /k/ + aim = came
   (a) /s/ + eyes = _____
   (b) /l/ + eight = _____
   (c) /w/ + ate = _____
   (d) /b/ + air = _____
   (e) /b/ + ache = _____
   (f) /p/ + oust = _____
   (g) /t/ + rue = _____
   (h) /r/ + owes = _____
   (i) /ʃ/ + eat = _____
   (j) /tʃ/ + it = _____

3. Read the following conversation that has six questions asked by X. Write F if they are spoken with a falling tone and R if they are spoken with a rising tone:
   Example: Are you a student?
   X: What is your name?
   Y: Sheela
   X: Where do you come from?

Y: Margao

X: Is that in Goa?

Y: Yes, that's right.

X: How long have you been here?

Y: For almost 10 years.

X: I see. Are you a teacher?

Y: No, I am an artist.

X: And do you like the place?

Y: Yes, very much.

4. Transcribe the following twenty words phonetically using IPA symbols:

   Houses, office, abyss, India, obscure, thereafter, era, oodles, conquer, thatched, women, manipulate, voice, caught, virtue, daunt, endorse, music, garage, pleasure

5. Write the following twenty words in ordinary orthography:

   /tʃɔːk/, /ðuθ/, /ðen/, /luːdikrəs/, krevin/, /kɔstik/, /mɔðð/, blænd/, /lisn/, /dʒuðri/, /ɔstið/, /rek/, /welfeð/, /indɔːs/, /iksentrik/, /ðust/, /ɔðːst/, /hiðriɔ/, /ʃuː/, /juːnɔvðːsðti/

6. (a) Mark the primary stress in the following words:

   Tunnel, deafening, teacher, yesterday, physiology, examination, interview, laboratory, economic, effectively.

   (b) Underline the word or words you would like to give emphasis in the following sentences:
   (i) He's a charming fellow.
   (ii) He wasn't by himself when I saw him.
   (iii) Give yourselves plenty of time.
   (iv) You won't hurt me, you'll only hurt yourself.

7. Underline the word or words you would like to give emphasis in the following sentences:
   (a) You must be back in time
   (b) I have been working all day.
   (c) He that is content is rich.
   (d) I enjoyed the trip very much.
   (e) Your shoes need polishing.
   (f) Her acting was excellent.
   (g) The play attracted a huge crowd.
   (h) This year the conference was held in Bengaluru for two days.
   (i) Yesterday, he came at 10 O'clock to the office.
   (j) At times, he will see even two or three movies in a day.

CHAPTER 6

Conversations and Dialogues

OBJECTIVES

You should study the chapter to know

- ○ how to apply both listening and speaking skills simultaneously in various contexts of communication
- ○ strategies for effective conversations
- ○ the essentials of a conversation such as beginning, ending, expressing opinions, paraphrasing, reflecting others' feelings, and involving everybody
- ○ how to listen, reflect, and speak while communicating with others
- ○ how to write situational dialogues

Introduction

Sender and receiver are the two most important components of communication—the sender writes and the receiver reads. Similarly, when the sender speaks, the receiver listens. Thus, speaking and listening activities complement each other. In many situations in our personal, academic, or professional life, we may have to speak in front of our friends and colleagues or listen to them. But how effectively we perform depends on our conversational ability. That is, our ability to listen effectively, process the information, consider the reaction expected of us as well as our expectation of the conversation, and respond appropriately. The audience's views, reactions, and responses may either encourage or discourage us, but it is important to remember that much of this reaction is a result of how we handled the situation, when it was our turn to listen or speak. For instance, as a listener, one needs to understand and respond to the speaker to engage his/her interests in the conversation. This would definitely help the speaker feel comfortable and encouraged, and thus respond positively.

Although in face-to-face conversations we can use our body language more effectively than in telephonic conversations, the same basic strategies for effectiveness apply for both the types. How we begin our conversation, how we listen to other participants in the conversation, how we carry on further, and how we finish our conversation—all are important facets involving our speaking and listening skills. Now that we have understood the basics of listening and speaking individually, in this chapter, we will learn how these two activities are combined in various forms of communication. To start with, the following section will enable us to understand the term 'conversation', its different types, and the techniques for effective conversations—both face-to-face and telephonic.

> 'You are awake 16 hours a day—5840 hours a year. You spend more time contacting other people than in any other single activity. How well you converse with these people is the magic key to whether your days will be pay-off days filled with personal and social popularity.'
>
> –*James A. Morris*

Conversations

Let us begin with an example. It is break time at a conference. You spot some participants with whom you would like to build a relationship. If you let them talk about themselves in detail while responding only with 'uh-huhs' and some occasional questions, you may come across to that person as being reserved. The best conversations are exchanges, not interrogations. Rule of thumb for effective conversations: talk about yourself enough to not appear withholding or reserved, listen carefully, ask questions, and make comments based on what the others are saying. You must talk almost 25 to 50 per cent of the time to make the conversation lively.

Conversation is a mental occupation and not merely converting casual thoughts into words. Everybody talks in a friendly environment. But mere talking may not be considered as a conversation. A person may talk for a long time without saying anything worthwhile, but even then his/her talk would be an interesting one if others in the company draw pleasure in listening to him/her. None of the professional talks, those of the teacher, the lawyer, the public speaker, the actor, the debater, may be considered as conversations per se. An uninterrupted talk, though being a good discourse, can never be considered as a good conversation. A good conversationalist is neither one who monopolizes the conversation nor one who keeps silent all the while. Conversation is an almost invisible art, efforts put into improving which are more felt than seen.

Everybody has the power of transmitting thoughts in some way. But an effective conversation is a mutual process, where the thoughts are transmitted not in isolation but in reactions to those that are received as well. Good conversational skills help us in both professional and social conversations.

Types of Conversation

There are six types of conversation. These are as follows:

Chat   The least formal of all conversations, e.g., two friends discussing about cricket and the latest fashion, or two women talking about their husbands.

Tête-à-tête   French for head-to-head, meaning a confidential/private conversation.

Dialogue   Conversations in a book, play, or film. Also, a formal discussion between two groups or countries, especially when they are trying to solve a problem, end a disagreement, etc.

Parley   A formal discussion between enemies regarding the terms of a truce.

Colloquy   The most formal of all conversations (*a colloquy on nuclear disarmament*); it can also be used to ironically describe a guarded exchange (a brief colloquy with the arresting officer).

Communion   A form of conversation that may take place on such a profound level that no words are necessary (communion with nature).

Strategies for Effectiveness

Each type of conversation serves a specific, well-defined purpose. By following certain simple strategies discussed below, we all can make our conversations more interesting, inspiring, and influential.

- Listening intently
- Arousing and sustaining interest
- Starting and ending conversations
- Developing ideas adequately
- Involving everyone in the group
- Using appropriate language

We have already understood the importance of listening intently in Chapter 4, including how one can encourage the speaker by clearly indicating one's interest in what the speaker has to say. In the following sections, we discuss the rest of the strategies listed here.

Starting and ending a conversation

While conversing with friends and acquaintances, we may not have any difficulty in starting or ending the conversation as we share many common events or experiences with them. However, with a stranger or with somebody whom we do not know very well, we should have some strategy to start and end our conversations. In such cases, an effective conversation may play an important role in shaping our career or developing a long-lasting relationship. Following are a few openers and closers for conversations we can use depending on different situations and topics.

Conversation openers   We can state a fact or ask a question to start a conversation. We can also start with a smile and some greeting word. In general, we may start by referring to a situation or to the person we are conversing with.

Starters for conversations with known people

- How did you like yesterday's programme?
- I did not like today's class on 'circuits and signals'.
- Your shirt looks new. Is there anything special today?
- Have you completed the assignments?
- The Mess food is getting better these days.
- I like your earrings. Where did you get them?
- The Government should not have introduced this scheme.

Starters for conversation with strangers

- Hello James, how do you do?
- Hello Sir, welcome to our college!
- Hi, Rohan, the weather is very nice. Isn't it?
- I am a second year student here. Are you a fresher?
- Hello Mr Gupta, I have heard a lot about you but meeting you for the first time.
- Mr Mahesh, nice to meet you.
- Good morning, I have been asked to appear for an interview at Sapient Corporation. Please guide me where I should submit my records.

> 'A conversation is a dialogue, not a monologue. That's why there are so few good conversations—due to scarcity, two intelligent talkers seldom meet.'
> —*Truman Capote*

Conversation closers   No matter how good and interesting our conversations are, we need to close them at some point. Signals such as boredom, restlessness, silence, etc., often indicate that our conversation should end. Moreover, if there is a time constraint, we have to end our conversation on or before time. In any case, we should end it emphatically. For instance, in an interview, you need to thank the panel at the end and tell them how you have enjoyed the interview. Similarly, if you are speaking to your professor on a project, you need to end your conversation by emphasizing what you have done so far and what would be your next step. The following samples would be useful in ending a conversation effectively:

- May be we can get in touch by next week. Hope to see you again.
- Thank you very much for this opportunity. It was great to attend this interview and I look forward to meeting you again soon.
- It has been nice speaking to you. Have a nice day/good night.
- I have some more things to discuss but I would like to take leave now. I will get back to you later. Thank you.
- I will not take any more time but it has been nice talking to you. Have a great evening.
- So let me confirm that we would take care of your investment plans. Thank you very much and have a nice day.
- It is my first visit to your organization and I would like to meet many others during my next visit. Thank you for making my first visit memorable.
- Ok, then I will email you the details you had asked for. See you soon.
- John, thanks once gain for your advice.
- Thank you Ma'am, I will certainly follow your guidelines and perform better next time. Have a nice day Ma'am!

Involving everyone

Assume that you are conversing with two of your teammates on an important class assignment on a team presentation:

You: Hello Mohan and Ramesh, I think we need to decide today on the topic for our presentation.

Ramesh: Yes, we know. What we thought of was ….

You: I think we can select 'Controversies on Moon Landing'. It will be interesting for others and we have adequate material.

Mohan: We have another topic in mind. Shall I …

You: As I already told you, the topic I said will be liked by everybody and there should not be any problem in presenting it.

In this conversation, you may realize that you have not involved your teammates in deciding upon the topic for the presentation.

A good conversationalist is not one who holds the listeners spellbound by his/her speech, but one who involves everybody into the conversation. We should not forget about the others in the group who may have something to contribute. Rather than individual contributions, participation of the group as a whole is more important in a conversation. Involvement of everyone makes a conversation more interesting, informative, resourceful, and effective.

Arousing and sustaining interest

Conversations become successful only when all the participants have interest in the discussion and participate in it. A conversation filled with questions, answers, views, expressions, statements, and information always turns out to be an interesting one. Good ideas from the various participants are often helpful to make the conversation interesting. Every participant may be different from the others in terms of their response and behaviour. Hence, it is necessary to arouse and sustain the interest of everybody while maintaining a balance among their different interests.

The following conversation will enable one to understand how one can arouse and sustain interest in the conversation partner.

You: Hey, what are you doing?

Counterpart: I am struggling to put up some web pages together.

You: What are they about? (You are encouraging your counterpart to share her problem.)

Counterpart: Figures of speech in American English, similes, metaphors, oxymorons, idioms, etc.

You: That sounds interesting! Could I have a look? (You are arousing her interest in speaking to you)

Counterpart: By all means. In fact, I wanted you to go through these pages.

You: Absolutely, just give me a few minutes, I will finish my coffee and be right back (You are sustaining her interest)

Counterpart: That is fine.

…

You: I think you are doing a great job of it. The content will be quite helpful for people who are trying to learn English, and also for those who want to add to their knowledge of the language. (You are appreciating her)

> 'Questioning is the mode of conversation among gentlemen.'
> *–Samuel Johnson*

Developing ideas adequately

Once we start the conversation and set the stage, we should have some strategies to develop our ideas. The best strategy is to build a knowledge reservoir that never lets our conversation run dry. Facts and information always make others enjoy the discussion. We can gather knowledge from reading, listening, watching television, and observations and experience. We can draw our examples from this gathered knowledge to make our conversations more meaningful and convincing.

Enjoy the following conversation between two friends on the T20 World Cup Cricket Tournament held in West Indies. Both the girls add value to their conversation by bringing in more matter from newspaper, TV, etc.

Shilpa: Hi Preeti, I feel very low today because India is out of semi finals.

Preeti: Yeah, me too. It is a shame on us.

Shilpa: Did you read today's newspaper? It says that Indian team played its match against West Indies very casually and it lacked in all fronts—fielding, batting, bowling, and captaincy.

Preeti: Yes, I know. I just watched the TV news channel. There are many controversies surrounding this match!

Shilpa: Like …?

Preeti: Like—the players did not have any rest after IPL T20, some of them were not fit to play, they were not focused on the tournament because of the overnight parties, etc.

Shilpa: It is indeed sad to watch our team perform badly.

Using appropriate language

This is the most difficult aspect of a conversation. We use language in conversations to convey our ideas and feelings, to paraphrase the presented thoughts, to appreciate others, to reflect implications and underlying feelings, and to invite further contributions. Given below are some practical hints to accomplish these purposes.

Practical hints   Our body language is very important to reinforce whatever we are speaking or listening to. For example, using gestures such as nodding our head while listening, or uttering uh-huh, etc., adopting suitable postures, maintaining purposeful eye contact, and facial expression would enhance the quality of our conversations. In addition, certain specific phrases that can be used in certain situations during our conversations are as follows.

Paraphrasing the presented thoughts or feelings

- What you are saying according to me is …
- As far as I understand what you mean is that …
- Do you mean to say that …
- So your feeling is that …

Making the others feel appreciated

- Fantastic! How do you keep coming up with such good ideas?
- What a memory! Wish mine was as good …
- You learn fast! I wish I could too …

Reflecting the implications (where the content is leading)

- So that might lead to a situation in which …
- Would that help with the problem of …
- I suppose if you did that, you would then be in a position to …

Reflecting the underlying feelings

- Had it happened to me I would have been rather upset.
- That must have been rather satisfying.
- I guess that must make you rather anxious.

Inviting further contributions

- Could you tell me a bit more about that?
- How did you feel when you learnt that you were promoted?
- What happened then?

Interrupting politely

- Excuse me, sorry to interrupt, can I add something here?
- Do you mind if I say something here?
- May I ask you something at this point?

Useful phrases   Given below are phrases and expressions that would be helpful in putting forth preferences and opinions, offering explanations, and disagreeing with others' views.

Opinions, preferences

I think …, In my opinion …, I would like to …, I would rather …, I would prefer …, The way I see it …, As far as I am concerned …, If it were up to me …, I suppose …, I suspect that …, I am pretty sure that …, It is fairly certain that …, I am convinced that …, I honestly feel that …, I strongly believe that …, Without a doubt …

Disagreeing

I do not think that ..., Don't you think it would be better ..., I do not agree, I would prefer ..., Should we not consider ..., But what about ..., I am afraid I do not agree ..., Frankly, I doubt if ..., Let us face it, The truth of the matter is ..., The problem with your point of view is that ...

Giving reasons and offering explanations

To start with, The reason why ..., That is why ..., For this reason ..., That is the reason why ..., Many people think ..., Considering ..., Allowing for the fact that ..., When you consider that ...

To sum up, conversing involves use of appropriate verbal cues besides other factors such as active listening, involvement of every member, understanding of implications, use of non-verbal cues, and so on. Choosing the right word at the right time to the right person is very important for a conversation to be effective. For example, while talking about the latest features of a database product to some database administrators, unnecessary superlatives should not be used to describe the product as that may exert an adverse effect. On the other hand, while narrating a sensational incident at a social gathering, one should try to re-create the scene by using sensational words and expressions. We always need to remember that acquiring conversational skill is the magic key to our professional success and our social popularity.

Conversation Practice

Now that you have learnt phonetics and strategies for effective conversations, practise the following short conversations with one of your friends or family members. Try your best to apply the rules of pronunciation, accent, intonation, and use your voice effectively to convey the message:

> 'Conversation is an art in which a man has all mankind for competitors.'
> –Ralph Waldo Emerson

In the class room

Good morning boys and girls. Let me talk to Sunil. Sunil, do you speak English at home?

Yes Ma'am, I speak English with my dad but I speak Hindi with my Mom.

Do you have any difficulty in speaking English?

Yes, at times, in using the correct tenses.

Do you need any help?

Yes, it would be nice if you can help me.

At the office during lunch break

What time do you get up every day?

At quarter to seven.

Are you able to complete all your household chores before going for work?

Yes, but with a tight schedule.

I know, it would be very hectic for you. When do you have your breakfast?

At around eight fifteen.

In the mall/at a restaurant/in the park

Do you like watching movies?

Yes, I do, but I do not find time to watch them in cinema hall. I watch them mostly on television.

I see, do you watch with your family?

Yeah, mostly when they are telecast in the evening.

That should be great experience, isn't it?

Yes it is, we even have our snacks and dinner while watching the movie.

Oh! I wish I had such an experience!

Have you ever visited Goa?

Yes, last December. In fact, we enjoyed Christmas in Goa.

That should be a satisfying experience!

Exactly. It was wonderful to visit a number of churches and attend a party. What about you? Have you been to Goa?

No, I yearn to. I have heard a lot about its beaches, people, art, etc.

Why don't you come with us this December when we go there?

Thanks. It is so nice of you to invite me. I will consult my family and let you know.

That is fine. Bye.

Bye!

Outside the exam centre

Hi Reena, how are you and how was your exam today?

It was okay … but I could have done better.

Never mind. You can do the remaining exams well. I know you are a good student.

Thanks uncle, for your encouraging words.

What are your vacation plans?

This year I have to do my summer internship at NELCO, Delhi.

That would be an enriching experience. Right?

Yes, my friends who have done there earlier say so.

I wish you good luck for your remaining exams and enjoy your internship.

Thanks uncle, I have to catch 43C to reach Gandhi Circle. Bye.

Bye bye.

Refer to sample conversations.

Argumentative Skills

Arguing to justify your point of view is required in your personal and professional contexts. For example, during your graduation period, you may have to appear in several competitive tests such as GRE, TOEFL, IELTS, IAS, etc. Similarly, you may have to prove your views on many occasions at your workplace. On most of these contexts, you may stand for and prove your point with strong reasoning garbed in apt statements and examples. Irrespective of the type of contexts—oral or written, it is necessary and important to be clear and effective in expressing your views/opinions/thoughts in appropriate words, expressions, and tone. In short, you need to possess sound argumentative skills.

An argument means a reasoning, or a reason given for or against a matter under discussion. It is not necessary to be angry during your argument. In fact, you should argue in such a manner that others agree with you. At times, as a team leader, you may have to justify your views so that your members accept them. For example, when you want to argue for the choice of a particular software, you may have to bring out both the pros and cons and prove that the benefits outweigh the limitations. Before arguing, (i) you should have done thorough research on the issue in hand. In other words, you should understand the ins and outs of the issue completely. For example, assume that you wish to argue for online education to a group of students and faculty members. You should read some scholarly articles, survey results, etc., and if possible, should talk to some students and faculty who have attended online classes; (ii) you should be able to critically analyse the research material you have collected and pick up the points for reasoning; (iii) you should look for appropriate words and expressions that you can use during your arguments. Look at the following sample expressions that may help you while justifying your views:

Some of you may disagree with this idea…but I would like to tell them….

Some of you may say that…You may be right. However, once again I would like to emphasize upon…

According to Prof…, who has done extensive research in the field of online education, '…'.

That's an interesting question Mr…. Let me answer this with an example.

In fact, there are three problems here: one, …; two, …; three, ….

The way I see it….

Precisely! You are very correct but at the same time, it would be better if you consider….

After listening to you, I agree up to some point. However, you can still reconsider your later point of your argument because….

As discussed above, research, critical analysis and speaking/writing skills are essential ingredients of argumentative skills.

Given below are a few guidelines to sharpen your argumentative skills:

- Prepare an outline for your argument (e.g.: Introduction to online education, Boons and Banes, Online platforms used across universities, Success stories of online education, Online vs classroom education, while arguing for online education)
- Organize your thoughts logically/chronologically as the context warrants.
- Think about a thesis statement that brings out your core idea.
- Present the examples as evidence to support your claim.
- Be prepared with a strategy to control disruptive members.
- Control your emotions, especially emotions such as anger, irritation, etc., while arguing.
- Summarize your argument with the most important points of your discussion.

Dialogue Writing

A dialogue is a two-way conversation between people, thus involving effective listening as well as speaking. Aimed at understanding and responding to the other person's opinions on a particular communication message, dialogues may involve opposing points of view. In several occasions we may have to participate in or initiate such conversations and, hence, it is important to know how to construct appropriate dialogues. For our dialogues to be effective, we must observe how people talk to each other, what words or phrases they use to provide information, how they ask questions, and how they convince others. Written dialogues always maintain a narrative flow in a

conversation if judiciously used. The dialogues should be written in such a way that the reader can discern the speaker's tone of voice and the state of mind.

Types of Dialogues

As already mentioned, dialogues are two-way communication between people. Both listening and speaking play a significant role in every oral conversation. Effective writing skill is necessary in written dialogues (e.g., dialogues in novels, plays, etc.). Conversations become successful when all the conversationalists get almost equal opportunity to express themselves. However, some conversations may have a few active participants and a few passive participants. Dialogues vary in tone, style, and diction according to the contexts or situations. We can categorize dialogues into four categories:

- Active dialogues
- Passive dialogues
- Internal/Inner dialogues
- External/Outer dialogues

Active dialogues according to their nomenclature are dynamic and play an important role in a conversation. In a play or novel, the dialogues happening between main/important characters fall under 'active dialogues'.

Examples

- A dialogue held between a father and a son regarding the latter's choice of a particular degree for graduation. Both the participants play an active role to argue for their views.
- The dialogue between Macbeth and Lady Macbeth in the Shakespearian play *Macbeth*.

Passive dialogues as against active dialogues discuss matters that are not so important. In novels and plays, they may be written for minor/weaker characters to keep readers from distractions from important details.

Examples

- A dialogue between a father and a son regarding which restaurant to go for dinner; as the issue is not very important, it may be short. Also, both may not participate equally in the dialogue; one may agree with the other quickly.
- A dialogue between two guards in *Hamlet*.

Internal dialogues are dialogues that happen within oneself. They are in fact thoughts that go on in a person's mind. In a play we call it monologue.

Examples

- Before discussing with his father, the son speaks within himself about the pros and cons of taking up a particular degree for his gradation.
- A monologue by the protagonist Hamlet in Shakespeare's play *Hamlet* (To be or not to be ....).

External dialogues are the most common dialogues. These are dialogues spoken on/written for various situations and hence they are also called situational dialogues. They can be either formal or informal. While a dialogue you have with a colleague may be informal, whereas a dialogue that you have with your manager is formal. They are categorized as formal or informal depending on the participants and situations.

Now, what are the differences between role plays and dialogues? We shall look at them in Table 6.1.

Table 6.1 Differences between role plays and dialogues

Role plays	Dialogue
Enacting/Assuming the role/character and speaking the dialogues	Only speaking the dialogues in a conversation
Generally used in plays	Used in novels and everyday conversations

(Contd)

Role plays	Dialogue
Happen in-person	Happen in-person, over e-mail, letter, etc.
Done in front of an audience	Done between/among the participants
Gives a better understanding of the dialogue because of the use of body language and movements	To be spoken with the right tone and right expressions for better understanding of other participants

The following section discusses situational dialogues with examples.

Situational Dialogues

There are several occasions or situations that require us to speak to another person for getting or giving information, making enquiries or a request, or getting something done. The dialogues spoken on such occasions are called situational dialogues. The following are some examples of such situations.

Tips for Writing Dialogues

1. Use quotation marks to show words spoken by an individual.
2. Start a new paragraph when the speaker changes. It is not possible to have the same paragraph for different speakers. However, several paragraphs can be used for the same speaker.
3. Ensure that the reader knows who is speaking.
4. Use punctuation, capitalization, and spacing correctly.
5. Vary the use and placement of speech tags to avoid monotony. Speech tags are used to convey many aspects. For instance, a dash (–) can be used as a tag for conveying a pause in the dialogue. Ellipses (…) can refer to stammering or stuttering during a dialogue. We can also bring in variety by using replied, remarked, suggested, asked, etc., instead of using 'said' many times.
6. Use narrative sentences to show the character's concurrent acts, thoughts, and perceptions.

At the tailor's shop

A customer (Ram) meets his tailor to get his clothes stitched:

'Good morning! I have got two shirts and a pair of trousers for stitching,' said Ram entering the tailor's shop.

'Let me take the measurements...Sir,' smiled the tailor.

'No, no, it is not necessary. I have got the sample shirt and trousers,' Ram handed over the samples in a hurry.

'Oh, that is good. Let me see them.' The tailor's voice expressed relief from taking the measurements.

'Do you want to have two pockets at the back of your trousers or only one as in this?'

'Please follow exactly the sample.' Ram got annoyed.

'Okay sir.'

'When can I get the shirts and trousers?'

'After a week, that is, on the 20th,' he assured.

At the library desk

A student finds it difficult to locate a book in the library and at this time the librarian comes forward to help her out.

'Good morning, what can I do for you?' asked the librarian with a smile to Leena who was struggling to locate a book.

'I need the book *Organic Synthesis* by Stuart Warren,' uttered Leena without looking at the librarian.

'Did you search through the online catalogue search?' asked the librarian gently.

'I did, but I could not succeed,' answered Leena disappointedly.

'Let me try,' said the libararian starting the online search.

'Do you need the details once again?' asked Leena expressing concern.

'If I am right it is a book on Organic Synthesis by …' said the librarian still searching online.

'Stuart Warren,' completed Leena.

'Here it is, the number is 201.15W,' said the librarian triumphantly. 'Please take this slip with you. It contains the number.'

'Thank you very much for your help.'

At the bank

A lady meets the manager in connection with understanding a scheme.

'Good morning sir, I am Swapna Sood from Goodwill Corporation,' said the lady to the bank manager.

'Good morning. What can I do for you?' replied the manager.

'I would like to understand the scheme XYZ-life offered by your bank,' expressed the lady in a hurried tone.

'You are most welcome Madam. But you know, as I am busy now, could you please come after half an hour or tomorrow?' asked the manager apologetically.

'I have travelled about 15 kilometres to reach this place and I cannot come again,' shot back the lady arrogantly.

'Please try to understand my problem,' requested the manager.

'It is your problem,' the lady replied, 'and it does not speak well of your bank if you turn down customers like this,' she added in an admonitory tone.

'Sorry, madam, I did not mean to discourage you. I understand your problem and I will put a person right now on this job.'

'Thank you,' said the lady curtly and went to the person referred to.

SUMMARY

Conversations—both face-to-face and telephonic—involving listening and speaking are the most frequently used form of oral communication. In order to be a successful conversationalist, one needs to adopt certain strategies that can be useful for both face-to-face and telephonic conversations. It is important to learn and practise how to start a conversation, how to carry it on effectively and efficiently, and how to end it smoothly. The other factors to keep in mind are listening intently, involving others, arousing and sustaining interest, using appropriate language, etc. In addition to these factors, arguments also arise in conversations and hence acquiring adequate argumentative skills is also necessary to become an effective conversationalist.

A dialogue performs the three-fold function of narrating a story, giving information, and describing the places or characters in the story. Dialogue writing requires great clarity of thought with respect to time, realistic characters, and places.

It is important to understand the various types of dialogues, the contexts in which they occur, and also how they differ from role plays. Such an understanding would enable you to be an effective conversationalist and dialogue writer.

EXERCISES

1. Given below are three situational dialogues that are jumbled. Arrange them in appropriate order by writing down the number in brackets as shown against one statement in Q. (a)

   (a) At the movie hall

   Neha: Three.

   Cashier: Sorry, how many?

   Neha: Excuse me, which screen?

   Cashier: Screen 4, it is on the first floor.

   Neha: Three tickets for Avatar please   (1)

   Cashier: Ok, here are your three tickets to Avatar. It is showing on screen 4.

   Neha: Screen 4 on the first floor, ok, thanks.

   (b) Discussing a movie

   Pinky: So you like at least something in the movie.

   Manisha: I think we have wasted Rs 500 on this movie.

   Pinky: Story and … the heroine.

   Manisha: hmm … then what was it that you did not like?

   Pinky: Was it so boring for you?

   Manisha: It is strange, I like only the heroine!

   Pinky: I like the locations, photography, and music.

   Manisha: Yes, but …

   Pinky: Leave it, let us have a good dinner at least.

   Manisha: Of course, it was; for you?

   (c) Discussing future

   Gyanesh: Yeah, I feel bad about leaving the campus.

   Guha: Thank you. Have a fruitful time in Matrix and do mail me.

   Gyanesh: I would like to work for two years and then go for higher studies in the USA. What about you?

   Guha: From one university and most probably I will join that.

   Gyanesh: Why did you choose Germany?

   Guha: Hi, Gyanesh, that is the end of our 4-year degree programme.

   Guha: I too. What about your future plans? Are you going to join Matrix Developers or pursue higher studies?

   Gyanesh: That sounds good. I am very happy for you and wish you good luck.

   Gyanesh: Congratulations! Have you been offered scholarship as well?

   Guha: You know that I was selected for a summer internship under DAAD and was working with a professor in Siegen University in Germany. I found the academic environment and research facilities great and hence decided to go for my MS there.

2. The following dialogue takes place between a space scientist and a research scholar after a special lecture on 'Climate Change' organized on the occasion of Antarctic Treaty Consultative Meeting in New Delhi. The special lecture was very informative on global warming and climate change and spelt out the importance of Antarctic research in continuous monitoring of climate variations. Complete the dialogue by framing suitable questions or statements.

   (a) *Research scholar*: _____?

   Space scientist: Yes, many observations indicate that the world's climate has changed during the twentieth century—the average surface temperature has increased by about 0.6 degree centigrade. Snow cover and ice extent have decreased. The sea level has risen by 10 to 20 cm.

   (b) *Research scholar*: ___?

   Space scientist: Climate has and will always vary for natural reasons. However, human activities are increasing significantly the concentration of some gases in the atmosphere, such as greenhouse gases (mainly CO_2), which tend to warm the earth surface and anthropogenic aerosols, which mostly tend to cool it.

   (c) *Research scholar*: _____?

   Space scientist: It is quite likely that the global mean temperature should increase between 1.4 and 5.8 degree centigrade. The northern hemisphere cover should decrease further, but the Antarctic ice sheet should increase. The sea level should rise between 9 and 88 cm.

   (d) *Research scholar*: _____?

   Space scientist: Take for instance the regional changes in climate, particularly increase in temperature, which have already affected some physical and biological systems. Both natural and human systems are vulnerable to climate change because of their limited adaptive capacity. This vulnerability arises with geographic location and time, as well as social, economic, and environmental conditions. Some extreme weather events and the damage, hardship, and death they cause are projected to increase with global warming.

(e) *Research scholar*: __?

Space scientist: Well, you see, the projected change in climate is expected to have both beneficial and adverse effects on water resources, agriculture, natural ecosystems, and human health. But the larger the changes in climate, the more the adverse effect should dominate. Human populations are expected to face increasing flooding and heat waves but reduced cold spells. The geographic range of infectious diseases should increase.

(f) *Research scholar*: ____?

Space scientist: Well, there are many techno-logical options of reducing greenhouse gas emissions, some at low or negative cost. Forests and agricultural lands provide signif-icant but not necessarily permanent carbon sinks, which may allow time for other options. There will be both costs and benefits associated with reducing green-house gases. With coordinated actions and international regimes, efficiency and equity should improve. Further research is required to strengthen future assessments and to reduce uncertainties.

(g) *Research scholar*: __?

Space scientist: Well, you see, it is not possible to link any particular event definitively to global warming. But as the world warms, more of some types of extreme events are expected, such as heat waves, heavy precipitations, blizzards, and droughts; for some other events, such as extra-tropical storms, there is little agreement between current predictive models.

(h) *Research scholar*: _____?

Space scientist: Ecosystems have a limited capacity to adapt to climate change; some might not be able to cope as they had done in earlier periods and are expected to suffer damages because the rate and extent of climate change is expected to be faster and greater than in the past and could exceed nature's maximum adaptation speed. Human activities and pollution have increased the vulnerability of ecosystems.

(i) *Research scholar*: _____.

Space scientist: The pleasure is mine and I hope to see you on some other occasion.

(j) *Research scholar*: _____
_____.

Space scientist: Bye and good day to you too.

3. Read the following dialogue and answer the two questions:

'What happened to my proposal on introducing a new centre for software development in our organization, Harish?' she complained. Come on, give me an answer. Surely there must be something you can tell me apart from "I will let you know"'.

'I am sorry, Ms Leela. The only aspect we are yet to decide is the budget. As soon as I have got anything concrete to say, I promise you will be the first to know.' Harish softened his words with a smile.

He turned to leave. But she pleaded, 'Listen, it is pending for long. Do something quickly and let me know.'

(a) What does it say about the relationship between the speakers?

(b) Who has the power in this exchange?

4. Find two crowded places, such as railway station, airport, restaurant, etc., and write down snippets of conversations that you hear. For instance, you may write down the conversation between two friends, two professionals, etc.

5. Take an episode from one of your favourite television shows. Analyse it carefully. What did it reveal about the characters? What was the mood of the scene?

Now write a scene of your own for the same containing dialogues.

Include all the four types of dialogues: active, passive, inner, and outer.

CHAPTER 7

Formal Presentations

OBJECTIVES

You should study the chapter to know

- ○ the importance of face-to-face presentations
- ○ how to be clear with the purpose in your presentation
- ○ how to plan, structure, begin, develop, end, and use effective body language, voice, and visual aids in your presentation
- ○ how to handle questions and criticism
- ○ how to control nervousness and stage fright while making a presentation

Introduction

Successful and inspiring speakers are remembered not only because they were eloquent, humorous, or had a good style, but primarily and principally because their messages and ideas caused a change in their audience's actions, attitudes, lives, or made the purpose clear to them. This is true for all types of presentations—professional presentations, business speeches, classroom lectures, and so on, especially in an age of instantaneous communication via telephone, computer, and fax. Face-to-face business presentations are enormous time consumers—from scheduling a date when everyone can attend, to making every arrangement necessary for the presentation, it takes much more time and effort than it would have taken to send the same message as an attachment in an email, in the form of a memorandum, circular, or notice. Nevertheless, presentations still play an important role in business for obvious and good reasons.

Throughout our career, we are bound to encounter innumerable situations that require professional presentations to be made. To mention a few, a team leader may have to present before the corporate body about a product that his/her team has brought out; a top administrator of an institution may have to present the goals, activities, and achievements of the institution to an important visitor; a project manager may have to present before a committee the results of a project recently undertaken by the company; a college student may have to attend seminars or may have to present project reports to fellow students and faculty members. These situations call for effective, memorable presentations. Although these circumstances differ in purpose, the strategies in making good presentations do not differ much.

The more successful our career, the more often will we be called upon to make presentations for a variety of situations and audiences. Constant practice is the key to acquiring this skill. The following points are to be attended to when preparing for a professional presentation.

'Half the world is composed of people who have something to say and can't, and the other half who have nothing to say and keep on saying it.'
–Robert Frost

- Planning: Occasion, audience, purpose, thesis, and material
- Outlining and structuring: Introduction, main body, and conclusion
- Choosing the mode of delivery

- Guidelines for effective delivery
- Body language and voice
- Visual aids

Planning

Preparing and delivering the first business presentation or public speech in our life can be daunting. We may find it difficult to decide what we want to say and how to say it, or perhaps the thought of speaking before an audience scares us. It is true that some people are naturally talented at public speaking. However, with some helpful guidance, anyone can prepare and deliver a successful speech that will be remembered for all the right reasons.

When preparing a presentation, the first instinct may be to sit down with a pen and paper and charge ahead into the first line of the speech. However, devoting some time to careful planning of the speech will save a lot of time and effort later on. Effective preparation enables us to answer all the questions and doubts about our speech before they arise. The contents of our speech, and how we deliver it, are based on five important factors:

- Occasion
- Audience
- Purpose

- Thesis
- Material

Occasion

Occasion refers to the factors such as the facilities available for our presentation, time, and context of our presentation. *Facilities* include the venue or locale along with the projection equipment, lighting, seating, ventilation, etc. Every location has its unique physical environment. We may present in magnificently large auditoriums or oppressively small conference rooms. We need to know the physical setting; find out whether we will have a podium or a table, whether we will have a public address system, and so on. Also, attention should be paid to the physical conditions prevalent in the venue such as seating, room temperature, and lighting. We should try to understand whether the audience will be seated on hard metal chairs for an hour in a freezing room, whether the lighting will be too powerful to render our slide presentation ineffective, etc. If we identify such problems in advance, we can either ask for alternative arrangements or modify our materials, visual aids, and style to suit the environment.

Time refers to both the time of the day of presentation and the duration of the talk. Straightforward and factual presentations may work well during the morning hours, but in case of an after-dinner speech, we may need to adapt our remarks to the occasion. Remembering the fact that most professional presentations are brief, we should present the important points in the first few minutes.

Context refers to the events surrounding our presentation. When we are presenting in a team, for example, we need to consider the team members. They might have left a positive or negative impression in the minds of the audience and, hence, we would need to adapt ourselves to the existing situation just before presenting our part. Besides these immediate events, the recent happenings in our company can also affect the presentation. For example, if you are about to present a new proposal on budget just after your company has suffered a financial loss, you should emphasize on those features of your budget that focus on reducing the costs.

The occasion dictates not only the content of our speech, but also the duration, the tone, and the expectations of the audience. For example, humour may be inappropriate during a serious sales presentation, while it may be welcome during a wedding speech, or a sports event. We should also be aware of our role and any observations that we might make during our speech. For example, if you are presenting the final report of your project to a group of professors or senior colleagues, you need to take care of the short duration, firm but polite tone, and also their expectations.

Audience

All audiences have one thing in common. They are at the receiving end of our communication. They may be our friends, clients, colleagues, sometimes unfamiliar faces, or a combination of all these. The nature of our audience has a direct impact on the strategy we devise for our presentation. Hence, it is necessary to have some prior knowledge of the audience.

- What are their interests, likes, and dislikes?
- Are they familiar with the topic?
- Is their attitude hostile or friendly?
- What is the size of the group?
- Age range? Gender distribution?

Adapt your speech to your audience.

For instance, people from a particular culture may feel uncomfortable asking questions or may not reveal their feelings through facial expressions. If we know in advance how our audience is likely to react, we can structure our presentation and adapt our style to help them feel comfortable. We are also less likely to feel distressed by their reactions.

> 'It is a remarkable observation that the more learned and respected the researcher, the simpler their talks often seem to be.'
> –Mike Grimble

If we are going to speak before an unknown group, we can ask our host or the organizer for help in analysing the audience and supplement their estimates with some intelligent guesstimates of our own. Whether we present locally or in a foreign country, we can expect at least some members to have linguistic or cultural backgrounds different from our own. Those who are not very conversant with English or with our accent will appreciate relatively slow speech and visual aids designed to aid their understanding. We should also adjust our style to accommodate cultural differences.

While speaking on a controversial topic, we ought to keep aside some time to tackle any opposition from audience. For example, if your topic is on 'Criteria for selection of projects' you may face a lot of opposition from those teams which do not conform to certain criteria. So, you need to be patient in listening to them and then only should react. Give the impression to your audience that you want to share your views with them.

The structure of a presentation can further be skilfully emphasized by pauses, through interactions with the audience, and through changes in delivery techniques.

If we are going to speak about something controversial or if we have to break some bad news perhaps, we can set aside some time before our presentation to chat with those who will be affected. This will help to:

- Build support
- Anticipate problems

- Consider strategies

Testing the waters beforehand, so to speak, will help to fine-tune the approach. Speak with confidence and conviction. Make your points crystal clear and easy to understand. Maintain an attitude of alertness and confidence. Encourage questions from the audience. Audience participation gives the opportunity to clear up any misunderstanding.

Tips for Creating an Impact on the Audience

- Before beginning your presentation, look at all the sections of the audience.
- Always begin with a smile and greet them in pleasant tone.
- Give the impression that you are not lecturing but sharing your views with them.
- Modify your tone/material according to the reaction of your audience. For instance, if you find them bored or not understanding your point, soften your tone, ask them if they have any difficulty and give one more clearer example.
- Choose examples that are familiar to the major section of the audience (e.g., if you are presenting on 'Meditation' you can give example of a student who has found a significant improvement in his power to concentrate by regularly practising meditation).

- Choose words as per your audience's background (e.g., if you are giving a technical presentation to your professors or classmates, you can use specialized terms. When you are presenting to non-technical audience, simplify or define terms such as 'lean manufacturing', 'mach 2', etc.).
- Do not get annoyed if there is a slight disturbance among the audience (say, two people at the back are whispering something).
- Concentrate on your ideas and be with the topic rather than thinking on what impression you are making in the audience's minds.
- Inform the audience at the start of your presentation whether you would prefer to answer their queries at the end or you would not mind being interrupted.

Purpose

There can be three different purposes of a presentation: to inform, to analyse, or to persuade. The purpose of a presentation not only decides the content and style but also affects the amount of audience interaction. For instance, when our purpose is to provide information or to analyse a situation, we generally interact with the audience in a limited manner. Examples of typical presentation forms with an informative purpose can be a presentation at the new employee-orientation programme or an explanation of our project status.

On the other hand, when our purpose is to persuade people to take a particular action, collaborate with them in solving a problem, or making a decision, the interaction would be more. We generally begin by providing facts and figures that increase our audiences' understanding of the subject; we may also offer arguments in defence of certain conclusions and recommendations. In addition, we invite them to participate by expressing their needs, suggesting solutions, and formulating conclusions and recommendations. However, this would need a lot of 'on-the-spot' thinking skills and in-depth knowledge of the subject. Sales presentations, speeches by political leaders during election, etc., come under the category of persuasive presentations.

At times, our goal may be to help the audience have a good time. When we welcome the gathering at a conference, we are cheering and gearing them up for the coming sessions. Likewise, when we give an after-dinner speech at a company gathering or an awards dinner, our purpose is to leave the group in a jovial mood.

Depending on the purpose, we should be flexible enough to adjust to new inputs and unexpected audience reactions.

Thesis Statement

The thesis statement is very important in a presentation because it spells out the subject and establishes its impact among the audience. It is also the central idea of a presentation. Using a question or a sentence fragment should be avoided. Simple language should be used to frame a complete, declarative statement. Let us look at the following versions of thesis statement written by a student for a presentation on *Choosing a reputed university for higher education*. We can observe that the first two are ineffective for the reasons mentioned in parentheses, while the third is an effective thesis statement.

- Why should we be careful in choosing a reputed university? (question: does not reflect the content)
- Choosing a reputed university for higher education (fragment: does not tell anything specific; repetition of the topic of presentation)
- Choosing a reputed university for higher education has five significant advantages (tells the audience that they will know these benefits after listening to the presentation)

Begin to formulate your thesis statement as soon as you select your topic and decide on your purpose. Then allow yourself enough time to explore and develop your ideas. Shown below are the steps to arrive at your thesis statement:

Topic: Choosing a reputed university for higher education
Topic area: Advantages of a reputed university
General purpose: To inform
Specific purpose: I wish to tell my audience about the benefits they will reap by choosing a reputed university
Thesis statement: Choosing a reputed university for higher education has five significant advantages.

Material

Once we complete formulating our thesis, we need to develop the information that elaborates it. Collecting material requires some research. For example, when we are explaining a process or procedure, the main text of our presentation will include a series of steps involved. Similarly, when we are giving a product presentation, besides the complete information about the product, we may have to collect information pertaining to the competing products and their features. For most of the professional presentations, we may have to consult the library, Internet, magazines, newspapers, organizational records, statistics, and publications. Sometimes, we may even have to collect information through surveys or interviews. We may also have to contact external organizations to procure information for some of our presentations. Once we finish collecting material and ideas for our presentation, we should assemble them at one place. We may list all the ideas on a piece of paper and then organize them.

Outlining and Structuring

An outline is a framework in which bits and pieces of the presentation material are fitted. It serves as a guide to show us the right path for our presentation. Hence, spending time in developing an outline never goes waste. In fact, we can use an outline as our 'script', but should be prepared to deviate in response to audience feedback.

Suppose we are planning to deliver a presentation on 'Graduate study in the USA' to the graduating students of a college. This may be a thirty-minute presentation about the steps and requirements to pursue an MS or PhD programme in the US universities. The outline can be in the form of words, phrases, or sentences:

- Introduction
- Decision-making
- Basics of US higher education
- Graduate study programmes
  – MS
  – PhD
- Application forms
- Admission procedure

- Requirements
  – Statement of purpose
  – Academic aptitude
  – Professional development
  – Personal qualities
  – Presentation skills
  – Recommendation letters
- Conclusion

We may have to revise the subheadings under each or some of these main topics. But, as already said, we should be ready to skip or add some topics if the audience wants us to do so. We will learn more about outlining in Chapter 13 on reports.

Structuring or organizing the material clearly is vital for an effective presentation. A well-organized presentation can make our messages more comprehensible, create the desired effect on our audience, and boost our image as a speaker. On the other hand, rambling or taking too long to get to the point, including irrelevant material in the speech, omitting necessary information, or messing up the ideas can lead to a chaotic structure. Even experienced speakers get into trouble if their material is not organized appropriately and end up confusing their audience.

The key to all these problems is to organize our ideas into a well-known pattern. First, we need to tell our audience what we want to tell them; then, we should tell them the ideas; and finally, we should repeat what we have already said. In other words, a presentation should have the following format:

- *Introduction* should grab attention, introduce the topic, contain a strategy for establishing credibility, preview the speech, establish rules for questions, and have a smooth transition to the main text.
- *Main body* should contain all the main points and supporting material; the entire matter should be organized into a logical sequence.
- *Conclusion* should contain signal, highlight/summary, closing statement/re-emphasis, a vote of thanks, and invitation to questions.

Introduction

Look at the following introduction to a presentation on 'Effective use of DDT-based Insecticides'.

Sample introduction

Good morning friends and wish you all a happy World Environment Day! (Greeting and reference to the day)

Before starting my presentation, let me ask you a few questions: How often do you use DDT-based insecticide in your homes or offices? Do you find them effective? Have you heard of their misuse? Are you aware of their harmful effects and how to control them? (Attention grabber)

Well, I am glad that you find them useful and you use them in your homes and offices. But today I am here to talk about the measures you need to take in order to optimize the benefits of such insecticides. Yes, my topic is 'Effective use of DDT-based insecticides'. DDT, a chemical compound present in insecticides is dangerous when misused, but you can prevent serious health problems by carefully following directions. (Revealing topic and thesis statement)

I would like to tell you that I have been doing research on various kinds of insecticides—their production, distribution, effects, hazards, remedies, etc., for the last several years and have come up with a number of recommendations for their effective use. (Credibility statement)

We will first define and discuss the effects of DDT, then the types of DDT-based insecticides and their effects, the inappropriate way in which people use them and finally suggestions to use them effectively. (Preview)

If you want to ask any questions or give any comments please do so when the presentation gets over. I'll complete my presentation in twenty minutes and you will have the next ten minutes for the question and answer session. (OR) Please feel free to interrupt me by raising your hands if you have any query. (Rules for Q&A)

So let me begin with the first point, what is DDT? (Transition from introduction to main body)

Compare this with an introduction that starts like:

Good morning ladies and gentlemen, today I am going to talk on effective use of DDT-based insecticides...

and then straight away goes to the first main point. Think over, which one, according to you, may catch the attention of your audience so as to persuade them to listen to the rest of your speech?

The introduction to a presentation does the job of the preface to a book. It catches the attention (attention grabber) of the audience, tells them the topic and purpose (topic and thesis), develops in them a trust for the presenter and the presentation (credibility), kindles their interest in what the presenter is going to speak in the minutes to come (preview), and takes them slowly into the main body of the speech (transition). Having gone through the sample introduction, let us look into its components in a little more detail:

Greeting

We can start with good morning/good afternoon, etc., or can begin by (a) extending a compliment to our audience—I feel good to present before an intelligent/august gathering like yours/It is refreshing to look at your bright faces this morning; (b) referring to the location or occasion—I hope you all had a good time at the river cruise last evening; On the occasion of World Education Day, let me wish all of you to have lifelong learning.

Attention grabber

This catches the attention of our audience and prepares them to listen to the rest of our presentation. Depending on the topic, we can use a question, a quotation, a startling statement, an anecdote, or even a video or audio clip to grab the attention of the audience. For example, as in the introduction to 'Effective use of DDT-based insecticides' given above, you can begin your presentation with a series of questions: 'How often do you use insecticides? Are you aware of their harmful effects?' or with the statement, 'You would be shocked to know that thousands of people die every day by the careless use of insecticides'.

Imagine that you are giving a presentation on the topic 'The Role of Emotional Intelligence in Developing Leadership Skills'. You can start your presentation with an anecdote similar to the one given below:

On Friday, when I was attending a meeting, I could barely control my open appreciation of Ms Veena, the Chairperson. You may wonder why. I will tell you now. She was able to resolve the conflict between two participants amicably simply by using intelligence to understand their emotions. Yes, friends, in today's business world, emotional intelligence plays a very important role in cultivating various skills—leadership skill is one of them.

After catching the attention, state clearly and precisely the purpose of your presentation. For instance, it can be as pointed as this—'One reason brings me here today—to inform you about our new performance appraisal system'.

Topic and thesis statement

As the topic is very important, we need to include it in our visual aid (PowerPoint slide or overhead transparency) and project the same. Then our specific purpose can be stated in the form of a thesis statement as previously discussed under the heading 'Thesis'.

Credibility

Many factors may help us develop trust in the audience's minds for us and our presentation matter. While the audience may believe us because of our power/status/experience, we may need to speak out explicit statements in order to establish credibility in their minds if we are young or inexperienced. We can achieve this goal by stating our interest in the topic, by quoting some relevant statement from a recent newspaper or magazine, or by informing them how much research we have done on the topic.

Preview

We can tell our audience what is coming ahead in our presentation. We can also give a brief idea about the issues we are going to cover in the given time. This can be done by showing the slide containing the main topics and subtopics in the presentation outline.

Rules for question and answer session

Good presenters always anticipate questions and prepare their answers as well during the preparation stage. It is better to inform the audience in the beginning itself whether they can ask questions during or after the presentation.

Transition

Before going to the slide containing the first main point, we can speak out a phrase or ask a question to provide a link between the introduction and the main body of the presentation—'now that you know what are the issues I am going to discuss today, let me begin with the first point, that is, ...'; 'So, what are the characteristics of a reputed university?'—It is better to adopt a uniform style to state the main points of a presentation; that is, all the points are in the form of phrases, questions, etc. Do not mix different formats.

A good introduction creates interest and leads the audience effectively into the main body of the speech.

Main Body

The main body, the discussion, or the text part follows the introduction and supports the aim or specific purpose of a presentation. The major points we highlight in our opening section will be expanded here. Depending on the topic, and the introduction part, we can choose from any of the following patterns to organize the main body of a presentation.

Chronological   This pattern can be used for organizing points that can be arranged sequentially (in the order in which the events occurred or appeared before us). The entire presentation can be arranged chronologically. This method is useful for topics such as 'the profile of our institute', 'the changing face of the earth', and 'history of sports'.

Categorical   This is one of the easiest and most commonly adopted patterns for many topics. The entire presentation can be divided into various topics and subtopics arranged on the basis of subordination and coordination. This can be used for topics such as 'the role of advertising', 'environmental protection', 'importance of professional presentation', etc.

Cause and effect   This method can be adopted whenever a 'cause and effect' relationship exists. Here, we have to illustrate and explain the causes of the situation and then focus on the effects. It is relevant for topics such as 'impact of cinema on children', 'Internet—boon or bane', 'global warming', etc.

Problem–solution   Here, we divide the presentation into two parts. In the first part, we describe and analyse the problem. After the analysis we move on to the main objective of the presentation to suggest or propose a solution to the problem. It is a very helpful and effective way for persuasive presentation. For topics such as 'population explosion', 'addiction to gaming', etc., this method can be used.

We can strengthen our argument or ideas by providing examples, illustrations, statistics, testimonies, analogies, or definitions.

Supporting material Solid ideas do not always impress our audience. We need to back up our well-organized points in a way that makes the audience notice, understand, and accept our message. In other words, we need to use plenty of supporting material or develop our core points adequately.

As demonstrated by the examples given in Table 7.1, supporting material not only clarifies the main ideas, but also makes them more vivid and meaningful to the audience. In addition, they help establish and prove our main statement. Let us now discuss the various categories of supporting material (Table 7.2).

Table 7.1 Examples of supporting material

Main statement	Supporting statement
• Replacing the lens in the laser projector is not as complicated as it seems	• Let me show a diagram that demonstrates how to do it
• We could increase sales by extending the store timings until late in the evening	• An article in *Business Today* cites statistics showing that shops that extended their working hours to 10 p.m. boost profits by more than 20% of the direct overheads involved with longer business day
• A reputed university always has an excellent placement record	• In an interview with ABC channel, the VC of XYZ university said, '99 per cent of our students have got offers from IT, manufacturing, consultancy, healthcare, and other industries'

Table 7.2 Kinds of supporting material

Type	Definition	Function	Speech occasions	Tips
Definition	Explaining difficult term(s) with the help of simple terms	To clarify	Used in informative/ technical presentations	Use easy and known terms
Example	A brief reference that illustrates a point	To clarify and add interest	Used in all types of presentations	Use situations with which your audience may be familiar
Statistics	Quantification of the main point	To clarify, prove, and add interest	Used widely in presentations where sales figures, survey results, etc., are to be explained	Round off the numbers, support with visuals, and explain adequately
Analogy	Process that shows how one idea resembles another	To clarify, add interest, and prove	Make the comparisons vivid: select familiar analogies	Used in business presentations involving products, processes, and procedures
Testimony	Opinion of experts, peers, or celebrities	To clarify, add interest, and prove	Used in sales presentations	Memorize/paraphrase/read verbatim, cite source, use sources credible to your audience, and follow up with re-statement of explanation

Definitions When we deliver a presentation on a topic that we feel that the audience is not too familiar with, we can use a definition to develop our idea.

For example, we can start the main body of a presentation on 'Artificial Intelligence' with the definition:

> Artificial intelligence (AI) is the intelligence of machines and the branch of computer science that aims to create it. John McCarthy at MIT coined this term in 1956.

Examples Examples are the most commonly used supporting material in presentations. They give life to our ideas and make them immediately comprehensible to the audience. Almost all effective speakers use examples in their presentations.

> Artificial intelligence (AI) has a wide range of applications such as computer games, neural networks, robotics, and many other areas of technology. For instance, by using AI, you can study the human behaviour. You can assign human characteristics to several characters, program them, allow them to behave in various situations, and then study them.

Statistics Engineers, scientists, and business professionals use statistics that represent numerical data relating to groups of individuals or experiments to substantiate their ideas and strengthen the understanding of their audience. Statistics help to make certain arguments more convincing. They are used in presentations related to sales trends, trends in epidemic, experimental results, size of market segments, and the like. Most statistics are collections of examples reduced to numerical form for clarity and easy comprehension of a complex idea. When handled well, statistics are an especially strong proof, because they are firmly based on facts and because they show that the speaker is well informed. Consider the excerpt from the speech of Aditya Vikram Birla, 'Let the Competition be Afraid of Us', delivered at a conference organized by *Euromoney* in New Delhi.

> We also have a vast bank of talent, with over 3.5 million scientific and technical personnel, trained in the English language, of a quality and at a cost unmatched. India has a well-developed capital market. We have 21 stock exchanges with over 2,000 actively traded scripts, compared with 220 in Indonesia, 354 in Thailand, 423 in Malaysia, 235 in Singapore, and 181 in the Philippines.

Statistics are best presented using visual aids. When presenting statistics, the statistics should be explained completely. Unless we are presenting statistics developed or collected by ourselves, we need to cite the source in the slide and also to mention it while presenting.

Analogy An analogy can make a point by showing how one idea resembles another. Analogies compare items from an unfamiliar area with items from a familiar area.

When we want to talk about the basics of electronics, we can compare the components with the elements of hydraulics. We can say, 'current is measured in amperes and is equivalent to the volumetric quantity of flowing water over time'. As 'flowing of water' is understood by everybody, this analogy may be understood by all.

Whenever we propose adopting a policy or using an idea because it works well somewhere else, we can use comparisons. Presenters mostly use literal comparisons that link similar items from two categories. Look at how N. R. Narayanmurthy, the founder and former Chairman and CEO, Infosys Technologies Ltd, uses comparison as a device to explain his idea pertaining to professionalism in India and the western countries:

> 'Yet another lesson to be learnt from the West is about their professionalism in dealings. The common good being more important than personal equations, people do not let personal relations interfere with their professional dealings. For instance, they don't hesitate to chastise a colleague, even if he is a personal friend, for incompetent work. In India, I have seen that we tend to view even work interactions from a personal perspective. Further, we are the most 'thin-skinned' society in the world—we see insults where none are meant. This may be because we were not free for most of the last thousand years.'
>
> (Excerpt from a lecture delivered at the Lal Bahadur Shastri Institute of Management, New Delhi, on 1 October 2002.)

The strength of an analogy lies in the choice of the points of comparison and the effectiveness with which we deliver it to our audience. An analogy should always be short and simple, and use of a wrong analogy that may confuse the audience should be avoided.

Examples of analogy   The use of analogy in the presentation can drive home the point more effectively and connect better with the audience. Analogies take much imagination with them. Some analogy examples are as follows:

1. While referring to creativity in advertising, an example of a baby diaper advertisement can be represented by showing a dam holding enormous water instead of getting it endorsed by crying babies or restless parents.
2. If a presentation is on 'Importance of humility', you can say that one should not become high headed as one grows professionally. One can always learn from a tree as it bends to the ground when it is laden with fruits. Similarly, when one becomes successful, one must be humbler.

Testimony   Testimonies are remarks made by others who are authoritative or articulate and could make a point more effectively than we could, on our own. For instance, when the chairman of Steel Authority of India Limited says something about the production technology of steel, people will accept it without a question because he is an authority in the field of steel production. Same is the case with Sachin Tendulkar quoting something on cricket. Therefore, to add punch to our talk on team effectiveness in cricket, we can quote Sachin Tendulkar:

'Isn't cricket supposed to be a team sport? I feel people should decide first whether cricket is a team game or an individual sport.'

(*Source*: http://www.brainyquote.com/quotes/authors/s/sachin_tendulkar_2.html)

Other testimonies include: remarks made by a celebrity who may be a non-expert in the field, an article written by a relatively unknown person in a journal or newspaper, some good arguments put forth by our colleagues/relatives/professors, etc. Testimonies also help build a persuasive case. Whenever we use a testimony in a presentation, we should cite the source and quote verbatim if it is short. If it is lengthy or confusing, we should try to paraphrase.

The temptation to include too many points in the body of a speech should be resisted. We should restrict ourselves to four or five main points. We can help the audience follow our presentation by summarizing the points as we go along. Every main point is a unit of thought and an essential part of the speech. Each point should be clearly stated, independent of the other main points. Hence, we need to balance the time devoted to each point accordingly. We should plan on bridging the main points so that we can move smoothly from one part of our presentation to the next. We can make use of transitional expressions such as 'therefore', 'because', 'in addition to', 'apart from that', 'on the contrary', 'next', etc.

Conclusion

The conclusion of a presentation provides yet another opportunity for us to impress the audience. Hence, the conclusion should be prepared and presented with the same interest as we take for the introduction. We can conclude our presentation by reviewing the main points. A signal such as *to sum up, to conclude, to review, in the end*, etc. to indicate the end of the presentation must be used. As we conclude, we should remind the audience briefly about the purpose of our presentation, which could be either to persuade them or to inform them. We should tell them what we want them to do, think, or remember based on the presentation. The temptation to wrap up in haste or add something new in this part of the speech should be avoided.

We can also conclude with a quotation or can recall the earlier story, joke, anecdote with which we commenced our presentation to bring it to a full circle. Some presenters bring in a change in the pace or pitch of their voice. They slow the rate and speak in a lower pitch so as to mark the difference between the main body and conclusion of their presentation.

Nuances of Delivery

All of us have listened to more than our share of bad presentations. We have sat through presentations that were delivered so haltingly that we could not care what was being said. We would also have come across presentations that were delivered smoothly but had practically put us to sleep, as the presenter droned on endlessly. However, if delivered effectively and efficiently, presentations can capture the audience's attention without the risk of being shuffled aside. We can reveal our enthusiasm to the audience better than any other means of communication and can address their questions or objections directly.

Modes of Delivery

What is it that makes our presentation hold the attention and interest of the audience? Our manner of presentation, our vocal inflections, our perfectly timed pauses, our facial expressions, and our gestures—all these are part of an expert delivery. Even a dull and drab topic will turn out to be more interesting if presented well, whereas a really interesting topic may appear to be dull because of poor delivery. So one thing becomes clear—*having something to say is not enough; you must also know how to say it*. Good delivery does not call attention to itself. It conveys our ideas clearly, interestingly, and without distracting the audience. Most audiences prefer delivery that combines a certain degree of formality with the best attributes of good conversation—directness, spontaneity, animation, vocal and facial expressiveness—and a lively sense of communication. The following discussion provides some suggestions so as to enable one to select the best mode of delivery for a presentation. There are four modes of delivery that can be used for making presentations:

- Extemporaneous
- Manuscript
- Impromptu
- Memorization

Extemporaneous mode

Extemporaneous presentation is by far the most popular and effective method when carefully prepared. When speaking extempore we must prepare the notes beforehand and rehearse our presentation. There is no need to learn every word and line by rote. Our presentation will sound quite spontaneous to the audience, as after thorough preparation, we are speaking while thinking. Careful planning and rigorous practice enable one to collect the material and organize it meticulously. Let us look at some of the positive and negative aspects of this mode of presentation.

Advantages

- As we have enough time to prepare for the presentation, we work hard on the theme/central idea. We can present the theme in the best possible structured way.
- Thorough preparation on our part makes us feel secure and we carry out our responsibility with self-confidence and assurance. Adaptation is also possible if the need arises. In other words, the language of any written text does not bind us. We can be flexible in our use of language.
- Supporting material helps to present our points clearly and also adds weight to our agreement. Appropriate selection of quotations, illustrations, statistics, etc., helps us to substantiate our point.
- Our delivery sounds natural and spontaneous to the audience as it allows us to establish a rapport with the audience through more eye contact.
- It enables us to move freely, with ease.

Disadvantages

- If preparation is inadequate, we can get lost and find ourselves uncomfortable.
- If we rely too much on note cards and start reading out from them instead of just consulting them for reference, then the speech will lose its spontaneity.

Manuscript mode

In manuscript presentation, material is written out and we are supposed to read it out aloud verbatim. We are not supposed to memorize the speech and then recollect it. It is there in front of us to read. But, we should be wise enough not to attempt to read a speech until we have become a proficient reader. Unfortunately most speakers are not good readers. They make it uninteresting by reading in a dull and monotonous way. However, we *can* overcome this problem with consistent efforts. We could maybe rehearse with a friend or colleague.

For effective use of this mode, we should go through the material several times beforehand until we become absolutely familiar with the text. We should strive to choose material designed to achieve understanding. We should know *what* is written *where*.

Advantages

- It is a permanent and accurate record of whatever we have to say.
- There is no chance of tampering with the facts and figures.
- The material is organized systematically. We just have to keep in mind the step-by-step development of main points.
- Language gets polished because we can write and rewrite our material until we feel satisfied on all counts.

Disadvantages

- Since we will be reading from the manuscript, we get less time for making proper eye contact, which is essential to feel the pulse of our audience.
- Since we will be reading to the audience, we cannot talk to them. There is not much scope either for non-verbal communication.
- Adaptation is rather difficult, if the need arises, to give a different twist to our material.
- In the absence of effective reading skill, we fumble over words, lose our pace, and miss punctuation marks, etc. This adds up to an uninteresting speech and loss of audience attention.
- Conversational flavour along with vocal inflection takes a back seat here, which is a great asset for a speaker.

Impromptu mode

The impromptu mode, as the word suggests, is what we use when we have to deliver an informal speech without preparation. For example, at a formal dinner party you may be invited to deliver a vote of thanks. Do not panic and babble something in an unmethodical way. Instead, calmly state your topic and then preview the points you are to make. Support your points with whatever examples, quotes, and anecdotes

Mr Naidu was called upon after dinner to give an impromptu speech.

you recall at that time. Then briefly summarize or restate your points and end with a smile. Remember, it is not difficult for you to anticipate certain occasions where you may be asked to speak a few words. Be as brief as possible during your impromptu presentations.

Advantages
- We sound very natural because we do not get enough time to make any elaborate preparation.
- We get a chance to express our thoughts irrespective of what others think or say about that particular topic.
- We are spontaneous as we say what we feel, not what we ought to say.

Disadvantages
- The presentation lacks organized development of ideas because of the shortage of time.
- There is no supplementary material (no data, no statistics, no illustrations, no figures) to substantiate the speech.
- Chances of rambling are very high. Various points may hang loose.
- There is frequent use of vocalized pauses.
- The presentation may turn out to be a failure if the speaker has inadequate proficiency in the language he/she uses.

Gaining a reputation for being a good impromptu speaker can do a great deal for our career aspirations. It has been shown that there is a positive correlation between communication effectiveness and upward mobility.

Memorization mode

This method of presentation is very difficult for most of us. Probably only a handful of us can actually memorize an entire speech. Usually we memorize only the main parts and are in the habit of writing key words on cards to help us out through the actual presentation. In some cases, if we wish to quote somebody or narrate an anecdote or a joke, it is better to memorize these for our presentations.

This type of delivery stands somewhere between extemporaneous and manuscript presentation. Speech is written out beforehand, then committed to memory, and finally delivered from memory.

Advantages
- It is very easy for such speakers to maintain an eye contact with the audience throughout the presentation.
- The speaker can easily move and make use of appropriate non-verbal communication to add extra value to the speech.
- It is possible to finish the speech in allotted time.

Disadvantages
- Memorization requires too much of time.
- There are chances of making it a dull and monotonous presentation because we go exactly by whatever we have memorized.
- Even our memory skills may fail us if we have not rehearsed adequately.
- No flexibility or adaptation is possible during the speech.
- The speaker gets flustered if he/she forgets a word, sentence, or a whole paragraph.

Among all the four modes of delivery, extemporaneous is the best because of its flexible nature and its effectiveness. Hence, it is always better to use this mode to make presentations more lively, effective, and memorable.

Guidelines for Effective Delivery

Success of any presentation also depends on the various elements—verbal, non-verbal (body language), vocal, and visual—used during a presentation.

Verbal elements

Word pictures We may give our speech a graphic quality by painting word pictures that allow the audience's imagination to take over. Specific details allow an audience to see the scenes we are describing. Our major job as a speaker is to tell somebody something. We should present our point clearly and just enough so that the listener clearly understands the intended message. The task is not merely to get words out of our mouth, but to transfer ideas into the listeners' minds.

Warm words Cold words leave us uneasy and unsure while warm words make us feel secure and comfortable. Words are powerful. They conjure images, evoke emotions, and trigger responses deep within us and we react, often without knowing the reason. In the early days of instant coffee, advertisers got off to a bad start by stressing words such as 'quick', 'time-saving', and 'efficient'. All these words are without warmth and feeling. Makers of fresh coffee fought back with warm, happy, and appetizing words such as 'aroma', 'fresh', and 'tasty'. The instant coffee industry learnt the lesson and its product became 'delicious', 'rich', and 'satisfying'. Sales soon boomed.

Words also suggest whether something is good or bad. We should use those words that strengthen our arguments and weaken those of our opponents. For example, look at the following words:

Good	Bad
• Independent	• Unaccountable
• Well-regulated	• Red tape
• Free-thinking	• Wishy-washy
• Appropriately rewarded	• Fat cat

Similes and metaphors Although technical presentations do not require the use of similes and metaphors, we cannot deny that they not only add flavour to a speech but also make abstract ideas imaginable. Reach for vivid comparisons your listeners can understand and remember. Try the following metaphors in your speeches:

- As inflexible as an epitaph
- As cold as outer space
- Building a business is like building an empire
- As profitable as a gold mine
- Delay is the deadliest form of denial

Impact words 'We' and 'you' are the most important words of all. We cannot stir the audience up if we do not address them directly and relate them to us and our topic. Remember the five-to-one rule: Every time you use the singular 'I', try to follow it with five plurals. Given below are some words that you may use in your presentations or speeches to get desired results:

discovery, guarantee, love, proven, safely, easy, health, vigour, money, results, save, protect, interest, challenge, opportunity, excitement, enthusiasm, flourish, progress, favourable, adaptation, circumstances

Smooth flow We can also make our speech flow smoothly and gracefully from beginning to end by using some transitional devices. They promote clarity, emphasize important ideas, and sustain our listeners' interest. Some transitional devices are discussed below:

- *Bridge:* A bridge is a word that alerts the audience that we are changing direction or moving to a new thought. Some examples are:
  (a) We completed the project in January. *Meanwhile* other developments were taking place.
  (b) That was bad enough. *However*, there was even worse to come.

- *Number item:* A number item keeps the listeners informed about where we are in a presentation, which covers several points such as:
     The *first* advantage of the new plan is …
     The *second* benefit of the plan is …

- *Trigger:* A trigger is a repetition of the same word or phrase to link one topic with another, such as:
     That was what the financial situation was *like* in March. Now I will tell you what it is *like* today.

- *Interjection:* An interjection is a word or phrase inserted in a commentary to highlight the importance or placement of an idea, such as:
     So what we have learned—*and this is important*—is that, it is impossible to control personal use of office telephones.
     Now here is another feature—*perhaps the best of all*—that makes this such a terrific plan.

- *Internal summary:* Internal summary helps our audience stay oriented by providing a one-sentence summary during the course of delivering the main text of our presentation, such as:
     Now, you can see that the problem grew from several causes: a shortage of parts, inexperienced maintenance people, and the overload of opening a new warehouse.

- *Internal preview:* An internal preview, like an internal summary, orients the audience by alerting them to the upcoming points, such as:
     You are probably wondering how all these changes will affect you. Well, some of them will make life much easier, and others will present some challenges. Let us look at three advantages first, and then we will look at a couple of those challenges I mentioned.

- *Signpost:* Signposts tell our audience where we are in our presentation. When we say, 'There are four advantages, as I had already said. Having discussed the first two of them, let me move on to the third advantage', the statement would serve as a *signpost* to tell the audience that we have completed two and there are two more to go.

- *Rhetorical question:* A rhetorical question can subtly change the direction of the discussion, such as:
     That is what a change of image can do to a company. *So how can we improve our image?*

- *Flashback:* A flashback is a sudden shift/reference to the past, and breaks what seems to be a predictable narrative. For example:
     Today, we are the market leader. However, *three years ago*, this was not the case.

- *List:* A list is a very simple way of combining apparently unrelated elements, such as:
     We made *four* attempts to solve the problem.

- *Pause:* A pause is a non-verbal method of showing our audience that we have finished a section of our speech and we are about to move on to another.

- *Physical movements:* Physical movements towards a visual aid, such as a black/white board, flip chart, or screen, suggest that we are moving on to something new.

- *Quotations, anecdotes, and jokes:* A quotation, anecdote, or joke can serve as an excellent link. We may see a joke like the one given below as a good link to the idea that one may wish to take up next:

The Chairman told me a story of a job applicant who said, 'I like the job, sounds fine, but the last place I worked at, paid more, gave more overtime, more bonuses, subsidies, travel allowances, holidays with pay, and generous pension schemes.'

The Chairman said, 'Why did you leave?' The applicant answered, 'The firm went broke.'

Non-verbal elements

Our appearance, facial expressions, eye contact, postures, gestures, and the space we share with our audience, all communicate our interest, enthusiasm, dynamism, intention, and confidence to our audience. Whatever the occasion, the following tips will help you to use body language effectively during your presentation.

- Wear a formal dress and use simple accessories; take care of your personal hygiene
- Use facial expressions to exhibit your enthusiasm and interest; do not show your irritation or anger even when someone interrupts you or asks a question that appears to be silly to you; be polite in answering them
- Make eye contact with all sections of the audience (avoid staring at somebody) to observe their reactions and also to show your sincerity and interest
- Use well-timed gestures; avoid monotonous gestures
- Stand tall and straight with shoulders upright; walk/ move swiftly; avoid too many and monotonous movements
- Do not come very close to the audience; maintain a distance of at least 4–12 feet
- While using blackboard, raise your voice and look at the audience in between
- Avoid fiddling with key ring or tie while presenting
- Avoid looking outside even if there is some external noise. Try to concentrate on your ideas and audience

What you are doing rings so loudly in my ears that I am unable to hear what you are speaking.

Vocal elements

Our voice can serve as an important tool to support our verbal message. How we sound is as important as how we look or what we say. Our vocal elements, namely the tone, pitch, rate, and volume, reflect our attitude about ourselves, our message, and our audience. Try the following tips to help you use your vocal elements to enhance the impact of your presentation:

- Speak with enthusiasm and sincerity
- Adjust the volume of your voice
- Avoid monotones or vocalized pauses
- Use your optimal pitch

- Avoid fast delivery
- Use silence and pauses effectively
- Articulate each word clearly

Ways to Improve Body Language

Try and videotape a part of your presentation. Play it back and identify one aspect of your body language you want to improve. After making conscious efforts for improvement, record this part again and see if there is any improvement. You can also practise in front of a mirror to improve on facial expressions. When you have worked through your entire presentation, and feel sufficiently confident, invite a friend to watch your performance. Ask your friend for comments on aspects of your body language that are good or that need improvement.

Visual elements

Our audience will remember facts easily if the ideas are connected to the right-brain stimulation. The way to stimulate the right side of the brain is to show pictures. Visual stimuli are more effective than verbal stimuli. We often recall the colour of the cover of a book rather than its title and subtitle.

Advantages

People find our message more interesting, grasp it more easily, and retain it longer when we use visual support along with our words. Besides increasing the clarity of the message, visuals make presentations more interesting. For example, investment brokers often use an array of well-prepared charts, tables, models, and so on, to add variety to information that would be dull without them.

Graphics can also boost our image in ways that extend beyond the presentation. They add a professional flavour to our presentation. Finally, our audience remembers a visual message longer than the verbal message.

Tips for Effective Presentations and Speeches

- Be clear with your purpose.
- Know your audience.
- Keep enough time for preparation.
- Develop interest in the topic; know more about it by reading books, newspapers, etc., listening to and discussing with people.
- Collect adequate material and then select what to present according to the purpose and time given.
- Organize and make an outline with the main points and sub-points.
- Structure your presentation into three parts: beginning, middle, and end.
- Prepare the PowerPoint slides with care keeping in mind the *one minute, one slide* rule.
- Keep animations to minimum.
- Prepare illustrations, such as graphs, maps, drawings, tables, etc., accurately. Ensure that they are visible to everybody in the audience.
- Familiarize yourself with the venue and the available equipment.
- Arrive early and check the arrangements and your PowerPoint slides.
- Be excited about your presentation. Think all positive qualities in you and feel confident.
- On reaching the stage, look at the audience for a few seconds before you start speaking.
- See to it that your introduction goes smooth. You have won half the battle if this is done.
- Use transitions effectively so as to provide a smooth flow to your speech.
- Give a feeling to your audience that you are not dictating but sharing information.
- Explain each slide adequately. Do not just flip slides. Give time for the audience to grasp its contents.
- Maintain eye contact with all sections of your audience.
- Exhibit your enthusiasm, excitement, sincerity, and interest through appropriate facial expressions.
- Use well-timed gestures to substantiate your points.
- Adopt postures that reveal your confidence. Avoid monotonous postures. Do not move excessively.
- If necessary you sit and present in front of a small group.
- If necessary distribute handouts (copy of your slides or any other material) at the right time.
- If you sit and present in front of a small group, adopt a straight posture on your chair and lean forward while presenting. Do not sit in a relaxed posture as it will reflect a casual attitude.
- Listen to questions carefully and answer them completely.

Scan the QR code for tips on giving spectacular presentation.

Quizzes and interjections

Quizzes and interjections help in reinvigorating the dullness of long presentations. When the audience start getting distracted, quizzes help to get them back. The expressions of ennui start reflecting through one's face. That is the time to present a quiz for audience involvement. Quiz is one of the ways to captivate the audience's attention to the presentation. While planning a quiz, take care of the following points:

- Ask questions that are easy to answer.
- Make it relevant to the topic of your presentation.
- Provide appropriate choices.
- Conduct the quiz in a dramatic manner like a TV artist.
- Involve each member of the audience.
- Give reward for right answers.
- Encourage even when you get a wrong answer.
- Use interjections sparingly.

Humour

Humour helps in engaging, connecting, and entertaining audiences spontaneously. You can bring laughter to the plethora of audience who have gathered to listen to your presentation. Good humour can create a connect with the audience. If it is used appropriately, it is an effective tool to bring smiles on their faces. The audience can be made more comfortable and receptive with the help of humour. It becomes easy to convince the audience by interlacing the ideas with humour. Laughter evokes a lot of positive energy, increasing the heart rate and thus making people receptive of new ideas. Even serious talks can make the environment light, providing the audience a much-needed respite from serious issues. Audience always remember ideas better as humour increases cognitive function and memory. The American poet Maya Angelou said, 'I've learned that people will forget what you said, people will forget what you did, but people will never forget how you made them feel.'

People enjoy humour when they are at the receiving end but while delivering a presentation, they are not quite sure of using it. But anyone can use humour with a little practice. It is a skill and the more you utilize it, the better you are in creating an impact. While using humour in your presentation make sure of the following:

- Don't try to emulate famous comedians.
- Go for already existing humour rather than creating your own.
- Relate the humour to your talk. Don't use humour just for the sake of using.
- Ensure it is befitting for the audience and the occasion.
- Keep yourself as the central point in the humour. It works.
- Narrate personal stories as they are the best source of humour.
- Test the humour on a smaller audience before using it with a real audience.
- You should yourself find your humour interesting.
- Don't try too hard to make your audience laugh. Proceed ahead.
- Don't repeat your humour.

Learning how to use humour is not difficult; therefore keep trying to use humour in informal talks and learn from other effective speaking platforms.

Controlling Nervousness and Stage Fright

Does the thought of speaking in front of an audience make you nervous? The symptoms of stage fright are racing heart, sweating, dry mouth, shaky hands and legs, knocking knees, blinking eyes, queasy stomach, and loss of memory. If your answer is 'no', you may be an experienced speaker. If it is 'yes' then you should feel happy to know that you have thousands and thousands of companions to share the same answer with you. You would also be glad to know that even many seasoned speakers feel nervous when they need to present a complex topic, to present before their superiors, etc. Hence, anxiety or nervousness is not a sin or a bad quality. It is common for almost everybody. However, as nervousness leads to stage fright, which in turn may affect the presentation, we need to know how to control them. The following discussion may be useful in this regard.

> 'All the great speakers were bad speakers at first.'
> –Ralph Waldo Emerson

Strategies for Reducing Stage Fright

Strategies in advance of presentation

- Develop an interest in the topic of your presentation.
- Reserve adequate time for preparation.
- Anticipate easy as well as hard questions and try to work out your answers.
- Practise your opening statement several times.
- Rehearse your entire presentation at least twice.

Strategies just before the presentation

- Arrive early; check the arrangements, equipment, and your PowerPoint slides.
- If you see some participants, look at them, greet them, and talk to them so as to ensure that you are not nervous while speaking and your voice is flowing freely.
- Take a few sips of tepid water.
- If you have time walk around or outside the venue.
- Concentrate on your ideas.
- Relax yourself by taking deep, even, and slow breaths.

Strategies when the presentation begins

- Feel good about your presentation and walk up to the dais taking a few deep breaths.
- Do not begin immediately or in a hurry. First look at the friendliest faces among your audience and smile.
- If your legs are shaky, lean on the lectern or table on the dais and hold it.
- Remember that the audience may not realize your nervousness as much as you feel it.
- Never comment on your nervousness during your presentation (some speakers say 'I feel nervous, let me have some water please').
- Do not show explicit signals, such as clearing throat, drinking full glass of water, wiping forehead, etc., which display your nervousness to the audience.

> 'The best way to conquer stage fright is know what you are talking about.'
> –Michael Mescon

Visualization Strategies

Positive visualization is a proven technique to reduce nervousness or stage fright on any occasion such as a presentation, an interview, a group discussion, etc. Try this: Visualize (you need to imagine and see things or people in front of you; not just think) that you have prepared very well and you are now standing in front of your audience. All the eyes are on you. Here you can imagine your friends or other known people sitting in front of you. Visualize their smile, clapping, etc. See in your mental image that many among them are appreciating and shaking hands with you after the presentation is over; you are very happy and enjoy that day with your friends in your favourite restaurant.

While practicing this technique, initially you may only think and not visualize. But after a few attempts, you will be able to visualize people and activities in front of you. The concept behind this technique is 'positive thinking'. Rather than thinking 'My presentation is going to be a failure as I do not have experience. All are going to mock at me ...', you should visualize positive things such as success, commendations, happiness, etc.

On-camera Techniques

When your presentation is captured in a video format, you may have to follow certain guidelines:

- If it is a presentation in front of an audience that is being video recorded, forget that there is a camera in front of you.
- Do not be conscious of the camera and behave naturally as you would do in front of your audience with all the guidelines you have learnt.
- Do not look into the camera, but you may not be able to avoid it if some people in the audience are sitting in its vicinity.
- If you know that your presentation is being video recorded, dress yourself in suitable colours.
- If you are delivering an impromptu speech for a television coverage (you must have seen the reporters asking viewers to speak on certain occasions, such as cricket matches, elections, etc.), do not look to be surprised. Listen to the reporter carefully and present your views very briefly.

Role and Importance of Visual Aids

- Enhance interest in presentations. In the absence of visual aids, presentations become monotonous and banal.
- Explain the concept in a more lively fashion.
- Make the data appear more organized and exciting. Listeners make more sense of data when presented through illustrations.
- Help create visual imagery; therefore, the information can be retained for a longer duration.
- Helpful for speakers as they do not have to bank on the written notes while giving presentations.
- Provide aesthetics to the presentation and make the presentation easy to understand.
- Illustrate key points
- Signal transition from one part of the presentation to the next
- Increase impact of message
- Help listeners retain information
- Help present ideas without depending on notes
- For those not familiar with our language or accent, turn the incomprehensible into something understandable

Visual Aids in Presentations

Spoken words are temporary; as soon as they come out of our mouth they evaporate into the air. Because of this limitation, speeches often need strong visual support—handouts, chalk boards, flip chart, overheads, slides, computers, charts, tables, film, etc. If a picture is simple, clear, and appropriate to its purpose and audience, it will deliver its message more accurately and quickly than a verbal explanation. Ours is a visually-oriented society and an audience likes to hear as well as see information.

We should choose only those visual aids that suit the style and content of our presentations. We should use visual aids for any point that sounds vague and requires discussion in detail. However, they must be well designed and professionally generated. We should not use visual aids as a verbal crutch for the speaker!

Table 7.3 provides some tips pertaining to the types of visual aids one may commonly use in one's presentations.

Table 7.3 Types of visual aids used in presentations

Type	Tips
Overhead transparencies	• Use larger fonts. Avoid decorative fonts. • Separate the transparencies using sheets of paper. • Keep transparencies uncluttered. • Show only the required information. • Do not add multiple colours or exciting backdrops to your slides. • Use pointer on the screen, or your pencil or pen on the transparency to draw audience's attention to a specific item. • Familiarize yourself with the operation of the overhead projector. • Be ready with your notes in case of power failures.
PowerPoint presentations	• Check the computer system/equipment before loading. • Familiarize yourself with the operation of the slides. • Transfer your file to the hard disk. • Be familiar with the operation of slide show. • Rehearse your presentation. • Keep a printed copy of the slides (6 slides on a sheet of paper) for use in case of computer malfunction.
Blackboard or whiteboard	• Clean the board well before starting and check the condition of markers. • Write in large letters. • Stand to the side as you write. • Do not face the board while talking to the audience. • Divide the board into columns and write legibly. • Keep contents which you may want to refer to again.
Flip charts	• Use different coloured markers. • Keep two pads of paper. • Write in large letters. • Use only one side of the chart. • Wait for the audience to grasp the contents before turning pages.

We should be judicious in our choice of aids. If we are going to deliver a lecture to illiterate people, we should not use tables or complex graphs; instead we can use aids that they can understand. For instance, if we are talking about the function of the heart, we need to show them some pictures, name all the parts, use a pointer to make it clear to them about which part of heart we are discussing. If we wish to use visual aids of other organizations rather than our own, we should confirm beforehand the availability of those aids that we need. Refer to Chapter 1 for details on the various kinds of visual aids that can be used with presentations.

> 'When speakers use a lot of numbers, the audience almost always slumbers.'
> *–Charles Osgood*

Guidelines to Make Effective Use of Visual Aids

- If you feel that the audience needs explanation for your aids explain to them lest they should misunderstand it.
- Organize the visual aids as a part of the presentation. Fit them into the plan.
- Emphasize the visual aids. Point to them with bodily action and with words.
- Talk to the audience, not to the visual aids. Look at the visual aids only when the audience should look at them.
- Avoid blocking the listener's view of the visual aids. Make sure that lecterns, pillars, charts, and such things do not block anyone's view.
- Refrain from removing the aid before the audience has an opportunity to absorb the material.
- Do not talk about the visual aid after you have put it aside.
- Use enough visual aids to make your points clear, but donot overdo it.
- Do not use too many lines or figures on one aid; make sure that it is visible to one and all from all the corners of room.
- It should not be very light that the audience finds it hard to see. Too small an illustration will not be visible to those sitting at the back.
- Keep them at an inconspicuous place if aids are too many, or else they may distract the attention of the audience.
- Be familiar with the basic operations of the electronic devices that would be in use for presentations.

Expert Technical Lecture

These lectures follow almost the same strategies discussed in this chapter (formal presentation); however, there is a stress on presenter–listener interaction. The term 'lecture' includes both plan and delivery of a presentation. A lecture comprises three parts: introduction, main body, and close. Each one is equally important; therefore, should be planned meticulously. The introduction is given special attention as the audience pays maximum attention and makes opinions about the speaker during this time. Just leave with a question which will possibly be answered after the end of the lecture. This technique can get the audience more curious to listen to the speaker.

For instance, 'By the end of the lecture, you will be able to know why paralinguistic features are important in presentations?'

Try to relate the topic to its utility in the current times.

For example, 'Many times you must have given presentations but did not realize why you are not able to create an impact. If you think hard, you will find that even when you have focused on the content of the presentation, you have missed on the vocal aspects of your speech.'

Connect the lecture with the previously delivered contents.

For example, 'You must be wondering if you had taken care of non-verbal cues appropriately, or your presentations will be successful; however, this is not the case. Paralinguistic features are equally important and they cannot be ignored.'

Explain to the students about the applicability of the lecture material.

For instance, 'Now that you know the various paralinguistic features, utilize them completely in your formal presentation as they enhance the delivery. Use of right pronunciation creates an impression on the listeners, which is one of the paralinguistic features.'

Let the students know acronyms and difficult concepts.

For example, 'You should know the etymology of paralanguage, what it constitutes along with the non-verbal cues. Stress, volume, articulation, modulation, pronunciation, pauses, etc. are all part of paralanguage.'

The body of the lecture is equally important as it occupies a major time of the lecture. There should be enough flexibility in the lecture to accommodate students' questions and comments. Map the key points to be dealt with and then plan the coverage as per the time available. Introduce a concept and then elaborate on it with an appropriate example. Do not stretch it beyond a point that students lose interest.

Organize the material in logical order as discussed in formal presentations. The organizational schemes depend entirely on the type of topic selected. It can be chronological, categorical, cause and effect, or problem-solution. In between, whenever the students' attention diverts, engage them by telling jokes or related anecdotes. Keep summarizing the topics in between so the continuity is maintained. Keep a tap on the non-verbal cues to get an idea about the students' understanding.

The closing of the lecture should be as effective as the beginning. Answer some of the questions which were raised in the beginning and then briefly summarize the lecture. You can also ask one of the students to provide a gist of the lecture. Before you close tell them how this lecture concept can be applied in real life. You can state what you will deliver in the following lecture.

SUMMARY

Professional presentations and speeches enable us to inform, persuade, or entertain our audience, and thus form an integral part of our academic or professional career. Hence, we need to understand the fundamentals of such forms of communication and aim for their effectiveness.

Planning and preparation, structuring, delivery, use of language, body language, voice, visual aids, and rehearsals are the key drivers for the success of a presentation. While planning, we must be aware of the occasion, audience, and purpose for the presentation. Thereafter, we need to work on the thesis statement, which is the central idea of the presentation. After this we start collecting appropriate main and supporting material to prepare the presentation.

Any communication consists of an introduction, a main body, and a conclusion. Likewise, when outlining presentations, we must ensure that these three elements are in place, and are performing their functions effectively.

Once, the presentation material is in place, we have to understand the nuances of effective delivery, which includes the mode and manner of presentation. These would involve paying attention to verbal, non-verbal, vocal, and visual elements during the presentation. Our aim should be to keep all these threads intact, neither too loose nor too tight. An important aspect to take care of is self-confidence while presenting publicly. We should strive to overcome stage fright and nervousness while presenting. Oral presentation is an art that requires careful planning, preparation, and a great deal of practice. This tool is both valuable and relevant. With care and practice, we can achieve wonders with our oral presentations.

EXERCISES

1. How important is it to have good presentation skills?
2. What are the five important aspects to be considered while planning for your presentation?
3. Discuss the contents of an introduction to a speech.
4. What are the ways in which you can develop your presentation contents?
5. How can you overcome stage fright during a presentation?
6. Do you agree that language plays an important role in presentations? Justify your answer.
7. 'Voice quality impacts your business presentation.' Discuss this statement with suitable examples.
8. You are heading the sponsorship and marketing team of the annual technical festival 'Quark' in your college. You want to project the best image of your festival in order to get enough sponsorship for various events. You decide on a 15-minute presentation that can be captured as a video and uploaded on your website. Now prepare a full text of your presentation, which should contain a catchy introduction, organized main body, and an emphatic conclusion.
9. Prepare a set of 15 PowerPoint slides for professional presentations on each of the following topics:
   (a) Recent trends in animation
   (b) Use of technology for effective communication
   (c) Significance of time management
   (d) Web advertising
10. Recall any presentation made in your class by one of your fellow classmates. Examine the non-verbal signals/cues sent by him/her. How do the speaker's gestures, facial expressions, eye contact, and posture contribute to the message? Do these non-verbal signals detract from your confidence in the speaker? Do you detect any signs of nervousness? Is there any aspect of his/her delivery that you think this speaker should work on to improve? Explain.
11. Write an outline and a catchy introduction to the following presentation topics and tabulate the elements attention getter, topic and thesis statement, credibility statement, preview, rules for Q&A session, and transition:
   (a) Sleep disorders
   (b) Special economic zones
   (c) The power of mass media
   (d) Interdisciplinary sciences
12. Think and write down an analogy to explain each of the following ideas:
   (a) A car smashed in from the front and from behind
   (b) All the campuses of the university working in consonance
   (c) A humid 40-degree in April
   (d) An aircraft gliding from the sky
   (e) Hira Singh was an extremely large man, who at 6 feet 5 inches and 260 pounds made his second-grade students look extremely small
   (f) The stock market was up one day, down the next and then up again the day after that
13. Prepare a technical expert lecture on 'Role of blockchain technology in online marketing'.

CHAPTER
8
Interviews

OBJECTIVES

You should study the chapter to know
- the importance of interviews as a psychological tool
- the various objectives, types, and modes of interviews
- the factors responsible for failure at interviews
- how to prepare for and participate in job interviews successfully
- about the various types of résumés and how to design them
- what media interviews and press conferences are and how to handle them

Introduction

An interview is a psychological and sociological instrument. It is an interaction between two or more persons for a specific purpose, in which the interviewer asks the interviewee specific questions in order to assess his/her suitability for recruitment, admission, or promotion. It can also be a meeting in which a journalist asks somebody questions to determine their opinions. It is a systematized method of contact with a person to know his/her views and is regarded as the most important method of data collection. In addition, interviewing a person gives an idea of how effectively the person can perform a particular task.

We may have to face interviews at different times in our life. If we consider an interview just as an interaction between two or more people, we may not feel nervous to face the panel members of an interview. However, the thought 'I am being observed and assessed by each member' often makes one nervous.

Although the nature of interviews may be different for different organizations, several rules are common for all. For example, for any job interview, one needs to prepare or update one's résumé, know the profile of the company, prepare answers for commonly asked questions, etc. This chapter throws light on how you can achieve success by adopting certain strategies before, during, and after an interview, especially a job interview.

Objectives of Interviews

Interviews may be conducted for various reasons. Generally, interviews are conducted to achieve some of the following objectives:

- To select a person for a specific task
- To monitor performance
- To collect information
- To exchange information
- To counsel

Types of Interviews

Depending on the objective and nature, interviews can be categorized into the following types:

- Job
- Persuasive
- Evaluation
- Conflict resolution
- Termination

- Information
- Exit
- Counselling
- Disciplinary
- Media

Each of the above types has a slightly different approach. For example, in a job interview *you* may have to convince the interviewer that you are the best person for the job, whereas in a termination interview your *employer* may have to convince you that your services have been terminated for reasons that are specific, accurate, and verifiable.

Comparing the involvement and contribution of the interviewer and the interviewee, an interview can be divided into three types: *telling, telling and listening,* and *problem solving.*

Telling   In a telling interview, the flow of communication is almost entirely one way—downwards. It is used most effectively in a directing, time-constrained situation; but it can cause hostility and defensive behaviour when the employee does not have the opportunity to participate.

Telling and listening   In a telling and listening interview, more feedback from the subordinate is allowed, but the interviewer still maintains control over the flow of communication.

Problem-solving   In a problem-solving interview the flow of communication is two-way. The bulk of communication is upwards, a genuine rapport is established, ideas are pooled, and exchange facilitated.

Job interviews

In job interviews, the employer wants to learn about the applicant's abilities and experiences, and the candidate wants to learn about the position on offer and the organization. Both the candidate and the employer hope to make a good impression and to establish rapport. In the initial round, job interviews are usually formal and structured. But later, interviews may be relatively spontaneous as the interviewer explores the candidate's responses.

Information interviews

The interviewer seeks facts that bear on a decision or contribute to basic understanding. Information flows mainly in one direction: one person asks a list of questions that must be covered and listens to the answers supplied by the other person, e.g., doctor–patient, boss–subordinate, etc.

Persuasive interviews

One person tells another about a new idea, product, or service and explains why the other should act on his/her recommendations. Persuasive interviews are often associated with, but are certainly not limited to, selling. The persuader asks about the other person's needs and shows how the product or concept is able to meet those needs. Persuasive interviewers require skill in drawing out and listening to others as well as the ability to impart suitable information, adapted to the situation and the sensitivities of the interviewee.

Exit interviews

In exit interviews, the interviewer tries to understand why the interviewee is leaving the organization or transferring to another department or division. A departing employee can often provide insight into whether the business and human resource is being handled efficiently or whether there is a considerable scope for improvement. The interviewer tends to ask all the questions while the interviewee provides answers. Encouraging the employee to focus on events and processes rather than on personal gripes will elicit more useful information for the organization.

Evaluation interviews

A supervisor periodically gives an employee feedback on his/her performance. The supervisor and the employee discuss progress towards predetermined standards or goals and evaluate areas that require improvement. They may also discuss goals for the coming year, as well as the employee's long-term aspirations and general concerns.

Counselling interviews

A supervisor talks with an employee about personal problems that are interfering with work performance. The interviewer is concerned with the welfare of both the employee and the organization. The goal is to establish the facts, convey the company's concern, and steer the person towards a source of help. Only a trained professional should offer advice on problems such as substance abuse, marital tension, and financial trouble.

Conflict-resolution interviews

In conflict-resolution interviews, two competing people or groups of people with opposing points of view, such as Smith versus Jones, day shift versus night shift, General Motors versus the United Auto Workers, explore their problems and attitudes. The goal is to bring the two parties closer together, cause adjustments in perceptions and attitudes, and create a more productive climate.

Disciplinary interviews

In disciplinary interviews, a supervisor tries to correct the behaviour of an employee who has ignored the organization's rules and regulations. The interviewer tries to get the employee to see the reason for the rules and to agree to comply. The interviewer also reviews the facts and explores the person's attitude. Because of the emotional reaction that is likely, neutral observations are more effective than critical comments.

Termination interviews

A supervisor informs an employee of the reasons for the termination of the latter's job. The interviewer tries to avoid involving the company in legal action and tries to maintain a positive relationship with the employee. To accomplish these goals, the interviewer gives reasons that are specific, accurate, and verifiable.

Media interviews

Most of us might have watched programmes such as *Walk the Talk, Meet the Entrepreneur*, etc., as well as press conferences organized by the government/businesses/industries on television. Many a time, reporters call up over the phone the head of an educational institution, an important person in the government, or the chief executive officer (CEO) of a company to ask about their success stories or their alarming anomalies. We might have watched the interview given by Mr Ratan Tata, Chairman of Tata Group, during the release of Tata Nano or the one given by Mr Shashi Tharoor, the former Minister of State for External Affairs, during the IPL Kochi Franchisee controversy. All these are media interviews, which are generally conducted to disseminate information to the public on the lifestyle and achievements of an individual/business or on the new policies introduced by the government. When there is an emergency, such as a terrorist attack, internal disturbances, etc., the media conducts interviews with the people in power and also with the experts in order to get their views, interpretations, and more information on the steps taken by the government. At times, we may give some news and the media may interview us over the phone to confirm some part of the message or to get more information on some issue. Thus, media interviews can help viewers to get quick updates on the issue.

In our professions, we may come across most of the types of interviews mentioned above. However, in this chapter, we will focus mainly on job interviews and résumés, and later on provide some tips for taking control in media interviews and press conferences. In the following section, we will discuss the various aspects of job interviews such as employer's expectations, certain critical success and failure factors, preparation, process, follow-up, and guidelines.

Job Interviews

Job interviews can be classified into four major categories as depicted in Figure 8.1.

Campus interviews

Campus interviews are the interviews conducted at the campuses of colleges. The companies inform the students well in advance through the placement department of the college that they would be visiting their campus to select students for jobs. Once the companies arrive at the campus they would deliver a presentation (known as *Pre-placement Talk*) to the interested students about themselves, the type of projects they carry out, the selection mode (aptitude test/group task/case study/technical interview/HR interview), etc., and also answer the students' queries if any. As a company has to conduct several rounds of interview in a limited time, it may be able to spend only a little time with students. Hence, these interviews will be brief and to the point.

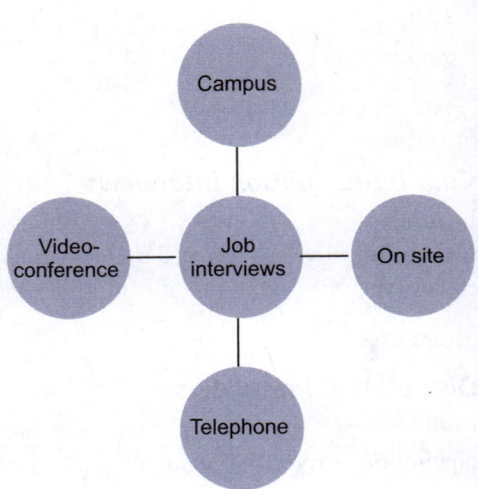

Figure 8.1 Categories of job interview

On-site interviews

On-site interviews are the interviews conducted at company premises. Many companies shortlist candidates after going through their résumés or talking to them over phone and call them to the company for a face-to-face interview (services such as www.placementindia.com, www.monsterindia.com, www.naukri.com, and www.bestjobsinindia.in give information about the job openings in various industries and forward the candidates' résumés to the companies). At times, shortlisted candidates of the campus interview may also be called for a final interview at their office. As the interviewers have more time on hand in this type of interview, they may be able to spend more time with each candidate. Hence, this type of interview may be more detailed than the campus interviews.

Telephonic interviews

Telephonic interviews are the interviews conducted by the companies over the telephone. This type of interview may be used for shortlisting the candidates by talking to them and verifying the details of the résumés that they have submitted. The interviewing company informs the candidates well before, seeks their convenience, and sends an email to confirm the date and time of the interview. Generally, a telephonic interview will be shorter than the face-to-face interview and may not be the final interview for selecting the candidates.

Videoconferencing interview

With hiring becoming increasingly global, many companies especially multinationals conduct videoconferencing interviews to select candidates for jobs. Generally when hiring for senior positions from countries across the world, companies may use this mode of interviewing. Nevertheless, it can be used for recruiting within the same country as well. If the interviewers inform the candidates about the videoconferencing facility they have arranged in the candidates' institute or campus, they can use such facility. Otherwise, the candidates need to go to a nearby agency that provides videoconferencing facilities. In either case, they will attend the interview in a professional setting as they do in an in-person interview.

Stages of Interview*

Appraisal of résumé   Some companies may shortlist candidates on the basis of projects they have completed, specific courses they have done, internships they have taken up, etc.

Tests   Companies conduct aptitude tests (written/online) comprising sections such as technical, quantitative, verbal, reasoning, psychometric, etc.

Group discussions   Most recruiters use this as the second stage after the aptitude tests. Some companies may conduct group activities as well to assess the candidates' personality, leadership skills, knowledge, communication skills, etc.

Presentations   A few companies may ask the candidates to present themselves in two or three minutes in order to shortlist.

Face-to-face interview   This may be the final round of an interview. Many companies these days give the candidate a short or long case containing a hypothetical problem in the business/industrial context, ask the candidate to present the case, and suggest a few alternative solutions in a limited time.

Videoconferencing interview   This could be a technical-cum-HR interview.

Negotiations   This stage comes after the candidate has been selected. The company makes a job offer out to the candidate after discussing the candidate's expectations about salary and other benefits.

Medical test   Once the job offer is confirmed, the candidates may be asked to go for a medical examination. This is done by the companies that offer the benefit of medical claims. Since the amount that can be claimed is substantial, the company ensures that the candidate is suffering from a major ailment. This is also done to ensure that the candidate is not suffering from any condition that would prohibit him/her from performing the duties required of him/her.

Face-to-face Interviews: Campus and On Site

A face-to-face interview for any job is a personal communication between the interview panel and the interviewee. It gives adequate scope for both the parties to know about each other and to get immediate feedback during the various stages of the interview. As this is the most commonly used form of interview, let us discuss how one can give a face-to-face interview whether it is conducted on campus or at the company site.

---

* Please note that all these stages are not mandatory. Companies may choose according to their practice.

Skills and attributes most employers look for

The following is a list of the skills and attributes most employers look for in prospective employees.

Technical skills: The candidate's subject knowledge suitable for the post he/she has applied for. For example, for a 'programmer-analyst trainee' the company may look for the candidate's ability to plan, develop, test, and document computer programs, and apply knowledge of computer techniques and systems. Interview Situation 4 in the Online Resource Centre demonstrates the importance of in-depth subject knowledge.

Analytical skills: The candidate's ability to examine and assess a situation, look at it from different perspectives, improve upon, and streamline it. For example, there may be a complex process that one may be asked to analyse critically.

Career objective: The candidate's goals and aspirations—what the candidate wants to pursue in his/her life and whether he/she is clear about it, whether the candidate's background and aptitude matches his/her career objectives.

Mental agility: The candidate's ability to quickly grasp things/mental alertness.

Communication skills: The candidate's skills in listening, speaking, reading, and writing.

Interpersonal skills: The candidate's skills to build relationships with colleagues, seniors, and subordinates, and ability to move with team members.

Flexibility/adaptability: The candidate's multitasking skills or ability to adapt himself/herself to the changing situations or environment and handle multiple concurrent projects. The candidate's ability to adapt himself/herself to culturally diverse work environment.

Management/leadership skills: The candidate's ability to plan, organize, motivate, inspire, manage, and lead the colleagues to achieve the organizational goal.

Creativity: The candidate's out-of-the-box thinking and ability to innovate. For example, if others suggest *imposing fine* for an employee for violating a rule and you suggest *appointing him* as the guardian of rules, and justify your solution, you are creative.

Positive/can-do attitude: The candidate's positive way of looking at things and people. For example, if one thinks of and projects good aspects of one's college/organization/parents/employers/job, one has a positive attitude.

Social skills: The demeanour in public or with strangers/employers or how a person conducts himself/herself with others—the way one meets and greets others, stands, sits or moves in front of others, shakes hands, reacts to opinions, eats or drinks, etc.

Honesty and integrity: The candidate's candidness and trustworthiness.

Determination/steadfastness: The candidate's ability to accomplish the given assignment despite several odds.

Professionalism: The candidate's maturity and fairness in handling business activities.

> Your attitude, not your aptitude, will determine your altitude.

Inclination for learning: The candidate's willingness to learn with an open mind.

Factors responsible for failure

There may be many reasons for failure in an interview. The following are ten common reasons for a candidate being rejected and some tips for handling them:

Arrogance It refers to overconfidence. If you exhibit overconfidence the panel may take you as arrogant. Avoid interrupting even before the interviewer completes his/her question. Do not display a facial expression that conveys, 'Why this silly question? I know the answer. It is very simple'. Do not lean on the back of your chair all the time airing arrogance.

Apathy   It refers to lack of enthusiasm or interest. Avoid *frozen* or *nil* expression on your face. A smile on your face, eye contact, confident posture, timely gestures, etc., may convey that you are enjoying the interview and that you have really come for an important occasion in your life.

Uninhibited nervousness   It refers to *explicit nervousness*. Though nervousness is common during an interview, you should try to control it rather than showcasing it through your clammy hands, dry lips, sweaty forehead, shaky hands or legs. If you are nervous, avoid keep clearing your throat/placing your hands on the table/wiping your forehead with tissue/handkerchief. Try to place your hands on the armrests of the chair and tell yourself, 'I am fine/all is well' and look confidently at the interviewers.

Equivocation   It refers to *evasion* or *beating around the bush*. When you do not know the answer to a question even after spending a few minutes on thinking, tell the interviewer politely, 'Sorry, I am unable to recall the correct point. However, can I make a guess?' Similarly, when you are unable to understand a question correctly get it clarified either by asking, 'Excuse me sir, could you please repeat the question?' or paraphrasing in your own words rather than answering incorrectly.

Lack of concentration   It refers to *inadequate focus because of poor listening, wandering mind or apathy*. The panel gets an idea of this quality when you give irrelevant answers or look elsewhere when the panel speaks to you. Remember to listen attentively exhibiting non-verbal cues and maintain eye contact. Listen completely and then answer.

Lack of crispness   It refers to *lack of precision, conciseness, and clarity in your communication*. Time is precious for everybody, and hence keep in mind that the interview panel is busy with many interviewees like you. If you are well-prepared you can be focused and clear in your answers. Avoid being verbose and sounding artificial. Preparing answers for certain anticipated questions will enable you to be concise and clear.

Lack of social skills   It refers to *using inappropriate/not following certain etiquettes during your interview*. Meet the interviewers with a firm handshake and a warm smile. If some snack or beverage is offered to you during interview, either refuse politely or take it exhibiting appropriate table manners. Use polite expressions such as, 'could you please…, sorry, pardon, excuse me, thank you, etc.' Thank the interviewers when you are offered a seat, speak softly but assertively, thank the panel before leaving the room, use positive and powerful words and be excited about your interview. Consider it as a learning experience.

Lack of firmness   It refers to *lack of determination consistency/decisiveness*. Do not keep on changing the areas of your expertise. For instance, if you have mentioned in your objective that you are specifically interested in computer programming, your answers should reflect the same. You cannot suddenly change your interest to some other unrelated area. Try to look into your skills and knowledge while preparing for the interview. Know well what you want to become in life and what your interest areas are.

Inadequate quantitative/qualitative skills   It refers to *inability to justify your answers* or *points of view*. Keep ready some examples to prove your skills or personal qualities. If you keep on speaking without adequate justification, the panel may not trust you.

Unsuitable personality   It refers to *a personality that does not match the job requirements*. For instance, if you are appearing for a marketing manager's position, you need to be an excellent communicator. If you need to handle a lot of employees you need to be cordial, patient, and a good listener. If your personality does not match the demands of the job you are applying for, the panel may not be interested in selecting you. Hence, it is important to know the job description well before you appear for the interview.

Preparing for interviews

The key to success in an interview is not one's experience, grades, extracurricular activities, but one's attitude. To rise above others with better experience, grades, or skills, a highly positive work attitude is needed. The way most employers differentiate among candidates at the entry level is by the candidates' attitude towards work. They look for those who have the 'can-do' attitude and are sincerely willing to put forth their very best effort. In the following paragraphs, we will touch upon the various aspects of preparing for interviews.

Preparation of résumé   A résumé is a written record of a candidate's education, and past and present occupation, prepared when applying for a job. This document enables the employer to judge the candidate's potential fit for the post. The résumé should be modified as per the requirements of the job and the organization. We will discuss how to prepare a job-winning résumé later in this chapter.

Personal attributes   One needs to analyse one's own hard and soft skills, strengths, weaknesses, attitude, likes, and dislikes. At least two unique strengths and weaknesses must be distinguished. For example, if you think of your *hard work and commitment* as your strengths, many candidates may have this. On the other hand, your *passion for a particular job, your strong foothold in certain areas of study and research*, etc., may differ from others. Likewise, *being very sensitive or short tempered* may be a weakness common with many candidates, whereas *taking many responsibilities at one time and struggling with the same* need not necessarily be. Hence, analyse yourself carefully and note down your important strengths and weaknesses. Think whether you look at things in a positive perspective or negative perspective. If you have a negative attitude, try to change yourself. Knowing yourself or introspecting your qualities and skills is a very important step in the preparation of your job interviews. Interview Situation 8 in the Online Resource Centre demonstrates a confident and convincing answer to the common interview question of strengths and weaknesses.

Mock interview   A mock interview is more than just a chance to work out the interview jitters. It is an opportunity to practise interviewing technique and answers live. It is also a chance to hear constructive feedback from someone who can guide towards improving the style and presentation during the real interview. Just one mock interview may bring about a marked improvement in the interview skills. Ideally, the mock session should be videotaped, and thereby one can have two opinions—the mock interviewer's and one's own. Go through at least one mock interview. For maximum effectiveness, review your answers and then go through a second mock interview. This will give you confidence in your first real interview.

Knowing the prospective employer   You need to know the company that you wish to apply/have applied for. You can collect information about the following factors:

- Age of the company
- Services or products
- Competitors within the industry—both national and international
- Growth pattern
- Reputation/where it stands in the industry
- Divisions and subsidiaries
- Locations/length of time there
- Size of organization
- Sales/assets/earnings
- Provision for career growth
- Ongoing projects
- Mission, culture, and values

You can collect the information through the company website, annual report, brochures, columns/articles in newspapers and magazines, personal contacts, if any, in the company, etc. The depth of information that is collected beforehand is far greater than that provided in the pre-placement talks or at the interview.

Awareness of job description   The nature of the job should be understood thoroughly. You can get an idea about the job profile from the company website. Also, you need to acquire a clear idea about the subject knowledge and skills that the job demands and also the knowledge of the type of activities you will be required to do. Such exercise will enable you to match the requirements with what you have in hand. If necessary, you can seek clarification from the person concerned at the company.

Subject fundamentals   You need to quickly go through the contents of basic courses done in the college. The job description generally gives an idea about which subjects you need to refresh in mind. Interviewers generally ask very basic questions (e.g., What is an array? What is the difference between RAM and ROM?). Knowing your main subjects well before an interview gives you immense confidence, which in turn leads to a better performance.

Examples corroborating skills   Besides testing the technical skills, the interviewers may also assess the candidate's team skills, decision-making ability, leadership skills, problem-solving skills, etc. Hence, pick up at least four or five such examples that show the above-mentioned qualities. Then prepare the narration of these examples using the STAR (situation, task, action, and result) approach. The STAR approach provides the outline for the answers. Preparing examples saves time and makes one feel more confident while answering behavioural questions. The STAR approach has been explained in the Online Resource Centre.

Appropriate dressing   Campus fashions and work fashions are two different worlds. You should be doing the talking, and not your clothes. Select conservative, good-quality clothes. They should be neat, clean, and ironed. Make sure your shoes are conservative, clean, and polished. Arrange all your documents systematically in a neat folder and carry it with you.

Questions That You can Ask the Interviewer

- Whom will I report to?
- Whom will I be working with?
- What training opportunities are there?
- What promotional prospects are there?
- When will you be making a final decision?

Questioning the employer   Interviewers expect you to come in with a working knowledge of the company as well as with a list of questions. When you have really done your homework you may not be able to think of any questions because you already have the company's history. Still, make up some questions ahead of time to ask during the interview. They can be based on the job that you are applying for or your prospects in that job.

Memorizing your résumé   Imagine the embarrassment if your interviewer asks you to elaborate on the project which you have done during your course of study, and you fumble. For every item on your résumé, try to have a paragraph's worth of information in addition to what is already said; even better, try to think of a way in which each item illustrates one of your particular strengths or weaknesses. If you are too nervous to remember everything, it is all right to hold a copy of your résumé in your hand to jog your memory.

Punctuality   The waiting room is your initial face-to-face connection point with your potential employer. Always arrive at least ten to fifteen minutes early. This will give you the time necessary to do a quick mental review before the actual interview. Have a glass of water to avoid the 'cotton mouth' syndrome. Check in with the secretary or administrator. Ask how long the interview is scheduled for, so that you have an idea of how much time you will have.

Relaxing the nerves   By the time the interview day comes closer, you should be set. Get a good night's sleep, eat well, and take a relaxing walk beforehand. And remember, it is just a job interview. If you do not get it, it is not the end of the world—take it as something better being in store for you in the future.

Knowing the possible types of interview questions   Interview questions may either be open-ended or close-ended. While open-ended questions allow one to give more information, close-ended questions restrict the responses to 'yes' and 'no'. For example:

Open-ended: How good a manager are you?

Close-ended: Are you a good manager?

There are basically six types of questions you may face during an interview:

1. Experience questions
   The main purpose of this type of question is to objectively measure the features of your background.

   > What is your C.G.P.A. (cumulative grade point average)?

   > How long were you at …?

2. Credential questions
   This type of question aims at subjectively evaluating the features of your background.

   > What did you learn in your network programming class?

   > What were your responsibilities in that position?

3. Opinion questions
   This type aims at analysing subjectively how you would respond in a series of scenarios.

   > What would you do in this situation?

   > What are your strengths and weaknesses?

4. Questions requiring innovative answers
   These questions are asked to find out if you are capable of an original thought:

   > Can you sell this pen to me in one minute?

   > What kind of animal would you like to be?

5. Behavioural questions
   The purpose of this type of question is to anticipate future responses based upon your past behaviour.

   > Can you give me a specific example of how you did that?

   > What were the steps you followed to accomplish that task?

Behavioural questions are gaining greater acceptance by the trained interviewers because past performance is the most reliable indicator of future results.

6. Tough questions
   Good interviewers often ask difficult questions to establish the weaknesses as well as the strengths of each candidate. They want to find out how you stand out from the other candidates who possess almost the same skills as you. Look at your curriculum vitae from the interviewer's perspective.

List out the gaps, weaknesses, and any problems you can see. If you were the interviewer, what would you ask? Work out your answer to each question.

What can you do for us that someone else cannot?

What do you look for when you hire people?

The interview process

In its simplest form, an interview consists of three distinct steps:

- Establishing rapport
- Gathering information
- Closing
- Using body language effectively

Understanding and successful completion of these basic steps are critical for one to reach the next step in the process, whether that be another interview or the actual job offer.

Establishing rapport   The rapport-establishing step is where the vital first impressions are formed. Some employers may claim to be able to make a decision about a candidate in thirty seconds or less. The truth is that you set the tone for the interview through your physical appearance and initial responses. When you enter the room, look around and establish eye contact with the people there. Smile warmly and greet them. Shake hands with a firm grip, if required, and sit when invited to do so. Address the panel members as 'Sir/Madam' or use their surnames if you know correctly. Do not call them by their first names unless they insist you to do so.

Interviewers will analyse you in reference to the company culture. Further, your initial responses will greatly affect how you are perceived in the eyes of the interviewer. It is not necessarily the words you say, but how you say them. This is where your positive attitude and confidence will establish the tone for the interview.

Information gathering   At this stage, the employers will ask questions and match your answers against their expectations. Your honesty and sincerity in answering the questions should be evident; remember that interviewers are experienced and can judge whether you are speaking the truth or telling a lie. Most interviewers are keenly aware of when they are being deceived or tricked. Questions in this step will usually be probing questions that drill deep into your background, attempting to get past the interview veneer. In fact, this is the stage in which you will need to consolidate the employer's view. You will be judged on attitude, work ethics (will you really work hard or are you just looking for an easy job?), intelligence, and honesty. Interview Situation 9 in the Online Resource Centre demonstrates undesirable and desirable answers to an ethics-based question.

Closing   If your interview has been successful, there will usually be an indication of what is to come next. You may be given further company information that is reserved only for the select few. You may get a hint from the interviewer's body language. No matter what your view of the interview is up to this point, it is important to personally close the interview by establishing continuity of the process. Ensure that you understand the next step and be prepared to follow up from your side. Always pursue each interview as if it were your last.

Using body language effectively   Various aspects of body language, namely personal appearance, facial expression, posture, gesture, eye contact and personal space—all need to be used effectively during a job interview as they communicate your confidence, sincerity, enthusiasm, interest, seriousness, social skills, etc., to the interview panel.

Personal appearance   Take care of your attire, accessories, and personal hygiene. Keep at least two sets of neat, well-ironed formal attire (men: pants, shirt, tie, belt, shoes; women: pants and full sleeved top/salwar suit/sari and blouse) specially for your interview. Clip your nails. Be well groomed. Avoid gaudy colours, clunky jewellery, and excess perfume.

Facial expressions   Your face is an excellent tool to communicate your interest, sincerity and enthusiasm about your interview to the panel members. Wear a smile on your face while entering and meeting the interview panel and use appropriate expressions while answering the questions. Even if you do not like some questions, try not to show your dislike on your face. Be happy and sporting and answer the questions patiently.

Posture   Do not sit on the edge of the chair and do not lean on the chair either. Sit in a straight posture in the beginning and after some time you can change the posture. Be natural but at the same time try to control nervousness if you have any by resting your hands on the arms of the chair.

Gesture   Use small gestures (e.g., if you wish to show two fingers to tell 'I have two points' keep the fingers close to you rather than stretching close to the panel) while speaking as there will be little space between you and the panel. Exhibit suitable gestures, such as nodding head, tilting head, shaking hands, etc., at appropriate times.

Eye contact   Maintain eye contact with all the panel members right from the time you meet them till you leave the interview room. While answering a question, look first at the member who posed the question and then at other members as well. Remember that if you do not look at the panel, you may appear to be diffident. Eye contact will also help you in getting feedback from time to time about how the panel members receive your answers, thereby enabling you to change your approach.

Personal space   As you will be sitting just on the opposite side of the panel, the space between you and the panel will be very less (a table may separate you and the panel). Hence, do not bend too much or stretch your hands on the table.

(*Note:* It will be helpful if you observe the interviewers' body language when they ask questions and also when you answer them. You can understand their intention and interest in asking the question and also their reaction to your answers.)

Interview Situation 1 in the Online Resource Centre demonstrates the importance of appropriate body language in interviews.

Answering techniques

Behavioural answering technique

- Talk about how you have done rather than how you would do.
- Be prepared to use examples from your work, classes, and extracurricular activities.
- Be ready to offer not just any example, but your own example.

Compelling story technique

- Expand your answers by developing the specific examples into compelling stories with personality, flair, and interest.
- Captivate your interviewer by providing the details and nuances that bring your story to life.
- Do not, however, be tempted into lengthy monologues that will stretch the interviewer's time.

Personality matching/mirroring technique

- Take your cue from the interviewer in terms of tone and approach. For instance, if the interviewers are using minimal gestures or facial expressions, you can also follow the same approach. If they speak in low tones, you can also do so. However, do not be too casual, even if the interviewer seems to be. Watch and learn.

- Bring under control the 'too much' area (too loud, too pushy, too confident, too egoistic, too formal, or too conventional) in your own personality.

Parroting technique

- Do not assume or make a 'best guess' of what the interviewer is looking for.
- If a question is unclear to you, it is absolutely appropriate to 'parrot back' the question in your own words to make sure you have understood the correct meaning.
- Use it as a temporary stall when you do not have a ready answer. You will get some time to think and answer.

Reframing technique

- Always attempt to answer the questions as straightforwardly as possible, initially.
- Reframe the original question to illustrate an area of your background that can further enhance your overall image. For example, if you are asked who your favourite professor is, you might give a short answer about a particular professor, and then reframe the question by explaining why that professor is your favourite—'... in fact it was her inspiration that encouraged me to participate in a two-week internship over the winter break, where I combined my classroom knowledge with practical experience in the field of ...' Thus, you can use this technique to your advantage in the interview.

Abraham Lincoln technique   Abraham Lincoln, while arguing in the court, would usually argue both sides of the case to the jury. He would first take the opponent's side of the issue and then his client's side.

Answering 'Problem' Questions

If you are asked an awkward question, you should try to turn this into an advantage. Do not evade the question or lie. Answer in a straightforward manner, dealing briefly with the negative aspects and move on, giving more time to detailing the positive aspects of the situation. Compare these two situations:

A *Interviewer:* Priya, you seem to have worked for just six months at CompuSoft. Why?

Priya: They were going through a financial crunch and I was laid-off.

Interviewer: Why?

Priya: I just told you. They were ...

Interviewer: No, no, I meant why did they lay you off, and not someone else?

B *Interviewer:* Deepti, you seem to have worked for just six months at CompuSoft before leaving. Why?

Deepti: They were going through a financial crunch and laid me off. That gave me time to do a course on web design which came in most helpful in my next assignment.

Interviewer: Was that at Worldcom?

Deepti: Yes.

Interviewer: Tell me more about that.

As you can see, Priya has painted herself in a bad light, whereas Deepti has turned a negative point into a positive one by emphasizing the new skill she has acquired.

Some More Interview Tips

- Ask permission if you wish to take notes.
- Remain calm if you sense prejudice or any preconceived notion on the part of the interviewer. Keeping yourself cool will keep the situation from getting out of hand.
- Turn off your cell phone. If it goes off accidentally, apologize and turn it off.
- Respond to both verbal and non-verbal cues of the interviewer.
- Ask relevant questions. Avoid becoming familiar or indulging in unnecessary chit-chat or gossip.

- Point to your strengths instead of making excuses for shortcomings.
- First speak well of others when the interviewer asks why you are lacking in a particular area (be it grades, work experience, extracurricular activities, etc.) or how you are better than others for a particular job.
- Then establish your own strength in the specific area.

Refer to the Online Resource Centre for details on the STAR approach. The Resource also lists 50 standard interview questions as well as sample answers to some frequently asked questions.

Overcoming nervousness

The interview is your opportunity to be at your best. If you allow your nervousness to control your presentation, that may be the dominant impression you have on the interviewer, blocking out any other positive aspects you may present.

Why do we get nervous? Because of fear of the unknown. In most cases, the fear of not getting approval makes us nervous, which in turn makes it more difficult to gain that approval. Uncontrolled nervousness can destroy our ability to perform effectively in the interview.

In the box given below, a simple technique is shown that you can apply to overcome nervousness in any interviewing situation. This is known as Rowboat technique and will help you overcome your fears and successfully meet with and speak to people you have never met before in the interviewing situation.

You will find your body completely relaxed. Even if you are not nervous, it is always a good idea to use this technique when you are waiting to meet your interviewer. If you feel nervous during the interview, you can still effectively apply this technique. Simply take in a deep breath through your nose, and then contract your abdomen muscles in the 'top-to-bottom roll' discussed above, as you slowly exhale through slightly parted lips. Hold it at the bottom, take in a deep breath, and you are ready to go. If you are overcome by nervousness while answering a question, simply pause, take a deep breath, exhale and contract, and then continue.

This technique is virtually unnoticeable to anyone nearby. Make it a habit to apply this technique several times before going on stage or for an interview, whether you are feeling nervous or not. The rationale behind this technique is that the muscle contractions prevent the introduction of chemical imbalances into the body system that can cause nervousness. The deep breathing helps to dissipate any chemicals that have already been released. It forces the body to prepare physically for the upcoming task. The body begins to produce endorphins (hormones that produce feelings of excitement), which will be needed for the anticipated rowing ahead. And, this exercise will help your mind focus positively on the interview.

Few interviewers may wish to check how good you are in handling stress during your professional career. They may suddenly turn a smooth interview into a stress interview for you by asking questions or

The Rowboat Technique

The Rowboat technique is a simple contraction of the abdomen in combination with rhythmic breathing that will allow you to fully overcome your nervousness in any situation. The steps are as follows:

- Sit forward in a chair, with your arms outstretched as if you are grabbing oars in a rowboat.
- Take a deep breath.
- Slowly pull back your arms and contract the abdomen muscles just below the rib cage.

- As you continue to let out air, roll the contraction of the muscle downward, just above your pelvic region, centering on your naval.
- Keep your muscles tight until all the air has been expelled.
- Count to three (do not breathe in yet), and then inhale deeply.
- Repeat this two/three times.

passing on comments. For instance, even when you give a correct answer they may say, 'Are you sure? I feel something is wrong in your approach' or they may ask, ' I think that you have done a very poor interview. What do you have to say on this?' (in fact, one student answered to this question like this: 'I think you are good in cutting jokes! ' and the employers enjoyed that answer!). Some interviewers may pose a puzzle before you and ask you to solve or ask you to tell a joke. Do not get nervous at all in these situations. Such questions or comments are included to check your presence of mind, creativity, ability to handle stress, etc. They expect you to be clear and consistent and not carried away by emotions or performance anxiety. You can handle such questions if you keep the evaluation aspect aside and perform the interview considering it as a learning opportunity. Interview Situation 6 in the Online Resource Centre shows four different answers to the stress interview question 'On a scale of one to ten, I rate you four. How do you rate yourself?' This technique can be used in a variety of circumstances in which we need to focus our mind and body: overcoming anxiety, anger, fright, tension, nausea, etc.

Attempt the static animation interview given in the Online Resource Centre. Answer the questions keeping the preceding discussion in mind, so that you can assess the effectiveness with which you will perform during an interview.

Intelligent listening at interviews

Assume that during your interview, you are speaking enthusiastically and your interviewer looks at the ceiling of the room or looks at his/her watch. How do you feel? You may get a hint that he/she is not interested in what you are saying. However, as an interviewee, you may not have the privilege to ignore the interviewer like he/she has done. You need to be an active and intelligent listener.

We have already learnt the difference between active and passive listening. Active listening consists of two parts: analysis of and response to the message communicated. In both these activities, we need to use our intelligence, knowledge, and power of concentration. To become an intelligent listener, we need to keep our eyes, ears, and mind open. The following guidelines may help you become an intelligent listener at interviews:

- Listen to the questions with an open mind. Do not get upset when the interviewer criticizes you.
- Keep aside your personal agenda (during interview it may be to get the job) and minimize your internal distractions such as thinking about something that happened on the way, at home, etc., while listening.

- Concentrate on the main issue emphasized in the question. For example, 'Tell me a time when you were under stress and the measures you had taken to control your stress.' This question is on stress and the strategies you have taken to do away with that stress.
- Learn to read between the lines. While listening to the words, observe the interviewer's body language as well. At times what has been left unsaid may have more impact on you than what is being said. For instance, when the interviewer comments, 'Do you think your answer is right? I don't think so', you may listen to these words, but at the same time try to read him whether he is planning to trap you. If you are an intelligent listener you may understand your interviewer's intentions along with his words.
- Intelligent listeners show their interest and sincerity in listening by exhibiting body signals. When your interviewers say something, you can nod, smile, lean forward, etc., to show that you are concentrating on what is being said. Listen not only with your ears but also with your mind and body.
- Listen patiently and completely for the whole message. Some interviewees do not allow their interviewers to complete the question and interrupt bluntly, maybe because they know the answer, and hence get very excited. You should never make such mistakes. Control your urge to respond when the interviewer is still speaking. In fact, you will be appreciated and will be considered as a mature person when you react after listening completely.
- During your interview, there may be some distractions such as somebody entering the room, a knock on the door, some noise outside, etc. Do not get perturbed by these disturbances. Listen with concentration on what is asked and what is to be told.
- When you implement intelligent listening, you may be able to ask good questions to the interviewer. For instance, when he speaks more about ongoing projects in a specific area, you can pose your question on projects; when he speaks more about career growth, you can ask question relevant to that domain.
- Show your curiosity while listening.
- Manage your feelings and emotions while listening to the interviewers. Let them not overrule you, thereby showing you in poor light.
- Whenever necessary, check whether you have understood the question correctly by paraphrasing (you can start with so, what you are asking is ...) and seeking clarification.

Follow-up

There are two simple steps you can take to make a lasting impression after your interview, which greatly increase your chances of success.

1. Call the interviewers to thank them for their time. If possible, you may want to add additional information which was not discussed in the interview. This phone call should ideally take place the same day. If you are unable to reach the interviewer directly, leave a voicemail message. But it is a good idea to assess the situation before the call.
2. Immediately write the interviewers a short note-mail, thanking them for their time and restating your interest in the position. These simple gestures of a phone call and a thank-you email can make a big difference in distinguishing you from your competitors. It has taken a great deal of effort to get this far. Take the extra time to make this final impression a positive one.

Go through the PowerPoint presentation given in the Online Resource Centre to get a comprehensive overview about the types, preparation, and process of a job interview.

Telephonic Interviews

The telephone is a very useful communication tool; as a basic business instrument it has proved to be very essential. Many people make the first contact over the phone, and this first conversation can leave a lasting impression over them. A little tact and attention to what we say and how we say it, we can use the phone as an effective tool in getting and keeping cooperation, sales, and goodwill.

Tips for Face-to-face Interviews

- Be well prepared.
- Brush up your subject and general knowledge.
- Prepare, update, and memorize your résumé.
- Know yourself.
- Know about the company.
- Dress appropriately. Unless advised otherwise, wear business attire. Limit make up, perfume/aftershave, and jewellery.
- Be smart, clean, and well groomed.
- Carry a briefcase or neat folder containing all relevant papers.
- Show up ten to fifteen minutes early. In case you feel you may get delayed, call up and inform.
- When you meet your interviewer(s), shake hands confidently.
- Stay calm, do not fidget or twiddle your thumb.
- Be polite.
- Never chew gum or smoke during the interview.
- Be yourself, be honest.
- Show a real interest in the job.
- Be aware of all the answering techniques.
- Do not answer a question you did not understand; ask for clarification first.
- Speak clearly using positive words and phrases, such as 'enjoy', 'enthusiastic', 'positive attitude', 'excellence', 'striving to be my best', 'passionate', etc.
- Use the following appropriately: non-verbal eye contact (shows interest and confidence), facial expression (tells about your delight and excitement), posture (reveals confidence and power potential), gestures, space (shows your respect to the interviewer and awareness about the organizational culture).
- In the end, restate your interest in the job.
- Smile and say 'thank you'.
- Tell them how you look forward to seeing them again.
- Shake hands firmly.
- Tell them how much you enjoyed the interview.
- When you leave an interview, you should leave the building as gracefully as you entered it. You should be as cordial to people on the way out as you were while coming in. Then, as you return, take time to review the interview while it is still fresh in your mind because an interview is a learning experience to help you in future.

Tips for Videoconferencing Interviews

If the interviewers want you to attend a video-conferencing interview, they will inform you well in advance the date, time, and duration of the interview.

- Confirm the date and time of the interview through email.
- Send all the necessary documents well before the interview.
- Assume that you are attending an in-person interview.
- Wear a formal dress as you would do for an in-person interview to give a professional appearance.
- Reach the venue early so that you get settled, familiarize yourself with the equipment, and use it with ease.

- Face the camera and speak a few words to test the focus and your voice. Use the picture-in-picture feature to see how you look on the screen.
- Do not clutter your table with papers. You may keep your folder containing copies of the relevant documents you had sent to the interviewer.
- Try not to be conscious of the camera in front of you. Just concentrate on the questions and your answers.
- Listen very carefully. If there is any technical problem, inform the interviewer and then seek assistance from the agency that has provided the facilities for the interview.

Many people do not take telephonic interviews as seriously as face-to-face interviews. A telephonic interview is also an interview and not just a phone call, and hence it has to be treated with all the importance given to a face-to-face interview.

There are three types of telephonic interviews:

1. You initiate a call to the hiring manager and he/she expresses express interest in your background. The call from that point forward is an interview.

2. A company calls you based upon a previous contact. You are likely to be unprepared for the call, but it is still an interview.

3. You have a pre-set time with a company representative to speak further on phone.

Preparation

One can prepare for the first and third type of interview call mentioned above. In these cases, you will have prior information regarding the date and time and possibly the duration as well of your telephonic interview. The following points will help you be prepared for the call:

- Keep all your documents within easy reach of the phone so that you can refer to them. In this respect, you have a major advantage in a telephonic interview that does not exist in a face-to-face interview.
- Have a note pad and pen ready to take notes.
- Keep a mirror nearby. Look into that mirror consistently throughout the phone call and smile. You will improve your telephonic presence by using this simple technique. This will help you sound friendlier, more interested, and more alert. If you feel self-conscious about seeing yourself in the mirror, you can use the mirror as an occasional checkpoint. But for most of us, seeing our reflection gives us the kind of feedback necessary to make instant modification towards a more positive presence.
- Always stand up when you are talking with a potential employer on the phone. It gets your blood flowing, improves your posture, and improves your response time. It helps give an action perception to your telephone call.
- Try to match your speaking rate and pitch with that of the interviewer's. Remember to stay within your personality range, but venture towards that portion of your range which most closely matches that of your interviewer.
- Place a 'Do Not Disturb' note on your door.
- Turn off your stereo, television, and any other potential distraction.
- Warm up your voice while waiting for the call. Sing an uplifting song to yourself.
- If your phone interview is at a set time, make sure you answer nature's call first.
- Have a glass of water handy, since you will not have a chance to take a break during the call.

Guidelines

Many interviewees feel that they can perform telephonic interviews better than face-to-face interviews, as they can have the details in front of them. However, the most obvious (and often most neglected) point to remember is this: during the interview, the interviewer has only ears to judge you with, and that is something you must exploit. Here are some tips.

Take a surprise call in your stride   If you receive a call as a result of a mailed résumé or a telephonic message you left, and you are unprepared, be calm. At times even the scheduled telephonic interviews may not happen as per schedule. The company may call you ten or fifteen minutes earlier or later than the specified time. Sound positive, friendly, and in control of your thoughts and feelings; take control of the situation like this: 'Thank you for calling, Pranay. Would you wait just a moment while I close the door?' Put the phone down, take three deep breaths to slow your heart down, pull out your résumé and a scratch pad to take notes on, put a smile on your face, and pick up the phone again. Now you are in control of yourself and the situation.

Be enthusiastic   During the start of the interview, when normal pleasantries are exchanged, greet the interviewer enthusiastically. Make a conscious effort to infuse enthusiasm and pep in your voice. Allow the company representative to do most of the talking. Keep up your end of the conversation and be sure to ask a few questions of your own that will reveal you as an intelligent person and provide you the opportunity to promote your candidacy. For example, ask what immediate projects the interviewer's department is involved in. When the interviewer answers your question, you will either have a clear picture of how to sell yourself, or you will ask a follow-up question for clarification. For example: 'What specific skills and personality traits do you think are necessary for a person to succeed with those challenges?'

Beware of giving yes/no answers   Yes/no answers give no real information about your abilities. Also, try giving answers that give details about you to the interviewers so that he or she can ask you more questions. Be factual in your answers. If a situation arises where you are forced to say, 'I do not know', do so gracefully and try and cover up your shortcomings with your strengths by saying, 'I do not know, but I can study that and I am confident that given the opportunity, I can master it in a short span of time.' Do not try to hide your shortcomings, but every time you acknowledge your shortcomings, do it in a positive manner.

Speak directly into the telephone   Keep the mouthpiece about one inch from your mouth. Do not smoke or eat while talking on the phone.

Take notes   Notes taken during a telephonic interview are invaluable. Towards the end of the interview, you will get an opportunity where the interviewer will invite you to ask questions, if any. You can make good use of your notes here. If, for any reason, the interview is interrupted, jot down the topic under discussion. When he or she gets back on the line, you can helpfully recap: 'We were just discussing ...'. That will be appreciated and will set you apart from the others.

The interviewer may talk about the corporation. A little flattery goes a long way (but do not overdo it): admire the company's achievements and you are, in fact, admiring the interviewer. Likewise, if any areas of common interest arise, comment on them, and agree with the interviewer when possible—people hire people like themselves. If the interviewer does not give you the openings you need to sell yourself, be ready to salvage the situation and turn it to your advantage.

The telephonic interview comes to an end when you are asked whether you have any questions. Ask any questions that will improve your understanding of the job requirements. If you have not asked before, now is the time to establish what projects you would be working on in the first six months: 'It sounds like a very interesting opportunity, Pranay, and a situation where I could definitely make a contribution. The project you just described sounds very exciting.' Once the details are confirmed, finish with this request:

'If I need any additional information, can I get back to you?' The company representative will naturally agree. No matter how many questions you get answered in the initial conversation, there will always be something you forgot. This allows you to call again to satisfy any curiosity—it will also enable you to increase rapport. Do not take too much advantage of it, though—one well-placed phone call that contains two or three considered questions will be appreciated, four or five phone calls will not.

In a Nutshell

- Use your voice effectively to express your genuineness.
- Always greet the interviewer with enthusiasm.
- Do not answer questions with one word. Try to give details of your area of expertise. Keep notes handy. If necessary, prepare a write-up on your responsibilities and refer to it during the interview to make sure that you do not forget anything.
- Do not use any words of your native language.
- If you have not understood the question, ask the interviewer politely to repeat/elaborate.
- Avoid repeating yourself.
- Do not raise your voice during the interview.
- Exhibit appropriate non-verbal cues while listening and speaking.
- Do not interrupt the interviewer while he/she is talking.
- When talking about your project, instead of trying to sell the product or your present company's capabilities to him, explain how you went about doing it and sell your capabilities to him/her.
- Even if the interviewer appears to be asking trivial or irrelevant questions, take all of them seriously. Maybe he/she is trying to check your communication skills.
- Do not ask the interviewer any personal questions.
- At the end of the interview, always thank the interviewer for his/her time.

Sample Interview Questions

Here are some sample interview questions:

1. Tell me about yourself.
2. Why do you want to join ABC Corporation?
3. What are your strengths and weaknesses?
4. Where do you want to be 5 years from now?
5. Which course in your undergraduate studies have taught you the real-life lesson? Explain.
6. Tell me an instance when you were in a team and things did not work according to your planning. How did you manage the situation?
7. Tell me how you built credibility in your first job.
8. What is the role of ethics in any organization?
9. Tell me what lessons you learnt from the mistakes you made in the last 6 months.
10. Did you like your previous job? Why or why not?

Let us now discuss the most important aspect of preparing for interviews—résumé.

Résumés

Résumés are technical as well as marketing documents that present the candidate's past and present performance to the prospective employers so that they can assess his/her future potential. In fact, a prospective employer forms his or her first impression of the candidate from the résumé. Of course a good résumé is not sufficient to get a job; but it can help the employer to shortlist the candidates to be considered. Employers usually have more applications than they can handle, and hence, they naturally look for ways of narrowing down the candidates to a manageable number. An effective résumé will put the candidate into that shortlist.

A cover letter is an essential accompaniment to résumés. Chapter 15 discusses covering letters in detail.

Résumé, Biodata, and Curriculum Vitae

'Let's hire this lady right away. Wait! This is my Résumé.'

Although the terms résumé, biodata, and curriculum vitae (CV) are synonymously used, they differ from each other in certain aspects. In French, résumé means summary. It is usually one page long, but may extend to two pages sometimes. It includes the gist of an individual's education, past employment, and skills for the new position. The features of a résumé include the following:

- Written in points
- Objective and formal in approach
- Written in third person
- Name and address of the applicant
- Summary of educational qualification
- Employment history
- Professional affiliation
- Skill sets

A résumé is suited for any position in an organization. Personal information such as age, date of birth, marital status, nationality, and gender are generally not included in a résumé. It is suitable for almost all types of organizations. It can also be modified according to the skill sets required by a particular job. For instance, if engineering students wish to apply for the post of a software executive post, they may highlight their skill sets in software.

A biodata is a shortened form of biographical data, and has now become an obsolete term. In this format, the emphasis is on personal details such as date of birth, nationality, marital status, gender and address. The applicant's hobbies may also find a place in a biodata. These details are followed by the educational qualification, work experience, and skills for the job.

A curriculum vitae contains all the elements of a résumé but it is more detailed in terms of the academic credentials. It is generally used for a position in a research organization or when the candidate applies for a research fellowship. A CV contains a detailed account of all the papers published, papers presented at the conferences, and research projects carried out. On the other hand, a résumé may just mention the number of conferences attended/ number of papers published/a brief summary of the projects carried out. We can say that a CV is more knowledge-oriented whereas a résumé is more skill-oriented.

Résumé Design and Structure

A résumé should present a brief summary of the candidate's personal details followed by details such as career objectives, educational qualifications, professional and technical skills, and extracurricular activities and achievements. It should not be very long, as the applicant will get the opportunity to present detailed information if shortlisted for the interview.

Appearance and elements

A résumé, like every important business document, should be impeccable. Any mistakes or sloppiness here could raise doubts in an employer's mind regarding the person's capability. The purpose of the résumé is to get called for an interview. It must be well-organized so that vital information is readily accessible.

A résumé should reflect the professional image that we want to create. It should be:

- Neat and error-free with no whiteouts or hand corrections;
- Legible and well-spaced;
- Printed on good quality paper of A-4 size; and
- Reproduced clearly on a high-quality printer or copy machine.

There is lot of debate on the ideal length of a résumé. The general notion is that the more the achievement in life, the lengthier the résumé. However, the ideal length for résumé is around one page, and it should never exceed two pages. Employers are often unimpressed with longer résumés that are hard to read and can seem padded, especially when they come from people with comparatively little job experience. A long résumé may even prompt your disqualification early in the selection process. All the details can be mentioned in brief. A concise, but complete résumé saves the reader's time and hence is more effective. If the employer needs further detail, it can be provided in the second round.

While résumés can be organized in more than one way, they will almost always contain the same basic information. Résumés are not autobiographies. The purpose is to gain the opportunity for an interview, and not to give a detailed history.

Personal information

The first thing an employer needs to know is who you are and where you can be reached. So include your name, address, phone numbers, email address, and website under this heading. Make sure that the information allows an interested employer to reach you easily. If you are currently employed, this can be difficult and delicate. Career specialists recommend that you proceed 'carefully and cautiously' and set up boundaries to keep your job search out of your current employment. You may not want to list your current business phone or business email. A personal email address and home or cell phone is preferable.

One might set up a separate email account especially for seeking employment. Ensure that it sounds professional and does not sound frivolous. An email ID like volcano2000@hotmail.com may be okay for personal use, but a prospective employer might not like it. A permanent postal address should be provided, indicating how long the address will be valid (i.e., until 'June 30').

Career/professional objective

This element is optional. However, most employers agree that a statement of professional objective should be included in a résumé. While stating the objective, make it effective by being as specific as possible about the requirement or aspiration. For example,

> Entry-level position in design and development of microprocessor circuitry; eventual advancement to position as project leader or technical manager.
>
> A software sales position involving international experience in a growing company.

Education/academic preparation

While applying for a job when one is about to graduate, educational qualification and experience are the highest selling points. Employers are usually interested in learning about the candidate's academic training, especially education and training since high school, degree earned, major and minor fields of study, courses or projects done, and also the practical experience gained during graduation.

Begin with the most recent education and work backward. If the information will be helpful and if space permits, we may consider listing notable courses taken. If the grade-point average of the candidate is impressive, it should be included. Finally, note any honours earned. If the individual has received awards for other accomplishments, all achievements can be listed in a separate section entitled 'Awards and Honours'.

Work experience/professional skills

A prospective employer would always be interested in a candidate's past work experience. When describing work experience, list jobs in chronological order, with the present or last one first. Include any part-time or summer internships or projects done, even if unrelated to the career objectives. It demonstrates the person's ability to get and hold a job—an important qualification in itself.

Each entry in this heading includes the name and location of the organization where one has worked or completed an assignment, the job title/designation, the duration of work, and also a brief summary of the work.

There is no need to use complete sentences; phrases will suffice. Be sure to use very concrete language, including technical terminology, to describe the work experience. Place this section either before or following the section on education, depending on which will be most important to an employer.

Activities, achievements/special interests, aptitudes, memberships

Most employers want to know about special abilities that will make an individual a more valuable employee. These include professional courses undertaken, community service/volunteer activities, languages known (written and spoken communication), knowledge of handling special equipment, relevant hobbies, and so on. The key here is to include only information that the employer will find useful, and that casts the candidate in a favourable light. Activities can be grouped into categories such as College Activities, Community or Social Services, and Seminars and Workshops.

Mention awards or honours received. Give details regarding the nature of the award, the activity for which the award was received, date or month and year of receiving, and also the authority from whom the award was received.

If we belong to any organizations in our field, those can be listed under 'Memberships'. Be sure to include any offices or committee appointments held.

References

This section should always be the last one in a résumé. For space and privacy considerations, one may simply include the phrase 'References available upon request' and supply the names only when and if asked for, as employers rarely investigate references until the candidate is under serious consideration.

If, however, the references are impressive enough to merit listing, follow these basic guidelines. Choose only the three or four people who combine the best elements of familiarity with the work and a credible position. A reference from a celebrity who barely knows you is not as good as one from an unknown person who has worked closely with you. In any case, do get permission beforehand from the people listed as references.

Media Interviews

Media interviews are an important aspect of public relations. Such interviews can do a lot to promote business or government policies, and create awareness and acceptance of sensitive issues concerning the public. These can be in the form of print, radio, or television interviews, each mode with its distinct pros and cons. Generally, media gets prior appointment for the interview and also informs about the topic focus.

However, if they call up without any notice, you can certainly ask for time—at least an hour or so. The following guidelines will help you successfully tackle interviews to the media.

- Be clear with your message and be ready with your points.
- Stay cool and smile.
- Ignore the camera and maintain eye contact with the interviewer/reporter.
- Correct errors in questions. For instance, if the interviewer asks, 'How many courses do you offer in your three campuses?' If the number of campuses is incorrect, you may interrupt and say, 'Please note, we have FOUR campuses'.
- State the most important information first and then provide the background.
- Do not exaggerate.
- Split complex/multiple issues in a question (For example, 'Can we have your views on reservation policy, disability issues, and the women's rights bill?') and then answer.
- Avoid saying things off the record/informal.
- Be honest. If you are asked about any negative information, try to explain what you are doing to correct it rather than hiding or refusing to give the information. For example, for the question 'Why your organization was ranked low last time by XYZ survey?', you can explain the measures you are taking to come up the list rather than getting annoyed over the question or refusing to answer.
- Do not be in a hurry to respond. Pause for a few seconds after the reporter completes the question and then answer.
- Do not take the reporter's mistakes personally. The reporters may err sometimes because of some communication gap.
- Maintain professionalism throughout your interview.
- Be assertive but do not be pushy.
- If a reporter asks you a question that you may not want to answer, explain why. ('We do not have enough information right now. We will inform you later on this.') Some people even say that they do not want to answer. Remember, you have the control when you are approached by media for an interview or you yourself are organizing it.
- Avoid using business/academic jargons.
- Finish the interview before the scheduled time if reporters misbehave.
- Take care of your appearance—dress, make up, accessories, etc.
- Use your body language and voice effectively during the interview.
- Do not try to force the information. If you spend time in pushing a particular point, you will lose your time for other issues.
- Never refuse a media interview.
- Stay on track with the message, and project enthusiasm through your messages.
- Make sure to track/monitor the results and get reviews of your performance.

Press Conferences

Press conferences (also known as news conferences) are voluntary interviews or presentations given by the governments, businesses, or other organizations to various media to get their stories or information across various television channels or newspapers across the nation. There may be more than one speaker in such a conference. Sometimes only questioning occurs; sometimes there is only a statement with no questions permitted.

The party initiating the press conference decides what information to be made public through the media and informs the media about the date, time, and venue of the press conference. The party starts the conference by presenting the prepared information and then answers the questions posed by the reporters invited from various media. Press conferences serve as powerful tools for organizations to publicize themselves in a new light that has not so far been covered by the media.

Press conferences are often held by politicians, sports teams, celebrities or film studios, commercial organizations to promote products, attorneys to promote lawsuits, and almost anyone who finds benefit in the free publicity afforded by media coverage. These are also often used as a tool to clear up any public doubt on any individual's or organization's actions. Every one of us might have watched several press conferences on the television channels during the deliberations on the nuclear deal between India and the USA.

'I am happy to say that both parties agree to disagree on the issue.'

Preparation

Guidelines for an effective and efficient press conference are as follows:

- Define your goal clearly: Why do you want to invite the media? To get publicity/to inform a new decision (your new admission process, your new mergers and acquisitions, etc.) taken by your organization/to convey the growth of your organization/to persuade people to accept your point of view.
- Ensure that the information you are going to convey through press conference has not been covered by any media so far.
- Decide upon and prepare the message you wish to convey.
- Prepare all necessary background material for the conference. For instance, if your press conference is about a new policy, you should have a clear idea about the existing policy, their shortcomings, why there is a need for change, etc.
- Prepare visual aids, handouts, etc.
- Set the venue, date, and time of the press conference.
- Inform the media at least one week in advance unless it is an emergency press conference.
- As your interview will appear on television, arrange the stage, public address system, etc., and prepare a good backdrop for the stage on which you and your colleagues will be sitting and addressing the media.
- Decide who else other than media representatives should attend the press conference. You may like to invite some senior officers from your sister concerns or other organizations as guests. Inform them the venue, date, and time.
- Choose/appoint your spokespersons for the conference so as to avoid too many people talking at one time. The person should be knowledgeable and should be a good speaker.
- Appoint a moderator who will be able to control the proceedings in case the reporters deviate from the issue to achieve their personal agenda, if any.
- Rehearse at least once with your team—the spokespersons, moderator, etc.

Process

The following points are guidelines to start, handle, and end the conference effectively.

- Arrive with your team at least ten minutes before the scheduled time.

- Invite the media personnel and distribute the material—a copy of the press statement.
- Register the contact addresses and names of attendees.
- Facilitate the media to set up their equipment and to be comfortably seated.
- Start the conference on time.
- Ask the moderator to welcome the gathering, introduce the speakers, and then allow the spokesperson to give the message.
- After the message is spoken, allow the reporters to ask questions. Answers from your side should be simple, brief, and to the point.
- Control the repetition of questions. Keep the conference short and crisp and do not stretch it too long.
- Be tactful in controlling the media. Keep yourself cool and do not get emotionally upset over any question.
- End the conference by thanking the invitees and inform them that they can contact your organization for additional information, if they need.
- If some media representatives do not turn up, you may send copies of your press statement to their offices.

SUMMARY

An interview is a major form of interpersonal communication. Conducted for various purposes, this form of communication plays a significant role in the lives of students as well as professionals. Interviews are conducted for recruitment, performance appraisals, information gathering and exchange, counselling, etc.

Although there are many types of interviews, the most common type of interview faced by almost everyone is the job interview. Job interviews vary in their nature and style from industry to industry and, hence, demand different types of skills depending on the specific job vacancy, level of experience, or industry. However, there are certain universal steps toward mastering the art of giving a successful interview. It is important to learn how to manipulate the various relevant factors, such as knowledge, personality, body language, leadership, and communication skills, to be successful in these interviews.

Job interviews can be conducted face-to-face, through videoconferences, or telephonically. This chapter provides guidelines on how to tackle these various modes. It lists the various steps towards preparing for an interview, such as brushing up subject knowledge, overcoming the jitters, answering techniques, and follow up. It also lists some standard questions and sample answers to help prepare readers for interviews.

Résumés are also considered to be an important aspect of preparing for interviews. The other interview types dealt with in this chapter are media interviews and press conferences. As an important aspect of public relations, these events help create awareness and acceptance of sensitive issues concerning the public. The chapter provides guidelines and tips to tackle these interactions effectively.

EXERCISES

1. Write short notes on each of the following terms with special reference to job interviews in about 100 words.
   (a) Mental agility
   (b) Unsuitable personality
   (c) Interpersonal skills
   (d) Apathy
   (e) Behavioural questions

2. Assume that you are one of the interview panellists who are going to a renowned institution for engineering and technology in India for campus interviews. Prepare the plan for your entire

interview process including all the types of evaluation components that are required to assess the candidates thoroughly. Invent necessary details.

3. Answer the following questions briefly:
   (a) If you wish to switch jobs because you cannot work with your supervisor, how would you explain this to a prospective employer?
   (b) What can you do to create a favourable impression when you discover that an open-ended interview has turned into a stress interview?

4. Read the following statements carefully and say whether they are true or false.
   (a) Job interviews are always conducted at the office of the employer.
   (b) Job interviews vary in style from industry to industry.
   (c) Preparing your résumé is all the groundwork necessary for an interview.
   (d) An interview is one sided if the interviewer asks questions and you answer.
   (e) Interviewers only ask simple yes/no questions.
   (f) Interviewers are interested not only in your résumé but also in your appearance.
   (g) There is no difference between a résumé and a biodata.
   (h) At a campus interview find out all about the organization from the interviewer.
   (i) Listing your skills and accomplishments is important to preparing for an interview.
   (j) Researching the employer allows you to play an active role in an interview.
   (k) Dressing in bright colours for an interview is appropriate.
   (l) One of the tools used to research a potential employer is magazines/newspapers.
   (m) Simply answer yes/no to questions in an interview.
   (n) Arrive for an interview on time.
   (o) It is not important to keep eye contact during an interview.
   (p) Relate your experience to the job you are applying for.
   (q) It is rude to ask an interviewer to repeat or clarify a question.
   (r) An interview is an opportunity for you to find out more about a prospective employer.
   (s) It is sufficient to thank the interviewer once at the end of an interview.
   (t) Call the interviewer by his/her last name unless otherwise specified.
   (u) An interviewer cannot ask you what salary you would like to earn.

5. Discuss with your friends and try to work out answer for each of the following questions that may be asked in an interview:
   (a) Imagine you are dead. You have lived a long happy life. What would your obituary say?
   (b) How will you want people to remember you when you have gone? (family, friends, etc.)
   (c) What is your ideal organization?
   (d) What is the one question you do not want to ask me?
   (e) What is the salary range that you expect?
   (f) What is the advantage of being a single child?
   (g) Why should we hire you?
   (h) Are you a quick learner? Give an example.
   (i) What motivates you?
   (j) Tell any three things you expect from our Company.
   (k) What is your biggest achievement/failure in life?
   (l) Who is your role model?

6. Assume that as the President of XYZ Global Services, Mumbai, you have called for a press conference to give information regarding the new collaborative ventures you have signed with ABC Global Corporation, Sydney. You are planning to introduce your Vice President and the Head, Public Relations and Communication, of your organization and then to brief the Press about the new collaboration. Prepare the text of your introduction and briefing in about 500 words.

CHAPTER
9

Group Communication

OBJECTIVES

You should study the chapter to know

- ○ the various forms of group communication, such as discussions, meetings, conferences, seminars, symposia, and negotiations
- ○ how to use effective body language in group communication
- ○ how group discussions (GDs) are conducted as a part of organizational discussions as well as the recruitment process
- ○ how to prepare for and conduct formal meetings and conferences
- ○ the strategies for effective and successful participation in GDs, meetings, and conferences
- ○ the written forms of group communication: brochures, bulletins, and newsletters

Introduction

As against dyadic communication (i.e., one to one) and mass communication (i.e., one to many), group communication implies a many-to-many communication. Teamwork and group communication form an integral part of most organizations that embrace the concept of an open organizational climate and participative management—in factories, corporate offices, research laboratories, universities, hospitals, law offices, government agencies, etc. This is largely because of the complexity of decisions that have to be made when we are dealing with groups as opposed to individuals.

Although teams are different from groups in that the former is process-based while the latter is function-based, they share common communication processes. For example, when a company sends one of its project teams to develop software for its clients abroad, that particular team carries out a process. However, when the company nominates a group to discuss the changes it may wish to bring about in its manufacturing unit, the group considers a function. In other words, a team's objective is set, and the members are assigned specific duties to achieve a specific goal; a group on the other hand is more involved in discussing and planning the achievement of long-term goals. However, groups develop into teams when their common purposes are clearly understood by all the members and their leaders are identified.

The members of an organization may communicate in groups to achieve any of the following purposes:

- Share and exchange information and ideas.
- Collect information or feedback on any project/policy/scheme.
- Arrive at a decision on important issues.
- Solve a problem concerning the organization as a whole.
- Discuss issues involving the group itself or for the benefit of a larger audience.
- Elicit feedback upon any work undertaken or research performed.

Discussions play a crucial role in relationship building and decision-making. In the following sections, among the various forms of group communication, we will study the various aspects of organizational as well as recruitment discussions as also the skills and strategies required for successful group communication.

Forms of Group Communication

Depending on the purpose, structure, and characteristics, group communication takes various names such as discussion, meeting, conference, seminar, group discussion, symposium, convention, etc. (Tables 9.1 and 9.2). Since all these forms of group communication involve oral communication, they require effective oral communication skills that would enable the members to present and discuss their or their organization's points of view on the topic of discussion convincingly.

Table 9.1 Major forms of group communication

Name	Purpose/objective	Structure	Characteristics	Procedure
Meeting	• To convey information to a group of people • To instruct, brief, make decisions, solve problems	• Two or more persons to several hundreds • Formal physical setting	• Punctuality • Presided by a chairman • Started with an introduction Problem-centred • Discussion-oriented • Information-centred • Fair chance to everyone • May be periodic • No side conversations • No hidden agenda • High degree of formality • Ends with a conclusion	• Notice given • Agenda prepared • Minutes recorded
Seminar	• To present the results of an original research or advanced study/evaluation of ideas • To share knowledge, and viewpoints	• Small groups of experts or well-informed persons	• Academic in nature • Close interaction with lead speaker • Free discussion	• Presentation of a topic • Discussion

(Contd)

Name	Purpose/objective	Structure	Characteristics	Procedure
Group discussion	• To exchange information • To solve problems • To convince • To make decisions • To assess or judge personality traits	• Seven to ten	• No named leader • Minimal rules • Free verbal interaction • Interdependent • Interdependent • Conclusive	• Comprehension • Discussion • Conclusion
Symposium	• To discuss different aspects of a problem for an audience	• Any number	• Formal • For larger audience	• Presentation of an aspect by each participant • Audience participation
Panel discussion	• To exchange ideas through conversation and cooperative thinking	• Small number of panelists (experts)	• Moderator present • Meant for public discussion • Includes programmes on radio and TV • Less formal than meetings	• Problem/topic put across as a question • Answers given by panelists
Conference		• Any number of participants • Wider subject matter • Several sessions		• Presentations • Discussion
Convention	• To discuss matters of professional interest	• Rigorously structured • Professional gatherings of companies, associations, societies, political parties, etc.	• Formal • Issues associated with the particular profession	• Sharing of views

Table 9.2   Characteristics of major forms of group communication

Forms	Intensity of structure	Degree of formality	Extent of use of body language	Level of interaction
Meeting	3	2	2	2
Seminar	2	2	2	3
Group discussion	1	1	3	3
Symposium	2	3	2	2
Panel discussion	2	3	2	3

(Contd)

Forms	Intensity of structure	Degree of formality	Extent of use of body language	Level of interaction
Conference	3	2	2	2
Convention	2	3	2	2
Very high = 3; High = 2; Low = 1				

Use of Body Language in Group Communication

Body language plays a very significant role in group communication, which involves more interaction among the participating members than any other form of communication. Each member of the group has an opportunity to use body language for effective communication, unlike in public speaking and professional presentations, where only the speaker's body language is noticed. In a group, each person exhibits and observes the facial expressions, postures, and gestures of others in order to understand the intentions behind what is being communicated through words. However, depending on the degree of formality of the groups, the use of body language may vary. For instance, while formal meetings at workplace may not involve much use of body language, GDs, which are less formal, may entail more use of body language as an aid in effective communication. Given below are certain general guidelines for the effective use of body language in groups:

- Maintain eye contact while speaking and listening to others.
- Adopt facial expressions that show interest and enthusiasm in participation. Restrain emotional expressions during an argument or disagreement.
- While communicating in a small group, use small hand gestures in order to avoid encroaching upon the personal space of others.
- As the members sit and communicate with each other, their movement may be restricted to a certain extent in a group. However, try to bring in some change in posture even while sitting, for instance, sitting cross-legged, placing one arm on the back of the chair, leaning on the chair briefly, etc.
- While setting up the venue for group communication, there should be adequate space between the seats so as to avoid members from sitting very close to each other.
- Observe the non-verbal cues of others and try to understand the intentions behind their verbal cues.

Table 9.3 gives a general idea about certain non-verbal cues frequently used in a group communication and their meanings. Group Discussion Situation 4 in the Online Resource Centre demonstrates the use of positive body language.

The following sections discuss the major forms of group communication that are widely used in academic and business settings and their features. We start with a general description of discussions and then explain the manner in which various discussion situations can be handled.

Table 9.3 Non-verbal behaviour in group communication

Body language signal	Intended meaning/interpretation
• Tilted head	• Interest
• Drumming fingers/tapping foot	• Impatience
• Drooping shoulders	• Lack of interest
• Open palms	• Sincerity, openness/candidness
• Folded arms	• Complaint/nervousness/feel threatened/ disagreement
• Slumped posture, looking away	• Suppressed anger or irritation

(Contd)

Body language signal	Intended meaning/interpretation
• Staring	• Attention seeking/aggression
• Hand-wringing	• Acute anxiety/worry/tension/stress/need for sympathy
• Pointing fingers at you while talking	• Complaint/aggression/anger/frustration
• Nodding while listening	• Agreeing/signalling understanding
• Head rubbing	• Puzzled/stumped by question
• Frowning	• Intense listening/potential disagreement/confused/need more detail
• Fiddling with accessories/jewellery	• Discomfort/embarrassment/stress/in need of second opinion
• Looking down to the left	• Having conversation with themselves
• Narrowing eyes	• Disbelief/anger
• Raised eyebrows	• Sarcastic/disbelief/smug
• Smiling/leaning forward/eye contact	• Friendly/interested
• Hand to cheek	• Evaluating/thinking
• Head resting in hands with eyes looking down	• Boredom, not interested
• Steepled fingers	• Authoritative/feeling superior
• Hands covering mouth or face/rapid blinking/little use of eye contact	• Shy
• Leaning back in chair with hands clasped behind head	• Overconfident/arrogant/wanting to belittle you

Group Discussions

In group discussions (GDs), a particular number of people (approximately three to eight) meet face to face and, through free oral interaction, develop, share, and discuss ideas.

Group discussions are widely used in many organizations for decision-making and problem-solving. They are also used widely as a personality test for evaluating several candidates simultaneously to select personnel for positions of responsibility, especially in the service sector, and also to select students for admission to professional institutes.

The Indian defence forces were the first to tap the potential of GDs and incorporate them into their battery of tests for recruitment of officers. Since then, GDs have become popular with various recruiting bodies because of their operational ease and effectiveness in terms of both time and cost. Most reputed business schools require students to participate in a GD before moving on to the final stage of the selection procedure, the interview. In a few cases, the GD may not be an elimination round, i.e., every person who participates in the GD also gets an opportunity to appear for the interview.

The preceding section on discussions familiarized us with the various functions performed by a participant during a discussion. A GD also includes all those functions, and hence the same guidelines apply. However, when we discuss issues as a group, we need to give importance to the characteristics of group behaviour—group orientation, orderly conduct, time-sharing, involving everyone, respecting others' viewpoints, cooperation, etc. In the following sections, we will discuss some strategies that can be used to perform well in group discussions.

Speaking in Group Discussions

Here we provide guidelines that would help one to speak effectively in a GD.

- Seize the first opportunity to speak if you have a good understanding of the topic of discussion.
- Listen patiently to others and then react to their viewpoints.
- Speak clearly and audibly so that everyone hears and understands.
- Be concise in your expressions. Do not repeat ideas just for the sake of speaking something.
- Ask for clarification, if necessary.
- Facilitate contribution from others.
- Use statistics and examples to justify a view point (see GD Situation 7 in the Online Resource Centre).
- Avoid talking to only one or two persons in the group.
- Assume an impersonal tone. Treat all members as fellow participants; none in the group is either a friend or a foe.
- Be assertive without being aggressive; be humble without being submissive (see GD Situation 2 in the Online Resource Centre).
- Raise your voice (just enough to be heard) and speak out a strong point in case of a fish-market-like situation in a GD.
- Conclude objectively by briefly presenting the important points of the discussion and any decisions taken (see GD Situation 10 in the Online Resource Centre).

Ambiguity can be dispelled by properly defining the concepts. In case we use a term that is not clear to the listeners, it should be immediately explained: for example, 'let me explain what I mean by honour killing'.

The language used must be accurate. Clarity and accuracy work in tandem. When we fall short of words, accuracy and clarity may suffer. Use words that correspond as explicitly as possible to reality. The choice of words must suit the discussion situation. For example, words and sentences can be less formal in some situations and more formal in others. If the problem being discussed is profound, it may require more careful word choices, definitions, and sentence constructions than simpler subjects. Avoid technical words, obsolete expressions, colloquialisms, etc. as listeners could be unfamiliar with them.

Connotative words should be used with caution. In fact, it would be advisable to be sensitive to the distinction between denotations and connotations; else we may not only fail in our communication, but also end up with tarnished interpersonal relations. Phrases such as the following may be regarded as inconsiderate:

- You sound preposterous.
- That is ridiculous.
- We will be mocked at.

- We are too small for it.
- Let us propose something feasible.

These statements may put the listeners off, and their reactions might be affected.

The other speech skill besides effective use of language is the effective delivery of messages. Speech should be made with intelligibility. It is important to be accurate with our articulation, e.g., with the words 'effect' and 'affect'. Listeners may misunderstand if some sounds are omitted or distorted.

We must sound natural as we would in an informal conversation. Stilted and unnatural voice puts off the listeners. Ensure that facial expressions, bodily actions, and gestures match the voice. It is essential to be direct in speaking as well as in making eye contact so as to hold the attention of and communicate effectively with the listeners. We must also speak with enthusiasm and conviction in order to engross the listeners. Use of vocal variety would be effective because speaking without modulations might make us sound dull and indifferent.

Discussing Problems and Solutions

In a discussion, identifying the problem can sometimes be a little difficult. The problem can be related to a personal, social, physical, or mental aspect, or it might be organizational—technical, managerial, or business related.

The problems related to an organization are easier to analyse and solve, because in an organization the number of possibilities of the cause of a problem and the consequences of the solution are relatively simpler to analyse than the problems related to a society.

Problems must be discussed in detail so that each participant of the team understands it. When beginning a discussion on a problem encountered, try to define the problem first. Defining a problem helps to understand what the problem is and also the nature of the problem, which in turn is crucial to solving the problem. For example, if a department is not working satisfactorily, then we should try to define the problem exactly, i.e., whether the problem is with the efficiency of the department or the output of the department.

Once the problem is identified, get into the discussion about finding a solution. Talking about problems in terms of systems makes the discussion easier. What are systems? Any mechanism put in place, which takes inputs from various factors and produces outputs, is a system. The management of a company or a pipeline for manufacturing a product can be regarded as a system. Having identified the problem means we have identified the system that is causing or facing the problem. Now all that needs to be done in order to find a solution to the problem is to identify the various inputs or factors that affect the system. Discuss elaborately the factors of the system. The problem should be discussed with participants who are experienced in the field so as to identify what factors are most likely to cause the problem. Each factor can then be individually checked to determine the root cause of the problem, which can then be directly addressed.

Once the root of the problem has been identified, there may be the problem of multiple solutions. Choosing the optimal one is essential because the solution should not only fix the problem within the system but also ensure that its consequences do not cause further problems in other systems. Discuss the pros and cons of the various solutions and take inputs from each member. Taking into account the consequences of the particular action taken for solving the problem helps obtain an optimal solution.

Creating a Cordial and Cooperative Atmosphere

It is said that ideas flow free in an atmosphere conducive to the participants. By creating a friendly and cooperative environment during a GD, we may be able to derive better solutions to our problems or create more ideas as the members' contributions are maximized. However, creating such an atmosphere is the responsibility of each member of the group. In an atmosphere that promotes friendliness and cooperation, we feel positive, optimistic, confident, and assertive and hence will voluntarily contribute to the discussion. The following strategies might help us understand how to create such an atmosphere in GDs:

- Listen to others' viewpoints with an open mind and interest.
- Respect others' ideas and try to understand the speaker's perspective.
- Develop mutual trust among each other.
- Avoid being too formal with others (e.g., knowing the names of others will help in addressing them by first names rather than using Mr/Ms; using I/you/we and active voice rather than impersonal passive voice—'Reena, I am unable to get your point. Could you please give some statistics and elaborate further?').
- Adopt a friendly attitude so that others put forth their ideas freely.
- Use body language effectively to convey interest and sincerity in the discussion.
- Avoid being friendly with only one or two participants in the group.
- Be objective and unbiased in the discussion.
- Allow others to speak.
- Recognize significant contributions by others and appreciate them.

Using Persuasive Strategies

Look at the following excerpt from a GD. Are the members persuasive?

Member 1: I feel that increasing the parking space is not advisable.

Member 2: Why do you feel so?

Member 1: Increasing the parking space is not beneficial to anybody, and hence we need not go for that.

Member 3: I don't think so.

Member 2: We have a lot of difficulty in parking. Hence, increasing the space will certainly be beneficial.

Member 1: I still feel that there is not much benefit.

We find that members of this discussion are not persuasive and the discussion does not include any meaningful points. The members are not courteous and they do not use any evidence to support their views.

Our ability to make others believe in what we say is important in GDs. In other words, mastering the art of persuasion or convincing others is crucial for GDs and the following tips may be helpful in this regard:

- Do your homework and be knowledgeable about the topic of discussion.
- Show maturity while reacting to others (by exhibiting appropriate body language and a firm tone of voice; by not interrupting somebody abruptly).
- Listen carefully and then react.
- Always use evidence (statistics/examples/testimonies) to justify your views.
- Establish mutual respect.
- Win the confidence of others.
- Be considerate. We may have a strong view on some issue. However, if others give evidences against these views, be willing to listen.
- Always be friendly and respectful during GDs.

Being Polite and Firm

As already mentioned, we need to be assertive but not aggressive in a GD and by being firm and polite we satisfy this requirement. When we feel that a particular idea suggested by another member may not go well with the organization, we need to express our view firmly, but politely. There are various expressions that can be used to express politeness with firmness. See the following sample expressions:

I understand that this idea may be liked by everybody. But, I am afraid it will not gel with our organizational environment.

I'm afraid this idea may not work in our system. (when we have to tell somebody something they may not like)

Has this idea failed? I am afraid so.

I wonder if I could have one more example on this point. (when asking somebody for a clarification)

Could you repeat that please?

Well, I am not convinced that the implementation of this idea is that urgent. / I am not sure about how urgent the implementation is.

I do not agree completely. Perhaps we should have more discussion on implementing this idea.

Turn-taking Strategies

Can I interrupt?

Would you mind me completing this idea?

Will you allow me to speak?

These are some of the utterances you might have heard in a GD because the members would not have used the turn-taking strategies effectively.

In a well-managed GD, members participating actively take turns through spoken and non-verbal modes in a GD in order to express their views. On the other hand, if the discussion is dominated by only one or two members, turn-taking is not used effectively. The term 'turn-taking' refers to a speaker giving a chance to others to comment on or question the point raised by him/her during a discussion. This process is repeated by the members of a group and if it does not go smoothly, the discussion will not be useful and can end up in an unpleasant argument.

> The open and/or subconscious behaviour by which an exchange of speaking turns is accomplished is known as a turn-taking mechanism.

It is often noticed in GDs that there are members who are quick thinkers, and hence would grab a turn to speak as soon as one of the others completes speaking. There are others, generally ineffective listeners, who may not allow anyone else to speak. On the other hand, there are members who give verbal/non-verbal signals to others for taking their turn. The following three events occur in turn-taking:

Taking a turn   Being ready to speak when one finishes is taking one's turn. For instance, if you respond to some member by agreeing, disagreeing, posing a query, or by paraphrasing what was said, you are taking a turn.

Holding the turn   There may be members in a group who do not wish to give turns to others to speak. They may like to keep the turn to themselves and continue further. Hence, they may suppress their verbal and non-verbal cues that are used to tell others to take their turn. Though they think that being able to dominate the discussion earns them points, they in fact lose by coming across as bad listeners.

Yielding the turn   When we give cues to indicate that we are concluding our remarks and others can take up the discussion further, we are yielding the turn.

Assume that a group is discussing the details of a project it is going to take up during next month. Given below are the statements spoken by some members during the course of this discussion. Read them carefully to understand the turn-taking strategies explained above:

> *Speaker 1:*   I don't think we need to use the project management software for cost control of our Mobile Medics project. The software is quite expensive and instead of procuring and using that we can think of some other alternative. Do you agree? (*By asking a question this speaker gives a verbal cue to others to take their turn; he is yielding a turn. He may also use non-verbal cues such as stretching his hand to someone particularly.*)

> *Speaker 2:*   I agree with you, but we should at least study the software available for project management. I've heard about them. They are efficient, no doubt, but... (*This speaker takes the turn but later on by pausing or by uttering an incomplete statement, the speaker gives the turn to others.*)

> *Speaker 3:*   You mean to say they are expensive? I don't think so. Some of my friends in other companies are using XYZ software for managing most of their projects. They are able to manage their projects very well in terms of planning, scheduling, resource allocation, cost control, issue tracking, etc. Such software may be economical and hence we should also go for them. (*In the beginning, even though the speaker asks a question, he wants to hold the turn for himself and hence continues.*)

The turn-taking process will yield good results when members have time to think before the discussionas they may have gathered many ideas about the various issues related to the topic of discussion. At times, groups are asked to go for an on-the-spot discussion of a topic. In such cases, participants who can think fast can take turns while others may not be able to do so and hence there may not be a well-balanced participation. These strategies can be used effectively by responding to questions, by making a suggestion, initiating interaction, paraphrasing, requesting the speaker to repeat, etc.

Effective Intervention

Interventions or interferences are necessary in a GD for reasons such as correcting an error, controlling unruly behaviour, adding some detail, or asking a question for clarification. Generally, members do not like intervention during their speaking turn. So, we should interrupt somebody only when there is a valid reason and also using appropriate phrases while doing so. If we do not use polite expressions while interrupting a member, the person might get annoyed. It is always better to use expressions such as *excuse me, sorry to interrupt, may I say something, can I add something*, etc., before interrupting in order to avoid confusion and also to exhibit a decorum during a GD. The following are some sample expressions for effective intervention:

> Excuse me for interrupting you, but it is of two months' duration, not one month.
>
> May I interrupt? Let me correct the time frame. It is two months and not one month.
>
> I would like to say something, if I may. The duration is two months and not one month.
>
> Can I just say that the duration is two months and not one?
>
> Sorry to barge in… but this idea has already been discussed.
>
> Can I add here that the duration also needs to be considered along with other factors?
>
> May I ask you a question at this point?

Reaching a Decision

Most of the GDs end with a decision—either final or interim. Whatever the type of decision, it can be arrived at only when members participate actively to explore the topic, contribute significantly to the discussion, and reach a consensus or an agreement. After thoroughly discussing various points involved in a problem by analysing their pros and cons, the group finally arrives at a few solutions. After ranking the solutions by considering their feasibility or practicability in their organizational environment, the members will accept one solution. When they decide on this, it will be presented by the leader of the group to everybody. However, before presenting the solution, the speaker will summarize the main points of discussion keeping in mind the goal of the GD. At times, because of various constraints such as time, inadequate participation, preparation, etc., the group may not be able to decide upon the given issue and it may decide to meet again. In such cases, some interim or tentative decisions may be taken. For instance, if the administrative heads of a college discuss a revamping of the curriculum, they may not arrive at a final decision, but may be able to take decisions on certain courses. They may continue their discussion later.

There are two main categories of GDs, namely organizational GDs and GDs as a part of a selection process. We will discuss these two formats in the following sections. Also, go through the PowerPoint presentation on group discussions in the Online Resource Centre to understand the various issues involved in a GD.

Organizational GD

In organizations, GDs are mainly used for group decision-making. Members of the interacting group take the responsibility of explaining their ideas and arriving at a consensus. GDs can help reduce many problems inherent to traditional interactive groups. The word 'traditional' here refers to an organizational set-up wherein hierarchy is given considerable importance. In such organizations, the group decision-making process may involve groupthink. A team suffering from groupthink will place so much value on maintaining loyalty, unity, and agreement that critical thinking and open enquiry are prevented. The participants may censor themselves and pressurize other group members into agreement. In order to minimize this, the following techniques may be used:

- Brainstorming
- Nominal group technique
- Delphi technique

Brainstorming

Brainstorming is a method for generating a variety of ideas and perspectives. It is as uncritical as possible because criticism inhibits the free flow of ideas. The people involved in brainstorming should ideally come from a wide range of disciplines and have divergent social and cultural backgrounds. The more diverse the group, the more likely it is to generate unexpected insights, ideas, and connections, and hence even unthought-of and novel solutions to problems. A typical brainstorming session follows the steps given below:

- A group of six to twelve people sit around a table.
- The group leader states the problem in a clear manner so that all participants understand it.
- Members then suggest as many alternatives as they can in a given length of time and write them down on a blackboard, whiteboard, flip chart, or a piece of paper.
- No criticism is allowed, and all the alternatives are recorded for later discussion and analysis.

There are two types of brainstorming techniques:

1. Storyboarding   2. Lotus blossom

In storyboarding, participants identify major issues and brainstorm on each of them. It is often used to solve complex problems. In the lotus blossom technique, a core thought is presented and participants provide eight ideas surrounding it like the petals of a lotus blossom. Then each of these ideas becomes a core thought to be surrounded by a further set of eight ideas and so forth, until participants can no longer generate ideas or until decision-makers feel that they have a good grasp of the problem and potential creative solutions.

However, brainstorming is mainly a process for generating ideas. The other two techniques—the nominal group technique and the Delphi technique—go further by offering methods to actually arrive at a suitable solution.

Nominal Group Technique

The nominal group technique restricts discussion or interpersonal communication during the decision-making process and hence the term 'nominal group'. In other words, in this technique, although group members are all physically present as in a traditional committee meeting, they operate independently, as described below. The problem is presented and then the following steps take place:

- Members meet as a group, but before any discussion takes place, each member independently and silently writes down his/her ideas on the problem.
- Each member takes his/her, going around the table and presenting a single idea until all ideas have been presented and recorded (typically on a flip chart or chalkboard). No discussion takes place until all ideas have been recorded.
- The group now discusses the ideas for clarity and evaluates them.
- Each group member silently and independently rank-orders the ideas.

The final decision is determined by the idea with the highest aggregate ranking. The chief advantage of the nominal group technique is that it permits the group to meet formally but does not restrict independent thinking, which an interacting group might do.

Delphi Technique

The Delphi technique is a more complex and time-consuming alternative in group decision-making. It is similar to the nominal group technique except that it does not require the physical presence of the group members. In fact, this technique never allows the group members to meet face to face. The following steps characterize the Delphi technique:

1. The problem is identified and members are asked to provide potential solutions through a series of carefully designed questionnaires.
2. Each member anonymously and independently completes the first questionnaire.
3. The results of the first questionnaire are compiled at a central location, transcribed, and reproduced.
4. Each member receives a copy of the results.
5. Based on the results, another questionnaire is prepared, and the members are again asked for their solutions, as the compiled results typically trigger new solutions or cause changes in the original opinion.
6. Steps 4 and 5 are repeated as often as necessary until consensus is reached.

Like the nominal group technique, the Delphi technique insulates group members from the undue influence of others. As it does not require the physical presence of the participants, the Delphi technique can be used for decision-making in geographically scattered groups. For instance, a company having branches in Tokyo, Brussels, Paris, London, New York, Toronto, Rio de Janeiro, and Melbourne can use the technique effectively to query its managers on the best global price for one of its products. By following this technique, business enterprises can avoid the cost of bringing their executives together at a central location.

The Delphi technique also has its drawbacks. As the method is extremely time consuming, it is frequently not applicable when a speedy decision is required. Additionally, the method may not develop the rich array of alternatives that the brainstorming or the nominal group technique does. Ideas that might be inspired by face-to-face interactions may never come up.

An organization can decide upon the type of GD required mainly on the basis of availability of time, urgency of the situation, location of the group members, and the complexity of the decision to be made.

GD as Part of Selection Process

A group discussion conducted for the selection of candidates for a job or for admission to a professional institution is a well-formulated tool for judging the personality of candidates, their communication skills, knowledge, and their ability to work as a team.

The group discussion is an important aspect of the recruitment process, especially for management trainees and executive positions. Employers look for candidates who have the potential to shoulder responsibility, work in a team, and also provide leadership. Hence, the objective of a selection GD is mainly to evaluate an individual's team-playing skills. As a team leader, one would be working with people. In such a setting, an independent or isolated worker approach does not always work. We have to understand the other person's point of view while making our point and ensure that the team as a whole reaches a solution or agreement that is both feasible and acceptable to all team members. To this end, the GD is a simulated managerial setting. Most GDs for selection purposes assess individual traits, group behaviour, and leadership qualities.

In the following sections, we will study the characteristics, evaluation, and analysis of group discussions, as well as understand how one should approach such discussions.

Characteristics

Typically, in GDs conducted for recruitment, candidates are given a topic or case for discussion. Normally groups of 8–10 candidates are formed into a leaderless group and are given a specific situation to analyse and discuss within a given time limit of about 30 minutes. They may be given a case study and asked to come up with a solution or they may be given a topic and asked to discuss it meaningfully.

Depending upon the infrastructure at the venue, the group is asked to sit in a circular, rectangular, or U-shaped arrangement. The group members may either choose their seats or be asked to take the seats allotted by the selection panel. This panel, which normally comprises the technical executives and human resources executives of the company, will observe and evaluate the members of the group. The rules of the GD—time limit, panel's expectations, etc.—are explained after the initial introduction of the panel to the participants. Thereafter, the panel assigns to the group the topic or case to be discussed, and observes the discussion either directly or from behind a screen.

The panel may, at its discretion, provide the group some time for thinking over the topic or case. Thereafter, upon directions from the panel, the discussion starts and carries on till they signal the termination time. Each candidate is supposed to voice his/her opinion and offer supporting and counter arguments as required. Although the panel specifies an approximate time for the GD, it may cut short or extend the time at will.

Evaluation and Analysis

The four components generally evaluated and analysed in a GD are as follows:

- Knowledge
- Communication skills
- Group behaviour (team spirit)
- Leadership skills

Knowledge

The depth and range of knowledge as well as analytical and organizational abilities of the candidate are judged. One should be able to grasp the situation and analyse it not just at a mundane level, but with a wide perspective. For instance, assume that the participants are asked to discuss the topic 'All tests and examinations should be abolished from the university education system'. Whatever be their stand, they should discuss not only the benefits or problems at their own institutions, but also those at the national and international levels. This approach will give a wider scope for the topic to develop and become more interesting. As a member of the group, one is expected to contribute substantially to the discussion.

The originality of ideas, knowledge and initiative, and approach to the topic or case contribute to one's success in the GD. Some knowledge of the topic concerned, supported by common sense, could win us laurels. The greater our knowledge of the subject, the more interested, enthusiastic, and confident we will be, the more fluent and forceful our contribution to the discussion will be. Of course, the contribution must be relevant, rational, convincing, and, above all, interesting and appealing to the selection panel. It is futile to beat about the bush or limit oneself to a particular perspective.

Aspects of Group Behaviour in a Group Discussion

- Consistency in participation
- Keenness in listening and observing
- Time sharing and orderly conduct
- Ability to handle turbulent situations
- Ability to cut excessively exuberant participants down to size

Being well-versed with the topic or understanding the case study well and being able to conceptualize it adequately will aid in taking the initiative. In other words, if we are able to form a concept or an idea of the given topic, it establishes the comfort level required to come up with many points for discussion. A lack of ability to take up an initiative may lead to a loss of opportunity to speak. Once the topic or problem is understood, it is easier to generate ideas as well as to organize them so that it is clear to all fellow participants.

Take a look at GD Situation 1 in the Online Resource Centre to understand how to initiate a group discussion and make a good impression on the panel.

One of the key skills looked for in any profession is problem solving. In a GD, it is vital to be creative and to come up with divergent and offbeat solutions. Do not be afraid to propose novel solutions. This is a high-risk, high-return strategy. Remember that we are evaluated on how we think and not what we think. GD Situation 9 in the Online Resource Centre demonstrates this point.

Communication skills

During a GD, an individual's skills will be assessed in terms of the following:

- Active listening
- Clarity of thought and expression
- Appropriate language
- Appropriateness of body language

Active listening Listening is as important as speaking in a GD. Unless we listen to others, we may not be able to continue the discussion in the right direction. As participants are interdependent in a GD, it is extremely necessary to listen very carefully. Only then will we be able to pick up the thread of discussion and continue. It is easy for the selection panel to identify poor listeners as their discussion lacks continuity. The selection panel may also pose questions at the end of the discussion to find out whether group members have been good listeners.

> *The importance of listening*
> Listening is a neglected skill. Good listening is difficult because we are usually more eager to state our own views than to pay attention to the views of others. When there is a difference of opinion, it is important to remain bjective by doing away with prejudices.

Clarity of expression The panel's perception of a candidate's personality and his/her ability to influence and convince others depends considerably on the correct use of tone, voice, and articulation. It is not sufficient to have ideas. They have to be expressed effectively. More than words, it is the tone in which they are spoken that conveys the right message. For example, a tone that orders is abrasive and may hurt the listeners' self-esteem. Hence, it is essential to cultivate an appropriate tone. Similarly, a lively and cheerful voice with appropriate modulations is likely to be appreciated.

Clarity of speech, achieved through phonetic accuracy, is very important. In addition, fluency of speech and good delivery are also expected. Slang, jargon, and an artificial accent are to be avoided in a GD. Given below is a brief list of the aspects of oral communication that we have to be aware of and develop to our fullest potential in order to be a successful communicator:

- Tone: Quality or character of the voice expressing a particular feeling or mood
- Voice: Correct projection of voice
- Articulation: Act of speaking or expressing an idea in words
- Fluency: Speaking or writing in an easy, flowing style
- Modulation: Variations in tone or volume of voice
- Good delivery: Ideas expressed fluently in the right voice, right tone, and right articulation

Some Useful Tips for the Effective Use of Body Language

- Look keenly at the speaker and nod your head to reveal that you are listening actively.
- When you speak, make sure that you do not ignore any of your group members. Try to look at everybody.
- Avoid overt gestures.
- Avoid pointing out fingers or raising your hands while speaking.

Apt language The language used should be accurate and free of grammatical errors. However, in case of an accidental grammatical error, do not pause to correct yourself because this might give others an opportunity to seize the discussion from you. The language used should be direct, clear, and precise, with the ideas flowing in an organized manner.

Do not use long, winding sentences. Rather, try to keep the language simple and unambiguous. Do not use jargon that everyone in the group may not be familiar with, as it may lead to a lack of clarity. Moreover, extensive use of jargon, high-sounding words, or ambiguous expressions may project the individual as a show-off and will certainly not endear the person to the group members. Remember, complication does not create impact, content does! Our main objective must be to create an impact in a positive manner.

Appropriate body language The scores in a GD depend not only on verbal communication skills, but also on non-verbal skills. Our body language says a lot about us. Our gestures and mannerisms are more likely to reflect our attitude than what we say. Panelists keenly observe the body language of the candidates and give due weightage to this aspect in their assessment of candidates.

Emotions such as anger, irritation, frustration, warmth, excitement, boredom, defensiveness, and competitiveness are all conveyed through body language. The selection panel observes an individuals's appearance, frequency of eye contact, postures, gestures, and facial expressions. A candidate who appears professional (i.e., not too overbearing) is more likely to be noticed favourably by the panel. A confident posture, appropriate facial expressions, and meaningful eye contact with other members will create a positive impression.

It is important to take care of our appearance and try to be as natural as possible. In order to appear self-confident, we must practise well so as to send the appropriate non-verbal signals during a GD. GD Situation 4 in the Online Resource Centre demonstrates the use of positive body language when discussing over a matter that one feels very strongly about.

Group behaviour (team spirit)

An individual's group behaviour is reflected in his/her ability to interact with other members of the group on brief acquaintance. Emotional maturity and balance promotes good interpersonal relationships. You are expected to be more people-centric and less egocentric. For example, when someone provokes us with a personal comment, do we keep our cool or do we react with anger? Does our behaviour come across as objective, empathetic, and non-threatening? These are the traits of a good team player.

Participating in a GD involves coordination and cooperation among the various members. The selection panel notes the differences in the amount of participation by members. There may be members who participate more and members who participate less. Some members may exhibit a shift in participation. That is, initially they might participate very actively, but as the discussion proceeds, they might keep quiet, or vice versa. Further, the panelists observe how the silent members are treated, who talks to whom, who keeps the ball rolling, etc. The successful candidate is the one who shows an active interest in the proceedings by being involved throughout the GD. Our success in a GD depends on how well we play the role of an initiator, informer, illustrator, leader, coordinator, and moderator. The ideal candidate will share time with others, listen, and react to their views. Hence, a good communicator is one who opens, rather than closes gates.

The ability to analyse a problem and persuade others to see it from multiple perspectives without offending group members is an important trait of a good communicator. While appreciating others' points of view, we should be able to effectively communicate our own view without obviously contradicting others' opinions. Also, do not speak just for the sake of doing so. Try to build up an argument from the point where the last speaker left. In case of a disagreement with what the predecessor just said, by all means feel free to disagree, but do so in a modest and amicable manner. We might benefit from keeping some facts ready to justify our point. A group can be persuaded by using valid arguments and appropriate verbal and non-verbal means of communication. The ability to convince others is an important trait during a GD. GD Situations 2, 3, and 6 in the Online Resource Centre, respectively, demonstrate how to be assertive instead of being aggressive, how to disagree amicably, and how to express agreement.

Leadership skills

The success of any team depends, to a large extent, on its leader. A group cannot carry on its assigned work effectively without a leader. Although there is no appointed leader in a recruitment GD, a leader usually emerges as the discussion proceeds. The candidate who possesses both functional and coordinating abilities will emerge as the leader. Functional ability involves knowledge, mental and physical energy, emotional stability, objectivity, communication skill, integrity, and emotional intelligence.

> 'Leadership is action, not position.'
> –Donald H. McGannan

Coordinating ability involves traits such as *group adaptability* and *group motivation*. Group adaptability means the ability to adjust with other individuals in the group, and to serve as a cohesive force that binds the group into a single unit rather than a collection of people. Group motivation means the ability to motivate and influence others, to bring out the best participation from the members, and to nurture cooperation, understanding, and team spirit among the team members.

Hence, leadership means influencing the proceedings by constructive participation, rational arguments, convincing other participants of a particular point of view, building support by working with supporters in the group, logically weakening the opponent's point of view, etc.

The selection panel observes the leadership style during the GD. Some leaders are *authoritative* in nature. They attempt to impose their will or values on the other group members or try to push them into supporting their decisions. They also pass judgement on other members and their views. Further, they block action when it is not moving in the direction they desire. Some leaders are *amiable*. They consistently try to avoid conflict or unpleasant feelings from being expressed and always try to maintain a peaceful environment during the GD. There is yet another type of leader who follows a *democratic* style. Such leaders try to include everyone in the discussion and express their feelings and opinions openly and directly without judging others. When there is a conflict, they try to deal with it as a problem-solving exercise. Obviously, the selection panel would prefer the democratic leader. Leaders should also know how to handle hostility and deal with the 'bulldozers', individuals who are high on lung power and low on logic.

A leader should be able to control the occasional fish-market environment that might arise during a GD. He/she should also be able to restrain exuberant speakers if they deviate from the topic or if they try to dominate the discussion without allowing others to speak. Further, the leader should try to include everybody in the discussion. GD Situation 5 in the Online Resource Centre demonstrates desired leadership behaviour.

Approach to Topics and Case Studies

When a topic is presented for a GD, take a minute or two to think about the topic with an open mind and note down the major issues that come to mind. Do not jump to any conclusions. Instead, arrive at a stand after examining all the issues in a balanced manner. Only then begin to speak.

While speaking, outline the major issues first and then state your stand. That is, give the justification first and the stand later. If we state our stand first, chances are that participants who disagree with our stand will interrupt to contradict before we can elaborate on the reasons. In this situation, the evaluator will only get an impression of what we think and not how we think. The guidelines given below will be helpful to handle topic-based as well as case study-based GDs.

GDs based on a topic

Topic-based GDs are generally more difficult to handle than case study-based ones as there is no starting point for a candidate's thought process, particularly when the topic is unfamiliar. The panel may or may not allow time for thinking. The dynamics in the first couple of minutes are generally chaotic. Ideally, to start with, some ideas have to be generated on the topic. These ideas must then be prioritized so that the presentation is coherent. At this point, there may not be much time to fully develop the ideas.

In order to pre-empt the possibility of other participants starting off first on the same ideas that we have thought of, we have to start speaking as early as possible. Not only must we develop the idea as we speak, but also think ahead for subsequent ideas. A weakness in any of these steps will lead to poor presentation.

As a rule of thumb, we should not speak unless we have content for a speech of at least one minute. Second, listening carefully to what the other participants have to say will trigger fresh ideas. A healthy discussion can take place only when there is an exchange of ideas and these ideas are subjected to analysis. Therefore, it is not necessary to keep on generating new ideas for the entire duration. It is also important to carefully examine each word of the topic, noting it down if possible, and checking that there are no words that can have different interpretations. If some ambiguity exists, it makes sense to define the terms first. GD Situation 8 in the Online Resource Centre demonstrates how one can deal with an unfamiliar topic.

GDs based on a case study

If an individual's analytical skills are good, then case studies are easier to handle than topic-based GDs, because there is a starting point in the form of a particular situation. Cases are discussions of situations (in business or other organizations) calling for an appraisal of past action, a decision on future action, or both. Virtually every case calls for both analysis and decision-making. Logical analysis and a firm grasp of the facts are crucial. Judgement is needed to sift through available information and find the relevant facts, and so is imagination for developing an action plan.

Tips for Success in GDs

The following is a list of tips for handling a GD successfully.

- Be thorough with current issues.
- Always enter the room with a piece of paper and a pen.
- Listen to the topic carefully.
- Jot down as many ideas as possible in the first few minutes.
- Try to dissect the topic and explore the underlying causes or consequences.
- Organize the ideas before speaking.
- Speaking first is a high-risk, high-return strategy. Hence, speak first only if there is something sensible and substantial to say.

- Try to contribute meaningfully and significantly every time you speak. Do not speak just for the sake of saying something.
- Identify supporters and opponents and allow the supporters to augment your ideas.
- Keep track of time and share time fairly.
- Have an open mind and listen to others' views.
- Maintain eye contact while speaking and listening.
- Do not indulge in parallel conversations.
- Use tact and wit. If you must use humour, do so judiciously so as not to hurt others or deviate from the topic.

- Display a spirit of cooperation and an accommodative nature.
- Draw out the silent members and encourage them to speak.
- If things get chaotic, take the initiative to  restore order by providing a fresh direction to the discussion.
- Attempt to arrive at a consensus although the ultimate aim is to reach a conclusion.

Within the specified time, the group may not be able to arrive at a consensus. However, working towards consensus will reveal the individual's capability and inclination towards being a good team player. GD Situation 10 in the Online Resource Centre demonstrates how to effectively conclude a group discussion.

A framework for a case analysis is provided below to ensure that the process is as orderly as possible.

- Understand the situation from different viewpoints.
- Work out alternative courses of action.
- Explore the pros and cons of each alternative.
- Make a decision.
- Work out an implementation plan.
- Work out a contingency plan to be used in case the first implementation fails.

In real life, the success or failure of any decision cannot be forecast. What can certainly be done is to have a logical decision-making process and a practical implementation plan.

Meetings

Meetings are an important facet of corporate life. A group of people performing different functions in an organization may come together during a meeting to work on a specific project. Major projects demand that knowledge and expertise from several sources be pooled for successful implementation. With the massive organizational, technological, and social changes taking place, there is an ever greater need for people in an organization to cooperate and share knowledge.

Meetings also serve as vehicles for individual advancement and organizational achievement. They act as a showcase for managerial talent, a forum in which employees get to audition before peers and senior management. They give participants an excellent opportunity to become opinion leaders in their organization, irrespective of their position in the hierarchy.

Beyond serving the professional purposes of individuals and organizations, meetings fulfil a more fundamental human need for socializing and communicating. They help employees to bond with one another and communicate the values of the organization, letting the employees know what behaviour is allowed and what is considered unacceptable. They reflect the organizational culture.

However, meetings are often poorly conducted, with many people considering them a waste of time. Due to a lack of open

'The major problem appears to be the lack of proper direction.'

communication among members, poor chairing of meetings, and ineffective use of time, the opportunity for people to share their knowledge and expertise and interact productively is lost.

Above all, meetings, if effectively managed, can become a potentially powerful tool for transforming the organization. This section discusses the purposes of meetings, proceeds to discuss the preparatory steps for a meeting, and finally presents an assessment form for gauging the effectiveness of a business meeting.

Purposes

Every meeting is called for a purpose and it is this purpose that decides the form of the meeting. In general, the purpose of a meeting is either informational or decision-making.

An informational meeting is called so because the participants share information and possibly coordinate action. This involves individual briefings by each participant or a speech by the leader, followed by questions from the participants. Examples of meetings for informational purposes:

- To communicate important or sensitive information (e.g., to inform about the rejection of a major proposal)
- To explore new ideas and concepts (e.g., to bring in a change in the work pattern)
- To provide feedback (e.g., to share the employees' reaction to a newly introduced bonus scheme)
- To present a report (e.g., accounts for the year's board meeting)
- To gain support for an idea or project (e.g., about the new product designed by the research division)

Decision-making meetings are mainly concerned with persuasion, analysis, and problem solving. They often include a brainstorming session, which is followed by a debate on the alternatives. Examples of meetings for decision-making purposes:

> Meetings and group discussions should not fall into the trap described by the person who said, 'A committee is a group of the unwilling, chosen from the unfit, to do the unnecessary.'

- To reach a group decision (e.g., to decide promotions for employees)
- To solve a problem (e.g., emergency meetings to solve crisis)
- To reconcile a conflict (e.g., to decide strategies)
- To negotiate an agreement (e.g., to decide the course of action before the actual negotiation)
- To win acceptance for a new idea, plan, or system (e.g., to explain the benefits of a new scheme to convince others)

Meetings can serve many other purposes. For example, they can be used to build morale, confer awards or recognition, plan projects and strategies, or provide training to employees, suppliers, and customers. The common denominator in any meeting is the group. If there is a clear need for a group of people to assemble, a meeting is called for.

Preparation

Before you call a meeting, ask yourself the following questions:

- Is this meeting necessary?
- What is my objective?
- Is the timing right?
- How much will the meeting cost?

If you have decided to call a meeting, then determine the following:

- Time
- Duration
- Meeting notice and agenda
- Participants
- Venue and set-up

Time

If a meeting is initiated by a crisis or an emergency, there may not be much choice in terms of scheduling. However, whenever possible, carefully consider the timing of a meeting because it can have a significant impact on the outcome. Given below are some scheduling guidelines:

- Choose a time during which participants are at their best.
  Do not call a meeting on Monday mornings or Friday afternoons when participants are likely to have little motivation. Do not schedule a meeting for the hour immediately after lunch when most people feel lethargic.
- Start at an unusual time and end at a natural break point.
  There is no law that says a meeting must start on the hour or half hour. If you select an unusual starting time, participants are more likely to show up punctually, especially if the meeting is scheduled to end at a natural break point, such as lunch or the close of the workday.
- Allow ample time for preparation.
  Make sure that participants have sufficient time to prepare. This will vary according to the nature of the meeting. The longer or more involved the meeting, the more the time required to prepare the related materials.
- Avoid surprise meetings.
  Nobody appreciates a surprise meeting. On-the-spot meetings interrupt the flow of the workday and allow people insufficient time to prepare, resulting in most participants entering the meeting room with a negative attitude.

The Four Ws

Successful, constructive meetings take place as a result of careful preparation. When preparing for a meeting, keep these crucial aims always in mind.

- Why is the meeting being held?
- Who should be present?
- Where should the meeting be held?
- When should the meeting take place?

Duration

The appropriate length for a meeting depends on the type of meeting and on the number, complexity, and sensitivity of the agenda items. The more complex or controversial the items, the fewer that can be addressed in a given time period. If there are more topics than can be effectively covered in a single meeting, schedule a second meeting rather than trying to cram too much information into the first.

The greater the number of participants, the shorter the meeting should be. In a small meeting, a higher level of interaction is possible. In a large meeting, interaction is restricted, and it is more difficult to maintain interest. Five people working on a problem may be absorbed for hours, five hundred people listening to a lecture can become distracted within minutes. When scheduling long meetings, allow sufficient time for breaks. For all-day meetings, allow at least an hour for lunch, but preferably an hour and a half or two hours. This gives participants time to recover from post-lunch fatigue and ensures more productive afternoon sessions.

Meeting notice and agenda

It is important to inform the participants well in advance about the meeting so that they can attend and contribute significantly to the meeting. This can be done by the following:

- Circulating a printed notice
- Informing participants over telephone
- Sending an email

> An agenda is the list of individual items that ensure that the meeting achieves its broad aims.

Few traditional organizations still follow the old method of circulating a printed notice and getting the signature of all participants, but the other two methods are widely followed today. When a meeting requires only internal participants, they can be informed through internal telephone lines. However, we must ensure that somebody at the other end receives the information if the particular person is not available at that time. Circulating the information by email is the other alternative. Any notice needs to include the following details:

- Date on which the notice is sent
- Details regarding the purpose (why); date, day, time (when); venue (where) of the meeting
- List of agenda items (what business is to be transacted)
- Signature
- Recipients (who)

A written agenda, which is the list of the individual items that need to be discussed in the meeting, should also be included with the notice. It is the single best predictor of a successful meeting. The agenda keeps the meeting on course and helps to ensure that the stated objectives are accomplished. It also serves as a planning tool for participants and a control tool for the leader. With the meeting objectives clearly defined, an agenda can be developed, keeping the following guidelines in mind:

> 'To get something done, a committee should consist of no more than three men, two of whom are absent.'
> –Robert Copeland

- Limit the number of agenda items.

Focus on a few critical items—three to six is a good rule of thumb. Separate the 'need to know' from the 'nice to know' and include only the former. The remaining topics can be summarized and included with the agenda as supplementary material or held back for a future meeting. Also determine the order in which items will be discussed, arranging them logically.

- Do not dwell on the past.

If a meeting is to be productive, it must focus on actions and decisions that will affect the future. The meeting agenda should reflect this future orientation.

- Present opportunities, not problems.

Keep the agenda upbeat. Instead of focusing on problems, recast them as opportunities. Every problem represents an opportunity of some kind—to improve a process or prevent a future problem, for example. If a problem is presented in this light, there is a better chance that it will be resolved.

- Allocate ample time.

An agenda should be well-organized, but not so rigid that it inhibits creativity or discourages full participation. Allocate some extra time for questions, creative discussion, etc.

- Include sufficient detail.

An agenda item should give participants the specific information they need in order to prepare for the meeting. It should enable them to understand exactly what is to be accomplished. For example, the agenda item 'advertising budget' may receive a more positive response if put across as 'allocate the advertising budget among direct mail, print, radio, and television'. With this level of information, participants can gather the appropriate data for the meeting and prepare their arguments for or against each advertising medium.

- Include the following details in an agenda: name of the organization, department/committee; first or follow-up meeting (this helps in maintaining a record); date, time, venue; items to be discussed.
- If possible, email the agenda along with the meeting notice to participants a week before the meeting. The more the preparation required for the meeting, the earlier the agenda should be sent out.
- If the agenda has less number of items, it can be included in the meeting notice itself. Otherwise, the agenda has to be circulated separately.

Exhibit 9.1 is a sample meeting notice that includes an agenda containing five items.

EXHIBIT 9.1 Sample meeting notice and agenda

Jupiter Group of Hotels Management Committee

23 March 2015

You are invited to attend the fifth meeting of the Management Committee of Jupiter Group of Hotels to be held at 6 p.m. on Wednesday, 29 March 2015, in the Venus Hall, 58, Park Avenue, New Delhi–110 005.

Agenda
5.01 Minutes of the last meeting
5.02 Recruitment of management trainees
5.03 Opening a branch in Nagpur
5.04 Innovations in travel packages
5.05 Any other matter

(Vivek Pai)
Secretary

Identify the Right People to Attend a Meeting

- Those who have enough knowledge of the subject to make a meaningful contribution to the meeting
- Those with the power to take decisions or approve projects
- Those responsible for implementing decisions made in the meeting
- Those who will be affected by the decisions made, or their representatives

Participants

One of the primary causes of unproductive meetings is not having the right people in attendance. An invitation to a meeting should always be based on purpose rather than on politics. In addition to inviting the right people, it is important to invite the right/optimum number according to the type of meeting. See Table 9.4.

Table 9.4 Optimal meeting size

Meeting type	Maximum no. of participants
Problem solving	3–5
Decision-making	6–10
Problem identification	6–10
Informational	20–30
Review or presentation	20–30
Motivational	No limit

Venue and set-up

'Where' people meet is as crucial as 'why' and 'when'. The choice of a meeting room has a significant impact on the overall quality of the meeting. Among other things, a meeting room can enhance or inhibit productivity, encourage or discourage communication, promote or stifle creativity, and make participants feel relaxed or tense. As the meeting room plays such an important role in meeting productivity, it is important to take great care in its selection and set-up. The following aspects of a conference room need to be considered in this regard:

- Size
- Lighting
- Ventilation
- Acoustics
- Electrical outlets

- Storage space
- Chairs
- Sound system
- Distractions
- Presentation equipment

- Projection screens
- Other amenities
- Special needs

Once the venue is decided, whether it is within the company premises or outside, it must be reserved early enough. In case the selected venue is not available, locate a similar site or consider postponing the meeting. The type of meeting affects not only the choice of the meeting room but also how it is set. Given below are some suggested layouts to improve access, heighten interaction, and increase visual contact:

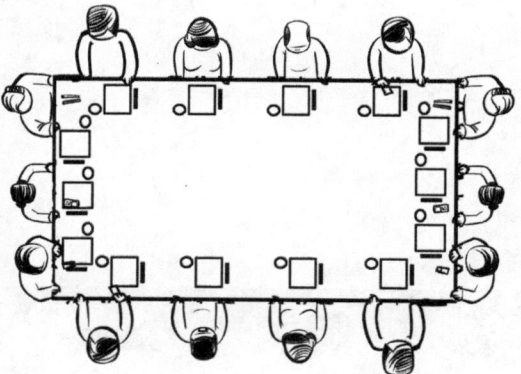

Figure 9.1 Boardroom style seating

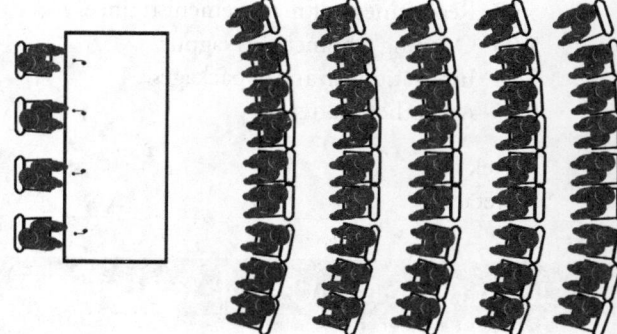

Figure 9.2 Conference style seating

Boardroom style This style is suitable for most meetings that involve sitting around a table. See Figure 9.1.

Conference or theatre style This style is suitable for annual general body meetings, or any other meetings that involve a large number of participants, presentation, and a question-and-answer session. See Figure 9.2.

A less formal layout is more appropriate for informal meetings. Consider eliminating tables altogether and sitting 'in the round'. However, please remember to provide some form of surface for placing laptops or notebooks for the

Figure 9.3 Teardrop extension to the meeting table

participants and the person taking down the minutes. A popular form of current office layout is to have a 'teardrop' (Figure 9.3) end to a standard desk around which two or three chairs can be placed, providing the meeting area.

Procedure—Conducting Effective Meetings

The quality of a meeting is largely determined by how well the players act out their roles. By playing the three primary roles of leader, facilitator, and participant, the chairperson can direct the meeting towards achieving its specified goal. Listed below are the responsibilities of the chairperson towards making a meeting effective:

- Getting the meeting off to a good start
- Managing conflict
- Encouraging participation
- Injecting humour
- Drawing silent types into the discussion
- Ending the meeting

- Joining the discussion
- Preparing and finalizing the minutes
- Managing emotions
- Evaluating the meeting
- Dealing with latecomers

Getting the meeting off to a good start   The chairperson/leader must establish control of the meeting from the start by setting the tone and stating the ground rules and objectives. A brief orientation speech by the chairperson can get rid of the needless talk that typically occurs at the start of the meeting. The chairperson must set up the meeting, state the problems and the general objectives and procedures, provide the relevant information base for the discussion, and note the boundaries and constraints of the discussion.

T–I–S Approach to Introductions

1. Introduce the Topic the speaker will address.
2. Clarify why the topic is of Interest or value to the group present at the meeting.
3. Explain why the Speaker is qualified to address this topic.

Encouraging participation   Encouraging participation is one of the leader's main obligations. Unfortunately, not all meeting leaders know how to go about this. Here are some tips for encouraging participation in a meeting:

- Do not monopolize the discussion.
- Do not exhibit verbal or non-verbal disapproval of ideas even if you disagree with them.
- Ask open-ended questions to stimulate discussion.
- Frame problems in positive terms (e.g., 'How can we achieve our goal?' rather than 'Is it possible to achieve our goal?').
- Identify the introverts and make a point of asking for their input.
- Do not let extroverts monopolize the discussion.
- Control participation by reminding people of the purpose of the meeting and by bringing discussions to a meaningful end. Sometimes, it is necessary to remind participants of the time, so that we can move on to the next topic.

- Recognize differences in individual styles.
- Change the seating arrangement (based on the previous experience).
- Rotate leadership.

Drawing silent types into the discussion   Silence can be divided into three categories:

- Silence of general agreement and no contribution
- Silence of diffidence
- Silence of hostility

Among these three categories, the first type is not of much concern; however, the other two may be a cause for disquiet. In the case of diffidence, the leader should gently bring the participant into the discussion and make it clear that the person's input is valuable. The silence of hostility signals that a person is detached from the whole proceedings and usually indicates that the person is experiencing some feeling of disrespect or humiliation. The chairman should try to get to the root of the problem and encourage participation.

Joining the discussion   In most cases, it is better for the leader to stay out of the discussion and remain in the neutral, facilitative role. If the leader wants to advocate a point, it is best for him or her to do so later in the discussion, when others have introduced the point, so as not to unduly influence the group's thinking. However, there is an exception: if the leader has relevant knowledge or experience, the group may feel the leader is obliged to share it with them. He/she cannot remain aloof from the proceedings as an impartial moderator, but can show himself/herself to be a fair-minded, even-handed participant who contributes significantly for the common good.

> 'Where all think alike, then no one thinks very much.'
> –Walter Lippmann

Managing emotions   The meeting leader should be conscious of not only the written agenda of the meeting but also the hidden and emotional agenda of the participants. The term 'hidden agenda' refers to the unspoken motives and aspirations of the group members, which could often conflict with the main goals of the group. Hidden agenda indicates what people really want as against what they say they want.

The leader should realize that meetings also have the goal of creating and maintaining positive emotions that promote working together effectively. This does not mean that the leader should ignore negative emotions. A person's anger over a budget cut, for example, may be a legitimate expression of concern. By exploring the reasons behind the anger, the leader can work with the group to address the problem productively. Acknowledging the existence of these emotions and creating trust will help people voice their concerns more openly. Apart from recognizing obvious displays of emotion, the leader must also be alert to subtle signs of emotional distress, using his own feelings as a guide.

Dealing with latecomers   The only way to ensure that a meeting starts on time is to start it on time. In addition, the leader should try to find out why certain people are habitual latecomers. He/she should make them realize their importance and their valuable contribution to the meeting. If people feel important and can anticipate achievement, they will be motivated to come on time.

Managing conflict   Some conflict is inevitable when people with varying views meet. This is not necessarily a negative feature. In fact, it is a mistake to discourage conflict because it can improve the quality of decisions, and stimulate creativity and innovation. Controversy promotes objectivity and enquiry, and enables thorough discussion and effective problem solving. The best solutions and plans often develop from contrary points of view within the meeting group. Of course, not all conflict is productive. Whether a given conflict is constructive or destructive depends on how skilfully the meeting leader manages it.

Injecting humour One effective way to dissolve tension in a meeting is to reframe a problem or conflict in a humorous light. Humour can be a valuable asset in many meeting situations. For example, humour can be used to help put people at ease, make bad news easier to accept, or introduce a sensitive subject. Personal anecdotes about funny, embarrassing, or ironic events that happened in college, on the job, etc. are a good source of humour. They often work better than jokes because the fact that they are real gives them added impact. In addition to using humour, it is important for the leader to acknowledge others' appropriate use of humour, and to discourage inappropriate jokes.

Ending the meeting When all items on the agenda have been covered, it is time to bring the meeting to a close, whether or not the scheduled time has run out. While closing the meeting, the leader should sum up what has been accomplished and what remains to be done.

> It is a good idea to close on a positive note.

He/she should recapitulate what action is expected of each participant. If another meeting on the subject is to be scheduled, the leader should set a tentative date and time. Even if the group has not reached an agreement, solved a problem, or met some other objective, the meeting leader can acknowledge any progress that has been made. This sets a positive tone for future meetings and helps to impart a feeling of control.

Tips to Manage Conflict

- Make it clear at the start of the meeting that it is fine to challenge ideas but that personal attacks will not be accepted.
- Make strategic seating arrangements. For example, the place to put a domineering individual is near the leader.
- Observe silence for a while if someone in the group becomes aggressive.
- Give an assignment (the bigger the better) to the participant who plays devil's advocate by continuously challenging everything the leader or other group members say.

Preparing and finalizing the minutes Traditionally, after the meeting is over, the secretary works on the minutes of the meeting. Minutes are a record of what happened at a meeting. They serve as a useful tool to remind the participants of the actions that need to be taken, show the process of arriving at certain decisions, and also serve as a repository of information for future reference.

> Minutes are a record of what happened at a meeting.

In general, the secretary takes notes during the meeting and then prepares the minutes to be read out and signed by the chairperson during the next meeting. However, the recent trend is towards managers who chair the meeting themselves preparing the minutes. Also, since participants most often have access to laptops, the information can be keyed in as and when points are raised and decisions are taken. After the meeting gets over, a fair copy of the minutes is prepared, with the contents arranged in an appropriate format. This is then circulated among the attendees in order to receive any comment. Once the comments are received, they are verified and the minutes are given a final shape by the Chairperson. It is these 'minutes' that are read as the first item of the agenda for the next meeting.

Minutes can be classified into two types and various styles as shown in Table 9.5. Depending upon the style, formal minutes contain the following information:

Table 9.5 Minutes—formal and informal

Formal minutes	Informal minutes
• Minimal	• Bullet point list
• Medium	• Table format
• Verbatim	• Narrative report

- *Minimal* The bare minimum of information, such as when and where the meeting was held and the decision arrived at
- *Medium* A précis of what was said at the meeting in the form of a structured report
- *Verbatim* A word-for-word account of what was said

Similarly, informal minutes contain the following details:

- A bullet point list of the decisions taken and the work progress to date
- A list in tabular format, giving the topic, the decisions made or progress to date, and subsequent action items
- A narrative report of a conversation that took place between several people

While preparing the minutes of a meeting as the chairperson/leader, it may be useful to check whether we have the following:

- A list of participants
- A copy of the agenda and all the accompanying papers
- A copy of any formal resolutions being proposed
- A copy of any presentations being given
- Two copies of the minutes of the last meeting, one for signature and the other for writing down amendments, if any
- Copies of any suggested amendments already received
- Notes of apology for missing the meeting from any participants
- A diary and a list of dates for future meetings

When putting together the minutes, we must ensure that we do the following:

- Draft the fair copy of minutes within a week from the date of the meeting.
- Decide the style of the minutes and prepare a template with all the headings.
- Keep a list of all the abbreviations and acronyms used and make a glossary of all these at the end of the document.
- Make an action point list, noting down all the things that were agreed upon, who is to do them, and any deadlines agreed upon.
- Make a note of all the items participants agreed to discuss at the next meeting and pass this on to the chair or committee secretary to include in the next agenda.
- Circulate the minutes to all the attendees for their comments; give them a deadline to send in their comments if they have any.
- Make necessary changes after verifying the comments.

Although we have mentioned the different styles for recording the minutes of a meeting, Exhibits 9.2 and 9.3 show the most commonly followed styles.

Evaluating the meeting In addition to paying attention to all the aspects discussed in the preceding sections and assigning follow-up tasks, it is important to evaluate the quality of the meeting so that future meetings will be more productive. Given below is a sample of a meeting assessment/evaluation form (Exhibit 9.4). By answering the questions given in this form, the group can measure the success of the meeting and highlight areas that need improvement.

EXHIBIT 9.2   Sample minutes in the formal style

Jupiter Group of Hotels Management Committee

Minutes of the fifth meeting held
at 6.00 p.m. on Wednesday 16 April 2019,
in Venus Hall, 58, Park Avenue, New Delhi-110 005

Present: Rakesh Chawla (President)
Navneet Arora (Vice President)
Vivek Pai (Secretary)
Lakshmi Deshpande (Joint Secretary)
Sunil Kishore (Treasurer)
Amritha Saxena (member)

Absent: Narasimha Rao (member)

No. of Minutes	Subject	Details of Discussion
5.01	**Minutes of last meeting**	Minutes from March 10, 2019 meeting were read, approved and signed by the the president.
5.02	**Recruitment of management trainees**	Navneet Arora stressed the need for recruiting five management trainees, 2 for the Delhi branch and 3 for the Bengaluru Branch. The committee discussed the matter at length. It approved the idea and decided that the secretary will place the advertisement in all leading details within a week.
5.03	**Opening a branch in Nagpur**	Lakshmi Deshpande presented the findings of survey conducted in Nagpur and opined that there is no need for a branch in Nagpur. The committee unanimously agreed with her.
5.04	**Innovations in travel packages**	Rakesh Chawla proposed that certain innovations in the travel packages for this summer will promote revenue arising out of tourism. The committee discussed the issue; a three-member committee consisting of Vivek Pai, Lakshmi Deshpande and Sunil Kishore was formed and was asked to work out on this matter and report to Rakesh Chawla latest by 25 April 2019.
5.05	**Complaints regarding room service**	Navneet Arora expressed great concern over the complaints that he received regarding room service in Chennai Branch. He read out the complaints and the committee accepted that out of the ten complaints six appeared to be genuine. Hence the committee decided to call for explanation from Sheela Tiwari, the Manager of Chennai branch

The meeting was adjourned at 8.00 p.m. The next meeting of the committee will be on May 6, 2019

EXHIBIT 9.3   Sample minutes in the informal style

- NA's proposal for recruiting five management trainees was accepted. Advertisement would be placed.
- LD reported that there is no need for a new branch in Nagpur. The committee accepted.
- To carry out RC's proposal for bringing in some innovations in summer travel packages, a three-member committee was formed.
- NA read out the complaints on Bengaluru branch room service. Committee decided to call for explanation from the Bengaluru Manager.

EXHIBIT 9.4 Sample of a meeting assessment form

Please fill in this form by providing the necessary details or encircling your preferred option:

Date: _____

Meeting Objectives _____

_____

	Not at all	1
	Not really	2
	Somewhat	3
	Pretty much	4
	Very much	5
	Not applicable	NA

1.	Was the meeting necessary?	1 2 3 4 5	NA
2.	Was the purpose clear?	1 2 3 4 5	NA
3.	Was the meeting manager prepared?	1 2 3 4 5	NA
4.	Were the participants prepared?	1 2 3 4 5	NA
5.	Were all the needed people present?	1 2 3 4 5	NA
6.	Were the participants motivated?	1 2 3 4 5	NA
7.	Was the control of the meeting adequate?	1 2 3 4 5	NA
8.	Were people involved in discussion?	1 2 3 4 5	NA
9.	Was the needed information available?	1 2 3 4 5	NA
10.	Were your resources and ideas used?	1 2 3 4 5	NA
11.	Were decisions reached on agenda issues?	1 2 3 4 5	NA
12.	Was commitment to decisions obtained?	1 2 3 4 5	NA
13.	Was the time used efficiently?	1 2 3 4 5	NA
14.	Were follow-up work assignments clear?	1 2 3 4 5	NA
15.	Were the room and facilities adequate?	1 2 3 4 5	NA
	Use the back of this form for comments and suggestions		

Conferences

A conference also is a type of business meeting. The level of formality of a meeting varies according to its size and purpose. When two or more people gather in an office to discuss a project, the meeting style will be much less formal than when 30–40 people assemble to learn about a new government ruling.

In general, the level of formality increases with the size of the group. However, the level of formality also is an indication of the objective and climate of the meetings. When a serious matter is being discussed by a small group, the meeting will be very formal. In a large meeting, called to announce record profits and thank all the employees, the atmosphere would be fairly informal.

The purpose of a conference is to confer with people having similar interests and to pool their resources, i.e., experiences and opinions. In this collaborative thinking process, a discussion generally results in a set of suggestions or recommendations on the topic/theme of the conference. The number of participants is greater and the spectrum of subject matter wider in a conference than that of a seminar or a symposium. There may be a wide range of activities such as formal lectures, exhibitions, and audio-visual presentations organized during a conference.

There has been a long tradition of organizing seminars and conferences in academic institutions. Business enterprises also utilize opportunities for learning and sharing new ideas and experiences. Often, academic institutions collaborate with business enterprises to organize conferences on various themes. These conferences serve as a forum for experts from various organizations to meet, discuss, and acquire new knowledge and insight into the theme of the conference.

> 'No grand idea was ever born in a conference, but a lot of foolish ideas have died there.'
> *–F. Scott Fitzgerald,*
> *'The Crackup'*

Significance

Conferences play a significant role in developing an analytical and questioning attitude among the participants. The participants in a conference make an attempt to define the subject of discussion and to ascertain its depth, scope, and related critical factors. They discuss all the factors and offer suggestions at the end of their deliberations. Since all the speakers are experts in their field, they suggest alternative solutions, which lead to the best decision.

Conferences have educational value in business and also in other fields where negotiation, collaboration, and collective thinking are essential. A corporate manager who participates in or leads a conference can develop his/her ability not only to define, analyse, and discuss a problem, but also to arrive at various solutions and take apt and sound decisions.

Within organizations, conferences are held to train employees. It may also serve the purpose of modifying attitudes, opinions, and feelings of participants. A conference within a business organization can provide necessary information on the policies, procedures, customs, traditions, and objectives of the organization to the conferees. New employees can share the practical knowledge and experience of seniors and superiors through conferences. The management can use conferences tactfully to correct mistakes and misunderstandings among groups although not to warn or reprimand the individual participants.

An employee who attends an in-house conference starts thinking in terms of the company as a whole. In a well-planned conference, information is collected, facts are assembled and studied, problems are defined and analysed, all the advantages and disadvantages of the alternative solutions are discussed, and then attempts are made to arrive at the best decision. This demonstration of a problem-solving approach is bound to have a positive effect on the employees who attend the conference, by boosting their morale and strengthening their confidence. Consequently, the employees tend to give more thought to the coordinated activities of the organization.

Conferences lead employees to think more effectively and more often about the objectives and challenges of the organization as a whole. For example, a weekly conference of the sales department can help salesmen to develop an effective sales strategy for the next month. In addition, the sharing of each other's views and experiences helps them to improve their performance. At some industrial conferences, experts and delegates with similar concerns are invited to discuss their common problems and prospects. Such conferences stimulate a creative flow of ideas and information, pooling of knowledge, views, and experiences. If the conference is hosted by an industrial concern, it enhances the host's prestige and goodwill. Industrial conferences can also promote public relations and direct the attention of the public as well as the government to the problems and prospects facing a particular industry.

Conference sessions can bring about a positive change in a conferee's attitude. An individual's attitude towards other persons and situations is governed by his/her past experiences, and the impressions caused by these experiences. In conference sessions, the individual listens to interpretations made by the other conferees about the same persons and situations. The different perspectives offered by the group broaden the individual's outlook and often have the positive effect of dissipating misperceptions and doubts. While this may not happen with every individual, the general tendency for the average individual is to be influenced by the group's attitude.

Planning and Preparation

Organizing a conference is a challenging task. It requires meticulous planning and systematic preparation. The following guidelines may be helpful in planning a conference:

- Decide on a broad area and then narrow it down to a specific theme after consulting other members of the organization.
- After deciding on the conference theme, identify related issues that have adequate scope for deliberations.
- Identify and prepare a list of prospective conferees.
- Decide the date.
- Identify a chief guest and get his/her consent.
- Work out the topics for various sessions and identify the chairperson, vice-chairperson, and lead speaker.
- Prepare a list of invitees.
- Estimate the budget for each session and fix the registration fee.
- Prepare a brochure containing the following details:
  - theme of the conference;
  - name of the sponsor/s, dates;
  - name of the organizer(s);

EXHIBIT 9.5   Sample covering letter accompanying a brochure

Birla Institute of Technology and Science, Pilani

April 22, 2019

_____

_____

_____

Dear Prof. / Dr/Mr/Ms _____

Sub.: International Conference details

'We are planning to hold an International Conference on 'Accessibility and Rural Development Planning' at our Institute during 25–26 May 2019. The International Forum for Rural Transport and Development, UK, supports this conference.

We are enclosing a brochure containing the details of the Conference.

Your participation would add value to the conference deliberations. Kindly confirm your participation by sending us the abstract of your paper and the duly filled in registration forms as per the deadlines given.

Thank you and regards

Yours truly

sd/_____   sd/_____

A.K. Sarkar   M.L. Dash

(Convenors)

- an introduction to the theme and to the host organization;
- related issues on which papers can be presented;
- travel accommodator;
- registration form, and deadline for the submission of abstracts of papers; and
- address (both postal as well as email, including phone number) of the conference coordinator.

Brochures are discussed in detail later in this chapter.

- Prepare the format for a covering letter to be sent along with the brochure giving all necessary details. See Exhibit 9.5 for a sample cover letter.
- Form an organizing and an advisory committee.
- Allocate work to individual task groups of the organizing committee so that responsibilities are shared.
- Plan well ahead (at least three months in advance) so as to give adequate time to the conferees to prepare.
- Meet the committee members frequently to learn the latest developments and to share any other information.
- Prepare a conference evaluation form.

Procedure

It is the duty of the convener of the conference to ensure that the required information reaches all the prospective conferees and invitees. After receiving replies from the participants, the relevant information has to be extracted and tabulated, and necessary arrangements made for the following:

- Transport
- Accommodation (ensure that the participants are well-received and have a comfortable stay)
- Reception
- Timings and duration of the sessions
- Lunch and tea breaks
- Invitations for inaugural and concluding sessions
- Venue
- Seating
- Projection facilities
- Rapporteuring (investigating a problem and reporting on it)
- Identifying volunteers
- Preparation of a souvenir containing
- Abstracts of all papers to be presented
- Production of required
- Materials to be distributed

At the inaugural session, the convener of the conference should welcome all the guests and delegates and introduce the theme of the conference, and also briefly mention the arrangements that have been made for conducting various discussion sessions.

Generally, the duration of a conference can vary from 1 to 3 days. If the number of participants is large, parallel sessions can be organized so that conferees get adequate time to present their views. During the closing session, the conference evaluation sheet may be circulated to be filled by the participants. This kind of feedback may enable the detection of any flaws and hitches that might have gone unnoticed, and also include any worthwhile suggestions for future conference planning. Once the conference is over, prepare a report containing the session details, names of participants, and summary of the discussion on each topic. This will help publish the conference proceedings later, which will incorporate all the papers that have been presented.

Symposia and Seminars

A 'symposium' is a formal meeting in which experts on a particular subject domain have discussions about certain topics in that domain. For instance, biologists and biotechnologists can organize a symposium on recombinant DNA technology and its applications in biotechnology wherein a number of experts will speak on various aspects of this topic and present their academic and research expertise. They deliver lectures and conduct laboratory sessions and provide the participants with the necessary notes. Participants may be faculty or students from academic institutions, R&D professionals from industries, etc. As against a conference, in which a lot of papers are presented, symposium presentations are restricted to few experts, but allows more in-depth discussion on the topic. The main aspect when organizing a symposium is to acquire contact with a good number of experts. Then the following steps may be carried out:

- Select a subject area and a topic (for instance, if the subject area in civil engineering is water resources, the topic can be water resource development and management).
- Divide the topic into various issues and identify one expert for each.
- Draft a proposal containing the objectives, theme, topic, issues, dates, duration, names of experts, budget, and sponsors and get the approval from the administration.
- Send proposals/requests for getting financial assistance from various agencies. The application procedures might be obtained on the respective websites.
- Prepare a brochure containing all details pertaining to the symposium.
- Prepare invitations to the experts and prospective participants. Attach the brochure.
- Identify the prospective participants and send the invitation through email or by post.
- Contact the experts and make necessary arrangements.
- Identify eminent persons for delivering inaugural and key note addresses.
- Form committees to take care of various functions such as technical sessions, reception, accommodation, hospitality, publication, publicity, etc.
- Ensure that the venue has all necessary facilities for conducting the symposium.

A 'seminar' is a meeting held for exchange of useful information by members of academia/business/ industry. It brings together groups of people from a particular sector for recurring meetings focusing each time on a particular topic. During a seminar, people present their research/new process/new technology in order to get the views of others. Hence, the participants of a seminar need to be very active and contribute significantly to the development of the research areas presented. However, as compared with the classroom lectures, seminars are less formal and involve more discussions. Students may have to present and participate in a number of seminars during their academic career (e.g., project seminars). Likewise, professionals also participate in seminars to enrich their knowledge on various topics (e.g., a seminar on 'recent trends in marketing').

Besides those organized for small groups, seminars are also organized for larger groups. In such cases, they are almost similar to conferences in terms of their preparation, planning, and procedure. Please refer to the section 'Conferences' in this chapter to understand the steps involved in organizing a seminar.

Now, we will discuss some of the documents that play a very important role in effective group communication.

These are brochures, bulletins, and newsletters. These are written forms of group communication that showcase an event or an organization in a few pages. Since the basic purpose of this form of communication is to advertise or communicate critical information to a wide variety of audience, it persuades the reader by using attractive language.

Brochure

A brochure, also referred to as pamphlet or leaflet, is used to advertise a company or organization and/or for informing the target audience about its goods and services.

Brochures are typically 8.5" × 11" or 8.5" × 14" trifold but can also have different dimensions and a number of folds. Usually a glossy paper is used, so that it does not get damaged on wide circulation. In a brochure, sentences are kept concise and readable; jargon and clichés are avoided; and repetition is totally avoided, if possible. Moreover, active voice is used to make the transmission of the message more effective.

A brochure serves multiple purposes. It usually provides explanation of the services offered by an organization. Moreover, it offers a platform for answering frequently asked questions about an organization or its products and services. A pamphlet is used by organizations like World Health Organization (WHO), who use them to disseminate health-related information and detailed instructions in a lucid manner. The flyer further provides paths and links to further information. A leaflet can educate people about a specific programme or event, though brochures are the most commonly used medium for announcement of conferences.

See Exhibit 9.6 for the general structure of a brochure. Please refer to the Online Resource Centre for more samples.

EXHIBIT 9.6   General structure of a brochure

Concept Note
Every culture has its implications on land and the way land is being used in various ways. In fact, cultural appropriation is done through modifications and manipulations of land and environment. In that sense, culture and cultural studies are closely related to ecology (interrelationships) and ecological studies but seen as different disciplines. But the term "ecoculture" brings these two disciplines in a single platform. Ecocultural studies has become a rumble in many western and eastern Universities and other

Call for Papers
Abstracts not exceeding 300 words should be copied to the filled in registration form (only on M.S. word document) and emailed to ecoethicsconference@gmail.com before **15 May 2014** for acceptance. The contributors will be informed of the acceptance of the abstract by **15 June**

Registration
Early Registration (till 31 July 2014)

Employees Category:	Rs. 3000 / $ 100 (with accommodation)
	Rs. 2000 / $ 50 (without accommodation)
Research Scholars:	Rs. 2000 / $ 50 (with accommodation)
	Rs. 1000 / $ 25 (without accommodation)

How to Reach BITS-Goa Campus
From Vasco-da-Gama Station:
The campus is 7 kilometers away from the station and can be reached by taxi and autorickshaw. Taxi fares cost around Rs.150 to Rs.200.

From Madgaon Station:
The campus is 28 kilometers away from the station and can be reached by taxi and autorickshaw. Taxi fares cost about Rs.500 to Rs.550.

From Dabolim Airport:
The campus is 5 kilometers away from the airport and can be reached by prepaid taxis which can be hired right at the exit of the airport. Taxi fares cost about Rs.200.

Important Dates

Deadline for Abstract and Registration form Submission: **15 May 2014**
Deadline for Abstract acceptance intimation: **15 June 2014**
Deadline for Early Registration: **31 July 2014**
Deadline for Late Registration: **31 August 2014**
Deadline for Paper Submission: **30 September 2014**

Log on to our website for recent updates of the conference:
http://www.bits-goa.ac.in/EcocultralEthics/index.html

Bulletin

A bulletin is a short report, especially one released through official channels to be broadcast or published. It is written in an easily legible format and usually divided in sections so that the information can be easily understood. However, a bulletin is larger in size in comparison to a brochure or a newsletter.

Bulletins are used for posting public messages, for example, to advertise items wanted or for sale. Bulletin boards facilitate addition and removal of messages for other people to read and see. In the field of journalism, bulletins containing brief description of events are published, based upon information received just before the edition goes to print. News bulletins are used for short news announcements of ongoing news stories.

One of the benefits of using bulletins is to make sure that the facts are disseminated in a quick and concise manner. They help in ensuring the basic information passed around the organization quickly, so that everyone has a clear idea of what is happening and what they are supposed to do. This saves a lot of time of the employer as well as employees.

Newsletter

A newsletter is a publication regularly distributed by an organization to disseminate information to select audience on a particular subject, or for circulation within the organization. It usually has one main subject or topic by one or more authors. Newsletters can contain information on upcoming events, conferences, etc. as well as articles on specific topics.

A newsletter is generally printed on a letter-sized paper, or published through other means such as newspapers or digests. Nowadays, e-newsletters are gaining a lot of popularity as they are environment friendly and easier to distribute. The content must be accurate, factual, brief, and impartial, and may contain technical jargon or specialized language not readily understood by a layperson. Review and proofreading is of prime importance in the preparation of a good newsletter. Newsletters do not have a word limit and the writer is free to express as much as he/she wants, though it should be relevant, to the point, and without redundancy.

Exhibit 9.7 shows snapshots from a newsletter.

EXHIBIT 9.7   Sample newsletters

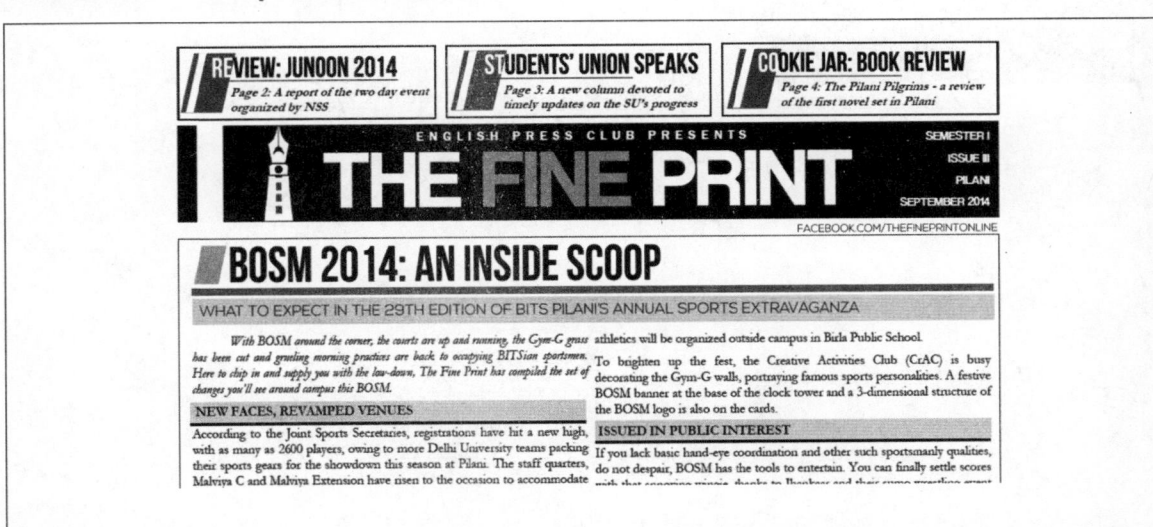

(Contd)

STUDENT AFFAIRS: THE WAY FORWARD

IN CONVERSATION WITH THE CHIEF WARDEN ON STUDENT MATTERS

We spoke to the Chief Warden, Prof. C. B. Das this week as a follow-up to the interview that appeared in the previous issue of The Fine Print. The conversation about disciplinary issues eventually digressed to other relevant topics.

AN UNEVENTFUL MUSIC NITE

Why the sudden stringent security measures at Music Nite? For one, students were admitted into the auditorium only on displaying their ID cards. Professor Das says this step was taken to prevent gate-crashers from other colleges who have been

SECURITY MEASURES IN THE AUDI

THE TIGHTENING SAFETY MEASURES AND THE FUTURE PROSPECTS

The day before Music Nite (Teacher's Day, in fact), all students woke up to a rather ominous e-mail from the institute. The mail declared that strict action would be taken on those who would arrive at the auditorium intoxicated/unable to speak the alphabet backwards (just ridiculing), and that any untoward incidents would lead to an immediate and indefinite ban on events in the auditorium. As Music Nite approached, breathalysers, not band names, became the major topic of discussion. All this was not without reason.

MUSIC NITE: A REVIEW

THE EVENT SAW A PACKED AUDI AMIDST STRICT SECURITY

One of the most awaited nights of the year, Music Club's Music Nite was all it promised. What started off with a half empty auditorium ended being one of the most attended nights of the year.

A seated audience soon moved to the front of the auditorium, where the space between the first row and the stage quickly turned into a makeshift dance floor. With people being thrown

JUNOON 2014

THE NSS INITIATIVE SPREAD OVER TWO DAYS WAS ALL-ENCOMPASSING

Junoon translates to passion, and that's what the folk at NSS have, the passion to help the physically challenged and specially abled kids to experience a decent life, and feel a sense of acceptance. They also strive to spread awareness about the conditions of such children and through *Junoon*, they try to help them in any little way they can. Aimed at providing a sense of equality and a stage for competition to the large population of specially-abled people living in our country, Junoon was organized on 6th and 7th of September.

The day began with an inauguration ceremony, followed by a free-for-all dance

SUMMARY

Group communication is essential for professional as well as personal progress as it enables us to share information with other members, make decisions, and solve problems in order to achieve our objectives. Group communication assumes various forms such as discussions, meetings, conferences, seminars, symposia, negotiations, etc., and all these forms require skills in expressing oneself, convincing others, adapting to a group environment, listening to others patiently, and leading and directing others. An understanding of these forms and also of the strategies that we can use to participate effectively in such discussions will enable us to achieve success professionally and personally. There are some written forms of group communication too such as brochures, bulletins, and newsletters, which showcase the event or the organization in brief manner.

EXERCISES

1. Based on your understanding of group communication, state whether each of the following statements is correct or incorrect. (Justify your answer in about 75 words.)
   (a) Surprise meetings are very important for the growth of an organization.
   (b) It is not necessary to strictly follow the sequence of agenda items while discussing them in a meeting.
   (c) A group discussion is a potential evaluation tool for assessing a person's attitude.
   (d) Conferences are useful in arriving at possible solutions to some problems.
   (e) Your success in a GD is decided by the number of times you speak.
   (f) You should never personalize your arguments during a GD.
   (g) Conferences organized in a business enterprise promote the executives' organizational and analytical capabilities.
   (h) It is always better to call a meeting in the early hours of the first day of the week.
   (i) Selection GDs have a named leader.
   (j) Non-verbal communication plays an important role in a GD.

2. (a) Assume that as the manager of a company, you are participating in a GD to decide upon the cost cutting measures to be adopted in your organization. Your organization has been providing transport facilities, refreshments,

subsidized meals, etc. to the employees. Now prepare at least four expressions for each of the following functions you may have to carry out during your discussion:

(i) Expressing your opinions
(ii) Disagreeing with someone
(iii) Raising questions
(iv) Interrupting someone
(v) Being polite and firm
(vi) Describing situations

2. (b) Organize group discussions on the following topics and evaluate the participants using the criteria mentioned in the evaluation form given below (if you are also a participant, you can assess yourself).

GD Evaluation Form

Candidate No.	Knowledge (25)	Communication Skills (25)	Group Behaviour (25)	Leadership (25)	Total (100)
1					
2					
3					
4					
5					
6					
7					
8					

Topics for Group Discussion
- Competition is ultimately more detrimental than beneficial to society.
- To be an effective leader, a public official must maintain the highest ethical and moral standards.
- The greatness of an individual can be decided only by those who live after them, not by their contemporaries.
- In any field of inquiry, the beginner is more likely than the expert to make important discoveries.
- Technology creates more problems than it solves.

3. (a) At the third meeting of the executive committee of the Staff Association of Nehru Institute of Technology, New Delhi, held at 5 p.m. on Thursday, the 21st of November 2019, in Gandhi Hall, the following business was transacted:
- Changing the duration of semesters
- Study leave for staff members
- Celebration of the new year
- Minutes of the last meeting

- Any other matter with the permission of the Chair
- Setting up a separate gym for the staff
- Arrange these agenda items in the most appropriate sequence and prepare the formal type of minutes as discussed in this chapter. Invent other necessary details.

3. (b) The Board of Directors of Bharti Steel Ltd, 20, Bapu Nagar, Nagpur, held its sixth meeting of the year on 23rd April 2019 at 11 a.m. on the company premises. The agenda was as follows:
- Minutes of the previous meeting
- Decrease in sales in the current year leading to losses
- Expansion of the annual capacity from 2 lakh tons to 4 lakh tons of steel
- Deputation of three mechanical engineers to the USA for a six-month training programme under Indo-US technical collaboration programme
- Improvement in the housing facilities for employees

- Any other matter with the permission of the chair

  As secretary to the board meeting, draft a notice to be circulated among the participants and write the minutes of the meeting.

3. (c) Consider yourself to be the Secretary in Attendance at the third meeting of the management committee of Gramin Cooperative Stores, Bhadpur. The meeting was held on 23rd March 2019. Arrange the agenda items given below in suitable order and write the minutes of this meeting:

   - Chairman's report
   - Purchase of books
   - Appointment of delivery staff
   - Minutes of the last meeting
   - Complaints regarding the quality of bakery products
   - Dealing with malpractices
   - Adding a new grocery section
   - Any other matter

3. (d) The executive council of the Students' Union of Sardar Patel University held its second meeting at the Students' Council building room number 20 on 3rd September 2019 to discuss the planning and preparation of OASIS, the 30th Annual Cultural Festival of the university. As the General Secretary, you have discussed the items of discussion with the President and are ready for the meeting. Prepare the agenda (six items) and minutes of this meeting.

3. (e) The sixth meeting of the managing committee of Tata Commerce College, Ajmer, was held at 3 p.m. on 20th February 2018 in room number B-45. The agenda was as follows:

   - Minutes of the previous meeting
   - Appointment of auditors for the financial year 2018–19

- Construction of a recreation centre for faculty and students
- Establishment of additional computer centre
- Any other matter with the permission of the Chair

  As the Secretary, write the minutes of this meeting.

4. (a) A national conference on 'Technology Forecasting Techniques' is being organized in Delhi under the auspices of Department of Science and Technology, Government of India. About 200 scientists, engineers, and technocrats from academic, research, and industrial organizations, both government and private, are expected to attend. The topics such as frontiers in biotechnology, energy prospects, telecommunication, approporiate technology and education for 21st century, are likely to come up for discussion.

   (i) Assuming yourself to be the Conference Secretary, write a covering letter to be sent to all the participants.

   (ii) Prepare a brochure giving details of the conference-theme, venue, time, sessions, president, etc.

4. (b) Prepare a conference brochure based on the guidelines given in this chapter for each of the following conference situations:

   (i) Apollo Medical College is planning to organize a national conference on 'Internet Applications in Medical Profession' during 10–11th September 2019. As the Head of the Department, prepare the conference brochure.

   (ii) Birla Institute of Technology and Science, Pilani, is planning to hold an International Conference on 'Meeting Communication Challenges at Workplace' during 27–28th July 2019. As the Head of the Languages Group, prepare the conferene brochure.

PART III

Reading and Writing

Chapter 10: Reading Comprehension

Chapter 11: Elements of Effective Writing

Chapter 12: The Art of Condensation

Chapter 13: Technical Reports

Chapter 14: Technical Proposals

Chapter 15: Formal Letters, Memos, and Email

Chapter 16: Research Papers, Thesis, and Technical Descriptions

CHAPTER 10

Reading Comprehension

OBJECTIVES

You should study this chapter to know

- ○ the techniques for achieving effectiveness in comprehending various texts
- ○ how to predict content and understand the gist of a particular text
- ○ the PQRST (Preview, Question, Read, Self-recite, Test) technique for meaningful reading
- ○ study skills to be applied while reading the texts

Introduction

According to Richard Steele, 'Reading is to mind what exercise is to body'. Akin to keeping one's body fit with regular exercises, reading helps to keep the mind active. Comprehending what we read is the most demanding aspect of our reading. Students and professionals need to read a variety of texts in order to understand concepts, enhance their knowledge and to know what is going on around them. However, if they are not aware of certain reading techniques they may feel that they are wasting time in reading texts. When they learn these techniques and practise them while reading, they may find that their reading is more meaningful, fruitful, and enjoyable.

Improving Comprehension Skills

Improved reading comprehension skills can positively impact many facets of a student's academic performance. Students who have effectively read and understood reading assignments are better prepared for class, leading to improved class participation and more accurate and complete notes. Performance in exams and quizzes could greatly improve as students become more proficient and effective readers. Their interest in a subject is often fostered when they understand the reading assignments. In addition, as students gain proficiency in reading, self-esteem improves. Similarly, professionals also read a lot of material related to their work, and it is important for them to have good reading comprehension skills.

The following are some of the ways in which comprehension skills can be improved:

- Read a variety of materials. Do not limit yourself to textbooks.
- Read a fairly long portion of the material. It would be difficult to assess reading comprehension based on one or two paragraphs. Try to read an entire section or chapter instead.
- Circle unknown or unfamiliar words as you read.
- After reading, recall as much of the information as possible. Jot down points if you like. Then check the accuracy and completeness of your recollections. If the main ideas are presented in a particular order, see if you can recall the structure.
- Consider how interesting the subject matter is and how much you already know about the subject.
- Answer questions about the material after reading it.

These strategies may help you in achieving the following purposes.

- To enhance understanding of the content in a text
- To improve understanding of how information is organized in a text
- To improve attention and concentration while reading
- To make reading a more active process
- To increase personal involvement in the reading material
- To promote critical thinking and evaluation of reading material
- To enhance registration and recall of text information in one's memory

> 'No matter how busy you may think you are, you must find time for reading, or surrender yourself to self-chosen ignorance.'
> –Confucius

Techniques for Good Comprehension

As we have already discussed, poor comprehension skills are usually attributable to one or more factors: lack of interest, lack of concentration, failure to understand a word, a sentence, or relationships among sentences, or failure to understand how information fits together.

In this section, we will look at the underlying aspects of text you need to be aware of and the techniques you should develop and practise to hone your reading skills:

- Skimming and scanning
- Non-verbal signals
- Structure of the text
- Structure of paragraphs
- Punctuation
- Author's viewpoint (inference)
- Reader anticipation: Determining the meaning of words
- Summarizing
- Typical reading comprehension questions

Skimming and Scanning

Before starting to read a text in detail, you should take a moment to preview the text. Read quickly, without pausing to study the details. This is called skim reading or skimming. You should understand

- for which audience the text was written (general public, professionals, laymen . . .);
- what type of text it is (report, informal letter, formal letter, article, advertisement . . .);
- what the purpose of the author is (to describe, to inform, to explain, to instruct, to persuade); and
- the general contents of the text.

After having skimmed the text, you can study it in more detail, reading more slowly and carefully and looking for specific information that you are interested in. This is called scanning.

Scanning

Non-verbal Signals

The meaning of a text is not only conveyed by means of words. All texts also contain non-verbal signs. Non-verbal signs may include certain style features, such as different fonts, bold print, underlining, or italics. The meaning of these style features can vary from one text to another. In one text italics may be used to emphasize a word, in another text italics can be used for subtitles.

Layout features are also non-verbal signals (Exhibit 10.1). For example:

EXHIBIT 10.1 Layout features

Heading ⟶ **GENERAL KITCHEN LAYOUT**

Kitchen is a busy place and cross-traffic can really hamper the operations. There are certain factors that one needs to keep in mind while planning a kitchen. This is usually done by 'facility planning department', which carefully plans the layout of the kitchens. Some hotels contract out these services but certain hotel chains, such as Oberoi Hotels and Resorts, have their own facility planning department which is responsible for planning and layout of all the kitchens.

Planning a kitchen entails much more than just placement of equipment in its place. A well-planned operation will always follow a systematic procedure (see Fig. 3.1).

Photograph ⟶

Figure ⟶ Receiving → Storing → Processing → Cooking → Serving

Figure caption ⟶ **FIG. 3.1** General workflow pattern in a kitchen

When we talk of design of a kitchen, it would generally mean the overall planning of the space with regards to size and shape of the operations. Layout would mean the detailed arrangement of the floor of the kitchen and allocation of places for the kitchen equipment to be placed where the specific tasks would be carried out. A cluster of such places is referred to as workstation.

A well-planned layout is not only important for the smooth workflow in the kitchen but it also adds to the profitability of the entire operation. Smooth workflow will ensure timely pick up of food for a busy meal period thus creating happy guest and good reputation. Such an operation is also directly linked to the motivation and overall morale of the staff.

Heading: A title printed at the top of a page to indicate the subject matter that is going to be discussed in a particular chapter, column, or section

Title: Tells you what the text is about

Subheading:	Presents you with a brief summary, introduction, or explanation
Photographs:	Pictures related to an article or a text
Captions:	Comments on pictures related to a text
Division into paragraphs:	Each paragraph is a unit and deals with one particular idea (see also the section on paragraphs)
Figures, graphs, bar charts, etc:	To visualize facts and figures presented in the text

Structure of the Text

Most texts start with a title and sometimes a subtitle. After that comes the introduction and the body, followed by a conclusion or summary.

An important aspect of reading is prediction. The better you can predict what you are going to read, the faster and more effective you will read. The prediction process begins with the title.

The introduction mostly informs you about what you can expect. The body consists of paragraphs. Each paragraph deals with one aspect of the subject matter. Paragraphs are linked in a logical way. The conclusion sums up and puts the subject matter in the right perspective.

Structure of Paragraphs

A typical paragraph consists of three parts. The first part is the topic sentence, which is the heart of the paragraph, which can figure either at the beginning or at the end of the paragraph. The topic sentence (also called *thesis sentence* or *key sentence*) contains the new aspect of the subject of the text. The second part of the paragraph contains sentences which develop support for the topic sentence. These sentences may contain arguments, explanations, details, examples, and other supporting evidence. The third part of the paragraph is often a summary of the paragraph or a linking sentence to the next paragraph.

In many well-written texts the reader will get a good idea of the contents by reading just the first sentences (i.e., the topic sentences) of each paragraph.

Punctuation

Punctuation is partly based on grammar. For example, commas are often used to separate clauses. If you understand the meaning and usage of punctuation marks, it will be easier to understand the grammatical structures.

Punctuation marks also indicate how the author wants you to interpret a piece of text. For example, if the author puts something in brackets, that part of the sentence should be interpreted as an addition to the sentence.

Author's Viewpoint (Inference)

In reading a text you make inferences or assumptions about the position of the author. Is the author neutral or does he/she have an opinion? The author often shows his or her opinion either by adding certain phrases or by adding a value to a word.

Authors can add words like *luckily* or *unfortunately* to show approval or displeasure, respectively. They can also add words which show their surprise, regret, or other emotions, for example, *surprisingly, to shock, unexpectedly, regrettably, pity, desirable, to be disappointed*, etc. They may also use words to indicate the level of certainty, for example, *certain, obviously, undoubtedly, naturally, always, often, likely, probably, maybe*, *unlikely*, *hardly*, *rarely*, and *never*. The author could also add words to comment on more or less objective facts. For example 'Only 40 per cent of the staff is female' has quite a different meaning from 'As much as 40 per cent of the staff is female'.

The author can also reveal his/her viewpoint by adding value to a word. For example, if something is big he/she may use the word *huge* or *gigantic* to indicate he/she is impressed. If something is small he/she may use words like *tiny* or *microscopic* to indicate that he/she is not impressed. If someone is afraid he/she may use the word *terrified* to add suspense; if something is good he/she may use the word *fantastic*.

Reader Anticipation: Determining the Meaning of Words

Even readers who have a wide and flexible vocabulary will encounter words whose meaning they do not know. You may have often come across a not-too-familiar word whose meaning you were able to guess accurately. This is possible if you understand the relationships between words in meaning and form.

You can develop this skill by following a method. Look for clues or indications that help you find the meaning. The following steps can help you:

Step 1   Determine the word class, i.e., a verb, a noun, an adjective, an adverb, etc.

Step 2   Determine its function in the sentence, i.e., how it is related to other words in the sentence.

Step 3   Analyse the contextual clues, i.e., the context in which we find this word positive or negative; what the context tells us about the word.

Step 4   Is there any relation between this word and a word that is familiar to you? Consider the word *aging* in *the population is aging rapidly*. We recognize the word *age* in *aging* and therefore we can easily derive its meaning.

Step 5   Try and derive the eventual meaning of the word.

Summarizing

It is very difficult to remember the complete contents of long texts. For that reason it is often advisable to make notes of essential information in the text. The result is a short outline of the text containing all its important aspects. The length of the summary largely depends on the density of the text. The average length of a good summary is about one-third of the original text, but summaries of detailed texts may be longer.

The following steps can be taken when summarizing a text:

Step 1: Familiarize yourself with the material

After you have read the text or a section of it, you can start summarizing. The length of the section you can easily read through at a time, in an attempt to summarize the text in parts, depends on the structure and the complexity of the text. With reports, it is often a good idea to read one section and then summarize. Newspaper articles are often best read as a whole before starting with the summary.

Some General Rules to Determine What is Important and What is Not

- *Important*: Author, title, reason why the text was written, purpose, theme, key words, link words, all major aspects (especially topic sentence of each paragraph), explanations, author's opinion.

- *Less important*: Introduction, summary, repetition, examples, anecdotes, analogies.

Step 2: Select important information
You should go through each paragraph, sentence by sentence, asking yourself which information is absolutely essential to the argument. You should write down complete sentences as much as possible. Jot down your points clearly or you will have to read the original text all over again. An alternative is to underline or highlight important sentences or phrases.

Step 3: Paraphrase the information
Paraphrasing, or rewriting from the original text in your own words, forces you to completely understand what you are noting down. You should try to condense long and complex sentences into much shorter ones. Use active sentences as much as possible and avoid adjuncts.

Practise economy with words. Avoid descriptions if they can be summarized in one word. For example, do not write 'The state exclusively controls and possesses the trade in stamps', but write 'The state monopolizes the trade in stamps.'

Step 4: Insert links between sentences and paragraphs
Make sure that the connection between sentences is clear and logical and that each group of sentences smoothly fits in one paragraph. This can be done by inserting link words such as *therefore, nevertheless, but, however, because, on the other hand,* etc. Sentences can also be linked by relative pronouns, for example, *who, which, whose,* and *that.* Using the correct links means that your summary becomes more logical and coherent.

Step 5: Adjust the length of the summary
If you have to write a summary as part of an assignment and you are assigned a maximum number of words, count the words after writing it out. Depending on whether your word count exceeds or falls below the given number, you can leave out unessential information or add important information.

Typical Reading Comprehension Questions

You should become familiar with the main categories of reading comprehension questions asked in standardized tests. This will help you to focus your attention while reading the passages.

Reading comprehension questions usually take one of three forms: questions based on the entire passage, questions based on sections of the passage, and questions based on particular words or sentences. Each of these is discussed and exemplified in Tables 10.1–10.3.

- Questions based on the entire passage
- Questions based on the entire reading usually target the main point of the text, author intentions, main ideas, and content. Ten types of questions based on the entire passage are given in Table 10.1 along with sample wordings.
- Questions based on sections of the passage
  In order to answer questions on specific sections of the passage, one must be able to identify and understand the main points in each paragraph. Look for cue words such as *advantages, disadvantages, similarities, differences, in contrast with, in comparison to, most importantly, primarily,* and *on the other hand.* Questions based on portions of the text usually deal with inferences, applications, and implications of the information. Six types of questions based on sections of the passage along with sample wordings are given in Table 10.2.
- Questions based on words, phrases, or sentences
  Here, specific details and pieces of information from the text may be the subject of test questions. The content itself is not usually the subject of questions. Rather, the reasons for using the information or the meaning of the information could be questioned.

Two types of questions based on words, phrases, or sentences are listed in Table 10.3 along with sample wordings.

Table 10.1 Sample questions based on the entire passage

Question type	Sample wording
1. *Main point*: What is the passage trying to tell you?	The passage is mainly concerned with . . .
2. *Primary purpose of author*: What does the author want to tell you?	The author's primary purpose in the passage is to . . .
3. *Mood or attitude of author*: What is the tone or attitude of the author?	On the basis of the passage, the author's attitude towards _____ can most accurately be termed as one of . . .
4. *Assumptions made by author*: What assumptions are made by the author but not directly stated in the passage?	Which of the following is an assumption made by the author?
5. *Implications of passage or author*: What does the author or the passage imply?	The author implies that _____ is . . .
6. *Applications of main ideas*: How can you extend the main ideas of the passage? According to the author, _____ would lead to . . .	The author provides information that would answer all of the following questions except . . .
7. *Summary of passage*: In a few words, how would you describe the passage? What title would you give the passage?	Which of the following titles best summarizes the content of the passage? Which of the following would be the most appropriate title for the passage?
8. *Content of the passage*: What is the passage really about?	Which of the following describes the content of the passage?
9. *Inferences*: What can you infer from the passage as a whole?	It can be inferred from the passage that . . .
10. *Statements with which the author would agree*: What could you say that the author would agree with, knowing the way he/she wrote the passage?	With which of the following statements regarding _____ would the author probably agree?

Table 10.2 Sample questions based on a section of the passage

Question type	Sample wording
1. *Inferences*: What can you infer from specific sections in the passage?	It can be inferred that the ancient atomic theory was primarily based on . . .
2. *Applications*: How can you apply information in specific sections of the passage to other areas? What precedes or follows the passage? What do you think was written right before the passage or right after the passage?	It can be inferred that in the paragraphs immediately preceding the passage, the author discussed . . .
3. *State ideas*: Can you find in the passage a specific reference to a stated idea?	According to the passage, blacks were denied entrance into anti-slavery societies because . . .

(Contd)

Question type	Sample wording
4. *Implications*: What is implied by a section in the passage?	The author implies that many Americans' devotion to the ideal of justice is . . . In describing American attitudes about the land (lines 7–8), the author implies that . . .
5. *Tone or mood*: What is the tone or mood of a section of the passage?	At the conclusion of the passage, the author's tone is one of . . .

Table 10.3   Sample questions based on words, phrases, sentences

Question type	Sample wording
1. *Reason for use*: Why are certain words, phrases, or sentences mentioned or used in the passage?	The author mentions Newton's Principia in order to . . .
2. *Meaning of a Word or Phrase*: What is the meaning of a certain word, phrase or sentence in the passage?	The enemy referred to in the last sentence is probably . . . According to the author, the words in the Declaration of Independence, 'all men are created equal', are meant to represent . . . By 'this skepticism' (line 35), the author means . . .

Predicting the Content

Predicting the content of a given reading passage enables you to quicken the process of comprehension. In other words, thinking ahead of the given information or anticipating the information yet to come in the reading passage, makes you understand the entire passage in its total perspective. In order to guess the information through the process of prediction, you need to develop rapid reading skills because an efficient reader is able to think ahead, hypothesize, and predict.

You may find several cues in the passage that may assist you in predicting the content. Your knowledge about the subject, the topic sentences of each paragraph, headings, subheadings, graphics (e.g., graphs, diagrams, and charts), and cohesive markers (e.g., pronouns, transitions, repetitions, and synonyms), all would enable you in the process of prediction.

A pre-reading survey of the passage is a necessary condition for predicting the content. This is the first function you may carry out when you need to comprehend the contents of a given passage. The following steps and the example discussed thereon may help you accomplish the prediction process:

- Familiarize yourself with the subject of the passage by glancing rapidly through it.
- Guess the information through the linguistic/graphic cues.
- Recall related information.
- Use background information related to the topic to accelerate the prediction process.

Example

Given the widespread acceptance of the image of the ocean depths as serene and undisturbed, it is unsurprising that researchers initially greeted the evidence for powerful deep-sea storms and dynamic currents with disbelief. The earliest arguments in support of the existence of such powerful deep-sea currents were based on models of ocean circulation derived from hydrodynamic theory. Because the density of cold water is greater than that of warm water, cold water tends to sink. Thus, near the poles, the chilly water sinks; in theory, this sinking should create powerful, regular deep-sea currents flowing from the poles toward the warm equatorial regions. Over time, investigators have gathered sufficient evidence to confirm the accuracy of this theory. Oceanographic studies have verified that deep-sea currents exist and have demonstrated that, on the western side of the great ocean basins, periodic underwater storms transporting masses of fine sediment scour the sea floor. Corroborating these studies, photographs of the sea floor have disclosed the existence of vast graded beds suggesting the active transport of large volumes of silt and clay. Further corroboration has come from an experimental program known as the High-Energy Benthic Boundary-Layer Experiment (HEBBLE).

- A quick reading of the passage tells us that the topic is about 'deep sea currents' and the related research
- Phrases such as *powerful deep-sea storms and dynamic currents, no surprise, disbelief, arguments, models of ocean circulation derived from hydrodynamic theory, oceanographic studies, over time, deep-sea currents exist, corroborating, further corroboration, HEBBLE*, enable us to guess that though initially the existence of deep sea currents were not believed by scientists, over the years, they started accepting it because of the various studies conducted. We can also predict that the author will talk about the HEBBLE experiment in the next paragraph.
- After identifying the phrases, we can recall the topic once again and try to paraphrase the details in our mind
- If we have some knowledge on the characteristics of liquids with different density we can understand the principle: *Because the density of cold water is greater than that of warm water, cold water tends to sink. Thus, near the poles, the chilly water sinks; in theory, this sinking should create powerful, regular deep-sea currents flowing from the poles toward the warm equatorial region.*

Understanding the Gist

When you try to get the general meaning from a text without concentrating on the individual words, you are reading for gist. In other words, when you read a passage, if you are able to understand its core meaning or main theme, you have understood the 'gist'. You might have observed that you could understand the gist of some passages in the first reading itself. For instance, when you read a passage on environmental pollution, you may understand the core meaning at the first reading itself, whereas when you read a passage on American Civilization or Nuclear Power, you might not be able to do so.

There may be several reasons for this ability or inability to understand the gist at the first reading. Your previous knowledge about the topic would certainly enhance your capacity to get the gist quickly. At the same time, if you are not familiar with the topic and the passage is complex, you may take more time to get the gist. However, for proceeding further with your reading activity, understanding the gist is very important.

Now, let us discuss how to go about understanding the gist of the passage. Each passage that you read will include some main ideas supported by adequate details. If the passage is a paragraph, it will have one main idea and various details to substantiate that idea. If the passage is long, it may generally talk about the main idea in the first paragraph and may contain the subordinate ideas in each of the remaining paragraphs. The concluding paragraph will once again emphasize the main idea. Each of the subordinate ideas will be supported by adequate details contained in a number of sentences. Hence, in order to understand the gist you need to identify essentially the main idea or theme of the passage and the subordinate ideas and also the relationship between the ideas and the supporting details. You can do this by posing questions such as what does the author want to say or what the topic is about.

Usually, you can find the main idea within the first one-third of the passage. The title or the main heading of a text, if any, can also give a clue to the content. You can also look for repetition of certain important terms or their synonyms that may help you identify the main topic. Similarly, the topic sentences of each paragraph can help in identifying the central idea. Each paragraph will have a subordinate idea that you can find either in the beginning of that paragraph or at its end. You may also glance over the beginning of the text to identify its logical organization as mentioned in the previous discussion on skimming.

With these strategies, you may be able to get the gist of your reading text or passage. Read the following passage carefully with a view to get its gist. Then go through the discussion that follows.

Communicating through Space

One way that people can communicate with one another is by manipulating the space between them. People have a very strong sense of personal space that surrounds them and are greatly discomforted when it is invaded. Crowded subway cars, for example, may be experienced as psychologically uncomfortable, and outbreaks of aggression are more likely in crowded situations.

Edward T. Hall studied attitudes toward physical proximity in several cultures. He found that people from different cultures vary in the degree of closeness that they will tolerate from strangers or acquaintances. Americans seem to require more personal space than any other people—a distance of at least 30 to 36 inches, unless the relationship is a very intimate one. American travellers to other countries find that the inhabitants stand almost offensively close. But people in these cultures are apt to consider Americans—who are always backing away when one tries to talk to them—disdainful and rude.

Hall suggests that there are four distinct zones of private space. Intimate Distance is the zone which extends up to 18 inches from the body. It is reserved for people with whom one may have intimate physical contact. Personal Distance is the zone which extends from 18 inches to 4 feet. It is reserved for friends and acquaintances. Some physical intimacy is permitted within this zone, such as putting one's arm around another's shoulder or greeting someone with a hug, but there are limits. Social Distance is the zone which extends from 4 to 12 feet. It is maintained in relatively formal situations, such as job interviews. There is no actual physical contact within this zone. Public Distance is the zone which extends for 12 feet and beyond, and is maintained by people wishing to distinguish themselves from the general public. Speakers addressing an audience, for example, maintain this distance.

(*Source*: Roberta Steinberg, *Complete Tools for the TOEFL Test*, p. 313, McGraw-Hill, New Delhi, 2006)

You might have identified that the main topic of this passage is 'communication can take place by using the space that is maintained between the speaker and the listener in various ways'. You would have done this by identifying the title (as this passage has a title) and also the first sentence. Then you would have got the gist of the passage by identifying the topic sentences (here the first sentence) of the second and third paragraphs. So, the gist of the passage according to you might have been as follows: 'Communication can take place by using the space that is maintained between the speaker and the listener in various ways. The space communication is affected by cultural variations and it is of four types.'

To sum up the discussion, we can say that you should carefully read the following in order to understand the central idea of a text:

- The title or the main heading
- The topic sentence of each paragraph
- The opening paragraph
- The last paragraph

PQRST Technique

Some of us read the texts several times to understand them and in this process, we lose time. To reap maximum benefit out of our reading in terms of time and content, we can make use of a reading technique called 'PQRST' wherein the acronym PQRST stands for

- Preview
- Question
- Read

- Self-recite
- Test

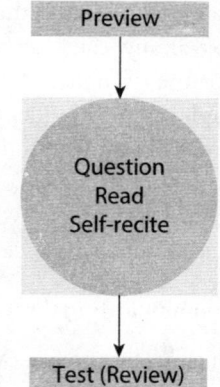

Figure 10.1 PQRST technique

George Bernad Shaw said, 'Question, Examine, Search' to understand things in their right perspective. On the same lines we can also apply certain steps while reading a text to understand and interpret the same with effectiveness and efficiency. The functions of PQRST technique (Figure 10.1) enable us to get an idea about the structure and organization of the text and show us the direction in which we need to proceed further. You might have understood from the above diagram that the first step, 'Preview', relates to the introductory part; the second, third, and fourth steps, Question, Read and Self-recite, pertain to the discussion or middle part of the text; and the last or fifth step, 'Test (Review)', corresponds to the conclusion or summary part of the reading text. When we read a text, we try to understand its gist first. Then we move on to the detailed reading of discussion paragraphs. Finally, we read the concluding paragraph to recall, review, and test our comprehension.

Now, let us investigate the constituents, namely P, Q, R, S, and T of the technique in some detail.

Preview (step 1)

Whether it's a speech, writing, or reading, a preview gives you an idea about the purpose and overall picture of the content. It's like a survey wherein you try to identify what the text is about and what the author aims to communicate through the text. It also conveys to you the various points or issues the author intends to discuss.

Just like we try to understand a problem before attacking it in order to find a solution, we need to get an overview of the text before finding out answers for the given questions. Get an overview of the text by

- Reading the title, contents, headings and subheadings
- Reading the chapter summaries
- Looking at the illustrations, their captions, etc.
- Making a note of highlighted words, sections, etc.
- Reading the introduction and conclusion

We can assume that surveying a text is more or less equal to scanning the text as detailed in an earlier section of this chapter. Surveying the text also helps us to judge our background knowledge related to the text.

In a nutshell, we can say that 'Preview' is nothing but a survey of what you are going to read further in the text.

Question (step 2)

Many a time we pose questions to ourselves to clarify something. For instance, when we look at our passport we ask ourselves, 'Is my name typed correctly? What is the expiry date? When should I apply for a re-issue?' Such questions enable us not only to get a clear picture of the given information but also enable us to check the details. Similarly, while reading a text, we should ask questions on its contents. Assume that you are reading the following paragraph in a text:

> The flow direction of refrigerant through a heat pump is controlled by valves. When the refrigerant flow is reversed, the heat exchangers switch function. This flow-reversal capability allows heat pumps either to heat or to cool room air.

After reading these lines, we could ask ourselves this question to check our understanding: 'What is it that allows heat pumps to heat or cool room air?' When we ask questions while reading each paragraph, we may be able to get the main points of the text. We may also grasp well the sequence of ideas presented in the text.

> 'Reading without reflecting is like eating without digesting.'
>
> –Edmund Burke

Read (step 3)

Once you preview and question, you should read the text focusing on the main points. We may have to slow down our reading speed in case of certain dense or complicated texts whereas we may read the simple texts fast. Making notes while reading may be helpful to understand and interpret the texts. Read the guidelines given in the section on note-making. Given below are certain quick guidelines to carry out this reading step of PQRST technique:

- Understand that each paragraph contains one main idea and the related supporting ideas. So, identify the main idea and note down.
- Note the punctuations, pronouns, transitional words/phrases, etc. carefully to get a clear idea of the discussion points. In other words, don't ignore them.
- Get an idea as to how each paragraph is linked to one another.
- Try to check whether you have understood each paragraph correctly by paraphrasing wherever necessary.
- Keep track of time. Do not spend too much time on note-taking, paraphrasing, etc.

Self-recite (step 4)

This step can also be called 'Recall'.

Just as we retain some information regarding places, people or products by recollecting the details when we had come across them earlier, we need to recall the main points and supporting points once we finish the reading and note-making processes. At this point, we may try to recall the supporting points for each of the main points and recite them/speak them aloud in our own words. We may even write down the points. But we should not copy from the passage but use our own words to recite or write. For instance, in the following text, you may see that there are two supporting points for the main point, which is 'getting more corneas', contained in the first two sentences.

> Getting more corneas for the blind is a difficult choice. Two types of solutions have been proposed for getting more corneas. One answer is to impose compulsory eye donation on patients dying in hospitals. The other answer is to educate the masses through vigorous campaigns for voluntary post-death eye donations.

While the questioning step enables us to probe further into the text, the next two steps, namely reading and self-reciting, assist us to retain most of the information in the text.

Test/Review (step 5)

After completing steps 1 to 4, we can move on to the last step, namely 'test or review'. At this stage, first, we need to test ourselves by covering the key points to see whether we remember them. If we are successful in this attempt, we can check the correctness by referring to the text. Otherwise, we can reread or expand our notes or discuss with our friends the main points. These steps will enable us to review the text and its salient points. In case you are reading a long text such as a book or report, you can also test your comprehension by writing section summaries. Go through the solved reading comprehension exercises in the Online Resource Centre (text supplements) and then assess your skills.

Study Skills

Besides employing various techniques while reading a text, you need to apply certain skills in order to make the most of your comprehension. These skills may be useful in answering your reading comprehension questions. Two important skills, namely note-making and sequencing of sentences are discussed in this section.

Note-making

Note-making is a systematic process of writing down important information, ideas, facts, viewpoints, and arguments contained in a written text for future reference. Note-making is necessary to recall the main ideas in what you have read as you may not be able to keep in mind all important information contained in a variety of material you may be reading. You would need to develop qualities such as quick comprehension, identification of main ideas, and recognition of their relevance to your needs, besides an ability to record them with quickness and precision. The following are some formal note-making methods:

- Outline/linear method
- Sentence/categorical method
- Schematic/mapping method

We will discuss each of these methods for making notes from the following passage on Indian economy.

Indian Economy

Just over a decade after the Indian economy began shaking off its statistic shackles and opening to the outside world, it is booming. The surge is based on strong industry and agriculture, rising Indian and foreign investment and consumer spending by a growing middle class. After growing just 4.3 per cent last year, India's economy is widely expected to grow close to 7 per cent this year.

The growth of the past decade has put more money in the pockets of an expanding middle class, about 30 crores strong, and with more choices in front of them, their appetites are helping to fuel demand-led growth for the first time in decades.

India is now the world's fastest growing telecom market, with more than ten lakh mobile phone subscriptions each month. Indians are buying about 10,000 motorcycles and scooters a day and 20,000 cars per month. Banks are now making two crores a year in home loans, with the lowest interest rates in decades and thus helping to spur spending on building. Credit and debit cards are slowly but steadily gaining momentum.

The potential for even more market growth is enormous, a fact recognized by multinational and Indian companies alike. For example, in 2001, according to census figures, only 31.6 per cent of the country's 18 crore households had a television, and 2.5 per cent a car, jeep, or van.

Outline/Linear method

In this method, you need to identify the main ideas and the corresponding main and subsidiary points from the reading text and arrange them in a table consisting of two columns. Use the first column for writing down the main ideas and the second one for the corresponding main and subordinate points of each main idea. You can distinguish between the main and subordinate points by using indentation or two levels of bullets, as shown below.

Boom in economy due to liberalization	• Liberalization of economy
	◊ strong industry
	◊ strong agriculture
	◊ rising investment
	◊ consumer spending by growing middle class
	◊ expected growth this year 7 per cent from 4.3 per cent last year

(Contd)

Growth spurred by demand	• Demand
	◊ demand-led growth because of consumer spending by growing middle class (30 crores)
	◊ more money
	◊ more choices
	◊ more demand
India—fast-growing market	• India—fast-growing market
	◊ 10 lakh mobile phone subscriptions each month
	◊ 10,000 motorcycles and scooters sold daily
	◊ 20,000 cars sold per month
	◊ banks earning 2 crores in home loans
Enormous potential for further growth	• Potential for further market growth enormous
	◊ in 2001 only 31.6 per cent households (18 crore) had a television
	◊ only 2.5 per cent owned a car, jeep, or van

Sentence method

In this method, also known as the categorical method, sentences are used to represent the main idea and subordinate ideas. Generally, the text contains one main idea/thought/fact in one paragraph and this is supported by several supporting points. You will have to identify the main and subordinate ideas and write them as short sentences. As mentioned earlier, you may be able to identify the topic sentence in each paragraph. All you have to do is to write the topic sentence in your notes, after editing it, if necessary. Then you should identify the sub-points and write them down below the sentence containing the main point. You might have observed that this method resembles the linear method, the only difference being the use of sentences in place of words or phrases, as shown below for the passage on Indian economy.

Ravi's notes are always complete. He never misses a thing the teacher says.

- There is a boom in Indian eco-nomy due to liberalization.
    - It is based on strong agriculture and industry.
    - There is an increase in Indian and foreign investment.
    - There is greater consumer spending, especially by growing middle class.
    - The growth rate expected this year is 7 per cent.
- The growth has been spurred by demand.
    - There is greater consumer spending by growing middle class (about 30 crore).
    - There are more choices, more money, and, hence, more demand.
- India has become a fast-growing market.
    - 20,000 cars are sold and 10 lakh mobile phone connections are issued every month.

- 10,000 motorcycles and scooters are sold everyday.
    Banks are earning about 2 crore as interest on home loans every month.
- There is enormous potential for further growth.
    - In 2001, only 31.6 per cent of households had TV.
    - In the same year, only 2.5 per cent owned a car, jeep, or van.

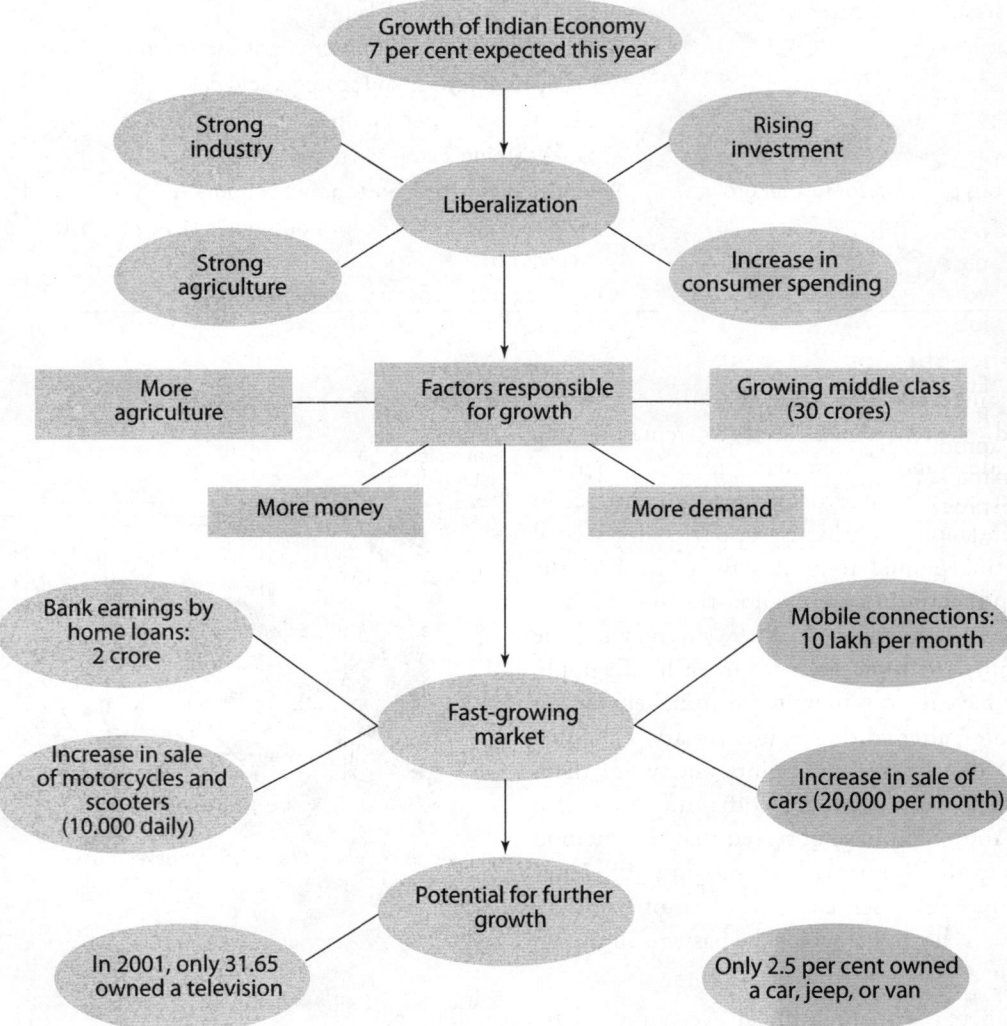

Schematic/Mapping method

You must be aware of mind-mapping technique used for categorizing your ideas on a particular topic where you use circles, blocks, and arrows to represent various main and subordinate ideas. The schematic or mapping method also is a semi-graphic representation of the contents of a text to show the relationship of each point or idea to other facts and ideas contained therein. In fact, this method has an edge over other methods as you can record a great deal of information in less space and also show the complex relationships among several ideas and facts. Given below are some tips on using this method for your note-making tasks:

- Represent the main ideas in a central box/circle.
- Depict the subordinate ideas as though they are radiating from this central image.
- Use branches attached to the higher-level branches to represent further minor points.
- Ensure that the branches form a connected nodal structure.

Practice 1

Read the following passage and make notes from it. You can use any of the three methods explained in the preceding discussion.

The march of scientific mechanization over industry, throughout the world, is unabated, so is the misery of the poor. Equally unchecked is the thriving of a few among the upper rungs of industrial hierarchy and the bureaucratic and political echelons. One feels proud of the scientific innovations in the manufacturing process where manpower is speedily being replaced by capital potential. The researches in the field are so swift that the labour factor is gradually being eliminated. Automation has withdrawn hazards from man but has offered him unemployment. In the progressing countries, where there is no check on population, scientific industrialization has become an eye sore on the job front. It has resulted in long queues at the employment exchanges.

Scientific industrial development has proved manna for the progressive countries of Western Europe, Canada, the USA, and Japan. They reap certain advantages over the backward nations. It has provided a luxurious life even to the common man there. Scientific researches result in mass production of commodities. Thus, the West has a huge surplus of finished goods to flood the world market and to reap profits. The resultant economic growth in these countries enables them to proceed with further scientific researches to make their industries more capital oriented. Thus they always have an advantage over the backward countries.

Notwithstanding the backwardness or poverty, scientific industrialization is responsible for an increase in the gross national product in progressing countries like India too. But it creates two-fold problems. The advantages of this increase in gross national product are reaped by the rich for it is they who have invested money. The growth of capital too is their prerogative. The profits and interest go to the coffers of the industrialists. The fruits of progress and mechanization do not go the poor. It is only a class—a group that benefits from the scientific development not all.

(*Source*: Bhatia,. M.P., *A Handbook of Applied Grammar and Composition*, 8th Edition, p. 138, M.I. Publications, Agra, 2005)

Critical Reading, Critical Thinking and Effective Googling

Critical reading

When you read a text specifically with a view to discover and analyse some information, you are reading 'critically'. In other words, you are not just satisfied with receiving the matter (what) presented in the text but you also apply your critical faculty or your power of judgement to find out the manner (how) in which what the text describes happens. For instance, when you read about a scientific discovery, you just try to get the information if you are a noncritical reader. On the other hand, if you are a critical reader, you go further and attempt to understand the factors behind the discovery and connect the ideas that lead to that discovery. As a critical reader, you try to understand the text from three different angles, namely the author's purpose, his/her tone and style, and his/her preferences/biases on the topic. However, you need to remember that intensive reading is mandatory for critical reading.

The authors of various texts have their own choice of content, language, and structure. For instance, the authors of scientific texts may describe a process using a matter-of-fact style and may follow a chronological order to discuss the steps, whereas a fiction writer may describe a natural phenomenon using figures of speech such as metaphors, similes, etc. As a critical reader, you need to examine the choice of content, language, and structure used by the authors to arrive at the correct understanding and assessment of the text.

As already mentioned, critical reading requires you to discover, analyse, and interpret the information contained in the text. Out of these three steps, analysis and interpretation are possible only when you think creatively and critically. Hence critical reading, and creative and critical thinking are always talked of together. They form the two sides of the same coin and hence are inseparable. When you read a report, for instance, presenting the results of a survey on the job preferences of engineering students over the past five years, your critical reading ability may enable you to discover how many students pursuing chemical engineering preferred a particular type of job over these years but at the same time you need to use your critical thinking also simultaneously if you need to analyse the trend, interpret the reasons behind such trend and evaluate their merit. In fact, it can also be said that only when you think critically can you also read critically because critical reading does not allow distortion of facts, and you can get the facts in the right perspective only when your critical thinking faculty works in order. While creative thinking enables you to generate innovative ideas, critical thinking helps you to evaluate what you are reading. We can sum up this discussion on critical reading as follows:

- Critical reading involves the discovery, analysis, and interpretation of the text.
- Non-critical readers satisfy themselves with only information but critical readers go further and try to identify the style of writing and the meaning of the text.
- Critical readers try to understand the purpose, tone, and style of the author and also his/her treatment of the topic.
- Creative thinking and critical thinking are an integral part of critical reading; without one the other cannot function.
- While reading a text, your creative ability enables you to come out with innovative ideas and your critical faculty leads you towards judging and evaluating the arguments.

Checklist for Critical Reading

- Whom will I report to?
- What is the author's main purpose?
- What is the scope of the text?
- Does the author discuss an idea/argue for or against/explain/evaluate and provide suggestions?
- What is the author's tone? Is it opinionated/argumentative/descriptive/factual/even handed?

- Has the author given substantial evidence to prove his/her ideas?
- Has the author provided necessary statistical/visual aids in support of his/her discussion?
- Has the author drawn reasonable inferences?
- Do you agree with the author and if not, why?
- Are you biased because of your prior knowledge of the topic?

Critical thinking

Assume that you wish to buy a new mobile phone. Would you go immediately to the shop and buy a phone? No. You would find about various options/types/brands, analyse them by looking into their positive and negative aspects, evaluate, decide, and communicate. Similarly, in your academic career, you may have to apply your analytical, interpretive, and evaluative skills to arrive at an appropriate solution to a problem. Such skills can be categorized under one term: Critical thinking.

Critical thinking refers to your ability to analyse, interpret, synthesize, and assess or evaluate various tasks in your personal, academic, and professional contexts. It involves understanding and addressing an issue with the available facts and information. You need to think deeply, organize your thoughts logically, interpret them correctly, analyse them using the available data, evaluate them critically without any prejudice and communicate them clearly. As already mentioned earlier, critical thinking is a part of critical reading. Without critical thinking skills, you may not be able to apply your rational thinking to any text or for that matter, to any problem.

You need to carry out the following steps of critical thinking to understand and manage/solve a problem:

- Observation
- Analysis
- Evaluation
- Communication

Assume that you, as the manager of an organization, must solve a conflict between two employees. Let us discuss the steps mentioned above in the light of this situation:

Observation involves identification and research, and it is the starting point of critical thinking. If you can observe well, you may get an idea about the reasons for the conflict. You must identify or know the ins and outs of the conflict and observe the conflicting parties to understand the problem thoroughly. You may even predict the causes of the conflict if you have good observation skills. Observing the background, behaviour data, etc. would enable you to understand the conflict very well.

Analysis encompasses having a deep insight into the problem and possible solutions. For instance, in the given context, the conflict might have occurred because of one main reason. But it is not enough to analyse only that reason. You may have to analyse the underlying issues or causes as well. A conflict may arise between two employees because one has not been promoted. But, with your critical thinking skills, you must analyse the various issues that led to this situation. You must also think about solutions and analyse them.

Evaluation refers to judgement. Once you have observed and analysed the possible reasons and solutions pertaining to the conflict, it is time for you to evaluate each solution thoroughly. Think about the positive and negative impacts of the solutions and evaluate them so that they may not create any untoward situation in future.

Communication plays a very important role in explaining and discussing the solutions with the employees. In the absence of effective communication, your other skills may lose their value. For instance, after evaluating and arriving at the solution/s, you may have to explain and discuss with your employees with clarity, correctness, and completeness so that they can understand the solutions and proceed further.

No matter what academic or professional career you take up, your critical thinking skills will always be relevant and will enable you to progress well and succeed. Whether it is to find answers to the questions in a reading passage, solve a mathematical problem, decide upon a suitable job or manage your project team—you would find your critical thinking skills mandatory.

Effective Googling

Today, we all live in the technology dominated globalized environment which requires several skills to succeed. Effective Googling is one important skill which enables us to be well informed on various issues that we may have to discuss and manage. It is one skill that assists us extensively in our personal, academic, and professional contexts. Though we may be using 'googling' for various purposes, the following discussion would make our googling more effective so that we can save our precious time and get the desired information with more accuracy.

'Effective googling', as the phrase suggests, refers to the effective ways in which we can use Google search for getting the information we need. It enables us to find information on innumerable issues. However, if we are not effective in our search, we may not only spend hours and hours in our search but also be frustrated with our failures.

Given below are few tips for effective googling:

- Use the tabs, namely, *All, Images, Videos,* and *News* appearing under the Search window to focus on what you wish to search. For instance, if you want to see the various types of cars available under a specific brand, you can key in the required words and click on *Images.*
- Use correct words or phrases within quotation marks for exact search. For example, if you type 'price of 1 g gold in India' or 'life in a desert', you may get the desired result.
- Exclude words if necessary to limit your search by using minus (–) sign just before the word you wish to exclude. For instance, when you want to search various universities in USA for graduate studies, and you do not want a particular state, say, New York, you can key in 'USA Universities for graduate studies –New York'. Remember to leave a space after the word 'studies' to avoid the minus sign looking like a hyphen.
- Use OR if you look for one or more terms of your input. For example, you can type 'autonomous universities OR colleges in Hyderabad', and you can find the matches for any one of these two.
- Use 'define' command if you wish to look for the definition of a term.
  Example: define:competitive advantage
- Insert an asterisk (*) mark to find out a missing word in a phrase, title, song, etc.
  Example: The Taming of the *
- Use the name of filetype [doc (Word), ppt (PowerPoint), pdf (Adobe Acrobat)], when you look for information under that particular filetype. For instance, when you wish to find information available in PPTs about 'English Intonation', just type the following:
  "English Intonation":.ppt
- Use your search to carry out mathematical calculations by typing the expression in the Search window.
  Example: 345678/24
- Use descriptive words to refine your search. In other words, use different words/question to get the required information when you don't get information in your first attempt. For instance, rather than searching for: 'How to attend a job interview?', (when you really want to know about your preparation), you should refine your search by typing 'How to prepare for job interview?'
- Look for correct statistics/numbers by using appropriate year or period in your search.
  For example, when you wish to know the novels by Anuja Chauhan after the year 2005, key in 'Anuja Chauhan novels..2005'

Besides these tips, you can also go through the following site to understand the nuances of Google search: https://support.google.com/websearch/answer/134479?hl=en

Understanding Discourse Coherence

The term discourse refers to a serious speech or a piece of writing on a particular subject and the term coherence refers to the logical development of ideas within a text and it is an important sub-skill for you to be aware of. A coherently structured discourse is assumed to be a collection of sentences, and in turn paragraphs, that are ordered such that they make good sense in some relation to each other. We usually expect written texts to be coherent, meaningful communications in which the words and/or sentences are linked to one another.

Incoherence can arise from too much or too little information as well as from incorrectly ordered sentences or thoughts. It is important to ensure that the reader understands the relation between the various thoughts expressed in a discourse. We normally use our own interpretations and perceptions of familiar situations and concepts while reading/hearing discourses of interest. Therefore, it must be ensured that there is no scope for ambiguity, especially in technical discourses.

To clearly express the relationship between the various segments of a discourse, and thereby effectively communicate the intended meaning, you can use a variety of expressions such as the following:

> but, yet, so, as, and, since, while, similarly/likewise/in the same way, on the other hand/on the contrary/contrariwise/whereas/ while/however/yet/although/nevertheless/nonetheless/whereas/unlike/despite/in spite of/even so, therefore/as a result/as a consequence/consequently/thereby/thus, because/due to/on account of/inasmuch as, though/although/even though, in addition/moreover/furthermore, either…or, neither…nor, of course, in fact/clearly/evidently/obviously/actually/indeed/ surely/certainly, so that/so as to/in order to, in other words/that is/to put it differently/to repeat/namely, for instance/for example, hence/in conclusion/to sum up/at last/finally

These words are known as transitional words or phrases that join one idea to another idea. You can use transitions within a paragraph as well as between paragraphs. They help your reader to follow from one idea to the next smoothly. These words/phrases are also known as connectors, markers, or signal words.

Discourse markers help you develop ideas and relate them to one another. They indicate how one piece of text is connected to the other. They show the connection between what has already been written and what is going to be written. But remember not to overuse these markers, as doing so would not only make the style of your text artificial but also affect its smooth flow. A few examples of discourse markers that provide coherence between various parts of the text are discussed below.

Example 1

The following expressions are used to focus on the contrast or contradiction between two ideas or issues

On the one hand …on the other hand, on the contrary, nevertheless, nonetheless, whereas, despite, in spite of, despite, however, yet, although, while.

By emphasizing the contrast, you can signal the switch of direction in your argument.

1. Most of the students enjoy doing chemistry experiments in the lab. *On the other hand,* they hate performing physics experiments.
2. Some people prefer to spend most of their time alone. *However,* some others enjoy most of their time in the company of their friends.
3. *On the one hand,* we have people who are without work and, *on the other,* we have countless jobs which are not being done.

Example 2

Markers such as *in addition to, furthermore, moreover, as well as, also, too, neither… nor, nor,* and *neither* are used to show addition. You can use them to signal the reader that you are about to add an additional reason or example to substantiate your claim. See the following examples:

1. First, they have a desire to conquer nature to achieve the most difficult and most impossible task. *Moreover,* their spirit of adventure and interest in exploring unknown places provide them the impetus to go mountaineering.
2. Minicomputers are the cheapest, smallest, and slowest. *In addition to* these advantages, they rarely occupy more than one room and are often used for a particular kind of work such as data analysis in a research laboratory.

Example 3

You can use expressions such as *as a result, as a consequence, consequently, thereby, thus, therefore,* and *hence* to show the effect or result of an action.

1. Scientists have worked hard to develop ways to decrease infant mortality rate and increase longevity. *As a result,* more people are living longer and scientists will soon have to develop some methods with which to control overpopulation.
2. The team completed the given assignment in record time and proved their efficiency. Consequently, the company appreciated their efforts and offered them few more prestigious projects.

Example 4

The expressions such as *regarding, with regard to, with respect to, as regards, as far as … is concerned,* and *as for* are used to draw your readers' attention to something that you are going to say next. Among these expressions *as for, as far as … is concerned,* and *as regards* would enable you to inform your readers in advance that there will be a shift in the subject. Look at the following sentences:

1. This manual provides you with all guidelines pertaining to the working of automatic grinders. *As for/as regards* the manual grinders, you need to look into the supplementary booklet.
2. This company is reputed for its purification process for various liquids. *As far as* the purification of water is concerned, it needs to resort to a different method.
3. *With regard to/regarding/with respect to* this new technology developed by our research division, I would like to specify that it could withstand high temperatures.

From the preceding discussion, we can say that a text is coherent when it flows smoothly in a clear direction and when all the sentences are logically arranged.

The following points need to be kept in mind in order to achieve discourse coherence:

- Avoid using too many short, choppy sentences.
- Organize your sentences in a logical sequence.
- Use transitions that suit your purposes, taking care to avoid their overuse.

Example of incoherent paragraph

Canadian software companies face several tough challenges in the new millennium because of 'brain drain', the weak Canadian dollar, and the monopoly held by their American counterparts. However, the Canadian dollar continues its downward spiral even today. 'Brain drain' is a bad thing. Our greatest resources are leaving. Microsoft is squeezing out its competitors. In comparison, if the 'brain drain' continues, Canadian companies will find it difficult to produce innovative software. The weak dollar will hurt us. It will help the Americans.

(*Source:* http://www.writingcentre.ubc.ca/workshop/tools/unity.htm)

The paragraph above is incoherent because it uses choppy sentences ineffectively, arranges the ideas illogically, and uses transition words in improper contexts. The following paragraph is an example of a coherent paragraph.

Example of coherent paragraph

Canadian software companies face several tough challenges in the new millennium because of 'brain drain', the weak Canadian dollar, and the monopoly held by American corporations. 'Brain drain' is a catchy new term for the practice of American companies enticing brilliant Canadian doctors, scientists, researchers, programmers, and business people to cross over the border into the United States. The US firms lure Canadian talent with the strong American dollar and the competitive research and business edge that many such companies have as a result of industry monopolies. Simply put, Canadian companies are being soundly beaten because they find themselves on a playing field that is no longer level.

(*Source:* http://www.writingcentre.ubc.ca/workshop/tools/unity.htm)

Notice how each sentence flows logically into one another. The paragraph stays consistent with both the content and the organization of the topic sentence.

Sequencing of Sentences

The process of sequencing the sentences in a text or, in other words, arranging the order in which the sentences need to logically appear in a text, is closely related to what you have learnt just now, that is, discourse coherence. If you are clear about the concept of discourse markers you may be able to identify the order in which the sentences need to be organized in a text. Look at the following paragraph:

Practice 2

The passage given below does not include adequate discourse markers for coherence. Combine/ rearrange sentences if necessary and use appropriate transitional words/phrases so as to make the text coherent:

The ancient Egyptians were masters of preserving dead people's bodies by making mummies of them. Mummies several thousand years old have been discovered nearly intact. The skin, hair, teeth, fingernails and toenails, and facial features of the mummies were evident. It is possible to diagnose the disease they suffered in life, such as smallpox, arthritis, and nutritional deficiencies. The process was remarkably effective. Sometimes apparent were the fatal affl ictions of the dead people: a middle-aged king died from a blow on the head and polio killed a child king. Mummification consisted of removing the internal organs, applying natural preservatives inside and out, and then wrapping the body in layers of bandages.

In Chapter 3, we report a study that found that subjects perceive as variable units only what the theory claims is a unit. Another series of crucial studies is the comparison and contrast experiments reported in Chapter 4, which shows that we do not distinguish complex concepts of different lengths as some current theories do. To a great extent, we have succeeded in showing our theory is valid. Our main concern was to empirically test the theory that forms the background for this work. Chapter 2 reports a study which shows that the rate of perceiving variations in length relates directly to the number of connectives in the base structure of the test.

Are the sentences logically organized? No. What has gone wrong and how can you rectify this mistake? Given below are some steps to arrange the sentences of the text in a logical and coherent manner:

- Read all the sentences and try to identify the topic sentence.
- Place it in the beginning of the text/paragraph.
- Look for discourse markers such as signal words, pronouns, repetitions, synonyms, etc.

Here is the revised version beginning with the topic sentence and containing the supporting sentences in a logical order:

Our main concern was to empirically test the theory that forms the background for this work. To a great extent, we have succeeded in showing our theory is valid. Chapter 2 reports a study which shows that the rate of perceiving variations in length relates directly to the number of connectives in the base structure of the test. In Chapter 3, we report a study that found that subjects perceive as variable units only what the theory claims is a unit. Another series of crucial studies is the comparison and contrast experiments reported in Chapter 4, which shows that we do not distinguish complex concepts of different lengths as some current theories do.

Practice 3

Rearrange the following sentences in a logical and sequential way so as to make a coherent paragraph:
- (a) The dual purpose bicycle has a permanent attachment and modified broad stand cum carrier.
- (b) There are two types of pedal power devices in existence today.
- (c) This modified bicycle is called a dual-purpose bicycle.
- (d) It can also be used to power a drill, a woodworking lathe, and a circular saw.
- (e) The second type of pedal power device is a modified bicycle, which can be used both as a vehicle for transportation as well as power production.
- (f) Some examples are paddy threshers, winnowers, groundnut shell removers, small water-pumps, and grinders.

(g) This rotary motion can be used to operate machines such as threshers, winnowers, pumps, woodworking lathes, and metal lathes.

(h) The first type is a stationary one, in which bicycle parts such as the frame, crank, chain, and free wheel are used to produce a rotating notion.

SUMMARY

How we are reading a text is as important as what we are reading. By skimming and scanning a text we may be able to get its gist. In addition, by applying certain other techniques such as summarizing, predicting the content, and anticipation, and by practising certain study skills, such as critical reading, critical thinking, effective googling, note-making, and sequencing sentences in a text, we will be able to improve the effectiveness and efficiency of our reading comprehension. These skills will also enhance our ability to understand and analyse information obtained through various sources. Besides using all these techniques and skills, we can also use the PQRST reading technique to enhance our reading comprehension skills.

EXERCISES

1. Read the following passage and then answer the questions that follow:

It is like a horror movie without an end. Scenes of death and devastation brought on by the 'worst ever quake to hit the country since independence' are now etched permanently in our memory. On the morning of January 26, 2001, the unstable earth under the Rann of Kutch in Northern Gujarat heaved and collapsed causing an earthquake recorded 6.9 on the Richter scale (China recorded 7.4 and the U.S. measured 7.9 due to different methods of calculation). But no scale can possibly measure the magnitude of the desolation and sorrow that the killer quake left behind in the villages, towns, and cities of Gujarat. Places like Bhuj, Anjar Bachau, and Sukhpar have been completely flattened. Buildings collapsed trapping thousands of people in the rubble and left the survivors with nothing to live for. We saw the grief-stricken faces of those who had lost their families, distraught men, women, and children huddled in the open in the cold night, and terrified villagers on foot desperately looking for places that might be safe.

What causes an earthquake? The surface of the earth is made of huge plates. They slowly move over, under, and past each other. Sometimes, the movement is gradual. At other times, the plates lock into one another unable to release the energy created by the movement. When this accumulated energy grows strong enough, the plates break free and snap into a new position. Vibrations make the structures around quiver and shake and fall. The fracture in the earth's crust is called a 'fault'. If all the stress has not been released, more tremors (aftershocks) can occur in the fault zone. The epicentre is the point on the earth's surface directly above where the quake is focussed.

Earthquakes can occur beneath the ocean floor. Then immense waves (tsunamis) as high as 15 metres caused by the freed energy travel across the waters at great speed and reach the shores. They engulf the coastal areas and cause severe damage. India has a grim history of earthquakes. Calcutta (1737, 300,000 dead) and Assam (1897) saw the worst of them. A series of tremors ravaged Udaypur, Uttarkashi, Chamoli, Latur, and Jabalpur in the last 12 years. The Rann of Kutch itself lost 2000 people in the 1819 quake and again in 2001 the death toll was close to 20,000. Dams built in the quake-prone areas, concentration of population, decrease in groundwater level can all be reasons for these disasters, say environmentalists. Dr R. Bilham of Colorado warns that because of the southward movement of the surface, 60 per cent of the Himalayas are overdue for a quake.

(a) Say whether the following statements are true or false:
- (i) Rann of Kutch suffered earthquake in the year 1897.
- (ii) Earthquake in the ocean causes huge waves up to 15 metres.
- (iii) The Himalayas are prone to earthquake.
- (iv) Calcutta and Assam saw the worst earthquake in the last 12 years.
- (v) Dr R. Bilham of Colorado is an environmentalist.

(b) Write short answers within one or two sentences:
- (i) What do you mean by fault?
- (ii) What is epicentre?
- (iii) State a few reasons for the cause of an earthquake?
- (iv) In correct chronological order, mention the places affected due to earthquake.

(c) Choose the most appropriate answer:
- (i) Which one of the following is not a cause for earthquake?
  - Dams built in the quake-prone area
  - Concentration of population
  - Decrease in groundwater level
  - Himalayan mountains
- (ii) The exact intensity of 26 January earthquake in Kutch was
  - 6.9 in Richter scale
  - 7.4 in Richter scale
  - 7.9 in Richter scale
  - 6.7 in Richter scale

(d) Complete the following sentences:
- (i) The worst earthquake that hit in dependent India is _____ .
- (ii) Tsunami is the name of _____ .

(e) Fill in the blanks with suitable words:
- (i) Because of the _____ movement, 60% of the Himalayan region is overdue for a quake.
- (ii) Another name for an earthquake is _____ .

2.(a) Scan the passage on 'Extraction of Gold' given below quickly to identify the types of mining and the ways of obtaining pure gold. Then fill the blanks in the following sentences:
- (i) Two types of mining are _____ and _____ .
- (ii) Three ways of obtaining pure gold are _____ , _____ , and _____ .

Extraction of Gold

The method of mining gold varies with the nature of the deposit. Two types of deposit can be considered here: one is placer deposit, which refers to the occurrence of gold in particles in the sand or gravel in the bed of a river; the other is lode mine, which refers to gold occurring as veins in gravel or rock. In placer mining, the separation of gold from gravel or other impurities is done by sifting. Hand panning is also common, in which water and gold-containing gravel are swirled in a pan. Gold, being heavy, settles down, and the gravel is washed away. In lode mining, shafts are dug into the rock following the veins of gold. Using explosives, the rock is broken and the ore is obtained. The ore is then transported to mills.

In milling, the ore is first crushed using heavy machines. This is followed by sluicing, that is, using water to wash the ore into sluices or artificial water-channels in which there are grooves which trap the gold.

There are three ways in which this gold is treated to obtain pure gold. They are floatation, amalgamation, and cyanidation. In the first method, a frothing agent is added to produce foam. A collecting agent is used to produce a film on the gold, which then sticks to the air bubbles. Gold is then separated from the top. In amalgamation, the ore, mixed with water to form a pulp, is collected on a copper plate covered with mercury. The mercury is then removed, partly by squeezing it out and partly by distillation. The cyanide process is now widely used. In this process, a weak solution of sodium, potassium, or calcium cyanide is used to dissolve the gold. The gold is then precipitated by the addition of zinc dust.

The gold thus obtained is smelted and cast into bars.

3. Read the following passage and answer the questions that follow as directed:

The Fight against Wound Infections

In 1865, the British surgeon Joseph Lister (1827–1912) succeeded in devising for the first time a truly antiseptic principle for treating wounds. Until the middle of the nineteenth century, surgery was not only a very gruesome trade—as there was no general anesthesia before that time—but also a dangerous method of treatment, which was always followed by a protracted and often fatal infection.

The patients died even after the slightest operation. All wounds suppurated, and in the hospital wards the sweetish smell of pus everywhere prevailed. The ancient doctrine that this was 'good and laudable pus', and must, therefore, be regarded as a sign of the favourable healing of the wound, was not yet dead.

In the hospitals of Lister's time, 'charpie' was used as a dressing. Charpie was made out of old linen cloth, which had become easily teased through frequent boiling, and attendants and patients who were not very ill teased out threads of varying lengths and thicknesses; and these threads were then brought together again to make a soft, absorbent material. Before it was used to pack a wound or as a dressing, the charpie was washed only with cold water without soap—and often it was not washed at all. The instruments and the sponges used to staunch blood were likewise washed only in cold water.

Like all surgeons in all countries Lister was troubled by the fact that a compound fracture—that is, a fracture in which one or both of the broken ends of the bone has pierced the overlying skin and soft tissues—never did well, and that in such cases, amputation of the limb nearly always had to be performed. Further, in all countries, the mortality rate from septic diseases after amputation varied between 30 per cent and 50 per cent. This was a dreadful state of affairs. As a preliminary to an attack on this problem, Lister had long been carrying out important work on inflammation and the behaviour of the blood during that process. He was led to the conclusion that wound suppuration was decomposition (or putrefaction) brought about by the effect of the atmosphere on blood or serum contained in the wound. But Lister was handicapped by the belief, widely held at that time, that putrefaction was due to oxygen in the air. He spent much time trying to exclude the air from wounds but not unexpectedly, these efforts were unsuccessful. Then in 1865 Lister learned for the first time about the important work of the French chemist Louis Pasteur (1822–1895) on that 'organized corpuscles' (i.e., living bacteria) are present everywhere in the air. This was the clue for which Lister was searching. He deducted that in the case of septic or 'putrefying' wounds, it was not the air itself but the organisms in the air which caused the sepsis.

Lister decided that these organisms must be killed before they obtained access to the wound. He tested the killing effect of a number of substances on bacteria; and after very careful experiments, he decided to use carbolic acid, not only as a wound-dressing, but in a systematic manner so as to prevent suppuration entirely in the part concerned. The hands of the surgeon and his assistants, and also all the instruments to be used, were soaked in a solution of carbolic acid. So also was the wound itself, and Lister did a lot of research in order to find suitable materials for dressings which would give off the carbolic acid slowly into the wound. For many years also Lister had the atmosphere in the operating theatre sprayed with a fine mist of carbolic acid, and the spray was also used during the change of a dressing. But it was later shown that the spraying of the atmosphere was not necessary.

Having, after these experiments, decided on the method he would employ, Lister tried out his new principle on 12 August 1865 in the treatment of a compound fracture in a patient in his ward in the Glasgow Royal Infirmary. A perfect result was obtained. Two years later Lister published a series of cases treated by the aid of his principle, and within a short time he was performing operations which previously, because of the danger of sepsis, would not have been undertaken by any surgeon.

Lister's methods soon found favour in Scotland, but English surgeons were very slow to adopt them. It is to the credit of German surgeons that they understood, perhaps more quickly and more thoroughly than any other, the revolution in surgery which Lister had effected. His antiseptic principle was enthusiastically supported in Germany and it led to the great technical advances of German surgeons.

(*Source:* Pollak, K. and Underwood, E.A., *The Healer*, Nelson, 1968)

(a) Complete each of the given sentences with a verb from the following list. Use each verb only once.

> supported, performed, undertaken, deducted, soaked, led, sprayed, followed, varied, troubled, obtained, effected

(i) In the middle of the nineteenth century a surgical operation was _____ dead.

(ii) Lister was _____ by the fact that in the case of compound fracture, the affected limb had to be removed.

(iii) The mortality rate after amputation _____ from 50 per cent to 30 per cent.

(iv) Lister was _____ to believe that pus was formed by the effect of the atmosphere on the blood or serum contained in the wound.

(v) Lister _____ that organisms in the air caused sepsis.

(vi) Lister _____ a perfect result when he applied the antiseptic principle on the treatment of a compound fracture in a patient.

(vii) Lister _____ all the instruments in a solution of carbolic acid before they were used for any surgical operation.

(viii) Lister ensured that the atmosphere in the operating theatre was _____ with carbolic acid.

(ix) In Lister's time, surgery was _____ in very bad conditions.

(x) Lister began to do such operations which no one would have _____ some years back.

(b) A doctor who performs operations is called a surgeon. Try to match the name of the specialist in the left column with his/her specialty in the right column.

	Name	Specialty
(i)	ophthalmologist	women's diseases
(ii)	neurologist	nature of diseases
(iii)	cardiologist	skin
(iv)	dermatologist	mental illness
(v)	gynecologist	illness of children
(vi)	pathologist	nerves
(vii)	psychiatrist	heart
(viii)	paediatrician	eyes

(c) Here is an outline of the main ideas used in the reading passage. Construct a summary of the passage from these ideas after reading the passage again.

Surgery before 1865 was very dangerous __ A compound fracture never did well __. Morality rate from these diseases varied between 30 per cent to 50 per cent __ suppuration was caused by the effect of the atmosphere on the blood __ suppuration was due to oxygen in the air __ Lister learnt from Louis Pasteur that it was not by the living bacteria __ He used carbolic acid __. On 12 August 1865, he successfully experimented with his new __ for treating __. His revolution was __.

4. Use the appropriate form of the words given in the following list to fill the gaps in the paragraph that follows.

> renowned, globe, maintenance, option, survival, drawback, estimate, objective, disease, contaminated

The India Mark II hand pump is _____ all over the world. Clean drinking water is the key to human _____. The United Nations _____ that 80 per cent of all _____ in the world are caused by water. The United Nations' _____ is to provide clean drinking water for every man, woman, and child around the _____ by the year 1990. They considered many but finally decided that a hand pump was the best solution. Older types of pump used before the India Mark II had a number of _____, for example, they did not last very long. The India Mark II is very successful but it has one problem: that is the _____ of the pump over a period of time.

5. The following bar chart provides the comparative figures of the major sources of information in the manufacturing organizations of three countries, namely, the United Kingdom, Canada, and the United States. Interpret this chart and write a paragraph of about 200 words including the comparative analysis.

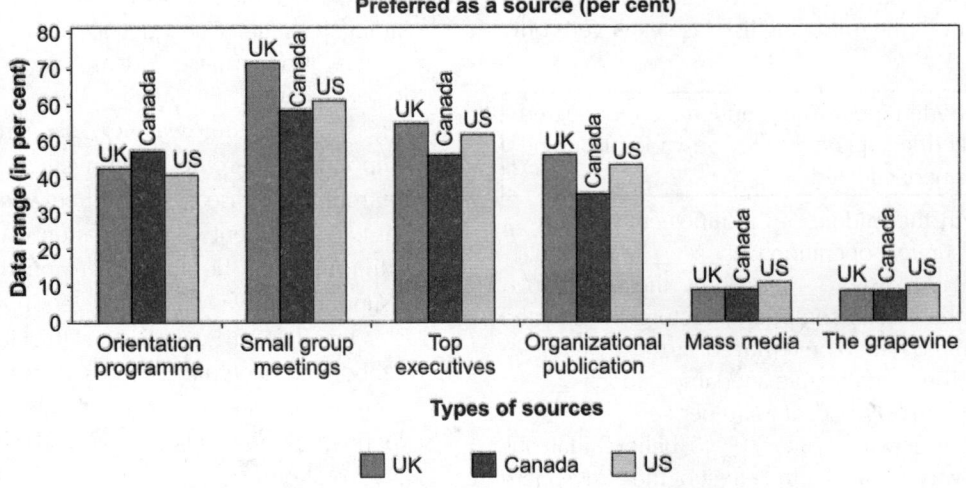

Preferred as a source (per cent)

6. Interpret the line graph given below. It depicts the training needs (in %) of entry level and the middle level managers in various areas in an organization. Including the information represented by all the three lines in the illustration, write a paragraph of about 150 words.

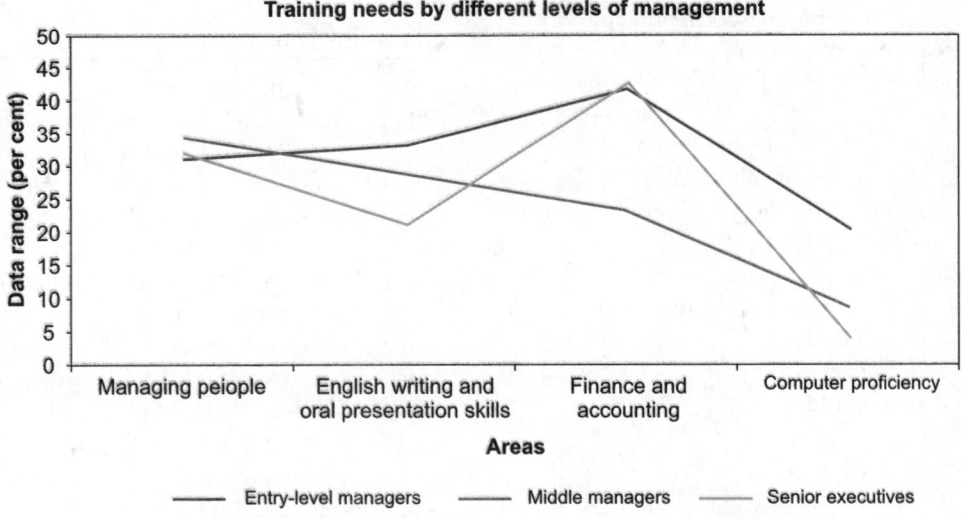

7. Read the following passage and make notes from it using the mapping method.

The use of heat pumps has been held back largely by scepticism about advertisers' claims that heat pumps can provide as many as two units of thermal energy for each unit of electrical energy used, thus apparently contradicting the principle of energy conservation.

Heat pumps circulate a fluid refrigerant that cycles alternatively from its liquid phase to its vapour phase in a closed loop. The refrigerant, starting as a low-temperature, low-pressure vapour, enters a compressor driven by an electric motor. The refrigerant leaves the compressor as a hot, dense vapour and flows through a heat exchanger called the condenser, which transfers heat from the refrigerant to a body of air. Now the refrigerant, as a high-pressure, cooled liquid, confronts a flow restriction which causes the pressure to drop. As the pressure falls, the refrigerant expands and partially vapourizes, becoming chilled. It then passes through a second heat exchanger, the evaporator, which transfers heat from the air to the refrigerant, reducing the temperature of this second body of air. Of the two heat exchangers, one is located inside, and the other one outside the house, so each is in contact with a different body of air: room air and outside air, respectively.

The flow direction of refrigerant through a heat pump is controlled by valves. When the refrigerant flow is reversed, the heat exchangers switch function. This flow-reversal capability allows heat pumps either to heat or to cool room air.

Now, if under certain conditions, a heat pump puts out more thermal energy than it consumes in electrical energy, has the law of energy conservation been challenged? No, not even remotely: the additional input of thermal energy into the circulating refrigerant via the evaporator accounts for the difference in the energy equation.

Unfortunately, there is one real problem. The heating capacity of a heat pump decreases as the outdoor temperature falls. The drop in capacity is caused by the lessening amount of refrigerant mass moved through the compressor at one time. The heating capacity is proportional to this mass flow rate: the less the mass of refrigerant compressed the less the thermal load it can transfer through the heat pump cycle. The volume flow rate of refrigerant vapour through the single-speed rotary compressor used in heat pumps is approximately constant. But cold refrigerant vapour entering a compressor is at lower pressure than warmer vapour. Therefore, the mass of cold refrigerant and thus the thermal energy it carries is less than if the refrigerant vapour were warmer before compression.

Here, then, lies a genuine drawback of heat pumps: in extremely cold climates—where the most heat is needed—heat pumps are least able to supply enough heat.

8. Read the following paragraph and write down which words indicate the author's opinion. Also explain what his opinion is.

There is the first problem with tipping: the more discretion you have in the matter the more unpleasant it is. Tipping is an aristocratic conceit—'There you go, my good man, buy your starving family a loaf'—best left to an aristocratic age. The practising democrat would rather be told what he owes right up front. Offensively rich people may delight in peeling off hundred dollar bills and tossing them out to grovelling servants. But no sane, well-adjusted human being cares to sit around and evaluate the performance of some beleaguered coffee vendor.

9. Read the following passage carefully and answer the questions in your own words.

The most frightening aspect of malnutrition is that it is likely to cause permanent damage to the brain. Experiments carried out on animals suggest that brain damage due to malnutrition is irreversible. This may not be true of human beings. Nevertheless, known facts as well as results of tests conducted so far point to alarming conclusions. The human brain grows very fast. Three months before its birth, the child's brain weighs one-fourth of the adult's brain. In one year, the brain of a child is already 66 per cent of the weight of the adult brain. At four years the brain weighs 90 per cent and at eight the child's brain is the size of an adult's and there is no further increase. Quite obviously the first four years are crucial for the development of the brain. And if the child suffers from malnutrition, the damage to the brain may well be permanent.

Fortunately and thanks to the research carried out by scientists in India and abroad and systematic attempts made by the National Institute of Nutrition, Hyderabad, to analyse the causes of malnutrition in this country, one need not wait for years to solve at least some aspects of the problem.

For instance, it was found that the addition of iodine in the lake-water salt consumed in the sub-Himalayan region would go a long way to controlling goitre. Again, experiments made by the Institute reveal that inexpensive green leafy vegetables are a good alternative to eggs and butter as a rich source of Vitamin A.

The Institute has also carried out considerable research into fortification of various foods. Modern bread is a case in point. But since bread is beyond the means of the poor the Institute has found some items of mass consumption which can be strengthened with certain proteins and amino acids. Fortification of common salt is considered the most promising possibility.

Questions

(a) What is the most serious harm that may be caused by malnutrition in childhood?

(b) Why is good nourishment so essential during the first four years of child's life?

(c) What suggestions have been made by the National Institute of Nutrition for controlling goitre in certain parts of India?

(d) List the other three suggestions made by the Institute for curing the ill effects of malnutrition.

10. Each passage in this group is followed by questions based on its content. After reading a passage, choose the best answer to each question. Answer all questions following a passage on the basis of what is stated or implied in that passage.

Passage 1

After inventing dynamite, Swedish-born Alfred Nobel became a very rich man. However, he foresaw its universally destructive powers too late. Nobel preferred not to be remembered as the inventor of dynamite, so in 1985, just two weeks before his death, he created a fund to be used for awarding prizes to people who had made worthwhile contributions to humanity. Originally, there were five awards: literature, physics, chemistry, medicine, and peace. Economics was added in 1968, just sixty-seven years after the first awards ceremony. (5)

Nobel's original legacy of nine million dollars was invested, and the interest on this sum is used for the awards which vary from $30,000 to $125,000. (6)

Every year on 10 December, the anniversary of Nobel's death, the awards (gold medal, illuminated diploma, and money) are presented to the winners. Sometimes politics plays an important role in the judges' decisions. Americans have won numerous science awards, but relatively few literature prizes. (9)

No awards were presented from 1940 to 1942 at the beginning of World War II. Some people have won two prizes, but this is rare; others have shared their prizes. (11)

(a) The word 'foresaw' in sentence 2 is nearest in meaning to
   (i) prevailed   (iii) prevented
   (ii) postponed   (iv) predicted

(b) The Nobel Prize was established in order to
   (i) recognize worthwhile contributions to humanity
   (ii) resolve political differences
   (iii) honour the inventor of dynamite
   (iv) spend money

(c) In which area have Americans received the most awards?
   (i) Literature   (iii) Economics
   (ii) Peace   (iv) Science

(d) All of the following statements are true EXCEPT
   (i) awards vary in monetary value
   (ii) ceremonies are held on 10 December to commemorate Nobel's invention
   (iii) politics plays an important role in selecting the winners
   (iv) a few individuals have won two awards

(e) In how many fields are prizes bestowed?
   (i) 2   (iii) 6
   (ii) 5   (iv) 10

(f) It is implied that Nobel's profession was in
   (i) economics   (iii) literature
   (ii) medicine   (iv) science

(g) In sentence 3, 'worthwhile' is closest in meaning to
   (i) economic   (iii) trivial
   (ii) prestigious   (iv) valuable

(h) How much money did Nobel leave for the prizes?
   (i) $30,000   (iii) $155,000
   (ii) $125,000   (iv) $9,000,000

(i) What is the main idea of this passage?
   (i) Alfred Nobel became very rich when he invented dynamite.
   (ii) Alfred Nobel created awards in six categories for contributions to humanity.
   (iii) Alfred Nobel left all of his money to science.
   (iv) Alfred Nobel made a lasting contribution to humanity.

(j) The word 'legacy' in sentence 6 means most nearly the same as
  (i) legend   (iii) prize
  (ii) bequest   (iv) debt

Passage 2

Ever since humans have inhabited the earth, they have made use of various forms of communication. Generally, this expression of thoughts and feelings has been in the form of oral speech. When there is a language barrier, communication is accomplished through sign language in which motions stand for letters, words, and ideas. Tourists, the deaf, and the mute have had to resort to this form of expression. Many of these symbols of whole words are very picturesque and exact and can be used internationally; spelling, however, cannot. (5)

Body language transmits ideas or thoughts by certain actions, either intentionally or unintentionally. A wink can be a way of flirting or indicating that the party is only joking. A nod signifies approval, while shaking the head indicates a negative reaction. (8)

Other forms of non-linguistic language can be found in Braille (a system of raised dots read with the fingertips), signal flags, Morse code, and smoke signals. Road maps and picture signs also guide, warn, and instruct people. (10)

While verbalization is the most common form of language, other systems and techniques also express human thoughts and feelings. (12)

(a) Which of the following best summarizes this passage?
  (i) When language is barrier, people will find other forms of communication.
  (ii) Everybody uses only one form of communication.
  (iii) Non-linguistic language is invaluable to foreigners.
  (iv) Although other forms of communication exist, verbalization is the fastest.

(b) The word 'these' in sentence 5 refers to
  (i) tourists
  (ii) the deaf and the mute
  (iii) thoughts and feelings
  (iv) sign language motions

(c) All of the following statements are true EXCEPT
  (i) there are many forms of communication in existence today
  (ii) verbalization is the most common form of communication
  (iii) the deaf and mute use an oral form of communication
  (iv) ideas and thoughts can be transmitted by body language

(d) Which form other than oral speech would be most commonly used among blind people?
  (i) Picture signs   (iii) Body language
  (ii) Braille   (iv) Signal flags

(e) How many different forms of communication are mentioned here?
  (i) 5   (iii) 9
  (ii) 5   (iv) 11

(f) The word 'wink' in line 7 means most nearly the same as
  (i) close one eye briefly
  (ii) close two eyes briefly
  (iii) bob the head up and down
  (iv) shake the head from side to side

(g) Sign language is said to be very picturesque and exact and can be used internationally EXCEPT for
  (i) spelling   (iii) whole words
  (ii) ideas   (iv) expressions

(h) People need to communicate in order to
  (i) create language barriers
  (ii) keep from reading with their fingertips
  (iii) be picturesque and exact
  (iv) express thoughts and feelings

(i) What is the best title for the passage?
  (i) The Importance of Sign Language
  (ii) The Many Forms of Communication
  (iii) Ways of Expressing Feelings
  (iv) Picturesque Symbols of Communication

(j) Who would be MOST likely to use code?
  (i) A scientist   (iii) An airline pilot
  (ii) A spy   (iv) A telegrapher

11. Read the following case carefully using your critical reading skills where you apply both creative and critical thinking. Answer the questions that follow.

The problem involved SBB Manufacturing, a large producer of non-woven materials such as clothing lining, automobile insulation, and hospital gowns. SBB identified a large market opportunity in supplying automobile repair shops with a new material for disposable shop rags—rags for cleaning and wiping oil and grease residue and spills.

The disposable rags replaced the reusable shop rags supplied by industrial launderers; they recycled cloth rags by cleaning them and returning them to the shops each week. The cloth rags were often returned torn and worn, and occasionally they had embedded metal shivs capable of gouging man and machine.

SBB, on the other hand, offered clean, soft, and strong rags. They had the feel of cloth and could be laundered, but they were less expensive for the mechanics. Another benefit was that each night they threw the SBB rags in the trash, instead of having to store the used cloth rags, which were full of oil, grease, and solvents, and created a fire hazard.

Everything was going well for SBB until the Environmental Protection Agency (EPA) created a new regulation. Suddenly, oil or grease soaked rags were considered hazardous materials, and companies were limited to the amount of hazardous materials they could dispose of each month. The amount of SBB contaminated rags that the large auto shops used far exceeded the 100 kilos per month that the EPA was allowing.

The auto shops had just five months to comply with the new EPA regulations, and they informed SBB that they would go back to cloth rags unless something could be done about the situation.

All of the ideas that SBB thought of either cost too much money or took too much time. As a result, we were invited by the marketing head to run a creative session to help SBB out of the mess.

Questions

(a) Write a brief (3–5 sentences) description of the problem.

(b) What have you tried to resolve your situation, and why do you think that resolution has failed?

(c) What other obstacles stand in the way of a successful solution?

(d) Summarize what needs to be changed in your situation in one sentence beginning with the words 'How to….'

12. Recall the skills that have been discussed in this chapter under critical reading. Read the following passage and using your creative and critical thinking choose the best option for each of the questions that follow:

Passage

Although the schooling of fish is a familiar form of animal social behaviour, how the school is formed and maintained is only beginning to be understood in detail. It had been thought that each fish maintains its position chiefly by means of vision. Our work has shown that, as each fish maintains its position, the lateral line, an organ sensitive to transitory changes in water displacement, is as important as vision. In each species a fish has a 'preferred' distance and angle from its nearest neighbour. The ideal separation and bearing, however, are not maintained rigidly. The result is a probabilistic arrangement that appears like a random aggregation. The tendency of the fish to remain at the preferred distance and angle, however, serves to maintain the structure. Each fish, having established its position, uses its eyes and its lateral lines simultaneously to measure the speed of all the other fish in the school. It then adjusts its own speed to match a weighted average that emphasizes the contribution of nearby fish.

(a) According to the passage, the structure of a fish school is dependent upon which of the following?

(i) rigidly formed random aggregations

(ii) the tendency of each fish to retain at a preferred distance from neighboring fish

(iii) measurements of a weighted average by individual fish

- II only
- III only
- I and II only
- I and III only
- II and III only

(b) Which of the following best describes the author's attitude toward the theory that the structure of fish schools is maintained primarily through vision?

(i) heated opposition

(ii) careful neutrality

(iii) considered dissatisfaction

(iv) cautious approval

(v) unqualified enthusiasm

(c) The passage suggests that, after establishing its position in the school formation, an individual fish will subsequently
  (i) maintain its preferred position primarily by visual and auditory means
  (ii) rigorously avoid changes that would interfere with the overall structure of the school
  (iii) make continuous sensory readjustments to its position within the school
  (iv) make unexpected shifts in position only if threatened by external danger
  (v) surrender its ability to make quick, instinctive judgments

(d) Generate at least five possible solutions to the problem posed in the following case, analyse each of them in the light of their positive and negative factors and present your analysis in about 300 words:

  A group of eight students from Erudite University plan to go on a motorbike expedition to the nearby hill station Mountpeak on a three-day break. On the way to the hill station, they have to go through a village hamlet. One of the two vehicles which were cruising ahead of the other two runs over a village kid and flees from the spot. Meanwhile, elders in the village come to know about the incident and decide to search out the culprits. For the students, this is the only route to get back to their home town. The group has four girls whose safety had to be taken care of. If you are the members of this group, what would you do?

(e) Supply the missing words from the choices provided. _____ (1) people breathe, pollutants in the air _____ (2) in the lungs or absorbed into the body. And polluted air can harm animals and plants _____(3) people. For this reason, our air supply should be _____ (4) watched and managed to assure _____ (5) good quality.
  (i) When, until, during, if, just as
  (ii) Have deposited, are depositing, had to be deposited, will have deposited, may be deposited
  (iii) As well as, in addition, even if, in spite of, supposing
  (iv) Alternately, previously, hastily, closely, furtively
  (v) Them, its, his, theirs, it

(f) Supply the missing word from the choices provided.
  Petroleum, or crude oil is one of the world's _____ (1) natural resources. Plastics, synthetic fibres, and _____(2) chemicals are produced from petroleum. It is also used to make lubricants and waxes. _____ (3), its most important use is as a fuel for heating, for _____ (4) electricity, and _____ (5) for powering vehicles.
  (i) As important, most important, so importantly, less importantly, too important
  (ii) Much, a lot, plenty, many, less
  (iii) Therefore, however, moreover, hence, rather
  (iv) Generated, to generate, being generated, generate, generating
  (v) Decisively, exclusively, especially, favourably, notably

CHAPTER 11

Elements of Effective Writing

OBJECTIVES

You should read this chapter to know

- ○ how to use the right words in the right context
- ○ the ways to achieve clarity and economy in writing
- ○ the various sentence patterns
- ○ the salient points of sentence construction
- ○ how to write effectively for the web
- ○ pointers on journal writing and collaborative writing

Introduction

Effective writing is an essential skill for any profession, be it technical or business. The purpose of writing any professional document is to communicate specific ideas, and everything about the document should contribute to this goal. Writing is a skill that can be learnt and practised, but to improve our writing skills, we need to be aware of the elements of effective writing. Choosing the right words and phrases, constructing grammatically as well as logically correct sentences, and well-developed and coherent paragraphs—all would contribute to enhancing the quality of writing. In this chapter, we will look into the elements that contribute to the effectiveness of writing.

Right Words and Phrases

Using the right words in the right context is one of the essential elements of effective writing. The following guidelines will help achieve clarity and economy in writing:

- Prefer simple, familiar words to obscure, unfamiliar words.
- Prefer concrete and specific words to abstract and general words.
- Use acronyms carefully.
- Avoid clichés.
- Avoid excessive use of jargon.
- Avoid foreign words and phrases.
- Avoid redundancy and circumlocution.
- Avoid discriminatory writing.

In the following paragraphs, we will delve deeper into these guidelines, explaining their effectiveness through examples.

Use Familiar Words

An important element in the use of words is the selection of familiar words—words that are commonly used in daily language. To do this, one must rely on one's judgement to decide which words are more familiar.

A rule of thumb is to avoid stiff and difficult words as far as possible. For example, use the word *destroy* instead of using the word *annihilate*, use *verify* instead of using *corroborate*, and use *outdated* instead of *antiquated*.

A sentence such as *The conclusion ascertained from a perusal of pertinent data is that a lucrative market exists for the product* would be more easily understood by writing it thus: *The data studied shows that the product has good demand*. Many writers fall into the trap of using high-sounding, pompous words in an attempt to make the work more impressive. On the contrary, such writing confuses the readers. There is no room for such writing, particularly in business and technical communication. The rapid pace of work life today demands that writing be informal in tone, brief, precise, and easy to grasp.

There is no formula for the appropriate selection of words. The writer must keep the target readers in mind and cater to the lowest common denominator of this group. The key is to use words suited to the intelligence level of the readers. The use of plain and familiar words, in contrast to pompous and unfamiliar ones, helps to convey the message in a better and more effective manner. Style does not consist in displaying the width of our vocabulary, but in how successfully we communicate the meaning to the reader. In general, simple writing is the key to effective communication.

> A rule of thumb is to avoid stiff and difficult words as far as possible.

> The writer must keep the target readers in mind and cater to the lowest common denominator of this group.

> 'A good style should show no sign of effort. What is written should seem to be a happy accident.'
> –Somerset Maugham

The following examples show how sentences can be simplified using familiar expressions.

Example 1

  Original: Antiquated machinery was utilized for experimentation.

  Revised: Old machinery was used for the test.

Do Not Insult Your Audience

Whereas using simple language and short sentences is the best way to communicate to all your readers, you could run the risk of irritating some of your audience who might feel you are 'talking down to them'. You will always have to fine-tune your judgement of what your audience needs and understands. You can do this by thinking about how well they know the subject and choosing words, especially technical terms, carefully. If you have to use unfamiliar words, try and build in an explanation of their meaning. Never ever patronize your audience.

Example 2

  Original: Company operations for the preceding accounting period terminated with substantial deficit.

  Revised: The company lost much money last year.

Use Concrete and Specific Words

Concrete words stand for things that exist in the real world such as chair, desk, automobile, flowers, the Taj Mahal, and so on. Abstract words, on the other hand, cover broad meanings—concepts, ideas, and feelings.

Concrete and specific words are always preferable to abstract words. Concrete words permit a limited interpretation and convey a more definite meaning. They tend to be forceful, direct, and to the point, while abstract words tend to be general and vague, and digress from the point.

Good business communication is marked by words that have clear meanings. Instead of saying 'The company suffered a *tremendous* loss in the second quarter of last year', it would make more sense to directly say 'The company suffered a 70 per cent loss in profits in the second quarter of last year'. Quantifying the loss in the above example makes the information more useful and creates a better impact. Otherwise, it hardly conveys much meaning.

It is essential to convey the message in clear terms; ambiguity results in reduced or no communication. Give special emphasis to sentence structure and construction in order to avoid ambiguity. For example, consider the sentence, 'If the child does not like spinach, try boiling it with milk'. Here, the word 'it' does not convey in clear terms whether the writer is referring to the spinach or the child. As another example, consider this obscure instruction written on a bottle top: 'Pierce with pin, then push off'. *Here too*, the ambiguity inherent in the sentence mars the communication. Hence, it is essential to be clear and precise in the choice of words.

Sometimes a sentence is correct in grammatical construction but can be interpreted in a number of ways. This is called *structural ambiguity*. For example, 'Lying at the bottom of the pool, I saw a five rupee note.' Similarly, ambiguity can result from the choice of words—the word selected may have multiple meanings. This is called *lexical ambiguity*. For example, 'The giant *plant* collapsed damaging many buildings'. Such words should be avoided in technical writing. If someone says, 'I was looking at the figures keenly,' the sentence is ambiguous. The words used should be very specific in their meaning in order to avoid alternative interpretations. Some commonly used ambiguous words and phrases a writer should avoid are *bimonthly, quite a few, tenement, quite a bit, table a proposal, slim chance, fairly good*.

Table 11.1 shows some examples of how abstract words and phrases can be substituted to sound specific and clear.

Table 11.1   Substitution of abstract words and phrases with concrete ones

Abstract	Concrete
An industrial worker	A welder
Furniture	An arm chair
Apparel	A shirt
Good attendance record	100% attendance record
The leading company	First among 3928 companies
The majority	73%
In the near future	By Thursday noon
Substantial amount	₹ 1,98,00,000

Use Acronyms and Abbreviations Sparingly

An acronym is a word formed by combining the first letters of the words that make up the name of something. For example, WHO is the acronym for World Health Organization. We are all familiar with the common acronyms FBI (Federal Bureau of Investigation), POTA (Prevention of Terrorism Act), IAF (Indian Air Force), and IBM (International Business Machines).

While working in a particular field, we learn the related technical terms and acronyms. In due course, we use these terms and acronyms freely in communicating with people in the same area of work. While acronyms help achieve brevity and are necessary in technical documents, excessive dependence on them leads to a loss of clarity. For example, one can talk to a friend in the field of electronics about re-routing OAM messages using UDP/IP in a CDPD radio. However, for someone who is pursuing a different field of study, these acronyms will be unintelligible. Therefore, do not use acronyms relating to technical fields when communicating with individuals unfamiliar with the subject.

Nevertheless, many acronyms are familiar to the layperson and can be freely used. For example, acronyms such as RADAR (radio angular detecting and ranging), NASA (National Aeronautics Space Agency),

IIT (Indian Institute of Technology), AIDS (Acquired Immune Deficiency Syndrome) are commonly used and may not be difficult to decipher.

Whether writing a technical document or sending an informal electronic mail, remember the following points:

- Introduce acronyms the first time you use them by defining them in full, parenthetically.
- Restrict the number of acronyms in individual sentences.
- When addressing non-expert readers, try to avoid using more than one acronym in a sentence.
- Use the full term for an acronym when beginning a new section or when the term has not been used for several pages in the technical document.
- When several potentially confusing acronyms are being used, they can be defined in a separate glossary (an alphabetized list of terms, followed by their definitions).

Given below are a few samples of acronyms that may cause ambiguity in a technical document or an email:

CIA	Cash in Advance
CPA	Certified Public Accountant
FICA	Federal Insurance Contributions Act
CIM	Computer Integrated Manufacturing
IRA	Individual Retirement Account/Irish Republican Army/Instructionally Related Activities

Acronyms are extensively used in email and sms communication. It is acceptable as part of informal communication, but should be strictly avoided in official mails. Some such acronyms are listed in Table 11.2.

Abbreviations on the other hand are short forms of words, such as 'St' for 'Saint', 'av.' for 'average', etc. The rules for using abbreviations are also similar to those for using acronyms. Some of the specific usages to avoid are as follows:

- Do not use phonic abbreviations like 'u' (for you) and 'tho' (for though) in formal or semi-formal writing. It creates the impression that the writer is not serious in his/her work.
- Abbreviations for English units of measurement use periods (doz., oz., ft., in., pt., yd.,) but abbreviations for metric units of measurement do not (g, m, cm, km, ha, MB.)
- Though this rule varies as a matter of style, abbreviations that include the last letter of the word being shortened should not take a period. For example, Dr (doctor), Mr (mister), but Prof. (professor), col. (column)
- Do not abbreviate words at the beginning of sentences.
- Do not abbreviate proper nouns.

Table 11.2 Acronyms in email

Acronym	Definition
Atw	At the weekend
Brb	Be right back
Btw	By the way
Cm	Call me
Cu	See you
cul	See you later
dur?	Do you remember?
f2f	Face to face
Gal	Get a life
gmtl	Great minds think alike
Gr8	Great
idk	I don't know
LOL	Laugh out loud
Ttyl	Talk to you later
w4f	Waiting for you
Wfh/wah	Working from home/working at home

Appendix B in the Online Resource Centre contains a list of commonly used acronyms and abbreviations.

Avoid Clichés

Clichés are overused words and phrases that have lost their appeal. Nobody likes to hear stale, once-fashionable phrases that add little or no meaning to what we say. Some well-known clichés include *food for thought, teeming millions, last but not the least, fishing in troubled waters,* and *to burn the midnight oil.* An indiscriminate use of clichés makes technical communication unimpressive and stale. However, there may be instances where a cliché conveys the idea so effectively that it becomes extremely difficult to improve upon it or think of a substitute. Hence, it is completely the writer's judgement on how often to use clichés and when to avoid them.

Example 1

Original: The report discusses threadbare the problem of air pollution in India.

Revised: The report discusses the problem of air pollution in India in detail.

Example 2

Original: The engineers left no stones unturned to construct the bridge in time.

Revised: The engineers worked very hard to construct the bridge in time.

Avoid Excessive Use of Jargons

Jargon is defined as words or expressions that are used by a particular profession or group of people, and are difficult for others (outside this group) to understand. Every profession has its specific jargon.

> Jargon is defined as words or expressions that are used by a particular profession or group of people, and are difficult for others (outside this group) to understand.

Doctors communicate with each other in medical terms that are seldom understood by the layperson. Similarly, software professionals resort to the use of computer-related jargon. The use of jargon is acceptable as long as the communication takes place among individuals belonging to the same field or profession. However, to sustain the interest of a large readership, it is necessary to minimize or avoid the use of jargon, since the reader may not be a specialist in the subject. The content has to be simplified to the extent possible when presenting it to a non-technical audience, and this requires the use of non-technical and easy words and descriptions.

Jargon is often used due to the false assumption that complex ideas cannot be expressed without the use of highly technical language. However, in effect, jargon only widens the semantic gap between the professional and the layperson. One of the hallmarks of effective communication is its ability to reach a population larger than what it is intended for. If such is the goal, excessive use of jargon has little role to play. Table 11.3 shows some examples of how technical jargon can be replaced by their corresponding simpler representations.

'The contract is legally sound now; all we need to do is flavour it with some jargon.'

Avoid Foreign Words and Phrases

Like jargon, some writers indulge in the use of foreign words and phrases. It makes little sense to use words such as *vide supra* or *raison d'être* in place of better understood English equivalents such as 'see above' and 'the justification for', respectively. Some foreign words are unavoidable and are also universally understood and accepted; for example, etc. (*et cetera*), i.e. (*id est*), e.g. (*exempli gratia*), a.m. (*ante meridiem*), p.m. (*post meridiem*), and so on. It is not wrong therefore to use such words or phrases as long as the reader is able to grasp the message being conveyed. Table 11.4 lists some foreign words with their meanings.

Avoid Redundancy and Circumlocution

We need to avoid the use of unnecessary words or resort to needless repetition of an idea. This makes the writing redundant. Redundancy is the part of a message that can be eliminated without loss of much information. Re-emphasizing the same point through the usage of extra words does not confer any extra worth to the idea.

Once unnecessary words are pruned, the idea stands out clearly and effectively. Some examples of redundancy are—the superfluous words are italicized—*basic* fundamentals, return *back, adequately* enough, resume *again, the month of* May, few *in number, detailed* perusal, true *fact, new* innovations, *future* prospects, and so on.

Superfluity in writing is also brought about by beating around the bush or writing in a circumlocutory manner.

> Redundancy is like killing a fly twice over.

In these cases, the entire sentence needs to be rewritten in order to make it crisp and precise.

Table 11.3 Avoiding technical jargon

Technical jargon	Familiar expressions
Accrued interest	Unpaid insurance
Annual premium	Annual payment
Assessed valuation	Value of property for tax
Charge to your principal	Increase balance of your loan
Conveying title	Signing and recording a deed
Maturity date	Final payment date
Per diem	Daily
Mach-2	Twice the speed of sound
Catalysis	Accelerated decomposition or re-composition of a substance

Table 11.4 Some uncommon foreign words

Foreign words	Meaning
raison d'être	The most important reason to exist
infra dig	Below dignity
teté-a-teté	Personal interview/person talk
status-quo-ante	As it was before
vide supra	See above
alma mater	Fostering mother university
bona fide	In good faith
quondam	Former
Éclat	Renown
sub rosa	Secretly

Rules that can Help while Using Jargons

- When a word or a phrase from the general vocabulary expresses a thought as well and as precisely as a specialized term, always use the general term.
- Even if the specialized term is more precise and economical than the general term, do not use it unless you are sure that the reader would understand it.
- When you have to use a specialized term repeatedly, define it clearly at its first appearance.

Examples

Original: Unemployment decline, which has continued over the past four months, indicates that the recession has bottomed and the upturn in the economy is a fact.

Revised: The strong reversal of unemployment trends indicates an improvement in the economy.

Original: It is not believed that the proposed design will meet all the required specifications based on the previous test experiences obtained in the laboratory.

Revised: Lab tests indicate that the proposed design will not meet all requirements.

Avoid Discriminatory Writing

In today's all-inclusive society, it is wrong on a writer's part to alienate a section of the masses through the usage of politically incorrect words. Such writing is offensive and does not speak well of the author's open-mindedness. All too prevalent are gender-biased words—words that differentiate people based on their gender. Although this form of discrimination can be against both genders, most instances involve discrimination against women, the reason being that a large number of words suggest male superiority and chauvinism. The repeated use of masculine pronouns—he, his, and him—creates a sexist image. A sentence such as 'A student in BITS does his homework well before coming to class.' shows gender bias, considering the fact that BITS is a co-educational institute.

The use of masculine pronouns can be eliminated in three ways:

1. Rewording the sentence: 'A student in BITS does the assigned homework well before coming to class.'
2. By making a dual reference: 'A student in BITS does his or her home work ...'
3. By making the reference plural 'Students in BITS do their homework ...'

Since language developed in a male-dominated society, it is common to find words such as 'man-made', 'manpower', 'chairman', 'firemen', 'manhole', 'batsman', 'layman', and so on. These can be made non-discriminatory by using non-sexist nouns such as 'of human origin', 'personnel/workers', 'chairperson', 'firefighter', 'layperson', and so on.

Words that stereotype by race, nationality, or sexual orientation must also be avoided. Such statements can severely damage the writer's image and lead to major controversies and embarrassments. Also, it is no longer acceptable to use words that stereotype by age or typecast people with disabilities. Terms such as 'deaf and dumb', 'mentally retarded', and 'handicapped' sound highly insensitive. Gentler and more humane words such as 'physically challenged' or 'mentally challenged' convey the same meaning but with a greater degree of sensitivity.

To summarize, the elements of style that enhance effective writing include a choice of simple words and phrases, precise expressions, and making a conscientious effort to express oneself plainly. Frills and verbal tricks, unfamiliar words, inflated or worn-out phrases, and needless repetition should be avoided.

> 'The writer does the most, who gives the reader the most information and takes from him the least time.'
> –*Charles C. Cotton*

Sentences

Expressing ideas in skilfully written sentences contributes considerably to the success of writing as a whole. The ability to construct effective sentences adds value to technical documents such as emails, letters, memos, reports, proposals, etc. On the other hand, if sentences are ineffective—choppy, loose, ambiguous, and awkward—they fail to create the impact one intends to make on the readers. Hence, it is necessary to learn to write effective sentences in order to construct technical or formal messages.

A sentence is a group of words that expresses a single, complete thought. While the basic units of a sentence are subject (main part of the sentence—noun/pronoun) and predicate (remaining parts of the sentence—

verb, object, compliment, and other elements), the larger units that make up a sentence are main (or coordinate) clauses, subordinate clauses, and phrases. Chapter 17 gives details on these individual elements. Look at the following examples to understand the basic parts of a sentence:

> A sentence is a group of words that expresses a single, complete thought.

Example 1

> Despite winning the lottery, my cousin seems to be unhappy.
>
> My cousin: subject (noun)
>
> Despite winning the lottery, seems to be unhappy: predicate
>
> Despite winning the lottery; unhappy, subject compliment (compliments the cousin/says something more about the noun)
>
> Seems to be: verb

A main clause, as the name implies, makes the main statement in a sentence, whereas a subordinate clause is a group of words that has a subject and a verb, but is dependent upon a main clause. In other words, a subordinate clause cannot stand by itself and make sense, while a main clause can stand alone and make sense.

Example 2

> After completing the filtration process, the students moved on to the next experiment.

From this example, you may be able to understand that 'the students moved on to the next experiment ' is the main clause and 'after completing the filtration process' represents the subordinate clause.

> A phrase is a group of words that does not have a subject and a verb, and that, like a subordinate clause, is dependent on a main clause.

A phrase is a group of words that does not have a subject and a verb, and that, like a subordinate clause, is dependent on a main clause; for example, after the game, seeing the danger, behind bars, above the ceiling, and so on.

Sentence Patterns

Based on the way in which the clauses are combined, sentences are classified as simple, complex, compound, and compound—complex sentences. A simple sentence contains only one clause—the main clause. The sentence can also have a compound subject as shown in the examples below:

> Gold is an expensive metal.
>
> The teacher along with her students completed the project.

A compound sentence contains two or more main clauses, which are called coordinate clauses, as illustrated in the following sentences:

> The teacher along with her students completed the project and they enjoyed the evening together.
>
> Gold is an expensive metal but many people still spend money in buying gold.

A complex sentence contains one main clause and one or more subordinate clauses as shown below:

> Although I wished to participate in the conference, I could not leave my home.
>
> The reports show that the process used for the purification of water by the National Fertilizers Ltd. who had won the best researcher award this year, was immaculate.

A compound–complex sentence or mixed sentence is made up of two complex sentences joined by the coordinating conjunction 'and', as shown in the following examples:

> Despite the fact that writing technical documents takes time, Mr. Kamal enjoyed such writing assignments   and I used to appreciate the enthusiasm that he had exhibited in preparing his technical documents.

Although roughly the same size as the Earth, Venus is generally much smoother despite being broken in one place by a mountain peak higher than the Everest, and it is blanketed by an atmosphere of carbon dioxide, ninety times denser than the Earth's, which raises its surface heat above the melting point of lead.

From the preceding discussion, we can learn that a complete sentence is one that contains a statement that will stand by itself, makes sense, and has a subject and a verb.

Salient Points of Sentence Construction

This section provides guidelines and examples to construct sentences with greater clarity and effectiveness.

Short sentences

Simplicity in writing can be achieved mainly by writing short sentences. Shorter sentences convey meaning better than longer ones. If too many ideas are stuffed into a sentence through words, the meaning is lost and the reader is left confused. Therefore, one should attempt to limit the amount of information in each sentence. Preferring shorter sentences does not mean that long sentences are to be completely avoided. When long sentences are required to convey a complex message, construct them meticulously so that clarity is achieved.

> 'Have something to say, and say it as clearly as you can. That is the only secret of style.' —Mathew Arnold

While giving a speech, long sentences can be used to good effect by using appropriate pauses, intonation, and body language. However, as non-verbal and paralinguistic features are missing in writing, one should be careful in framing long sentences. The best sentence is that which a reader can understand in the first reading.

As we have learnt in the preceding section, there are various types of sentences on the basis of their structure: simple, compound, complex, and compound–complex. All these types are normally required while preparing a technical document. However, one should remember to use a judicious mix of long and short sentences, as they not only arouse interest in the reader but also break the monotony of reading the same type of sentences. Sometimes, the completion of a thought requires a compound or complex sentence. In such situations some short sentences can be inserted either before or after these long sentences to provide some relief to the readers.

Long sentences are usually harder to understand than the shorter ones, as readers can absorb only a few words per glance. Hence, they may skip some words and not grasp the full meaning. Nevertheless, long sentences are well-suited for certain purposes such as summarizing, combining ideas, and listing. It is appropriate to use medium-length sentences (about twenty words) for linking ideas. Although there is no rule about the length of a sentence, most effective formal/business writing has an average length of twenty words per sentence or fewer. The following sentence illustrates how a long sentence can be ineffective in communication:

> The members of the management of Bharat Textiles Ltd are unable to see their way through the crisis that has engulfed the company, though they are keeping it as a well-guarded secret, not even revealing what ails the company, as they think this necessary for the protection of the image of the company.

It may not be possible to immediately grasp the main idea from this sentence when we read it once. The miscommunication in the sentence is not due to the words used but because of the length of the sentence, the number of words, and their relationship. A complex sentence can be simplified by breaking very long sentences into two simple sentences and/or using words economically. Try revising the complex sentence given above after going through this chapter.

Break the sentence

When a sentence is loaded with a lot of information, it becomes very difficult to understand the meaning. It is difficult to break a sentence only when the ideas are very closely knit. Otherwise a sentence can be split into two or three shorter sentences to bring in clarity. Table 11.5 illustrates the advantage of short sentences over long sentences.

Table 11.5 Long and short sentences

Long and breathless sentences	Short and clear sentences
• We might further mention that we would be glad to furnish any of these whistles on a trial basis to the extent that if the smaller size is not adequate enough, it could be returned in lieu of the purchase of a large size depending upon actual operation and suitability of requirement for signal distance and audibility.	• We would gladly provide any of these whistles for trial. In addition, we would replace the smaller size whistles with the larger ones, if you feel that they may help you hear your signals more clearly at a specific distance.
• We can see from the above list that though the average number of customers per day is small, the turnover is maximum because at other restaurants the main items of sale are tea, coffee, and snacks, whereas at this restaurant meals are also available.	• The above list shows that this restaurant has the maximum turnover though the average number of customers per day is small. This is because besides tea, coffee, and snacks, it sells meals also.
• Due to the fact that the production of reports involves considerable cost to our organization, it can easily be seen that the reduction of the time spent in writing and reading them, a shortening of the reports themselves, would represent an appreciable gain in reducing our general operating expenses, although the matter of the length of the report should naturally be considered in relation to the complexity of the material and its adequate coverage keeping in mind the requirements of the specific situation.	• The production of reports involves a large cost to our organization. If we shorten the reports keeping in mind the complexity, adequate coverage relevant to specification and contents, we can reduce the time spent on reading and writing them. By doing so, we can reduce the operating expenses.

Long and circuitous sentences are difficult to understand while the shorter ones convey a message more easily. Shorter sentences in fact are more precise in expression.

However, using too many short sentences also makes the writing appear jerky and irritating. It will also give an impression of elementary writing (such as writing for children). It is, therefore, a writer's job to bring in the right mix of simple and complex sentences.

Look at the following sentences:

Original: The tinder must be some soft, inflammable material. This may be dry grass, leaves, or wood shavings.

Revised: The tinder must be some soft, inflammable material such as dry grass, leaves, or wood shavings.

Original: The experiment was over. I completed the report. I shut down the power supply. I submitted the report to the instructor. I left the laboratory.

Revised: I completed the report as soon as the experiment was over. After shutting down the power supply, I submitted the report to my instructor.

Original: The logs are fastened to a chain running up an incline. When they reach the floor of the mill they are rolled onto the carriage. The carriage is about 40 feet long by 15 feet wide. It moves the logs toward the saw after each is cut by means of an automatic feeding device.

Revised: The logs are fastened to a chain and hauled up an incline to the floor of the mill, where they are rolled onto the carriage. The latter, which is about 40 feet long by 15 feet wide, is provided with an automatic feeding device by means of which the logs are moved towards the saw after each cut.

The original versions sound choppy and disjointed whereas in the longer sentences, the sequence of thoughts is smoothly linked.

Economy with words

A sentence can be shortened by being economical with words. One idea can be expressed in a number of ways; one way can be shorter than the other. Usually, a briefly worded sentence saves the time of the reader apart from being clearer and more interesting.

The following substitutions will help one achieve economy with words.

Avoid cluttering phrases   Sentences often become long because of cluttering phrases. One can replace these phrases with shorter wording, without loss of meaning. Let us look at these examples:

> In the event of procrastination and dilatory action, the operations will be shunned.

The phrase 'in the event of' is uneconomical and can be substituted by 'if' without loss of meaning. Similarly, 'procrastination' and 'dilatory action' can be replaced by 'delay in action'.

> If there is delay in action, the operations will be cancelled.

Similarly, the following sentence begins with an unnecessary phrase that adds to the length.

> I take this opportunity to tell you that you are an excellent leader.

The better substitute will be

> You are an excellent leader.

Avoid pleonasm or redundant phrases   To write with simplicity and clarity, avoid using excess words that do not contribute any meaning to the sentence. Sometimes the sentence will have to be reconstructed after eliminating the extra phrases, while at other times, it would suffice to delete the extra words. Look at some of the examples of pleonasm taken from technical writing:

What do I do for effective communication?

> It has been glaringly noticed from the records of the accounts that the company faced great loss in this fiscal year.

The initial words in the sentence do not add anything substantial to the meaning of the sentence. Therefore they can be dispensed with, and the sentence can be reconstructed thus:

> The records of the accounts indicate that the company faced great loss in this fiscal year.

Here is another example:

> The company is not prepared to expand at this point of time.

The phrase 'at this point of time' can be replaced by one word, 'now', and the sentence can be rewritten thus:

> The company is not prepared to expand now.

Some commonly used cluttering phrases that should be avoided are mentioned in Table 11.6. The sentences shown in Table 11.7 prove the point further.

Table 11.6   Avoiding cluttering phrases

Cluttering phrases	Better substitute
Owing to the fact	Because
Under the circumstances in which	When
For the reason that	Since
In the light of fact	Because
On the occasion of	When
Under circumstances in which	When
It is necessary that	Should
Has the opportunity to	Can
There is a chance that	May
It is important that	Must
It is necessary that	Should
In the meantime	Meanwhile
In very few cases	Seldom
With a view to	To

Table 11.7   Avoiding pleonasm

Sentences with pleonasm	Better option
• I am of the opinion that the company managers should be admonished for their misconduct.	• The company managers should be admonished for their misconduct.
• In the light of the fact that Mr Bansal has worked with effort to build this website, we must give him the contract.	• Since Mr Bansal has worked with effort to build this website, we should give him the contract.
• It is essential that there be no construction of houses in the area designated as the sanctuary for wildlife.	• There should be no construction of new houses in the area designated as the wildlife sanctuary.
• The antique dealer who is on Mukherjee Road has a pair of silver candlesticks that were designed by Punit.	• The antique dealer on Mukherjee Road has a pair of silver candlesticks designed by Punit.
• Many of the riders were boys with skinny frames and bold spirits.	• Many of the riders were skinny, bold-spirited boys.
• In the period between October and December, the business did well.	• Between October and December, the business did well.

Avoid circuitous expressions   While it is possible to write a sentence in a number of ways, some ways are more direct than others. Let us observe this sentence:

> If there are any points on which you require explanation or further details, we shall be glad to furnish such additional details as may be required by telephone.

A better substitute is

> If you have any questions, please contact us over telephone.

The examples given in Table 11.8 will illustrate the point.

Table 11.8   Avoiding roundabout sentences

Roundabout sentences	Direct sentences
• It is important that you shall read the notes, advice, and information detailed opposite, then complete the form overleaf prior to this and immediately return to the council by way of the envelope provided.	• Please read the notes given overleaf before you fill the form. Then send it to us by tomorrow in the envelope provided.
• The table is intended to assist investors in understanding the costs and expenses that a shareholder in the fund will bear directly or indirectly.	• This table describes the fees and expenses you may have to pay in connection with an investment in our fund.
• The following summary is intended only to highlight certain information contained elsewhere in the prospectus.	• This summary highlights some information already included in the prospectus.
• Persons other than the primary beneficiary may not receive these dividends.	• Only the primary beneficiary may receive these dividends.

Table 11.9 lists certain phrases that add wordiness to a sentence, and how such phrases can be substituted by more direct words to give a crisper, business-like effect.

Table 11.9 Using direct words

Phrases	Substitute
is aware of	knows
has knowledge of	knows
is taking	takes
are indications	indicate
are suggestive	suggest
considering the fact	consider

Avoid needless repetition   Repetition of a word or idea unnecessarily lengthens the sentence, without adding to the meaning. Most of the time, it serves no purpose at all as can be observed in this sentence:

> I was born in summer, the month of July.

A more direct version, *I was born in July*, conveys the same meaning. Another example:

> He remarked that he believes that I am a consummate speaker.

By not repeating the word *that*, the sentence reads much more crisp:

> He remarked that he believes I am a consummate speaker.

Redundant words also add up to the sentence length. For example, expressions such as past memories, various differences, true facts, future plans, past history, sudden crisis, and free gift are illogical and cannot be justified. Look at the following sentence:

> Before I finalize the schedule, please let me know your *future* plans.

This can be better written as

> Before I finalize the schedule, please let me know your plans.

You may have heard this expression often:

> Please return *back* my book tomorrow.

The sentence can be rewritten without loss of meaning like this:

> Please return my book tomorrow.

Given in Table 11.10 are more examples of redundancies along with the better versions.

Table 11.10   Avoiding redundancy

Redundancy ridden	Sans redundancy
• We will all *assemble together* for the condolence meeting.	• We all will assemble for condolence.
• This stick is *limited* in length.	• This stick is short.
• My *basic fundamentals* of Physics are not clear.	• My fundamentals of Physics are not clear.
• The speaker invigorated his presentation by narrating *humorous jokes*.	• The speaker invigorated his presentation by narrating jokes.
• At the *present time*, I am training two engineers.	• I am training two engineers.

Right ordering of words and proper emphasis

In order to convey the exact meaning of what we wish to say, words must be placed in the right order. A lot of information goes into a sentence, but all of it is not necessarily of equal importance. We now know that the sentence length affects the emphasis. Short sentences carry more emphasis than longer ones. It draws attention to its contents, giving a single uninterrupted message. This is well-illustrated in the following passage:

The community-type hybrid solar cooker was designed bearing in mind not only cost and time as the main factor, but also a special feature of heat energy being trapped by flat collectors that can be supplemented by burning coal or other conventional fuels, making it appropriate for traditional cooking in rural areas. The steel tray, which is fixed inside the box, is painted black and covered with glass sheets, accounting for the quick heat trap; however, this black paint and insulation apart from preventing the heat from escaping also let the cumulative heat build up inside the wooden box to accelerate the cooking of food items kept in aluminium vessels. This project was found to be commercially viable.

Longer sentences that contain many ideas get confusing, as the emphasis gets diluted across the contents. To highlight varying emphasis, let us take this example:

The report was completed in time. The completion took place in spite of difficulties.

Here equal emphasis is given to both 'report was completed in time' and 'difficulties'. In the second option, where these two sentences have been combined, the emphasis changes.

Although we faced difficulties, the report was completed in time.

Here the second clause has been given greater emphasis. Given in Table 11.11 are some more examples to show how proper emphasis alters the meaning of the sentences.

Table 11.11   Use of emphasis

Less emphatic	More emphatic
• Two of our members heard you speak in Delhi and praised you highly for your dynamic presentation when they returned.	• Two of our members heard you speak in Delhi and when they returned, praised you highly for your dynamic presentation.
• We feel we are missing some patients, and therefore losing revenue, by using this system.	• By using this system, we are missing some patients and therefore losing revenue.
• The primary force behind most stress relief and exercise programmes is the executives who are prone to stress and heart attacks, as shown by medical reports.	• The medical reports show that the primary force behind most stress relief and exercise programmes is the executives who are prone to stress and heart attacks.

The illustrations of Table 11.11 show how the construction of sentences affects emphasis. All clauses or ideas can be given equal importance, or any of them can be de-emphasized. However, care must be taken to place the key message either at the beginning or at the end of the sentence, rather than bury it in the middle.

Active versus passive voice

Economy of words is one of the hallmarks of good writing, and using the active, rather than the passive voice, is one way to achieve economy. Passive constructions often result in vagueness, and they are longer since they need helping verbs (such as has been). Passive voice can be used occasionally, especially when we want to avoid the use of personal pronouns; otherwise it is preferable to use active voice. Passive voice can be used in the following instances:

- When the doer of the action is anonymous or is not very relevant. For example: If kerosene is exposed to fire, it ignites immediately.
- In some standard phrases, for example, someone to be reckoned with or match was abandoned.
- If the attention is already on the object of action 'for example' Those black stones seem to augur bad luck—last night they were desecrated by the crowd.
- Sometimes to add variety, to help augment a point, or to add some stylistic effect.

A few examples are given in Table 11.12. Note from these examples that active voice produces more forceful and livelier writing. Since the emphasis is on action, it generally saves words.

Table 11.12   Passive and active voice

Passive	Active
• The concentration by the Training Division of its time, money, and other resources into a new programme is not advisable from our point of view.	• We do not advise the Training Division to concentrate its time, money, and other resources into a new programme.
• Our implementation of this new procedure is required by the board of directors.	• The board of directors requires us to implement this new procedure.
• My first visit to your organization will always be remembered.	• I will always remember my first visit to your organization.
• In completing the tasks, the planned ₹ 70,000 budget for June and July was exceeded.	• In June and July, we had to exceed the planned budget of ₹ 70,000 in order to complete the task.
• A weekly meeting is recommended as a way to reduce problems generated within the Training Division.	• We recommend a weekly meeting as a way to reduce problems generated within the Training Division.
• There are many positive aspects of cheque collection that are not brought out in this report.	• This report does not bring out many positive aspects of cheque collection.
• From these findings it is indicated that none of you has yet notified me.	• These findings indicate that none of you has yet notified me.

Avoid ambiguous sentences

Ambiguity is a hindrance to clarity, which in turn results in communication going awry. Faulty construction of sentences gives rise to ambiguity. Ambiguity arises mainly because of misplaced modifiers in sentences. A modifier is a word/phrase/clause that tells us something more about a verb, noun, or adjective in the sentence. The position of a modifier in a sentence is important. Whether it is a word, a phrase, or a clause, it should be placed as close as possible to the word it modifies. Given below are three examples where misplaced modifiers lead to ambiguity. In the improved version, these modifiers, once placed in the correct position, make the sentences clear and easy to understand.

> A modifier is a word/ phrase/clause that tells us something more about a verb, noun, or adjective in the sentence.

Example 1 (a word as a modifier)

Original: The delay in transit nearly drove the manager frantic.

Revised: The delay in transit drove the manager nearly frantic.

Example 2 (a phrase as a modifier)

Original: Quick Information Systems has bought new computer chairs for the programmers with more comfortable seats.

Revised: Quick Information Systems has bought new computer chairs with more comfortable seats for the programmers.

Example 3 (an elliptical clause as a modifier)

Original: While on a tour of India, my expensive watch was stolen.

Revised: While we were on a tour of India, my expensive watch was stolen.

Table 11.13 shows a few more ambiguous sentences along with their corrected versions.

Table 11.13 Avoiding ambiguous sentences

Ambiguous	Revised
• He noticed a large stain in the rug that was right in the centre.	• He noticed a large stain in the centre of the rug.
• The seniors were told to stop demonstration on the campus.	• The seniors on the campus were told to stop demonstration.
• New York's first commercial, human sperm bank opened on Friday with semen samples from eighteen men frozen in a stainless steel tank.	• New York's first commercial, human sperm bank opened on Friday when semen samples were taken from eighteen men. These samples were then stored in a stainless steel tank.
• You can call your father in Delhi and tell him all about Srimant taking you out to dinner for just ₹ 50.	• For just ₹ 50, you can call your father in Delhi and tell him about Srimant taking you out to dinner.

Journal Writing

We have heard of medical journals, academic journals, online journals, etc. These journals are magazines that deal with subjects or professions. For instance, a journal on English Language Teaching, a journal on Experimental Psychology or a journal on Indian Culture, etc. deal with the respective subjects mentioned in their names. However, a journal also refers to the written record of what we see, do, or experience every day or in a particular period. While the former type of journal involves academic/technical/political/sports subjects, the latter deals with our own reflection on various issues we come across every day and hence can be called a reflective writing. You would also have come across *blogs* which are a type of journal. They are also known as online diaries, personal blogs, etc., mainly written by people to keep in touch with friends and family.

Instead of keeping things in our memory, the events we attended, the places we visited, the friendship we cherished, the films we watched, etc., and trying to recollect them, it is better to record them in a journal so that we can refer anytime later. A journal can also be written in a classroom when your professor asks you to write about a recent event you have attended or other such instances. The following write-up discusses the guidelines for writing a class journal:

Look at the following journal prompts:

- Write about your family.
- Write about any two places you visited last year.
- Write down what you can do to make your environment greener.
- Write about the national leader you admire the most.
- Write about a novel you have recently read.
- Write about your new bicycle.
- Write about your career plan for future.

To write a journal on any of these topics, you need to reflect upon your personal experience and express the same in an appropriate style. You need to be expressive and creative so that your journal makes an interesting read. Though you may be presenting facts about various aspects, such as, the places, the food, the scenes, etc. you had experienced, you should garb them with attractive words, phrases, and expressions. For example, rather than starting your journal on a *nature walk* with 'I enjoyed my nature walk in the Aravalli hills two weeks back.', you can start with 'Right now, I am feeling the gentle breeze and the mesmerizing sunset at the Aravalli Hills in Rajasthan.' By creating a visual picture in your writing, you can develop interest in your readers to read on your journal.

The structure of a journal has the same elements as that of an essay: introduction, illustration, conclusion. While the introductory part gives a preview of your topic, the illustration and conclusion respectively include specific examples to support your topic and the summary of what you have already said. Find below a few guidelines for writing class journals:

- Start with a hook (question/quotation/anecdote/feeling)
  Example: Have you ever experienced a sunset standing on a hill?

- Include a sentence in introduction that serves as a preview.
  Example: My experience on the Aravalli Hills was filled with magnificent events and was unparalleled.

- Give adequate instances to depict your magnificent experiences.
  Example: The myriad sights, smells, and other sensory perceptions we had experienced at various points of the hills were breathtaking and unforgettable.

- Conclude with a clear statement that summarizes your experiences narrated in the journal.
  Example: In a nutshell, I can say, 'Visiting Aravalli Hills is a once-in-a-lifetime experience, and we would cherish this forever.'
- Use personal pronouns in your journal.
- Adopt a friendly tone and avoid didactic tone.
- Be as creative as possible.

Writing for the Web

With the phenomenal increase in the use of smartphones, tablets, kindle, etc., online readership has been witnessing a tremendous growth. Online readership includes various social networks, wikis, blogs, websites, etc. Though print material is widely used by all of us for academic and professional purposes, time is not far when online media will take over print media. Hence, it has become important for us to know not only the various types of online writing but also the style in which we need to formulate the web content.

Writing for the web is different from that for the print media. Think for a few minutes how we read a newspaper? Do we start reading from the first line of the front page till the last line on the last page? No. We may read headlines and some important columns on page one and then directly go to the sports section if we are interested in sports or go the editorial page if we wish to know the newspaper's views on certain topics of importance. Similarly, many readers who read on the web also skim through certain sections and quickly complete reading the content. Keeping this behavioural attitude of the readers in mind, print media try to attract and retain their attention by catchy visuals, colour scheme, and headlines. Likewise, when we write for the web, we need to understand the various platforms available and follow certain guidelines in order to grab and sustain the attention of our readers.

Social networks are forums that enable people to get connected to each other. Started initially for sharing more of personal information, these networks are now being widely used for sharing professional messages as well. Facebook, Twitter, and LinkedIn are three most popular social networking sites.

Wikis are collaborative websites where multiple authors add and edit content. They are also used by organizations to manage content, and interact with each other and their users.

Blogs are web pages created by individuals on a particular topic. Though personal blogs are online journals that can be created in MySpace or LiveJournal, the professional blogs engage the readers by writing on topics that are of interest to others. A blog on photography or on origami, for instance, may be of interest to many readers.

Websites are pages on the Internet to inform the readers about a person or an organization depending on whether they are personal websites or professional/corporate/university websites.

Clarity of purpose, audience awareness, planning content, using effective style of writing, formatting, and proofreading are the concerns of those of us who wish to write for web.

Theme Clarity

When we think of creating content for the web, we need to ask ourselves the following questions so that we are clear with our purpose:

- What are we achieving by publishing this content?
- What information should we include in this?
- Does my objective understand the readers' needs?
- Have I read the other web contents on the same topic before I set my objectives?
- Have I published the same topic on any other web forum and, if so, how does this content differ from the earlier one?
- Does my content have a single purpose or multiple purposes? If it has many purposes, which one is the most important one?

Audience Awareness

Readers read our web content with specific task in mind. For instance, when we like to know about a job profile in 'knowledge management' area, we surf the web specifically for that content. Similarly, readers go to the web with a specific demand. Hence, it is important to understand their needs. In fact, being aware of the recent trends in job market, industries, sectors such as academia and industry, etc. would enable us to understand the readers' requirements. To get increased readership for our web content, it is not only important to write what we know about, but also what readers want.

Information Design and Development

Designing and developing information is of utmost importance in web-writing. While the term '*design*' here refers to the appearance (or look) and feel of the information on the website, '*development*' involves its functionality (how it works). *Development* should happen both at the front end and at the back end of the web. In other words, *design* includes the layout, speed, colour, images, graphics, fonts, navigation, etc., and belongs to the front end which is developed by means of appropriate computer coding languages and other technological tools. *Development* is carried out both at the front end and back end. For instance, when you get web feedback from the customers, you receive lot of information. They fill out forms and provide rating, etc. The data thus received is stored in the back end (server). Based on the data, you can develop better feedback forms and apps for future use.

The input information should be designed in such a way that the users should find it easy to access whatever part of information and related link they want to. In fact, more and more users would like to visit the website only when they get the required information and when it provides all details in attractive layout with clarity. However, the use of images, text, etc. is decided by the type of content. For instance, if the website is of an educational institution, you can focus more on your text than on the other features. You need to organize the information logically/chronologically/spatially/psychologically according to your target audience/customers. Of course, pictures, speed, navigation, font, etc. are also important for all websites.

Development encompasses research, plan, organization, structure, and style. Before preparing any text for the website, you need to engage yourselves in all these steps. Think critically what the content should be and look for the same from various documents and sources. Then plan the design with which the text would be presented. Think about the pictures, graphs, photos, colour, layout, fonts, etc., to present your text. Organize the material in the most appropriate order with catchy introduction, illustrative discussion component, and impressive conclusion. The style of writing should be suitable for your target audience. In other words, choose words, phrases, expressions, figures of speech, etc., which your users can comprehend easily.

Design and development complement each other. It's difficult to separate them. However, if you learn and get trained in design and development to have a firm grasp and grip, you will find them helping you in all web writing throughout your career.

Effective Style of Writing

Though the fundamentals of effective writing may be same for print and online writing, the latter differs in certain aspects. For example, we should use first person and second person, rather than using third person. We should get to the important point right away without spending too many words on the introduction because if the readers don't find the required information soon, they would go for some other site. We should be friendly in our tone and speak directly using active voice.

Formatting

Our web content should be easily readable and scannable, and in order to achieve these, we can follow the following tips:

- Use appropriate font type, size, and colour; use bold or italics type to catch the important points of the content.
- Use catchy headings and subheadings that reflect the content under each.
- Provide easy navigations for any links or other sources, etc. It is better to connect the links to the appropriate words in the content rather than using the words 'Click here'.
- Bullet lists may be helpful in grabbing the attention.
- Use videos, photos, and other multimedia aids for enriching the text.

Proofreading

Our web content should be free from typos, incorrect spelling, and grammatical errors. Such lapses would put the readers off and keep them away from using any of our web writing. Just as we proofread our print content, we should proofread carefully the web content too. Carefully edited, revised, and proofread content would have better chances of getting more readership.

The following links would provide an idea of the various types of web writing:

- https://www.linkedin.com/in/sumedh08
- http://educationalwikis.wikispaces.com/Examples+of+educational+wikis
- http://www.baghaescup.blogspot.in/
- https://twitter.com/example

Collaborative Writing

When two or more people think, plan, organize, write, and produce one document, we can say that they are involved in *collaborative writing*. It is that type of writing which involves the collaborative effort of a set or group of people, probably with varied skills to achieve one common goal or bring out one document. It is like how collaborative effort produces better results in various assignments such as projects, interior designing, complicated medical treatments, etc., which require different skill sets. The old saying, 'Two heads are better than one', in fact, turns out to be true in a *collaborative writing* endeavour because the writing is enriched by more and more thoughts and ideas coming from different authors. In fact, the primary aim of *collaborative writing* is to bring out the best document by using the diverse ideas and skills of several authors. This type of writing can be used to produce research articles, books, websites, etc., wherein ideas from various authors are combined to bring out an effective text which benefits more and more readers/users. Certain steps and guidelines for successful *Collaborative Writing* are discussed below:

Steps in collaborative writing The steps include team formation, work allocation, document preparation, revision, and production.

Team formation refers to the selection and making of the collaborating team of authors after deciding and researching the document content. Depending on the requirements, the leader needs to identify suitable members for various tasks such as collecting information, writing, looking for correct visuals, revising, editing, proofreading, producing, etc. After forming the team, members are introduced and become familiar with each other's skills. The leader discusses and sets the deadlines for each task.

Work allocation involves further planning, discussion, and decisions on various assignments such as dividing the content into components, who will take care of each component, does any component require more than one member, what are the interim goals, what is the time limit, when should the progress be reported, etc. Division of work is the most important aspect as no team member should feel that he/she is overloaded as compared to others. Even if the team is very small and has less number of people, division of labour should be almost equal.

Document preparation comprises the actual writing or preparing suitable images, etc. Once authors start their work, they need to be in touch with each other to ensure that they are proceeding in the right direction. Once they finish their part, all of them should meet and prepare the first proof of the document. Again, discussion follows to refine the document. The leader, with the help of others, work on the first draft of the entire document.

Revision and production are the processes that the first draft undergoes for revision of content, editing, proofreading, formatting, checking the images, etc., before final production. Please refer to the chapter on Technical Reports for more details on editing, revising, etc. The final content, organization, structure, style, visuals, etc. should be approved by the team before the document is sent for production. Once it is produced, the team can send it for approval by the management/administration.

Guidelines for Success

To be successful, *collaborative writing* should be effective, efficient, and harmonious. Carefully go through the following guidelines to accomplish your collaborative writing:

- Realize that each author (team member) is important and plays a crucial role in the collaborative writing project. Whether it is the leader, consultant, writer, reviewer, editor, or creative expert—everyone is important in collaborative writing.
- Understand that collaborative writing involves shared responsibility primarily to generate more and more ideas and develop a healthy relationship among the collaborating authors.
- Be aware that communication and coordination play a key role in the success. Both the communication in the document and that among all members are vital. Coordination is extremely important at every stage of collaborative writing to produce a harmonious document.
- Keep in mind and meet the given deadlines for each component.
- Be ready with some strategies to resolve/manage conflicts if any during the various stages of preparation, discussion, and decision.
- Take criticism positively and work wholeheartedly. Never be overtly defensive.
- Avoid ethnocentrism when you work in a team with members from varied cultures.

Collaborative writing is becoming increasingly important in bringing out research papers. When you go through reputed journals, you may see that almost all articles are jointly authored. Similarly, many people come together in planning and producing the organizational websites which present the product of *collaborative writing* to the outside world. If you master the art of *collaborative writing*, your skills will pave way for a successful professional career.

Creating Indexes

When you read a book, you see a list containing words/phrases along with page numbers over the last few pages of the book. This list is titled as Index. ***Index*** of a book/document refers to the list of terms, topics, authors, etc. that are discussed in it. It is also known as BoB (Back of the Book) index. You would also have heard of the term *Citation index* which is an index of all the citations. In other words, it gives an idea to the reader about which documents have cited the earlier works. Another type of index, namely the *Periodical Index*, provides an index of the articles published in journals, periodicals, etc.

An index is an alphabetical list of main terms and associated terms of the document and enables the readers to identify/search the location of a particular term/author/concept quickly and easily. We can call it a roadmap to the document. Prospective buyers of a book, for instance, can have a quick look at the index and get an idea of the topics/authors discussed. Similarly, it can greatly help researchers who look for specific terms, concepts, authors, etc. in a document. Please look at the given screenshot example (Exhibit 11.1) to see how an *index* appears in the document.

EXHIBIT 11.1   Sample of an index

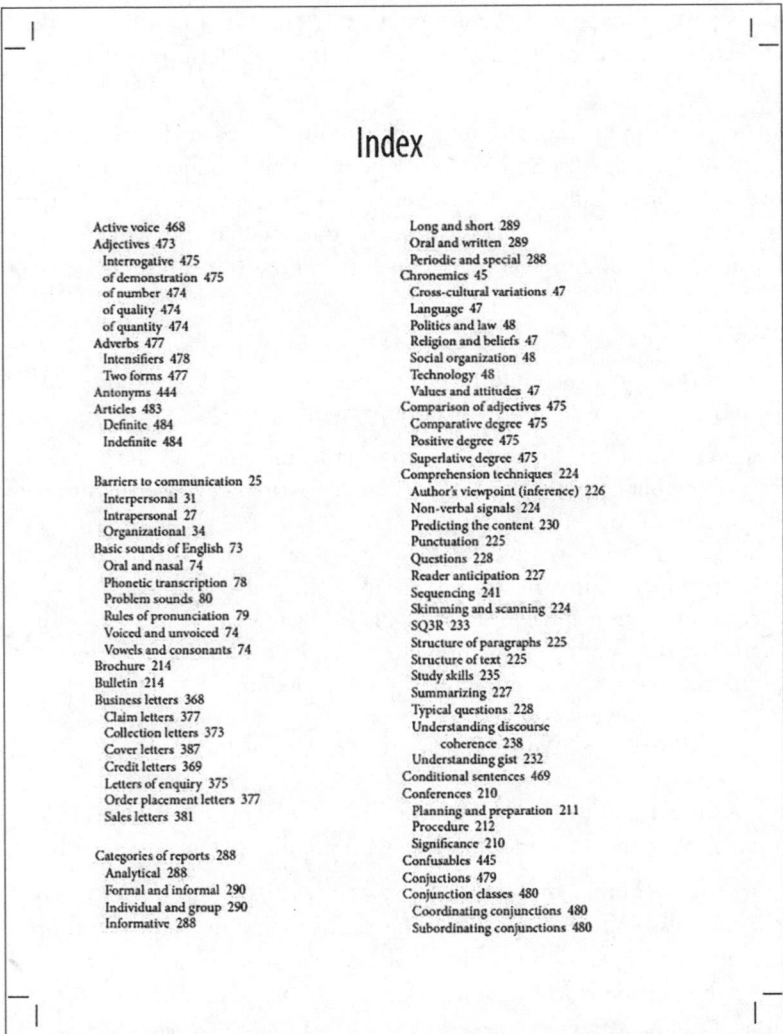

Active voice 468
Adjectives 473
 Interrogative 475
 of demonstration 475
 of number 474
 of quality 474
 of quantity 474
Adverbs 477
 Intensifiers 478
 Two forms 477
Antonyms 444
Articles 483
 Definite 484
 Indefinite 484

Barriers to communication 25
 Interpersonal 31
 Intrapersonal 27
 Organizational 34
Basic sounds of English 73
 Oral and nasal 74
 Phonetic transcription 78
 Problem sounds 80
 Rules of pronunciation 79
 Voiced and unvoiced 74
 Vowels and consonants 74
Brochure 214
Bulletin 214
Business letters 368
 Claim letters 377
 Collection letters 373
 Cover letters 387
 Credit letters 369
 Letters of enquiry 375
 Order placement letters 377
 Sales letters 381

Categories of reports 288
 Analytical 288
 Formal and informal 290
 Individual and group 290
 Informative 288

Long and short 289
 Oral and written 289
 Periodic and special 288
Chronemics 45
 Cross-cultural variations 47
 Language 47
 Politics and law 48
 Religion and beliefs 47
 Social organization 48
 Technology 48
 Values and attitudes 47
Comparison of adjectives 475
 Comparative degree 475
 Positive degree 475
 Superlative degree 475
Comprehension techniques 224
 Author's viewpoint (inference) 226
 Non-verbal signals 224
 Predicting the content 230
 Punctuation 225
 Questions 228
 Reader anticipation 227
 Sequencing 241
 Skimming and scanning 224
 SQ3R 233
 Structure of paragraphs 225
 Structure of text 225
 Study skills 235
 Summarizing 227
 Typical questions 228
 Understanding discourse
  coherence 238
 Understanding gist 232
Conditional sentences 469
Conferences 210
 Planning and preparation 211
 Procedure 212
 Significance 210
Confusables 445
Conjuctions 479
Conjunction classes 480
 Coordinating conjunctions 480
 Subordinating conjunctions 480

To create an index, the authors first identify the terms to be included and highlight them in the document which is then passed on to the indexers. The indexers need to be familiar with the indexing tools and software and also with the complete document to ensure that the concepts, terms, etc. are easily accessible by the readers.

Please visit the following link to understand how to create an index using Microsoft Office: https://support.microsoft.com/en-us/office/create-and-update-an-index-cc502c71-a605-41fd-9a02-cda9d14bf073

SUMMARY

Effective writing is essential in every profession. This can be achieved by maintaining clarity and brevity in writing. Words, being the smallest unit in writing, play a crucial role in effective communication of a message. Clarity can be achieved by using simple and familiar terms, concrete and specific words, by using acronyms carefully, and by avoiding clichés, jargons, and foreign words. Effective and good writing has to be able to reach out to a large audience. This requires the writer to maintain an impersonal and indiscriminatory tone, while being genuine.

Although it is essential to retain the correct order of words to achieve the desired emphasis as well as grammatical accuracy, variety can be added to the writing by altering the sentence types effectively. Other key factors to be remembered by a writer are a correct mix of active and passive voices and avoiding ambiguity.

Journal writing enables writers to reflect upon themselves and ponder over their experiences. On the other hand, web writing helps writers to present various activities and achievements of themselves and their workplaces to the public. Whether it is writing for web or for print, the respective writers find collaborative writing a boon. However, the authors always need to keep in mind the theme clarity, audience awareness, planning, etc.

EXERCISES

1. Substitute the unfamiliar words in the sentences given below with familiar words.
   (a) If *liquidation* becomes mandatory, we shall dispose off the mammoth buildings first.
   (b) The *boffins' propensity* to assimilate new instruments is *insatiable*.
   (c) Mr Bose's *idiosyncrasies* supply adequate justification to *oust* him out of his office.
   (d) The *exiguous* diet intake is reflected by the *gaunt* figure of the patient.
   (e) The *unanimity* of current forecast is not *inconvertible* evidence of an *impending* business.

2. Identify and replace the general words with more concrete and specific words.
   (a) His GRE score was not high.
   (b) The institute will buy some new equipment soon.
   (c) This room is large enough for the crowd.
   (d) Kota has many engineering and medical colleges.
   (e) Our company has made a significant profit this year.
   (f) I will visit your organization shortly.
   (g) He is interested in an occupation in the line of engineering.
   (h) Let us first find out what his angle is.
   (i) From the point of view of economy of operation, the small car is superior to a big car.
   (j) The institution needs to spend a considerable amount of money to set up this new laboratory.

3. Substitute clichés with simple English words.
   (a) *Last but not the least*, I thank the typist for doing a commendable job.
   (b) *By and large*, the company is making exorbitant profit.
   (c) *It goes without saying* that I had a wonderful experience by being the member of your coterie.

(d) If you give me this opportunity to serve your organization, I promise I will *leave no stone unturned*.

(e) It was a *blessing in disguise* to spend the holidays in this arid land.

4. Dispense with the jargon to make the following sentences more effective.

(a) The biota showed hundred percent mortality response.

(b) The oedema in the leg indicates that the leg might be amputated.

(c) The shareholders without any procrastination should sell the junk bond.

(d) We are designing a plane to fly at speed above Mach 2.

(e) The research division has taken up a detailed study of the meteorological effects on microwave propagation.

5. Correct the malapropism in the following sentences.

(a) Geometry is a science which deals with the earth's crust, its strata, and their relations and changes.

(b) What a supercilious knowledge of ethics you possess!

(c) An epigram is what we say about a man after his death.

(d) Herrings swim in the sea in shawls.

(e) She instructed them in geometry so that they might know something of the commercial countries.

(f) Don't attempt to extirpate yourself from the matter.

6. Rewrite the following sentences by avoiding needless repetitions.

(a) Such an act of kindness could be expected of any *living mortal*.

(b) Her *future prospects* are full of promise.

(c) He *again* made *another* attack.

(d) It was a fine *bright sunny* morning.

(e) She went home full of a *great many* serious reflections.

(f) I understand that Mohan and Rama *both agreed* upon this point.

(g) A *rich and wealthy* person looks down upon the *indigent and the poor* with an eye of contempt.

(h) It is *all meaningless nonsense*.

(i) *Superfluity of unnecessary* words and phrases is altogether improper.

(j) He has the *entire monopoly* of the liquor trade.

7. Remove redundancy and rewrite the following sentences to make them direct, precise, and effective.

(a) Although this option may be safer and less risky, there is an initial one-month period of time of great uncertainty at the start of negotiations.

(b) Disclosures made in the footnotes of the financial statement indicate that the company has suffered heavy losses due to its involvement in legal proceedings.

(c) Many companies simply cannot gain adequate access to the information they need to direct their market efforts efficiently, as our study shows.

(d) A list of all supplies that are needed by each centre in order to carry out the studies was prepared.

(e) The definitions of objectives and goals must be systematically made in order to ensure a successful training programme.

(f) For the maximization of the achievement of our corporate goals, abate your personal altercations and make sufficient efforts to realize our mutual cooperative ends.

(g) An analysis of the problems in this department and some recommendations that are straightforward, simple, and easily evaluated are presented in this report.

(h) The writing is on the wall: a tight monetary policy will have disastrous consequences.

8. Change the following sentences to avoid discriminatory language.

(a) Any worker who flouts the company rules will lose his increment.

(b) The Executive Committee constitutes of an engineer, a businessman, and a lady doctor.

(c) An efficient salesman has good time management skills.

(d) Handicap seats in the local buses are strictly for the use of the crippled.

(e) The Annual Conference on *Prevention of Earthquakes* was represented by three of our Chinese engineers.

9. The following extracts contain short, choppy sentences which make the style of writing appear immature. Rewrite after combining the sentences appropriately.

   (a) The raw data was collected. A questionnaire was used for collecting this data. The group analysed the data. It found that most people preferred air conditioners to air coolers.

   (b) Ruby and Roberts saw infinite possibilities in cyberspace. They were determined to turn their vision into reality. They wanted to build a simple Internet service for ordinary people. Their friends scoffed at their ideas. But Roberts and Ruby doggedly pursued their dream. They analysed other online services. They also assessed the needs of their customers.

   (c) The interview ended. Meera got up. She thanked the interviewers. They wished her good luck. Meera observed their facial expressions. She believed she was selected.

10. Break each of the following breathless sentences into two or three short sentences so as to make the sentence readable.

    (a) There is no particular gadget which will produce excellent writing but using a yardstick like the Fog Index gives us some guidelines to follow for making our writing easier to read because its two factors remind us to use short sentences and simple words.

    (b) We might be able to replace a significant portion of our interdivisional travel with electronic meetings that utilize videoconferencing, real time document sharing on PC screens, and other alternatives not with a view to reduce employee or team effectiveness, but to act in unison with other companies which are using these new tools to cut costs and reduce wear and tear on employees.

    (c) The company provides Phase I and Phase II environmental site assessments, preparation of site closure repairs, installation of groundwater monitoring wells and testing of soil and groundwater samples of environmental contaminants, which are possible because of the facilities available in the company's testing laboratory, which is certified by ISO, copy of which certification is enclosed.

11. Identify the cluttering phrases in the following sentences, change them into single words and rewrite the sentences.

    (a) This study is being conducted for the purpose of determining the material durability and is expected to be completed in the near future.

    (b) Due to the fact that the lesson on cloning is quite interesting but difficult, the instructor decided to brief the concept prior to the start of the lesson.

    (c) In view of the fact that around 20 per cent of the information carefully collected by NASA on Jet Propulsion Laboratory during the course of the year 1976 has been lost, the Defense Department can no longer read certain records.

12. Remove the redundant words from each of the following sentences so as to make it concise.

    (a) In all probability we are going to use an aluminium plate with the glass honeycomb collector.

    (b) It is essential to understand thoroughly, fully, and completely, the basic fundamentals of the physical characteristics of the machine.

    (c) These evidences are adequate enough in order to terminate their services.

13. Remove the ambiguity and rewrite the following sentences.

    (a) Early failures tended to discourage further investments in robotics technology of time and money.

    (b) Despite the intensive training computer failures are sometimes difficult for maintenance technicians.

    (c) Double tracks are found in the main entry to the mine, similar to the street car tracks in the city.

14. Change the passive constructions in the following sentences into active constructions so as to make them direct and concise.

    (a) The damage was caused as the material was exposed to sunlight.

    (b) Many electronic and mechanical processes are mostly explained by students by using block diagrams.

    (c) Besides digitized images, a wide variety of graphics that supplement written material are created by writers.

15. Make the following sentences more emphatic by rearranging the words or phrases:
    (a) This machine costs ₹5,00,000 but saves our maintenance costs by ₹2,000 annually.
    (b) Although Printwell Publishers enjoyed record sales in the year 2002, it lost money.
    (c) India, which has the lower wage rates, was selected as the production point for electrical goods.

16. Rewrite the following passages so as to make it more clear and effective
    (a) As the General Manager occupied his coveted chair, he witnessed an outrageously turbulent scene in the meeting. The Purchase Officer and Works Manager, at loggerheads with each other right from the word 'go', made the General Manager feel why in the first place had he called both in a meeting considering the fact that there was mutual animosity for each other. Both of the officers breathing fire, it became difficult for each item in the agenda. He tried to intervene but in vain. Both the Purchase Officer and Works Manager were charging in, firing on all cylinders. The General Manager was in utter dismay. Once he even tried to impose his authority by speaking vituperative words but both the officers in duel were least bothered. With every minute passing, the precious time getting lost without much fruitful discussion, the General Manager wondered how to bring these two recalcitrant subordinates under his control. What he lacked most was the authority; he searched for a symbolic gavel that decides the judge's authority in a court. But he was not a judge. He was a managing engineer—an engineer with managerial responsibilities. But as things went on in the meeting, it appeared that his unruly juniors, who were much too overt and trenchant to be tackled, managed him.
    (b) To begin with, in the development of your communication strategy, you assess your credibility from your perspective of the transmitter of the message. What your audience knows about you will determine your credibility from their perspective. Your credibility is not fixed but is mutable. On occasions, if you have had no former contact with your audience, they can merely assess your credibility from what amount of meagre information they preserve about you. In short, you can impart more information about yourself through your message, which enables your audience to accommodate their view of your believability. Your before and after levels of credibility are what Munter calls 'initial' and 'acquired' credibility.

    The aim then should be made to increase your acquired credibility to doing a self-analysis and self-introspection of your credibility. Your audience always envisages your knowledge about you from the content of communication. If you are delivering an oral presentation on sales projections for your company, exempli gratia, the audience will discern your status quo and your arena. If you are grouching about the miserable service in your bank branch, your audience will realize you are a regular customer and may have some information about your dealings in the bank. In either case you have some prima facie credibility and have the opportunity, by taking a strategic approach, to acquire more credibility.

    (c) Enthusiasm is one of the most vital elements that human beings can make use of to become successful and ultimately turn all their defeats into victories which they can boast of in their life. In order to be quite effective in your life, you need not simulate or pretend, but shedding aside all these simulations or pretensions, you are supposed to be quite sincere and genuine in your efforts to convert your defeats into victories. When you pretend to be what in reality you are not, then everything about you seems quite ostentatious but hollow, and when you are genuine, people have to appreciate you by all means. So you should leave no stone unturned. In short, try really hard to become what actually you are. It is very uncomfortable to remain your original self, but it is very hard or rather quite an exacting task to pretend. As far as enthusiasm is concerned you should try to discriminate between real and fake enthusiasm. If you shout, shrink, howl, cry, and indulge yourself into something boisterous, then it is apparent that this type of enthusiasm is not pure but fake. Truly genuine enthusiasm can be inculcated on the basis of your awakening experiences. When you know things and have the experiences, then only can we bask in the glory of enthusiasm. In short, you should have the ability to discriminate between real enthusiasm and fake enthusiasm. We can be enthusiastic about things only when we wield complete knowledge about something. So, wake up to the power of enthusiasm!

(d) It is a well-stated fact that those we recruit are not inchoate but have the innate ability to solve profound technical problems. What we are unaware of is how exactly they can explain and identify the technical problems, how dexterously they can manage, and how eloquently they can communicate. We endeavour to find all those pertinent things out before we hire them. We invariably ask for a short piece of writing, usually an answer to some technical problem that we are quite sure they will not fail to solve. This brings to fore their communicative skills, which enables us to gauge their potential, and when we bring them for an on-site interview, we make an earnest request to them to make an oral presentation on some technical area in which they are experts. Though, to tell you the truth, we are not really interested in what they have to say. The candidate is completely oblivious of this fact. They zero in on their technical knowledge, and to their consternation, all we want to know is how well they communicate.

17. Assume that you are assigned by your organization the task of preparing the web content for the awareness campaign pertaining to 'water conservation'. Write down the document keeping in mind the important ingredients of web writing.

CHAPTER 12
The Art of Condensation

OBJECTIVES

You should study this chapter to know

- ○ the importance of being concise and relevant while preparing technical documents
- ○ the various steps involved in condensing a text
- ○ how to prepare an effective précis of the given text

Introduction

In today's fast-changing, technology-driven world, acquiring the art of summarizing and essential writing skill is crucial, as it saves time and promotes effective communication. This skill is helpful in preparing synopses, abstracts, and executive summaries, which are essential elements of various technical documents.

The art of summarizing is also known as *précis writing*. A précis can be defined as an abridged form of the text or a piece of writing: *Précis writing involves summarizing a document to extract the maximum amount of information and conveying this information to a reader in the minimum number of words.*

Précis writing does not merely involve paraphrasing (or writing in one's own words) from the original document. The summary produced must be capable of 'standing alone' and making sense—the précis must be complete, informative, and easy to read.

There is no fixed length defined for a précis, but normally it is written in one-third of the number of the words in the original passage. The following qualities are essential to a précis:

Completeness   The précis must have the essential contents of the original passage without omitting any important fact or idea.

Compactness   All the ideas reproduced from the original document should form a compact whole. The words and sentences should convey a sense of unity with each other.

Conciseness   Conciseness is achieved by the process of sifting essential from unessential information, by avoiding repetition, and by omitting ornamental phrases. However, conciseness should not be achieved at the cost of clarity. All the original essential ideas should be presented clearly but in brief.

> 'Clarity precedes success.'
> –Robin Sharma

Clarity   This can be achieved by getting rid of circumlocutions and unclear statements. It should be intelligible to even those readers who have not gone through the original document.

Coherence   All sentences and ideas in a précis should follow a reasonable and natural order of development. Thus, the précis should not look like a collection of disjointed sentences, but a well-connected whole.

Steps to Effective Précis Writing

Given below are the step-by-step instructions for summarizing any given piece of writing.

Identify the reader and purpose of the précis

This step determines how much detail should be included and how formal the précis needs to be. For instance, the précis from a textbook chapter for study purposes need not be as carefully refined as the executive summary of a formal report for an important client.

Read the original document

Skim-read the document to get an overview, then read it again slowly to identify the main themes and to distinguish the key ideas and concepts from the unimportant ones.

Underline the key ideas and concepts

Each paragraph should have one key topic, which the rest of the paragraph clarifies, supports, and develops. Write down a title that sums up the theme of the passage.

The title or heading is the précis of a précis and indicates what is to follow. Hence, it should reflect the central idea of the passage. It can be a word, a phrase, or even a short sentence. A suitable title must be provided even if it is not asked for.

Write a note-form summary of each paragraph

The passage should be summed up in the form of points. This should always be done in one's own words as far as possible. All irrelevant material should be omitted.

Write the précis

The original text should be paraphrased to express the summarized points more concisely and to develop them into coherent sentences, expressing all important points in a generalized form. Any repetitions or irrelevant details should be eliminated. Only the third person and indirect speech should be used. No personal comments should be added.

Review and edit

Compare the précis with the original document and make sure that it emphasizes the same points. Ensure that the précis is readable, concise, and coherent.

Although we have discussed these tips or techniques in the earlier chapters, some examples are given in Table 12.1 as a ready reckoner.

> 'There is great power in words if you don't hitch too many of them together.'
> —*Josh Billings*

Guidelines

The following are guidelines for effective précis writing:

- The theme of the passage must be determined very carefully.
- A précis is not the reproduction of important sentences.
- Conciseness is good but not at the cost of clarity.
- The précis should be intelligible even to a person who has not read the original passage.
- The original passage must be condensed in the précis-writer's own language.
- A précis is always written in the third person.
- Statistical information should be treated according to one's own discretion.
- All the main ideas in the passage should be given almost equal treatment in the précis.

Table 12.1 Guidelines for effective précis writing

Type of ineffectiveness	Original version	Revised version
Redundancy	(a) The analysis was thoroughly and wholly complete. (7) (b) The Chairman wants to meet those employees who are working in the production department, so that he can discuss the difficult crisis immediately. (23)	(a) The analysis was complete. (4) (b) The Chairman wants to meet the employees of the production department to discuss the crisis immediately. (16)
Circumlocution	(a) Objects, on our first acquaintance with them, have that singleness and integrity of impression that it seems as if nothing could destroy or obliterate them, so firmly are they stamped and riveted on the brain. (35) (b) The reason why the technicians were so upset was because their boss seemed so angry with them. (17)	(a) Our first impressions of objects are the most lasting. (9) (b) The technicians were upset because the boss seemed so angry with them. (12)
Passive voice	(a) If no satisfactory results are obtained, another study is carried out, but this time the data obtained from the first trial is included. (23) (b) After the robot was installed, a series of problems were faced by the company. (14)	(a) If you do not get satisfactory results carry out another study including the data of the first trial. (18) (b) After installation the company faced a series of problems with the robot. (12)
Wordiness	(a) The secretary's proposal was adopted with the full agreement of all the members. (13) (b) It is not fair or just in the interest of the consuming public that any commercial concern should acquire the sole power of trading in some particular article to the exclusion of its rivals. (34)	(a) The members unanimously adopted the secretary's proposal. (7) (b) Monopolizing a particular trade by any commercial concern is unfair to the consuming public. (14)

- Personal ideas should not be introduced.
- The original source must not be criticized or changed.
- The précis should be limited to no more than one-third the length of the original passage; the number of words must be indicated.

Tips for Condensing Longer Sentences

- Try to replace clauses by phrases and phrases by words.
- Make use of one-word substitutions.
- Avoid all unnecessary repetitions.
- Link various sentences.
- Omit examples, comparisons, contrasts, digressions, and minor details.

Samples

In this section, we give three examples to demonstrate précis writing. The number of words in each original text is indicated. The original text is followed by a vocabulary list, which gives the meanings of the more difficult words. This list is followed by a listing of the main points of the original passage. Thereafter, the précis has been written, constructed using the main points, indicating the final number of words after condensation.

Original Text 1

A vital factor in the success of any business is the right selection of its administrative staff. In this matter, the management has the inescapable function not only of making the right selection, but after having made it, also of providing the fullest scope for legitimate ambition and individual advancement. A management which is so petty as to be jealous of the powers and authority of the officers will naturally select staff of the submissive type, docile men who are accustomed to obey without question. It would not take the risk of engaging able and aggressive/self-motivated employees for the fear that it may one day be supplanted by them. A really go-ahead management which understands the mood of the present times will do the direct opposite. It will seek out men capable of evolving policies within their own sphere and will train them to rise to the highest position.

In connection with most administrative bodies, management is self-perpetuating and is responsible for its own succession. Even the ablest management cannot foresee the future with certainty; its decisions, for the future, are at best intelligent guesses. What, however, it can and must do is to make available the enterprise to the men who will be capable of taking the decisions of the future and who are qualified, trained, and tested during the present to do so. No management can rely upon a constant supply of geniuses. It must so train its staff that, during normal times, the enterprise is capable of being run effectively by men of not much more than average ability and with a robust sense of purpose. (270 words)

> 'Never be so brief as to become obscure.'
> –*Tryon Edwards*

Vocabulary   vital—important; legitimate—proper; docile—submissive like a slave; supplanted—removed, replaced; evolving—finding out; self-perpetuating—continuing itself; foresee—guess; rely—depend; geniuses—very intelligent and able persons; robust—strong and healthy.

Points

1. Success in business depends upon right selection of staff.
2. The management must foster the staff's proper ambitions and individual advancement.
3. A petty, jealous management selects submissive people for fear of being replaced by them.
4. A progressive management tries to find decision-making, creative geniuses, and trains them for the present.
5. Good management prepares its own succession for the future; it can only guess intelligently.
6. As the supply of geniuses cannot be assured, the staff must be trained to run the enterprise with normal ability and a strong sense of purpose in normal times.

Précis of the original text 1 is as follows:

Efficient Management

Success in business depends upon the right selection of staff members and fostering their proper ambition and individual advancement. A petty, jealous management will select docile people for fear of being replaced by them, while a progressive management appoints decisive, creative, intelligent persons and trains them in anticipation of its own replacement. Even the ablest management can merely guess the future intelligently. As the supply of geniuses cannot be assured, the staff must be trained to run the enterprise with normal ability and a strong sense of purpose in normal times. (91 words)

Original Text 2

For all industrial development we need power, and the ultimate restriction on power is the fuel from which it is extracted. Is there enough fuel to satisfy our ever-growing hunger for power? For conventional fuels, such as wood, coal, and oil, the answer is quite clearly 'No'. The world's known stock of oil is only sufficient to last sixty years at the present rate of consumption and the rate of consumption keeps going up and up. We are burning too much wood already, and the earth's known fuel-wood forests will be consumed soon. Coal is still in fair supply, but in some areas—notably England—it is becoming increasingly difficult to mine, and it is therefore uneconomical.

Besides fuel as a source of power, there is the device for harnessing energy from rapidly flowing water. Few sources of water power remain untapped, and the power they yield meets only a fraction of our total need. Moreover, it is not very dependable, because storing water in reservoirs depends on rains, which are sometimes freakish.

Conventional fuels release energy by combustion, but fission makes use of another kind of fuel, remarkable for its concentration of power. All fissionable material is extracted or manufactured from two elements, uranium and thorium, and the world has plentiful stock of them. But even so they will not last forever. There is probably enough to last for several centuries. Fission in the techniques known up till now converts only one-tenth of one per cent of its fuel into energy. Complete conversion of fissionable fuels into energy is known at present at laboratory level only. If it can be harnessed into a practical power device, one pound of fissionable fuel would be equivalent to three billion pounds of coal. Now the scientists' quest is to find out some more efficient process for using these fuels outside the laboratory on industrial scale. But after even fissionable material is gone, what then? There is no reason to despair. The sun is continually pouring solar energy on earth—we have only to gather and harness it. Those who think that humans will one day be left without any source of power are not far-sighted enough. (364 words)

Vocabulary ultimate—final; untapped—not tried; freakish—whimsical; fission—division of cells; quest—search.

Points

1. Power, which is dependent upon fuel, is needed for all industrial development.
2. The conventional type of fuel is not going to last for a very long time.
3. There is enough stock of uranium and thorium in the world, and fissionable material, which can serve as fuel for a very long time, is extracted from it.
4. Scientists are trying to find some efficient process for the use of fissionable material as fuel for industrial purposes.
5. Besides this, the energy that we get from the sun can also be gathered and harnessed for our purposes.
6. It is only short-sighted people who think that there will be no source of power left for humans in the future.

Précis of the original text 2 is as follows:

Inexhaustible Sources of Power

Power, which is extracted from fuel, is essential for all industrial development. The fear is that conventional types of fuel are not going to last for a very long time. Fortunately, we have enough stocks of uranium and thorium, and all fissionable material, which is a great source of energy, is extracted from these elements.

Scientists are busy researching an efficient process for the use of fissionable material for industrial purposes. The solar energy that we receive can also be gathered, harnessed, and used for our purposes. So people who imagine that in the foreseeable future humans would be left without any source of power, only display their short-sightedness. (109 words)

Original Text 3

There is no doubt that people are growing more and more interested in the seas, and that there is a great need for that interest. Humans have long tried to probe the secrets of the oceans to gain knowledge for its own sake, but there are other practical reasons for doing so. The sea can provide us with many things that we need in everyday life. Future generations will probably depend more on the seas for their food, and not only food in the form of fish. Minerals necessary for modern industries are also there, when we can find out how to extract them.

We have explored and mapped most of the land, and we are quickly exploring the air. The seas present a greater difficulty because we cannot yet, and probably never shall be able to, set foot on the deep ocean floor.

The aim of the extensive ocean-going expeditions, of the marine biological stations around the coasts, and even of those who simply study the shore uncovered by the tide, is to build up our knowledge of this vast and unfamiliar world beneath the waves. In some cases the knowledge gained can be put to practical use, but much of it is for interest only.

For the very early mariners, interest lay in the currents, and especially those at the surface that carried their ships along. They were also interested in the weather over the sea. Yet, even these hard-bitten seamen were not immune from a curiosity about the animals and plants that lived below the waves. Their first impulse may have been to seek trade overseas, or to fish for food, but over and above this, anything strange or beautiful, whether caught up in their nets or cast ashore by the tides, caused them to wonder. So, from the earliest time, the pursuit of practical everyday things went on side by side with the inquiry that springs from a desire to know more. Bit by bit grew the knowledge of the physical features of the seas, of such things as currents, waves, and winds, as well as of the biology, the knowledge of animals and plants. (359 words)

Vocabulary probe—delve deep, find out, explore; extract—derive them, set them; expedition—journey; beneath—below; immune—to have no effect; impulse—feeling, emotion.

Points

1. Humans' interest in the seas is increasing and it is a healthy and useful sign.
2. Future generations are likely to draw more of their food from seas.
3. Exploration of seas is a difficult job.
4. The aim of all marine exploration is to know the world beneath the waves.
5. Even the earliest mariners had interest in exploring the seas for the sake of knowledge.

Précis of the original text 3 is as follows:

Human Interest in Exploring Seas

Humans' interest in exploring the seas has increased and it is a healthy sign. Probably, future generations are to depend far more on the seas for their food. Though people have been able to map the entire land seas offer difficult prospects. The aim of all adventure has always been to know about the world. Some knowledge thus gained may have practical utility, but most of it is for the sake of interest. Even the earliest mariners, though their primary interest was to seek out trade routes, had the curiosity to study the animals and plants that lived below the waves. This curiosity, of course, helped them in gaining knowledge. (110 words)

For more solved samples of précis writing, please refer to the Online Resource Centre.

SUMMARY

Writing abstracts, summaries, and synopses is a necessary part of academic and professional work. Preparing such professional/technical documents requires concise, clear, and relevant writing. Learning to write the précis of the given text is a significant step in achieving the essential qualities of effective writing. By understanding the various steps involved in précis writing given in this chapter and practising them meticulously, we will be able to condense any given text with effectiveness and efficiency.

EXERCISES

Summarize each of the following passages in about one-third of the total number of words. At the end, write the number of words in your précis.

(a) Experiments have shown that in selecting personnel for a job, interviewing is at best a hindrance, and may even cause harm. These studies have disclosed that the judgements of interviewers differ markedly and bear little or no relationship to the adequacy of the job. Of the many reasons why this should be the case, three in particular stand out.

The first reason is related to an error of judgements known as the halo effect. If a person has one noticeable good trait, their other characteristics will be judged as better than they really are. Thus, an individual who dresses smartly and shows self-confidence is likely to be judged capable of doing a job well regardless of his or her ability.

Interviewers are also prejudiced by an effect called the primacy effect. This error occurs when interpretation of later information is distorted by earlier connected information. Hence, in an interview situation, the interviewer spends most of the interview trying to confirm the impression given by the candidate in the first few moments. Studies have repeatedly demonstrated that such an impression is unrelated to the aptitude of the applicant.

The phenomenon known as the contrast effect also skews the judgement of interviewers. A suitable candidate may be underestimated because he or she contrasts with a previous one who appears exceptionally intelligent. Likewise, an average candidate who is preceded by one who gives a weak showing may be judged as more suitable than he or she really is.

Since interviews as a form of personnel selection have been shown to be inadequate, other selection procedures have been devised which more accurately predict candidate suitability. Of the various tests devised, the predictor which appears to do this most successfully is cognitive ability as measured by a variety of verbal and spatial tests. (300 words)

(b) A recent phenomenon in present-day science and technology is the increasing trend towards 'directed' or 'programmed' research, i.e., research whose scope and objectives are pre-determined by private or government organizations rather than researchers themselves. Any scientist working for such organizations and investigating in a given field therefore tends to do so in accordance with a plan or programme designed beforehand.

At the beginning of the century, however, the situation was quite different. At that time there were no industrial research organizations in the modern sense—the laboratory unit consisted of a few scientists at the most, assisted by one or two technicians, often working with inadequate equipment in unsuitable rooms. Nevertheless, the scientist was free to choose any subject for investigation he/she liked, since there was no predetermined programme to which he/she had to confirm.

As the century developed, the increasing magnitude and complexity of the problems to be solved and the growing interconnection of different disciplines made it impossible, in many cases, for the individual scientist to deal with the huge mass of new data, techniques, and equipment that were required for carrying out research accurately and efficiently. The increasing scale and scope of the experiments needed to test new hypotheses and develop new techniques and industrial processes led to the setting up of research groups or teams using highly complicated equipment in elaborately designed laboratories. Owing to the large sums of money involved, it was then felt essential to direct these human and material resources into specific channels with clearly defined objectives. In this way it was considered that the quickest and most practical results could be obtained. This, then, was programmed research.

One of the effects of this organized and standardized investigation is to cause the scientist to become increasingly involved in applied research, especially in the branches of science which are likely to have industrial applications. Since private industry and even government departments tend to concentrate on immediate results and show comparatively little interest in long-range investigations, there is a steady shift of scientists from the pure to the applied field, where there are more jobs available, frequently more highly paid and with better technical facilities than jobs connected with pure research in a university.

Owing to the interdependence between pure and applied science, it is easy to see that this system, if extended too far, carries considerable dangers for the future of science—and not only pure science, but applied science as well. (419 words)

(c) Beyond all the hoopla involving fancy terms like 'embryonic stem cell research' and 'gutted egg reprogramming' lay the nasty little fact that technology can xerox you. That is the nub of the whole hassle. Because who in his or her right mind is going to object to a sure-fire drug for diabetes? Come to think of it; nobody has any problem with a safe and permanent cure for cancer either. Also notwithstanding all the cacophony of national and international ethics committees around the world, mortality thumping and accusations of playing God, everybody in their heart of hearts knows that 50 or 75 years from now, cloning will be as common as warts.

The discord and friction is not whether the benefits of cloning outweigh the possible social consequences, or that its abuse can unleash powerful forces which can be exploited to produce horrendous results. It is not whether it will help to produce discoveries that would hugely affect the study of human genetics, development, and growth or it is an affront to religious sensibilities by interfering with the natural process. It is not the subconscious fear that men would no longer be needed for reproduction. What it is about is not being able to come to terms with an impending social upheaval, the likes of which human society has not seen in its entire ten-thousand-year-old history. In other words, 'What does your cloned child call you?'

By circa 2045 you could walk into the local clinic and get yourself cloned as easily as eating a pie. Nine something months later you get delivery of what? Your child, your brother/sister/yourself?

Strictly speaking though, one reason the child is your sibling is because it has the exact mix of your parent's genes as you have. That is because the process of cloning is fundamentally the same as what happens when a fertilized egg splits into two at a very early stage of development to produce identical twins sharing the same genetic blueprint. Usually there is always an age difference between twins that is measured in minutes; the only difference is, in your case it would be measured in decades.

Protests of losing diversity of genes and interfering with the course of Darwinian evolution are objections that will turn out to be peanuts compared to the magnitude of this interpersonal problem we will soon have to deal with. Of course we will resolve it, but the resolution will probably demand a jettisoning of a whole lot of received wisdom and will completely metamorphosize the way we live with ourselves and others. (440 words)

(d) Sweet and cold, with a wonderful mouth feel, ice cream is an American favourite, but far from the soft, icy product produced by hand-cranked freezers, today's commercial ice cream is a complex product designed and engineered for the best attributes.

'There are a variety of formulas which are used to derive recipes,' says Dr Robert Robers, associate professor of food science and director of the Penn State Ice cream Short Course, the nation's oldest and best-known educational program on ice cream manufacturing.

Legally, ice cream must contain no less than 10 per cent milk fat, and no less than 20 per cent milk solids. In general, most ice creams contain 10 to 16 per cent fat and 9 to 12 per cent non-fat milk solids with 11 to 15 per cent sucrose or equivalent for sweetness. Then, of course, there are the flavourings and the emulsifiers and the most important and often a forgotten component, air. Choices within these ranges produce economy, premium, and super-premium ice cream.

'Many people think that the higher the quality of the ice cream, the higher the fat content since fat makes the ice cream feel unctuous and creamy', Roberts told attendees at the annual meeting of the American Association for the Advancement of Science in Boston. 'Fat is also a cold insulator and is involved in trapping air and perhaps most importantly, it tastes good,' he notes. In essence, ice cream is frozen foam. During the freezing and shipping process, proteins in the ice cream mix encircle the air bubbles incorporated in the liquid and then the fat stabilizes the bubbles. 'Protein traps the air, but cannot hold it, much like skim milk foams', says Roberts. 'The fat in ice cream partially destabilizes and traps the air. In ice cream, in contrast to other products, emulsifiers are added to destabilize the fat, allowing partial agglomeration and air cell stabilization.' During ice cream mix manufacture, the ingredients are measured by weight, and then mixed, pasteurized, and homogenized.

'The pasteurization process is required by law to destroy any potential pathogens and make the product safe for consumption,' says Roberts. 'Homogenization, a high pressure process designed to reduce the size of the fat globules and increase whispability is very important. Without homogenization, the mix might over-destabilize during the freezing process leading to a defect known as buttery, which is definitely not what people want in ice cream.' The pasteurized, homogenized mix is cooled and allowed to age for at least four hours to create some fat crystals. Allowing time for the fat to assume the appropriate form is a critical step. 'The surface area of the fat in a quart of mix is equal to about 1,200 square yards', says the Penn State scientist.

After aging, the mix is ready to be frozen by commercial ice cream freezers, though much larger, operate on the same principle as hand-crank machines. The outside wall of the freezer gets cold and a series of blades remove the ice crystals from the wall and move them towards the centre, also incorporating air. Roberts and others have looked at the speed at which the dasher moves to determine if an optimum setting exists.

Contrary to conventional understanding, about 50 per cent of the energy removed by the refrigeration process is due to the frictional heat created by the dasher scraping the freezer wall. While freezing under agitation, only about half of the water in the ice cream mix freezes, leaving the other half liquid. The proteins, salts, and sugars in the mix lower the freezing point enough to require further freezing. The hardening stage, when the rest of the mix solidifies, must be done rapidly to avoid the formation of large ice crystals. (620 words)

CHAPTER 13

Technical Reports

OBJECTIVES

You should study this chapter to know
- O the importance, objectives, and characteristics of reports
- O the different categories of reports
- O various formats in which reports can be presented
- O prewriting steps followed for preparing reports
- O how to structure a report
- O the different types of reports
- O how to draft, edit, and proofread the report

Introduction

We see many examples of reporting in daily life. A nurse at the hospital reports to the doctor in the inpatient ward about the condition of each patient. A supervisor, at the end of the day, reports to the manager the progress of the work carried out in his/her supervision. Similarly, the manager of a bank sends a periodic report to the head office on the state of deposits, advances, overdraft limits, etc., during the period. Another manager posted in a remote rural area reports to the head office about the difficulties faced in sanctioning loans to farmers. A publishing firm keen on introducing a new series of paperbacks into the market has to ask its marketing research team for a report on the current reader preference. The managing director of a bank would like to get a report on the efficacy of the measures introduced by the bank to promote efficiency in the branch offices.

From time to time, the Government sets up committees and commissions to report on various issues of social, political, and economic importance. There are various Parliamentary committees, such as Committee on Human Resources, Committee on Agriculture, Committee on Industry, Committee on Science and Technology, Committee on Environment and Forests, and so on, which are set up to discuss, analyse, and report on various matters pertaining to the respective Ministries. Such committees meet from time to time, work out a detailed plan to conduct surveys and collect data from various sources, and finally submit their findings and recommendations in the form of a technical report.

Reports are a part of our lives—starting from school progress reports through news reports on television and radio to the many kinds of reports we are required to submit in the course of our work. But what is a report? A report is usually a piece of factual writing, based on evidence, containing organized information and/or analysis of a particular topic. It is a major form of technical/business/professional communication.

The word 'report' is derived from the Latin *reportare*—'to bring back'. Over a period of time, it has evolved to mean that the gathered information is unbiased and evidence-based.

A report is usually a piece of factual writing, based on evidence, containing organized information on a particular topic.

A person transmits certain facts, ideas, or suggestions useful for another person through a report. It can also be the description of an event or a condition.

Importance of Reports

A report is a basic management tool used in decision-making. Hence, it is extremely important for all organizations, especially for large-scale organizations that are engaged in different activities handled by different departments. Their top executives cannot keep a personal watch over all these activities. Hence, they have to base their decisions on the reports from the heads of various departments. Reports bear a lot of significance both to the organizations in which they arise and to the organizations they are submitted to. The following list will give an idea of the importance of reports and report writing to the organizations and the individuals.

> Remember that all reports carry legal responsibility. They can be used as legal instruments. Your activities, as an employee, and your competence at work are reflected through reports.

- *A report is the only tangible product of a professional.* All the efforts of engineers, academicians, and researchers culminate in reports that convey to others the efficiency with which they carried out their assignments.
- *Reports enable decision-making and problem solving in organizations.* Based on the information presented, analysis discussed, or the suggestions rendered, administrators can make important decisions and solve serious problems.
- *Reports help the authorities in planning new ventures and in evaluating men and material.* If an organization wants to open a new branch in a nearby locality, it can plan for the same more effectively after going through the feasibility report prepared for this purpose. Similarly, if the organization wants to assess the qualities and capabilities of a person, it can do so by going through the annual assessment form submitted by the supervisor of that person. It can also evaluate a newly introduced machine or product by listening to or reading the report prepared for this purpose.
- *Reports are an important means of information dissemination within and outside the organization.* Many of the routine reports, such as inspection reports, inventory reports, or annual reports, transmit information across and outside the organization.
- *Reports serve as a measure of the growth, progress, or success of an organization.* The success of any organization depends on the quality and quantity of information flown through its personnel in the form of oral or written reports. For instance, an organization focusing on research can bring out reports from time to time to provide information on the progress being made, which serve to prove its worth to prospective clients. Also, a manager can measure the progress his/her department is making by going through monthly reports on the status of all the projects.
- *Reports serve as a valuable repository of information.* Organizations of various kinds preserve reports of importance and value for a long time so that they can refer to these reports whenever needed. For example, an academic institution can refer to the previous assessment reports it had received from some committee in order to improve its performance.
- *Reports reveal gaps in thinking.* A report can give the recipient an idea about whether the writer had thought and proceeded logically and whether he/she had conducted an in-depth study of the topic.
- *Reports develop certain skills in the writer.* Reports not only help organizations but also help the writer to develop certain skills, such as the ability to organize, evaluate, and communicate with greater accuracy.

Objectives of Reports

The purposes for which reports are written vary widely. Some important purposes of reports are to

- Present a record of accomplished work (project report)
- Record an experiment (primary research report/laboratory report)
- Record research findings or technical specifications (a report on the details of a new product)
- Document schedules, timetables, and milestones (a status report on a long-term plan)
- Document current status (an inspection report)
- Record and clarify complex information for future reference (a report on policies and procedures)
- Present information to a large number of people (annual report)
- Present organized information on a particular topic (a report describing the working of various divisions of an organization)
- Recommend actions that can be considered in solving certain problems (recommen-datory report)

Characteristics of a Report

In this section we will study the important characteristics that all reports must have.

> In a business report, the purpose can be to inform, persuade, motivate, or express feelings—or a combination of these.

Precision

Precision gives unity and coherence to the report and makes it a valuable document. Effective reports must clearly reflect their purpose. The purpose should direct the investigation, analysis, and recommendations. The purpose determines the format, content, sequence, and word usage in the report. Hence, first the objective should be defined clearly, considering the expected response from the audience.

Factual details

The report should be very detailed and factual. It should meet the audience's expectation. The scientific accuracy of facts is very essential to a good report. Since reports invariably lead to decision-making, inaccurate facts may lead to disastrous decisions.

Relevance

The facts presented in a report should be not only accurate but also relevant. While it is essential that every fact included in a report has a bearing on the central purpose, it is equally essential to see that no relevant information is excluded. Irrelevant facts make a report confusing; exclusion of relevant facts renders it incomplete and is likely to mislead.

Reader-orientation

A good report is always reader oriented. While drafting a report, *it is necessary to keep in mind the person(s) who is (are) going to read it.* A report meant for the layperson will be different from one meant for technical experts.

Objectivity of recommendations

If recommendations are made at the end of a report, they must be impartial and objective. They should come as a logical conclusion to investigation and analysis. They must not reveal any self-interest on the part of the writer.

'The report is done, but who am I to give it to?'

Simple and unambiguous language

A good report is written in simple, unambiguous language. It is a document of practical utility; hence, it should be free from various forms of poetic embellishment such as figures of speech. It should be clear, brief, and grammatically accurate.

Clarity A good report is absolutely clear. Clarity depends on proper arrangement of facts. Report writers should make their purpose clear, define their sources, state their findings, and finally make necessary recommendations. They should divide their report into short paragraphs with headings, and insert other suitable signposts to achieve greater clarity.

Brevity A report should be brief. Brevity should not be achieved at the cost of clarity. Nor should it be at the cost of completeness. Sometimes the problem being investigated is of such importance that it calls for a detailed discussion of facts. Then this discussion should not be evaded. Brevity in a report is the kind of brevity one recommends for a précis. *Include everything significant and yet be brief.*

Grammatical accuracy The grammatical accuracy of language of a good report is of fundamental importance. It is one of the basic requisites of a good report as of any other piece of composition. Remember that *faulty construction of sentences makes the meaning vague and ambiguous.*

Special format

The technical report uses a rather involved format including cover, title page, table of contents, list of illustrations, letter of transmittal, and appendices. These have to be prepared according to a set standard, which will be presented later in this chapter.

Illustrations

Most technical reports contain illustrations, which may be tables, graphs, maps, drawings, charts, or photographs.

Homogeneity

A report should deal with one main topic. All the sections of the report should focus on that topic.

> Irrelevant facts make a report confusing; exclusion of relevant facts renders it incomplete and is likely to mislead.

Documentation

Technical reports acquire more value when adequately documented by acknowledging sources of information in an appropriate style.

A good report is

- Precise and brief
- Factual
- Unambiguous and accurate
- Relevant
- Reader-oriented
- Objective and homogeneous
- Detailed and documented

Categories Of Reports

On the basis of purpose, frequency, mode of reporting, length, approach, and target audience, reports can be classified as follows:

- Informative, analytical (purpose)
- Periodic, special (frequency)
- Event report
- Oral, written (mode of presentation)
- Long, short (length)
- Formal, informal (approach)
- Individual, group (target audience)

Informative Reports

An informative report, as the name suggests, entails provision of all details and facts pertaining to the problem. For instance, it could be a report that attempts to trace the growth of Company X in the automobile industry. In a report of this kind, the presentation of all details that led to the growth of Company X should be listed in a chronological order.

In a report of this kind, the introduction is followed by a presentation of information or facts and a conclusion thereafter, where all the details are collated in brief as a recap of earlier sections. Recommendations do not arise in this type of report.

The main purpose of an informational report is to present the information in an objective, factual, and organized manner. Supporting data, appropriate order, and good presentation style are important in writing an informational report.

> Informative reports focus on documenting new information; analytical reports assess information in order to propose a course of action.

> The main purpose of an informational report is to present the information in an objective, factual, and organized manner.

Analytical Reports

An analytical report is also known as an *interpretative* or *investigative* report. An analytical report analyses the facts, draws conclusions, and makes recommendations. For instance, a report presenting production figures for a particular period is informative. But if it analyses the causes of lower production in that period, it becomes analytical, interpretative, or investigative.

An analytical report comprises stages in which there is a proper identification of the problem, analysis, and subsequent interpretation. Recommendations or suggestions are then incorporated in the report, depending upon what is required by the report writer. Thus, in a problem-solving method, the steps observed are as follows:

- Drafting problem statement
- Evolving criteria
- Suggesting alternatives and evaluation
- Drawing conclusion(s) and making recommendations

Periodic and Special Reports

Periodic reports are either informational or analytical in their purpose. As they are prepared and presented at regular, prescribed intervals in the usual routine of business, they are called *routine reports*. They may be submitted annually, semi-annually, quarterly, monthly, fortnightly, weekly, or even daily.

Generally such reports contain a mere statement of facts in detail, in summarized form, or in the layout of a prescribed form, without an opinion or recommendation. Progress reports of various kinds, inspection reports, annual reports, and sales reports come under this category.

At times the routine reports can be analytical or interpretative. For example, when the heads of various divisions in an organization submit annual assessment reports of their employees to the higher authorities, they assess the data and give their recommendations so as to enable their superiors to take certain decisions.

Special reports are related to a single occasion or situation. A report on the feasibility of opening a new branch, a report on the unrest among staff in a particular branch, or a report on the causes behind the recent fire incidents in a factory are special reports. Special reports deal with non-recurrent problems.

Event Report

Events reports are short reports that provide an overview of the event that has taken place. The readers get an idea of what to look for while organizing an event. The success of the event mainly lies on the positive feedback received from the attendees. One of the essential features of event report is to offer the stakeholder the data on the event's success. The structure of an event report varies as per the type of event and organization conducting the event. However, its main elements are title of the event, name, telephone number, e-mail, conference summary, introduction, objectives, inauguration, sessions, feedback, and conclusions.

The language of event reports should be easy to understand, simple, precise, and matter of fact. As the audience is not defined, avoid the use of jargon and cliches. However, identify the target audience to know what is essential for them in the report and focus more on that aspect. Present the data in a readable form so that readers can easily interpret and make sense. One should know the motivation of the readers and align the event report with the language they speak.

Exhibit 13.1 shows a sample event report.

EXHIBIT 13.1   Sample event report

Event reports contain all the information about the event data and the value generated for the business. This report is important for the project management stakeholders to identify and finally decide on conducting the similar future events. The various elements of the event report include:

Title of the event: International conference on Media, Communication and Ethics
Date of Event: 9–10 April, 2021
Location of Event: New Academic Building, KIIT, Kanpur
Number of Persons Attending: 150
Contact Name: Rashmi Bhatnagar
Telephone Number: 9024568500
E-mail: rashmi@kiitkanpur.ac.in

Conference Summary
The International Conference on Media Culture and Ethics that was held at KIIT, Kanpur Campus on 9–10 April 2021 unfurled to be a great success. As many as 105 delegates from different parts of the country and abroad, including the US, UK, and Dubai, attended it and made a significant contribution to the deliberations. In addition, a galaxy of 20 distinguished speakers with diverse backgrounds delivered erudite talks on issues related to the theme of the conference, sharing their views and experience on a common platform. A collection of 84 full-length research papers presented by participants and published in the form of a book was released on the occasion. We received feedback from several participants expressing their appreciation for the excellent arrangements made by KIIT Kanpur for the conduct of the conference.

Introduction
Media and popular culture percolate in all aspects of our waking time. The unrelenting exposure predominantly guides our perception of reality, the formation of our values, beliefs, and attitudes, and above all, defines self and society. This has become an extraordinarily powerful educating agent among the majority of the population. The speed with which it is influencing society has blinded us. Hence, it becomes imperative to have the complete and true reflection of the cultures in which the stories are set. Authenticity and ethical guidelines also need to be employed seriously to propagate content through these different forms of media. As professional practitioners, it is of utmost importance to know their content and relevance in today's context and be good human beings, ethically cognizant of their responsibilities for others and the world.

Culture is the belief system of a particular group, region, or nation, for instance, the way content is circulated through films or other essential forms of the media can have both positive and negative impacts on different segments of society. Hence there is an urgent need to study the nuances associated with the information being distributed and shared through different forms of the media. This conference would provide a platform to discuss the challenges and possibilities of arriving at certain conclusions regarding the present status quo.

Objectives:
- To explore the Media and its influence on society and culture
- To understand the role of media in development, communication and nation-building
- To identify the extent to which media engineers and distorts our perceptions
- To find out the impact of news channels on the mind of the general masses
- To understand different formats of the print and electronic media
- To unravel the ethical dilemmas in different cultures and media
- To draw guidelines of cultural behaviour based on academic deliberations

Inauguration

The International Conference on Media Culture and ethics was inaugurated yesterday morning.

Ms Anu Hassan welcomed our delegates to the conference. Mr Sunit Tandon, was the chief guest and Ms Vasanthi Hariprakash was the guest of honour. Prof. Subrajeet Sarkar, director KIIT Kanpur, and Prof. Pankaj Chauhan, Head of Humanities and Social Sciences Department, KIIT Kanpur presided the inaugural ceremony. Dr. Rashmi Bhatnagar, convener of the conference introduced the theme, media culture and ethics.

Prof. Sarkar introduced the journey of KIIT, how it started as a school in 1950, and today it is one of the top institutions of the world. He emphasized how the Media is a means of communication and how it reflects and creates culture.

Ms Vasanthi Hariprakash, founder of pickle jar platform, enlightened us with the importance of the initiatives that function with an aim to reach to the remotest of areas and empower individual voices that are unheard and are subsumed under the mainstream. NWM, Navodayam, Voicebook and Gaon Connection and Community radio bring tribal and rural narratives and stories to the forefront, thereby bringing out the true India.

Mr Sunit Tandon, director of IIMC,emphasized the rise of social media that results in democratization of information but is vulnerable to a debate between what is true and what is fake. These debates will dictate how our culture evolves.

The ceremony concluded with the release of the abstract book and conference proceedings followed by a vote of thanks by Dr Suman Shekawat, the co-convener of the conference.

There were two plenary sessions on the first day, followed by four technical sessions.

Plenary Sessions

Session I

The first plenary session was based on clarity, consistency and constancy: Does brand communication really matter? It was chaired by Ms Vasanthi Hariprakash and co-chaired by Prof. Devendra Ghule. The speakers for this session were Prof. Falguni Oza, Prof. Paurav Shukla and Prof. Hemant Gaule.

Prof. Falguni Oza told us how companies could use their advertisements to communicate and entertain the audience. She showed us various advertisements and emphasized how low involvement products can use metaphors to represent themselves. She also spoke about how corporate social responsibility (CSR) can lead to brand positivity.

She concluded by saying that brand communication matters, and the brands must continuously change and adapt to the changes in consumer behaviour.

Prof. Paurav Shukla highlighted the shift in how media communicates to its audience. From newspapers and books, we now read and access information on our electronic devices through the internet, predominantly through social media. He raised the concern of how our interests and choices are assessed and predicted by AI systems.

Prof. Hemant Gaule shared his experience of working for our Prime Minister, Narendra Modi's Campaign, and how it was important to convey the desire of the Modi government very clearly to people. Analysing the US President Donald Trump's speech, he emphasized how the choice of words and *the consistency of words can communicate effectively.*

Session II

The second session was based on the *Role of Communication in creating Compelling Corporations & Communities.* It was chaired by Prof. Paurav Shukla and co-chaired by Dr Anupama Goel. The Speakers for this session were Prof. Sangeeta Bharadwaj, Prof. Ravindran, Prof. Sankar Srinivasan. Prof. Bharadwaj spoke about how media has become a platform that is not just for the privileged but now is a medium to express ideas and thoughts irrespective of socio-cultural barriers.

Prof. Ravindran defined corporate communication and the perception of corporate communication wherein the truth may always not matter; however, it is crucial to propagate transparency and honesty such that truth triumphs over fake news. Prof. Sankar Srinivasan, Senior Technical Lead, Federal Reserve Bank of New York explained the significance of corporate communications in reputation building.

A total of five insightful technical sessions: Films and Literature, Journalism & Political Communication and Organizational Communication, were conducted on the first day of the conference. Day one culminated in the mesmerizing ambience of the royal Alsisar Fort that gave us a glimpse into the culture and tradition of Rajasthan.

Session III

The second day of the conference began with two technical sessions followed by a plenary session on "Media and its role in Shaping Narratives and Public Agenda", It was Chaired, Prof Sunil Singh. The Speaker of this session was Ms. Vasathi Hariprakash. She spoke about - Evolution of journalism: the extinction of proofreading with the advancement of technologies.

The ethics of reporting has changed with the times and now focuses on sensationalizing news rather than bringing truth and justice to people. Today, the news has to be perceived exceptionally carefully. One is constantly challenged to distinguish between fake news and the truth. The true essence and power of journalism are such that it can instigate transformation and change for the betterment of society.

The plenary session was followed by a panel discussion on Media and Communication courses for Engineering Students: Opportunities & Challenges. It was moderated by Prof. Neel Upadhyaya. The participants were Prof. Ravi Kumar B, Prof. Naveen Gupta, Prof. Gyan Singh Shekhawat, Prof. Ravi Prakash, Prof. Supriya Agrawal, and Prof. S.S. Bhakar.

There were two more technical sessions which were followed by the valedictory ceremony. Prof. Nirupama Prakash spoke about the role of media, culture, and ethics in today's society. Dr Rashmi Bhatnagar, convener of the conference, thanked all the delegates and team members who had worked for the conference. The director, Subrajeet Sarkar, director KIIT appreciated the conference and expressed his desire to see more such events in the future. Some delegates appreciated the efforts of the organizing team and spoke about their wonderful experience at the conference. Dr Anupama Singh, co-convener of the conference, proposed the formal vote of thanks.

Feedback and conclusion

MCE 2021 ended with this ceremony, and all associated members and delegates have indeed learned and added experience and knowledge to their former understanding of Media, Culture, and Ethics. The feedback was collected from the participants to decide the future course of action.

One of the participants said, 'I am thankful to all of you from the bottom of my heart for organizing such a wonderful and enlightening International conference. It was indeed a mixture of several academicians, budding researchers, and working journalists. I found it academically very profitable'.

Another one said, 'Thanks for all the things like: hosting a fantastic conference, providing grand hospitality and organizing Alsisar trip. I have never seen such caring faculty and supporting researchers like you all. Although the conference duration was two days, the memories related to it will remain forever in my academic life'.

Oral and Written Reports

Reports can be oral or written depending upon the mode of presentation. When you rejoin duty after attending an international seminar, you meet your officer and report about the deliberations of the seminar. This type of reporting comes under oral reporting.

An oral report is simple and easy to present. It may communicate an impression or an observation. While oral reports are useful, written reports are always preferred as they enjoy several advantages over the oral ones. Table 13.1 compares oral and written reports.

Table 13.1   Oral and written reports

Oral reports	Written reports
• Immediate feedback is possible	• Immediate feedback is not possible
• Do not add to the permanent records of the organization as the information/facts can be denied	• Contribute to the permanent records of the organization
• Audience needs to comprehend quickly as and when these are presented	• Audience can ponder over these reports and understand at its own pace
• May be encumbered with irrelevant facts and overlook important ones	• Are more accurate and precise as the writer will be careful in putting down the facts in writing
• Cannot be referred to again and again	• Can be edited, reviewed, stored, and retrieved
• Have less professional value	• Have more professional value

Long and Short Reports

Reports are classified based on the frequency of appearance, length, degree of formality, and purpose. As is evident, long and short reports are classified based on length. When there is a definite purpose, short reports are suitable as they are precise and concise. Their structure is not very elaborate and the focus is not on format as well. Sometimes the format is provided by the organization. Short reports include memo and letter reports. A short report usually begins with an introduction, followed by the information and then conclusion. However, the format may vary depending on the situation.

In contrast, long reports place a lot of emphasis on the format. It has got elaborate structure and consists of abundant information. Moreover, it is properly organized and analysed. It is usually produced after detailed research and data collection. These can be further classified as informative, analytical, and recommendatory reports. Long reports also have a lot of illustrations to support the discussion.

Formal and Informal Reports

The nature and format decide if the report is formal or informal. A formal report is the result of proper survey and investigation and is presented in a prescribed format. It is prepared as per the requirement of the organization. The language is also very formal. The formal reports have all the elements of reports and follow the laid down rules of writing reports. The length may vary from a few pages to large volume. These reports include annual reports, auditor's reports, policy reports, interpretive reports, etc.

An informal report is usually transmitted from person to person. It can vary from a few lines to several pages. Sometimes an informal report may include raw data that can be used at the time of need. It is generally targeted at a few people. The memo report is an example of an informal report.

Individual and Group Reports

When a report is classified based on the target audience, it is classified as individual and group report. A report presents the information in an organized manner in the most usable form to the set of people. It may describe the series of event to the concerned people. For example, it may present the government expenditure of the entire year to all the citizens of the nation. Another report on the admission pattern in the engineering college can be presented to aspiring engineers and their parents. These two are the examples of group report.

However, some reports are of concern only to an individual and are not meant for the general reader. These include the attendance report of an employee or an individual's progress report. These reports are of more interest to the top authority than anyone else. These reports are called individual reports. The format and design of these two reports may not vary much, but for the individual report the formalities can be dispensed with, and a personal touch can be added.

In an organization the marketing head sends a group report showing next year's projections to all the other functional (line) heads, and the production line head sends an individual MIS report to another head of the organization. Thus, in the first case, one individual is sending his/her report to a group of individuals, whereas in the second case an individual is sending a report to only one individual.

Formats

A report may have any one of the following formats:

- Manuscript
- Letter
- Memo
- Pre-printed form

Manuscript format   This is the most commonly used format for reports and is generally used for formal reports. The length of such reports can range from a few pages to several hundred. Further, the manuscript form can be used for all types of reports—informational, analytical, or routine. As the length increases, these reports include more elements such as abstract, summary, appendix, glossary, and so on. All the elements of a manuscript report are discussed in detail later in this chapter under the heading 'Structure of Reports'.

Memo format   A report sent to somebody within the organization will be in a memo format. The analysis, conclusions, and recommendations are included in the main text part of the memorandum, the other parts being the same as the inter-office memorandum discussed in Chapter 15. A memo report (Exhibit 13.2) is written on the letterhead of the organization. Inside address or salutations are not required. The main body of the memo report includes headings appropriate to the discussed matter. Although there is no complimentary close or signature, sometimes the memo report is signed or initialled at the end. Most organizations have a printed format for memos in which a memo report can be submitted.

EXHIBIT 13.2 Sample memo report

MODERN INSTITUTE OF TECHNOLOGY
JAIPUR (RAJASTHAN)

INTER-OFFICE MEMORANDUM

To: Dean, Students Welfare Division
From: Chief Warden
Date: 15 October 2021
Subject: Negative effects of Internet facilities

Please refer to your letter No. SWD/IM/2003 in which you have asked me to study the negative effects of Internet facilities provided by the Institute. I would like to present my findings and recommendations.

Findings

The data for the report was collected by interviews with wardens, Mess Managers and the Chief of Information Processing Centre. Also, the medical records of the students were collected from the office of the Chief Medical Officer.

Physical Problems

A preliminary look into the medical records shows that about 75% of the students owning a computer and regular users of the IPC have complained about some physical problem or the other during the past one year. Dr R.K. Sen, Chief Medical Officer, told me that these were the symptoms of Carpal Tunnel Syndrome (CTS), a deadly disease that affects many computer users all over the globe. He also explained that these were due to excessive time spent in front of the computer, improper sitting posture, and the lack of physical exercise amongst the students. Also lack of proper sleep is a cause of this fast growing disease.

Academic Performance

Another disturbing trend has been the decline in academic standards of the student. Most of the wardens and teachers have complained about the declining academic output of the students since the facility was provided to the students. From the talks with Dr T. Bansal, Chief, IPC, I have concluded that most of the students make improper use of the Internet facility. The most common use is for chatting and watching movies over the Web. Though listening to music is also attributed as a problem, one may say that music is good for the students as it has a soothing effect. This abuse of the facility hampers the mental development of the student. Many have got so hooked to it that they live in a virtual world and the only friends they have are chat friends. This is an alarming fact and the trend must be stopped from growing.

Food Habits

The food habits of the students have also been a cause of concern lately. Most Mess Managers agree that the attendance for meals has gone down considerably since the last year, after the introduction of the computer centre. Also, the growing use of junk food by the students is a cause of concern, according to Dr Sen.

Recommendations

Given below are a few suggestions which may help in tackling the problem in question:
- The amount of time spent by a student in the computer lab should be fixed to no more than 4 hours. This can be easily implemented according to the IPC Chief.
- Undesirable sites should be blocked.
- Awareness should be raised among students about CTS and also how to prevent it.
- Considering Dr Sen's advice, chairs in the computer centre should be changed to specially designed ones for more comfort.
- Student participation in sports and cultural activities must be encouraged to shift their attention from computers.
- The computer lab must be shut down between 12 PM and 9 AM to give students proper rest.

I hope that this report will give you an insight into the nature of the problem and also its possible solutions. I would be glad to provide any clarification or additional information required in this regard.

Amit Goyal
Chief Warden

Letter format   While sending short reports of a few pages to outsiders, one can opt for a letter format. Besides all the routine parts of a letter, these reports may include headings, illustrations, and footnotes. The letter report is one of the most personalized forms of reports, but the degree to which it can be personalized depends upon the relationship between the writer and the audience, and the circumstances under which it is used. Exhibit 13.3 shows a sample letter report.

Pre-printed form   Reports containing routine matter and which are periodical in nature may be written in a form prescribed by the organization. All one needs to do is to fill in the blanks in a pre-printed form (see the sample inspection report in Exhibit 13.4). For instance, a report presenting the performance assessment of an employee, an interim report reflecting the progress of a project, or a report informing the condition of equipment in a laboratory may be presented in a printed form wherein the reporter needs to fill in certain details against the details asked for.

Prewriting

Before actually beginning to draft a report, various tasks need to be undertaken, which may be referred to as the *preliminary steps to writing a report*. The effectiveness with which one carries out the tasks involved in these steps decides the effectiveness of the technical report. In fact, the planning stage is the most crucial one. Enough time should be spent in collecting material, synchronizing details, and ensuring that nothing has been left out. If the planning is done in a detailed manner, there are very few chances of errors creeping in at the final stage. In fact, *planning for a report is as important as the process of writing itself.* The various steps involved in report planning are as follows:

- Understanding the purpose and scope
- Analysing the audience
- Investigating the sources of information
- Organizing the material
- Making an outline

Purpose and Scope

Purpose refers to the objective of the study, while scope refers to the depth or extent of coverage. Assume that you, as the senior engineer of an organization, have been asked by your department head to study why the recently constructed flyover did not receive the anticipated feedback from the users. You are also required to suggest some measures to modify the same. Unless you are clear with the objectives of the task your report involves, for example, (a) identifying the causes of dissatisfaction and (b) suggesting remedial measures, you will not be able to proceed in the right direction. It is the *purpose* of the report that enables you to decide the amount of data to be collected, the data collection method to be used, the quality and quantity of information to be included in the report, and the methodology to be adopted in analysing the situation and arriving at a solution.

Further, it is essential to understand the nature of the report—informative or analytical. In an informative report, one may stress factors contributing to collation of information at the time of stating the purpose. However, in an analytical report the writer would need to prepare a problem statement, the analysis of which becomes the thrust area of the report. Depending on the type of report to be written, there is bound to be a difference in the definition of the problem and purpose.

Audience

The audience for a technical report—or any piece of writing for that matter—is the intended or potential reader. For most technical writers, this is the most important consideration in planning, writing, and reviewing a document. One 'adapts' his/her writing to meet the needs, interests, and background of the readers. Lack of audience analysis and adaptation is one of the root causes of most of the shortcomings of professional, technical documents—particularly while writing instructions, where inadequacies surface most glaringly.

EXHIBIT 13.3 Sample letter report

SINLEY DISTRIBUTING COMPANY
3204 Jawaharlal Marg, New Delhi

Mr S.S. Moondra
September 27, 2021
Akshay Supermarket
Vidya Vihar
Pilani, Rajasthan

Dear Mr Moondra:

Subject: Advantages of Fully Stocked Shelves

As inquiries are increasing from several supermarket executives concerning grocery and drug shelf stocking, I have undertaken an investigation to determine the effect of fully stocked shelves on sales. This survey has been made considering representative grocery and drug products, with attention given to percentage increases through mass stocking.

Effect of Diversification

Seven supermarkets were surveyed, with several brands of products checked for a two-week period under normal shelf-stocking conditions, and then for two more weeks under fully stocked shelf conditions. Enclosed is the complete result of the survey: below is a simple breakdown:

Table 1: Sales in Relation to Number of Items Stocked

	On Total Grocery Product Sales	On Total Drug Sales	On All Products
Number of items checked	128	69	197
2 weeks' unit sales under normal conditions	8,404	607	9,011
2 weeks' unit sales when shelves were kept fully stocked	10,287	902	11,189
Change in percentage	+22.4%	+48.5%	+24.2%

If you notice the change in the percentage of sales resulting from fully stocking the shelves, it is obvious that this procedure is of tremendous value:

Grocery product sales ………………………………… 22.4 per cent increase
Drug product sales …………………………………....... 48.5 per cent increase
All products sales …………………………………....... 24.2 per cent increase

Margin and Turnover

We all know that it is the desire of every supermarket to offer goods at the lowest possible prices. This can be accomplished only by reducing markup and increasing stock turnover. Now, if you can increase sales on all products by 24.2 per cent merely by fully stocking your shelves, it is apparent that you will be able to reduce markups and offer merchandise at lower prices. By your giving maximum exposure to different commodities, the consumer has the opportunity to see more and as a result is motivated to purchase something that would never have entered his mind if certain brands had not caught his eye.

The rise in the general standard of living has caused a proportional increase in the demand for service. By our very nature, we cannot offer personalized service; therefore, we must do the next best thing—give intensive exposure to a large variety of brands. That is, substitute displays and printed selling appeals of various manufacturers for personal selling. The consumer is still our livelihood, and the more he sees, the more he will buy.

Recommendations

I suggest that you keep your shelves fully stocked at all times to increase sales of merchandise. It has always been our policy to sell through our retailers, which has been brought to light by the survey.

Yours sincerely
M.K. Hingle
President

EXHIBIT 13.4   Sample inspection report

EXCEL Technovation Pvt. Ltd
Ph. No. 377919 Fax: 37978-0141
CUSTOMER CALL FEEDBACK REPORT

Call Registration Number: _____ Date : _____
Customer: _____
Location: _____

Sys. Model:	Sl.No.:		Peripheral/Add-on Model:	Sl. No.:	
Service Type	Warranty/AMC/IRB/ Chargeable/Others		Product	Home PC/Desktop/Server/ Sun/IBM/Datacomm/ SW/Peripheral/Others	
Call Type	Ins/CM/PM/Proj/Upj/Upg/Siteinsp/Others			Call Category	HW/SW

Problem Reported: _____

Event	Date	Time	Event	Date	Time
Call Reported			Start of Service		
Call Assigned			End of Service		
Travel Time			Engineer Hands on Time		

Action Taken: _____

Call Status: ☐ Closed   ☐ Pending for Spares   ☐ Pending for Customers
              ☐ Pending for Others

Part Replaced: ☐ Yes   ☐ No   ☐ Under observation

	Part Number	Part Description	Quantity	Part Serial No.
Part Replaced				
Part Removed				

For Customer's Use: Please rate this call by ticking an option:

☐ Extremely Dissatisfied   ☐ Dissatisfied   ☐ Neither Satisfied nor Dissatisfied

☐ Satisfied   ☐ Extremely Satisfied

Customer's Feedback: _____

User Name :	Engineer Name :
Email ID/Tel.No :	
Signature :	Signature :
Date :	Date :

One of the first things to do while analysing an audience is to identify its type (or types—it is rarely just one type). In general, the audience can be categorized into three types: *experts*, *executives*, and *non-specialists*.

'Adapt' your writing to meet the needs, interests, and background of the readers who will be reading your writing.

Experts are the people who know the theory and the product inside out. They designed it, they tested it, and they know everything about it. Often, they have advanced degrees and operate in academic settings, or in research and development areas of the government and business departments. More often, the communication challenge faced by the expert is communicating to the technician and the executive in simpler terms.

Executives are the people who make business, economic, administrative, legal, governmental, and political decisions on matters that the experts and technicians work with. If it is a new product, they decide whether to produce and market it. If it is a new power technology, they decide whether the city should implement it. Executives may sometimes have as little technical knowledge about the subject as non-specialists.

Non-specialists have the least technical knowledge of all. The non-specialist readers are least likely to understand what the experts are saying, and also have the least reason to try. Their interest may be as practical as the technicians', but in a different way. They want to use the new product to accomplish their tasks; they want to understand the new power technology enough to know whether to vote for or against it in the upcoming union election. Or, they may just be curious about a specific technical matter and want to learn about it, but for no specific, practical reason.

It is important to analyse the audience in terms of the characteristics explained in the following sections.

Background—knowledge, experience, and training   One of the most important con-cerns is just how much knowledge, experience, or training we can expect in our readers. For example, imagine you are writing a guide to using a software product that runs under Microsoft Windows. How much can you expect your readers to know about Windows? If some are likely to know little about Windows, should you provide that information? If you say no, then you run the risk of customers getting frustrated with your product. If you say yes to adding background information on Windows, you increase your work effort and add to the page count of the document (and thus to the cost). Obviously, there is no easy answer to this question—part of the answer may involve just how big or small a segment of the audience needs that background information.

Needs and interests   While planning a report, we need to know the audience's expectations from the report. For example, imagine you are writing a manual on how to use a new microwave oven—what are your readers going to expect to find in it? Imagine you are under contract to write a background report on global warming for a national real estate association—what do they want to read about, and, equally important, what do they not want to read about?

Other demographic characteristics   The various characteristics of our readers might have an influence on the design and writing style of our document—for example, age groups, type of residence, area of residence, gender, political preferences, and so on.

Audience analysis can get complicated by at least three other factors: mixed audience types for one document, wide variability within an audience, and unknown audiences

Mixed audience   A report can be often meant for more than one audience. For example, it may be seen by technical people (experts and technicians) as well as administrative people (executives). Hence, one must write all the sections so that all the audiences of the document can understand them. Otherwise, one can write each section strictly for the audience who would be interested in it, and then use headings and section introductions to alert the audience about where to go and what to stay out of in the report.

Wide variability in an audience   Even though the audience fits into only one category, there may be a wide variability in its background. This is a tough one—if we write to the lowest common denominator of readers, we are likely to end up with a cumbersome, tedious book-like thing that will turn off the majority of readers. But if we do not write to that lowest level, we lose that segment of our readers. What to do? Most writers go for the majority of readers and sacrifice the minority that needs more help. Others put supplementary information in appendices or insert cross-references to beginners' books.

Using the following guidelines, you can make a report comprehensible to the non-specialist reader:

> If you write to the lowest common denominator of readers, you are likely to end up with a cumbersome, tedious book-like thing that will turn off the majority of readers. But if you do not write to that lowest level, you lose that segment of your readers. What to do?

- Add information required by readers in order to understand the report
- Omit information your readers do not need
- Add examples to help readers understand
- Change the level of your examples
- Change the organization of your information
- Use more or different graphics
- Add cross-references to important information

Unknown audience   At times the writers do not know who all will be the readers of their reports. Even in such a situation, it is better for the writers to anticipate a hostile audience and try to make their reports as justifiable as possible with adequate supporting details. They must organize the points of justification or evidence in the most convincing manner. In addition, they need to adopt a neutral and unbiased tone in discussing the topic of their reports.

Sources of Information

To accomplish the objectives of a report, we require facts and ideas. We may find them in company records, reports, bulletins, pamphlets, and periodicals; we may use library sources to look for information; we may observe some incident and collect the facts or ideas; we may conduct personal interviews with people to get information; or we may circulate questionnaires to get data for our report.

Investigating the sources of information is a kind of research. It must be done right in the beginning. The extent of investigation will, of course, depend on the length and importance of the report.

The two types of data that one can collect are primary and secondary data. Primary data are what a researcher gathers for the particular problem being addressed in the report. The important means of collecting primary data for a report are

- Personal observation
- Personal interviews or telephone interviews and experimental data
- Surveys (preparing and circulating questionnaires)

Secondary data are data gathered for some purpose other than the problem at hand. Common sources of secondary data are

- Internal records
- Published material such as directories, guides, statistical data, government publications
- Databases such as bibliographic and numeric databases
- Censuses
- Syndicated sources (information services provided by research organizations)

@ Please refer to the Online Resource Centre for details on primary and secondary data collection.

Organizing the Material

Depending upon the topic, purpose, and audience, we can organize the material/data collected for our report in either of the following:

- The order of occurrence
- A combination of orders
- The order of importance

Order of occurrence

Order of occurrence is otherwise known as *chronological order*. Here, the data or events are presented in the sequence of their occurrence in time—none of the parts are uniquely important. For example, the history of a transaction or the procedure for manufacturing or installing equipment.

Order of importance

When the matter/data collected for the report is not of uniform value, we may have to organize the information in descending or ascending order of importance. Generally, the descending order of importance is valid for informational reports, as the reader is interested in looking for the most important information first. For example, feasibility reports (in which immediate needs are more important than future needs) or a feedback report on a conference recently attended.

Combination of orders

This order is particularly useful in reports involving a double assignment. A combination of order of occurrence and order of descending importance is fairly common. Examples are a report on the appraisal of a situation and the recommended changes, and a report on a problem and the suggested solution.

To arrive at the correct order, examine the data as a whole, consider them for completeness, their relationship to the purpose of investigation, their total significance to the problem at hand, and then organize in an appropriate order.

Interpreting Information

Once the data are organized, the findings can be presented. It must be presented the way the reader wants. If the reader wants only in the form of topic and subtopics, mere organizing the data will be sufficient, but if the reader wants the application of it, some more efforts are required. Although interpreting the data is a sole mental activity, the following points should be kept in mind:

- Present the facts as they are
- Do not give your conclusions
- Interpret only the available data
- Only analyse comparable data
- Be conscious of unreliable data

Be aware of the statistical tools available for the interpretation. There are different ways of dealing with qualitative and quantitative data. Even the smallest calculation should be explained to make the content easy and comprehensive for the readers.

Making an Outline

It is extremely important to develop an outline of the report prior to commencing work on the report. The formatting of the report should be carried out only after completion of the outline.

An outline is a mechanical framework into which the information collected for the report can be fitted in bits and pieces. It shows the direction in which one needs to proceed in writing the report.

In school days when our teacher had asked us to write an essay, we used to think about some points, organize them in our mind, and then jot them down in some order. We also revised them and modified the order.

Then, keeping those points as guidance we started writing on each. Thus, we could produce a well-organized essay. Report writing also follows similar steps. Thoughts generally do not appear in our mind in a desired order. They usually come randomly, and we must organize them appropriately and systematically before we begin writing.

Preparing an outline requires a considerable amount of time. However, the time spent on a systematic outline is always beneficial as this helps save time in writing. On the other hand, a report written without an outline may lack organization and structure and may end up in confusion. The outline indicates the main topics and subtopics for the report in the form of words or phrases.

Go through the following sample outline and the discussion thereafter to make your outline more effective.

To prepare an impressive outline, you may follow the tips given below.

Education System in India

1. Introduction
2. Status before Independence
3. Present status
   3.1 Present status
       3.1.1. Primary
       3.1.2. Secondary
       3.1.3. Higher secondary
3.2. College education
3.3. University education
4. Status of women's education
5. Merits and demerits
6. Financial support
7. Employment prospects
8. Conclusions

Use words or phrases

The outline that is prepared with careful planning finally turns out to be the Table of Contents. Various parts of the outline are used as headings and subheadings to the sections of the report. Hence, the words for constructing the headings of an outline should be chosen carefully. We can choose short constructions (topic headings), which frequently consist of one or two words that merely identify the topic of discussion; we can also choose longer constructions (talking headings), which frequently include prepositions that not only identify the subject matter covered but also summarize the material they cover. Table 13.2 shows some examples of topic headings and talking headings as they may appear in a segment of an outline.

Table 13.2 Topic heading and talking heading

Topic heading	Talking heading
4. Marketing strategies	4. Creating strategies for marketing
4.1. Feedback	4.1. Collecting feedback from relevant areas
4.2. Promotional efforts	4.2. Designing strategies for promotional efforts
4.3. Distribution network	4.3. Setting up of new distribution network

Use parallel grammatical constructions

Ideas that are parallel in thought must also be parallel in grammatical construction. Therefore, corresponding parts of an outline, being of equal significance, must be stated in the same grammatical form. In other words, equal-level headings should be parallel in structure so as to show similarity. Through parallel headings we can show such equal-level divisions consistently.

For example, if we use a noun phrase to express an idea under one subdivision, we must state the other parts of the same subdivision in noun phrases only. Compare the parallel and non-parallel constructions given in Table 13.3 to understand the effectiveness of parallel constructions in an outline.

Table 13.3 Parallel and non-parallel constructions

Non-parallel construction	Parallel construction
1. Advantages of computers	1. Advantages of computers
1.1. Internet browsing	1.1. Internet browsing
1.2. Promotes sales	1.2. Sales promotion
1.3. Analysing data	1.3. Data analysis
1.4. Managing finance	1.4. Financial management
1. Introduction	1. Introduction
2. Computers mark the beginning of a new era	2. Computers: The beginning of a new era
3. What are applications?	3. Applications
4. Advantages of computers	4. Advantages

Avoid needless repetitions

Repetition of words should be avoided as it will result in monotonous writing, thus making the outline dull and drab. Look at the original and revised versions of a segment of an outline taken from a report on Education System in India given in Table 13.4.

By simply eliminating the repetition of the word 'education' in all the subdivisions, we have avoided sounding monotonous.

Table 13.4 Avoiding monotony in an outline

Original	Revised
3. Present status	3. Present status
3.1. Present status	3.1. School education
3.1.1. Primary education	3.1.1. Primary
3.1.2. Secondary education	3.1.2. Secondary
3.1.3. Higher secondary education	3.1.3. Higher secondary

Follow the principles of coordination and subordination

The outline indicates the level of each heading in the report, that is, how important each heading is. The closer a heading is to the margin, the greater is its importance in the report. Likewise, as the headings move away from the margin, they become less important.

According to the principles of coordination, all the main headings have the same importance with respect to their relation to the subject and also the seriousness of discussion. So, while organizing data, main headings of equal importance should be selected. Table 13.5 shows the principles of coordination.

Table 13.5 Coordination of headings

Illogical coordination	Logical coordination
2. Computers: Beginning of a new era	2. Computers: Beginning of a new era
3. Applications	3. Applications
4. Education	3.1. Education
5. Industries	3.2. Industries
6. Advantages	4. Advantages and limitations
7. Disadvantages	

As per the principles of subordination, subheadings should be appropriate to the main heading under which they are listed. Table 13.6 clearly demonstrates the principles of subordination.

Table 13.6 Subordination of headings

Illogical subordination	Logical subordination
3. Applications	3. Applications
3.1. Education	3.1. Education
3.2. Industries	3.2. Industries
3.3. Advantages	4. Advantages and limitations
3.4. Limitations	

Follow a suitable numbering system

Either the conventional numbering system or the decimal numbering system can be used to mark the levels of headings in outline. The conventional system uses Roman numerals to show the main headings and the letters of the alphabet as well as Arabic numerals to show the subheadings, as shown here:

I. First-level heading
   A. Second level, first part
   B. Second level, second part
     1. Third level, first part
     2. Third level, second part
       (a) Fourth level
         (i) Fifth level
II. First-level heading
   A. Second level, first part
   B. Second level, second part
   ....

The decimal system uses whole numbers followed by decimal digits to show main sections and subsections. Thus, the digits to the right of the decimal show each successive step in the outline. The following is an example of this system:

1.0   First-level heading
   1.1 Second level, first part
   1.2 Second level, second part
      1.2.1 Third level, first part
      1.2.2 Third level, second part
         1.2.2.1 Fourth level, first part
         1.2.2.2 Fourth level, second part
2.0   First-level heading
   2.1 Second level, first part
   2.2 Second level, second part
   ....

Structure of Reports (Manuscript format)

Various elements combine together to structure a report, and knowing them will help us in writing better reports. Although twenty elements are listed after this discussion, all of them need not be used in a report.

While some of them may be included in all reports, some may find a place only in reports that are to be published.

The elements of structure can be selected considering the following parameters: usefulness, terms of reference, and existing practice.

Usefulness refers to the need for including any particular element. For example, when you write an informational report, you do not require a section on recommendations.

The terms of reference (refer to the earlier section on prewriting in this chapter) specify the objectives of the report. These are limits that are set on what an official committee or report has been asked to do. If the terms of reference demand suggesting some measures for improving a situation, a section on recommendations should be included. Otherwise, we can end the report with a section on conclusions.

The elements that constitute a report are also determined by the *existing practice* in an organization in terms of producing reports. That is, if an organization does not require an abstract or summary for a report, these elements can be omitted.

The standard twenty elements of structure of a full-fledged report are as follows:

Prefatory parts	Main text	Supplementary parts	Optional elements
Cover page	Introduction	Appendix/appendices	Frontispiece
Title page	Discussion	References/bibliography	Letter of transmittal
Certificate	Conclusions	Glossary	Copyright notice
Acknowledgements	Recommendations		Preface
Table of contents			Summary
List of illustrations			Index
Abstract			

Prefatory Parts

Just as we catch the attention of our audience, introduce our topic, and give them a preview during the introductory part of our speech, the report writers use the prefatory parts of their report to get their audience ready for reading the report. They reveal the topic, author, contents and also introduce the objectives, results, significance very briefly in these parts. Now let us look into the various parts of the preliminary section of a report.

Cover page and frontispiece

The cover page of the report not only gives it an elegant appearance but also protects it from damage. It also serves as a quick reference to the readers to know the topic and the author of the report. The classification of the report (secret/top secret) and report number, if applicable, may also be mentioned on the cover page on the top left corner and right corner, respectively.

To stir the curiosity of the reader, one may superimpose on this cover page some illus-trations, such as photographs, drawings, and diagrams, provided these illustrations reflect the contents of the report. For example, a report on 'Designing Webpages' can have a webpage superimposed on the cover. Any such illustration is known as the frontispiece, which is an optional element, and generally finds a place in reports that are published.

Many organizations have standard covers for reports, imprinted with the organization's name and logo, as shown in Exhibit 13.5.

EXHIBIT 13.5 Sample cover page

Confidential   Report no. 115

**Growth of Medical Facilities
in India (2005–2015)**

**Prepared by
Anuj Singhania
Secretary**

Medical Council of India

August 2015

Title page

The title page (Exhibit 13.6) is the first right-hand page of a report. This page is more or less similar to the cover page, except that it contains the following additional information:

- Name and designation of the intended audience
- Name and designation of the approving authority, if any. (In some organizations the report does not directly reach the recipient. It may require the approval of an intermediary before reaching the recipient.)

If needed, the organization's emblem can be included just above its name given at the bottom (similar to the cover page, as shown in Exhibit 13.5).

Certificate

Certain reports, such as project reports and research reports require a certificate vouching the original contribution of the report writer. Generally, the certificate (Exhibit 13.7) contains the statement testifying the original work, place, date, and signature of the project supervisor or guide.

EXHIBIT 13.6   Sample title page

Growth of Medical Facilities
In India (2005–2015)

Prepared For

Gautam Chaudhary
Director
Medical Council of India

by

Anuj Singhania
Secretary

Approved by
ABC

Medical Council of India
August 2015

Letter of transmittal

An optional element of reports, the letter of transmittal (or memo of transmittal) conveys the report to the audience. This serves the same purpose as that of a preface in a published document. As the letter of transmittal is the written version of what we would say if we were handing the report directly to the person who authorized it, its style is less formal than the rest of the report. For example, the letter would use personal pronouns (you, I, we) and a conversational tone.

The transmittal letter usually appears right before the table of contents. The contents of a letter of transmittal are

EXHIBIT 13.7   Sample certificate

Certificate

This is to certify that the project entitled Growth of *Medical Facilities in India* (2005–2015) embodies the original work done by *Anuj Singhania* under my supervision.

Date: 26 August 2015   Prof. Ashok Joshi
Place: New Delhi

- Objectives/terms of reference
- Scope
- Methodology adopted
- Highlights of the analysis
- Important results

- Significance of the study Suggestions
- Any other details that may enable the audience to understand the report better
- Acknowledgements

This type of letter typically begins with a statement such as 'Here is the report you asked me to prepare on …'. The rest includes information about the scope of the report, the methods used to complete the study, and the limitations that became apparent. In the middle section of the letter, we may also highlight important points or sections of the report, make comments on side issues, give suggestions for follow-up studies, and offer any details that will help readers understand and use the report. We may also wish to acknowledge help provided by others. The concluding paragraph is a note of thanks for having been given the report assignment, an expression of willingness to discuss the report, and an offer to assist with future projects.

If the report does not have a synopsis, the letter of transmittal may summarize the major findings, conclusions, and recommendations. This material would be placed after the opening of the letter.

Acknowledgements

This is a list of persons whom we may like to thank for their advice, support, or assistance of any kind. It is not only customary but also necessary to acknowledge even the smallest help rendered by people. While writing the 'Acknowledgements', the following guidelines should be remembered:

- Categorize the audience: Courtesy, real help, emotional support, secretarial assistance, etc.
- Vary the expressions: Do not begin each sentence with 'I thank'; such a monotonous beginning will reduce the impact of the acknowledgements. Here are a few openings:

Thanks are due to …	Inadequate to express my sincere thanks to …
Our sincere appreciation to …	I am extremely grateful to …
We owe a lot to …	My heartfelt thanks are due to …
Mere thanks in few words would be highly …	I acknowledge with thanks the support rendered by …

- Avoid clichés such as:

First and foremost	Firstly ……… secondly ……… thirdly
Last but not the least	I take this opportunity to thank …

- Avoid listing the names
- Divide the content (if long) into well-structured paragraphs

Table of contents

'Table of Contents', generally titled simply 'Contents', helps the reader locate a specific topic easily and quickly. However, Contents need not to be included in a short report (ten pages or less). This table is actually the report outline we discussed earlier in its final form, or simply the frozen outline with page numbers. It indicates, in outline form, the coverage, sequence, and relative importance of information presented in the report. The Contents page is especially helpful to readers who wish to read only a few selected topics of the report.

The Contents lists all the three parts—prefatory, main, and supplementary—of the report along with their page numbers. It also lists the illustrations used in the report. However, if the report has more than five illustrations, a separate 'List of Illustrations' page can be included after the Contents. Although the outline can have subheadings up to any level, the Contents page should include headings up to a maximum of three levels only. Depending upon the length and complexity of the report, this page may show only the top two levels of headings or only first-level headings. See an example of a Table of Contents in Exhibit 13.8.

List of illustrations

As mentioned, this page serves as the Contents page for all the illustrations that appear in the report. Except tables, all other visual aids (graphs, maps, drawings, and charts) are grouped under the heading Illustrations or Figures. The List of Illustrations gives the titles and page numbers of all visual aids. When tables and figures are numbered separately, they should also be listed separately. These lists would enable the reader to quickly locate any specific illustration.

Abstract

An abstract gives the essence of the report. In business reports it is known as the synopsis. The length of an abstract is generally 2 to 5 per cent of the report. The length and readership of the report decide whether an abstract is to be included. If the report is less than ten pages it does not require either an

EXHIBIT 13.8   Table of contents

Table of Contents

Acknowledgements	ii
Abstract	iv
1. Introduction	1
2. Growth of medical facilities—An overview	4
3. Hospitals and dispensaries	11
4. Beds	16
5. Health centres	21
5.1. Community	
5.2. Primary	
5.3. Sub	
6. Conclusions	30
7. Recommendations	32
Appendix	34
References	41

abstract or a summary. (While an abstract is more appropriate in specialist-to-specialist communication, a summary is meant for all readers, and is longer than an abstract.) A report of 10–50 pages should have an abstract. If it is a long report, more than fifty pages, it needs both an abstract and a summary. An abstract is especially relevant in specialist-to-specialist communication where the reader would be expected to have some background knowledge of the subject. It does not allow abbreviations, acronyms, or illustrations. It tells the reader the following:

- Objectives
- Main findings or accomplishments
- Significance

Exhibit 13.9 shows a sample abstract written for an informative report on the Growth of Medical Facilities in India.

Summary

While an abstract is more appropriate in specialist-to-specialist communication, a summary, which is an optional element, is meant for all readers. It is longer than an abstract, 5–10 per cent of the length

EXHIBIT 13.9   Sample abstract

Abstract

Medical facilities in India have grown considerably over the years. With the primary objective of providing a clear view of this growth over the years 2005–2015, this report analyses in detail the various facilities related to hospitals, dispensaries, and centres in our nation. Essentially the study would provide a comprehensive view of the medical facilities, which in turn would enable the authorities to decide upon improving the same in future.

of the report. Business reports name 'Summary' as 'Executive Summary'. It presents the entire report in a nutshell. Summaries may contain headings, adequately developed text, and even visual aids. A well-written summary opens a window into the body of the report and allows the reader to form an impression of how well the topic of the report has been dealt with. Generally, the summary of a report presents information from various parts of the report in the same sequence as they appear in the report. After reading the summary, the audience should know the essentials of the report and be able to make a decision. Later, when they find time, they may read certain parts of the report to obtain additional detail. Table 13.7 presents the differences between an abstract and a summary.

Table 13.7   Differences between abstract and summary

Abstract	Summary
• Essence of the report	• Entire report in a nutshell
• 2–5 per cent of the report	• 5–10 per cent of the report
• More relevant in specialist-to-specialist communication	• Meant for all readers
• Information is qualitative	• Information is both qualitative and quantitative
• Does not include illustrations	• May include certain illustrations

Preface

The preface, an optional element, is the preliminary message from the writer to the reader. It is quite similar to the letter of transmittal, except that it does not formally transmit the report. It seeks to help the reader appreciate and understand the report. The contents of a preface are as follows:

- *Factors leading to the report* (what was the prevailing situation or earlier studies carried out and hence the need for this study and report)
- *Organization of the report* (what do the various sections contain)
- *Highlights* (important observations and findings)

- *Significance* (how the report would enable the readers in further study or research; how best the study in the report would help them)
- *Acknowledgements* (Frequently, the acknowledgements can be dispensed with the preface. In the preface itself one can express his/her indebtedness to those who helped in the study.)

Generally, the preface appears only when the report gets published.

Copyright notice

Copyright is a form of protection that covers published and unpublished literary, scientific, and artistic works, whatever the form of expression, provided such work is executed in a tangible or material form. Simply put, it means that if we can see it, hear it, and/or touch it, it may be protected.

Copyright laws grant the creator the exclusive right to reproduce, prepare derivative works, distribute, perform, and display the work publicly. For example, a copyright statement that reads '© Copyright 2015, OUP, India' means that the Oxford University Press, India has legal monopoly over the work, which was produced in 2015.

There is one thing that must be clarified, though. The actual *intangible idea* may not be copyrighted. What is copyrighted is the tangible result of the idea, which would be in the form of the published work.

An example:

© Copyright 2015 by Maxima Institute of Technology. All rights reserved. This material may not be duplicated for any profit-driven enterprise.

Main Text

The main text consists of the introduction, discussion, conclusion, and recommendations. This is the main body of the report, which gives the details of the study such as the method adopted, data collection methods, and the constraints under which the study was carried out.

Introduction

The function of an introduction is to put the whole report in perspective and to provide a smooth, sound opening for it. It presents the subject or problem to the readers and gets their attention. A good introduction must furnish the readers with sufficient material concerning the investigation and problem, to lead them to an easy comprehension of the rest of the report. It should also give the readers a general view of the report before they plunge into the details. In other words, the introduction is a section where a broad, general view of the report, rather than a specific and analytical one, can be presented.

> A good Introduction must furnish the readers with sufficient material concerning the investigation and problem, to lead them to an easy comprehension of the rest of the report.

An introduction includes the following information:

Background of the report Conditions/events giving rise to the project or survey need to be discussed here. Details of previous investigations and studies can also be included if there is significant time gap.

Purpose and scope The background will logically lead to the purpose of the report. If the investigator has received the Terms of Reference, they can be presented verbatim. Otherwise the objectives have to be clearly specified. The purpose statement describes the objective as well as boundaries of the work. Stating which issues will be covered and which issues will not be covered is especially important in the case of complex, lengthy investigations.

Authorization   The authority who has assigned to do the project/conduct the survey should be mentioned. In other words, the recipient's name and designation should be mentioned.

Basic principles or theories involved   The important theories and principles used for analysing the data should be outlined.

Methods of gathering data   The methods used or the sources consulted for collecting the data should be mentioned.

General plan in developing the solution (brief outline of methodology)   The methodology adopted in the analysis should be outlined.

General structure of the report (organization of various sections)   The Introduction may contain subheadings such as objectives and procedures, which serve as significant guideposts for the readers.

Ending the Introduction with an explanation of the general plan of the report will provide a logical transition to the next section of the report. It can also be ended with the discussion of the procedures used to gather data; and then the next section would present the results. A brief outline of the results or main conclusions may also provide a logical transition to the first part of the Discussion section of the report.

Discussion

The discussion section, the lengthiest part of the report, contains the information that supports the conclusion and recommendations, as well as the analysis, logic, and interpretation of the information. Here, information and data are presented, analysed, and interpreted. The writer must decide between pertinent data to include in the text and less important information to omit or relegate to the appendix. Meanings, ideas, and facts are made clear to the reader. Comparisons are made, facts are evaluated, and significant relationships are drawn. The solution of a

> The Discussion should lead the readers through the same reasoning process the author used to reach the conclusions and show them that they are sound.

problem may be given with an explanation of its advantages and disadvantages. Tables, charts, and other media for presenting figures and data are used. Other illustrative material may be included (the various types of illustrations that can be used are discussed in Chapter 1). Emphasis is on the results and their interpretation. The discussion should lead the readers through the same reasoning process that the author used to reach the conclusions and show them that they are sound.

Opposing contentions should be considered to show how the data prove otherwise. The writer should not assume that the reader agrees with a concept, unless it is generally accepted. Simple, straightforward statements of facts should be used, as these are most easily understood. Different aspects of the problem are treated in the discussion section. Major subject headings are used to guide the reader. Points may be arranged to suit the subject and reader.

Conclusion

The conclusion is that section of the report where all the essential points developed in the discussion are brought together.

The function of this section is to bring the discussion to a close and to signal to the readers gracefully that they have reached the end. It also refers to the logical inferences drawn, the judgements formed on the basis of analysis of data presented in the report, or the findings of the investigation. This section has the following characteristics:

> 'Tell your audience what you are going to tell them. Tell it to them. And then tell them what you have told them!'   –Hilaire Belloc

- Uses decreasing order of importance
- Can be narrative (in paragraphs) or tabular (in points)
- Uses narrative type when there are few conclusions
- Uses tabular form when there are more conclusions
- Both narrative and tabular forms are acceptable but the latter is better for quick comprehension.
- Contains only opinions and never suggests future actions to be taken by the reader
- Does not introduce any new idea not previously introduced into the report

Conclusions are the result of reasoned analysis and judgement of the data in the report and serve as a basis for recommendations growing out of the study. They may be summary or analytical in nature. Thus, the conclusion section is a recapitulation of the significant points developed in the discussion section. Concluding statements are supported by the facts in the discussion section.

An example of tabular type of conclusion is the following, taken from a consumer-preference survey made by Swift & Company for the moulded pulp egg carton versus the regular self-locking egg carton:

1. The moulded pulp carton is decidedly preferred, both by those having used it (77 per cent of them) and those who have not used it (68 per cent of them).

2. Protection is the principal reason of preference for the moulded pulp carton.

3. 'Hard to open' and 'can't see eggs' are the principal reasons for disliking the moulded pulp carton, given by those who have used it.

4. 15 per cent of those preferring the moulded pulp carton had opening troubles—69 per cent of those preferring the regular type found the pulp carton difficult to open.

5. 45 per cent of those preferring the moulded pulp carton liked to look at eggs before buying, compared to 74 per cent of those preferring the regular type. 'Want to see size' and 'colour' were given as reasons why.

Here is a sample of the narrative form of conclusion.

Until the final comparative analysis of sales, there was relatively little to choose between the two cities. Both were certainly well adapted to a location for conducting the presenting surveys of a small national-scale advertising agency, although at this point Kansas City has probably shown itself to be slightly more representative of the nation than Cincinnati. The analysis of retail sales, however, completely changed the picture.

Recommendations

Recommendations pertain to the action that is to be taken as a result of the report. They are supported by the conclusions, and they are aimed towards accomplishing the purpose of the report. If the purpose of a report, for instance, is to alleviate employee grievances over wage incentive plans, the recommendations will suggest ways in which this can be done. Conclusions and results of investigating the problem will support the recommendations.

Like conclusions, recommendations may take the form of a formal, long report. Recommendations generally follow the conclusions. They do not, however, always appear at the end of the report. They may be given first, especially in recommendation reports. They are also sometimes treated briefly in the letter of transmittal, preface, and separate summary section. If the reader is already familiar with the data or is chiefly interested in the action to be taken, then the recommendation should be presented first to avoid reading through a lot of material. If the reader is likely to react unfavourably to the recommendation, then it should be given at the end; the report can prepare the reader for it.

Recommendations, depending upon the relationship with the reader, can assume any of the following three types:

- Tentative (temporary solutions that may have to be reconsidered in future)

- Conciliatory (suggestions that may be accepted by the recipient)
- Aggressive (recommendations that are mandatory and are to be implemented immediately)

Supplementary Parts

While prefatory parts precede the main body of a report, supplementary parts follow it. However, they have a lesser role to play that the prefatory parts because they are not mandatory for a report. For instance, many reports may not contain documentary evidences such as references, bibliography, etc. Nevertheless, their importance cannot be ignored as they provide certain additional details to the interested readers to enhance their understanding of the topic of the report. The following discussion throws some light on the various supplementary parts of a report:

Appendices

This section of a report is used for information that has some relevance to the report but cannot be easily fitted into the text. It is a convenient way of presenting detailed information particularly of a descriptive nature, which, if inserted in the main body, would interrupt the smooth flow of the narrative. Hence, before including any material in the appendix, you should:

- See whether the material is related to the subject of your report
- Check whether it would interrupt the theme if included in the main body

An appendix should contain (a) material not strictly related to the main argument of the report but which nevertheless is of interest and (b) material that readers can safely omit but can consult if they want to examine the details, and also to carry out further study.

Generally, appendices contain materials such as sample documents, detailed calculations, experimental results, statistical data tables and graphs, specimen questionnaires or samples of forms used in investigations, summaries of results achieved by other organizations, etc. An appendix also helps to present recent work or data added at the last moment. It is better to put these in an appendix rather than completely rewriting the report.

If there are many appendices, they should be named as Appendix A, Appendix B, and so on. Also, an appropriate title should be given to each of them. However, lengthy and numerous appendices should be avoided, as they reveal the writer's poor organization.

Bibliography/references

A bibliography is an alphabetical list of the sources—books, magazines, newspapers, CD-ROMs, Internet, interviews—that have been consulted in preparing the report. This list is used to

- Acknowledge and give credit to the sources of words, ideas, diagrams, illustrations, quotations borrowed, or any materials summarized or paraphrased
- Give the readers information to identify and consult the sources
- Give the readers an opportunity to check the sources for accuracy.

This section may be named as References or Works Cited if it includes works that were consulted or directly quoted for particular pieces of information, and are mentioned within the report in the form of citations.

Please refer to the Online Resource Centre for referencing and bibliography for all kinds of technical documents.

Glossary and index

The glossary is a list of technical words used in the report and their explanations. If small in number, the terms are explained in the form of footnotes. The decision whether to include a glossary or not depends upon the readers. If they are likely to be unfamiliar with the context, it would be better to include them.

The index is intended to serve as a quick guide to locate the material in the report. The readers can locate a topic, subtopic, or any other important aspect of the report quickly and easily. This element is generally used in bulky reports where the Contents do not serve the purpose of locating a particular issue. The index is arranged in alphabetical order (please see the index of this book) and is extremely helpful in cross-referencing.

Types of Reports

There are many different types of reports. The basic format and elements remain the same but they vary in terms of the purpose and extent of formality. We will discuss some common types in this section.

Introductory reports

As the name suggests, these report introduce some topics but do not delve deep into any aspect. They just skim the surface of the issue, to give the audience a preliminary feel of it. Therefore, these are short reports and do not need subsections and subdivisions. The contents are brief and to the point.

Progress reports

These reports give regular updates about the progression of a particular ongoing project. It states the expected or set deadlines and then describes the work done till the date of writing the report. In case of delay, it specifies the reason for it and the support that can be provided to finish the work. For instance, in the case of a construction project, a progress report can help in keeping the track of the progress. Progress report format can vary as per the project undertaken. It can be very formal in case of the report on the progress of the building or it can be informal in case of students' thesis work progress and if the report is through email to the supervisor. In some situations it has a pro forma to be filled out.

Incident reports

These reports look more like an article than a report. It is used to describe an event or an accident without distorting the facts to a person who could not witness the scene. It is very important to maintain accuracy and truthfulness. The order of presentation of information is important. In general, the events constituting the incident are presented in chronological order.

Feasibility reports

While undertaking a new project or starting an establishment, the possibility of launching it should be assessed. The pros and cons of it and the cost, gains, glitches should be thoroughly studied. The report studies the problem, opportunity, and plan for taking action. The conclusions are very important as they indicate whether the project being considered is feasible, not feasible, or partially feasible, and hence directly helps decision-making.

Marketing reports

These reports are persuasive in nature and begin with marketing objectives, stating the available resources, plan of action, and goals. It is similar to feasibility report in terms of style, length, and content. This report is prepared by the marketing department while promoting or launching the product or while assessing the effectiveness of existing strategies. It contains the details about segmentation, targeting, and positioning. A market report can be about the global, domestic, or regional market. It would also include the customers' perspective about a particular product vis-à-vis competition. The report should also include the effectiveness of the promotional, pricing, and distribution strategies.

Laboratory test reports

These reports document the various experiments conducted in the laboratory. The observation, calculation, and results sections of these reports must be prepared with utmost accuracy and precision. These reports have the following sections:

* Aim
* Theory
* Apparatus
* Procedures
* Observation
* Calculations
* Results
* Conclusion
* Discussion
* Inference
* Possible sources of errors
* Precautions

Project reports

A project can be defined as a sequence of unique, complex, and connected activities having one goal or purpose and that must be completed by a specific time, within a budget and according to specification. At the end of a project the person or the team who has accomplished it writes a report explaining the details. For instance, as a part of their curriculum, students of technical and professional courses may undertake some projects of theoretical or practical nature under the guidance of professors. While some projects may last for a semester, some others may be completed in fifteen days or one month. But only when they submit their project reports their project is considered as complete. A student carrying out a project on the topic 'Effective use of Internet on campus' may conduct a survey among the users of Internet on campus, interpret, and analyse the data collected and suggest some measures to improve upon the effective use of Internet on campus. When he/she prepares the project report, he/she will choose appropriate elements and write the report. Such reports generally include the title page, certificate, acknowledgements, contents, abstract, the main body consisting of three or four sections, appendices, references, glossary, etc.

Writing the report

Any project or study ends with a report. After deciding and formulating a topic, the writers carry out various activities such as data collection, analysis, discussions, etc. Then they start writing the report. So this process is the culmination of all the hard work put forth by the report writers. As the report is the only tangible evidence for their efforts, they need to take utmost care of its structure, layout, and style. They need to understand clearly the elements to be included and in what order, which parts have to be drafted first, how much background material to be included, etc. For instance, most writers prepare the main body of the report first and then the other parts. In academic or research reports which include references or bibliography, the writers prepare these parts while writing the main body. Even experienced writers may not be able to write their reports in one attempt. Generally, reports go through at least two or three drafts.

First Draft

Once the work is completed, the writer starts preparing the first draft of the report. Writing the first draft is a step between the preparation and the editing of the report. As already mentioned, a writer generally prepares the main chapters of the report, then the introduction and finally the other parts of the report. Once the main chapters (topics) are written, the writer can get a comprehensive view of the main chapters and then he/she can write the introduction and abstract. Even at this stage any modification in the outline can be made and once he/she freezes the outline, it can appear as the contents page in the report. Given below are certain guidelines in preparing the first draft of a report.

- Start writing Chapter 2 and go on till the concluding chapter. Then write the Introduction and the Abstract
- Prepare the references/bibliography as and when you prepare the main topics
- Prepare the prefatory parts and the appendices, if applicable
- Keep in mind the elements of effective writing discussed in Chapter 7
- Ensure that the analysis on all main topics has been included; the trends discussed and suitable illustrations have been added at appropriate places (Refer to Chapter 1)
- Refer to the characteristics of each element of the report discussed in the previous sections of this chapter while writing your first draft

Revising, Editing, and Proofreading

After completing the first draft of the report, it should be reviewed carefully. The type of the report, its utility, the time available, and also the effectiveness of the first draft determine the amount of revision needed.

The first draft is reviewed for the following:

- Accuracy of facts
- Clarity of expression
- Overall organization
- Adequacy and appropriateness of the contents to meet the set objectives
- Suitability and conformity of illustrations
- Effectiveness in expression
- Grammatical accuracy
- Correctness of layout

Basically, editing is the same as revising because its purpose is to correct and improve the report, and to prepare and check the final copy. Although someone other than the author generally does editing, nowadays a number of writers edit their own material. Most of them even do the drafting and revising together, using word-processing software.

The report, revised from all aspects mentioned above with additions, deletions, or corrections, is again read to check and improve it before submission. If the report is to be published, editing also involves carefully preparing the final copy for the printer. First, the manuscript should be checked for its correctness, completeness, and clarity of its contents. Then it should be checked for overall and sectional organization, and finally for the correctness in the mechanics of style and form. In most instances, after revising meticulously, all one needs is to correct a few mechanical errors that remain.

Proofreading is done at each stage of preparing the first draft to ensure correctness. The proofreaders should read the copy as a whole, and then check it for form, layout, and mechanical accuracy. It is not necessary to read for content at this stage. The proofreaders should read each word and phrase carefully and thoroughly check the punctuation marks. The final copy needs to be checked against the first draft to ensure that all the corrections are incorporated. There are standard symbols available for proofreading (refer to the Online Resource Centre). These symbols are marked in the margin to indicate any corrections required in the text.

Almost everybody uses word-processing software these days, and hence tends to proofread the soft copy rather than the printout. Nevertheless, it is necessary to proofread the print version, especially for longer documents as the human eye may overlook errors while reading from the computer monitor.

Please refer to the Online Resource Centre for several samples of technical and professional reports.

SUMMARY

Reports are an indispensible part of academic and professional work. They are important as they are the tangible product of any project or study, and many problems are solved and decisions are based on reported recommendations or conclusions.

Reports put an investigation, analysis, and recommendation together in an organized manner, and can be used as permanent records as well as legal instruments.

Reports must be precise, based on facts, relevant, reader-oriented, objective, easy to understand, well formatted, and well illustrated. There are various categories of reports such as informative and analytical, periodic and special, oral and written, formal and informal, and group and individual. They can be presented in manuscript, letter, memo, or pre-printed format.

Identifying the scope and purpose, knowing the audience, identifying the sources of information, organizing the material, interpreting the data collected, and drawing an outline are the prewriting steps that lead to the preparation of an effective, convincing report.

The structure of a report can be broken down into prefatory, main text, and supplementary information segments. Some of the constituents of these segments are optional sections, such as the frontispiece and the copyright.

Once we have understood the characteristics, categories, format, and structure of reports, we choose the most appropriate type of report based on the purpose at hand. Using the material prepared during the prewriting stage, the first draft of the report is prepared. This draft is then edited and refined to create the end product, which is a high-quality technical, business, or professional report.

EXERCISES

1. Indicate the difference between the following reports:
   (a) Oral and written
   (b) Formal and informal
   (c) Individual and group
   (d) Long and short
2. Discuss the various types of reports. Give an example of each.
3. What is a project report? Write a project report on the recent project you have done in your college.
4. A company is considering a proposal to establish a new factory in your town. The Managing Director has asked you to write a report on suitability of the place for the establishment of this factory. For this report, an outline is provided below. Study it carefully and rewrite it in accordance with the principles of co-ordination, subordination, phrasing, numbering, ordering, etc.
   Establishment of a New Factory
   Outline
   1. Introduction
   2. Fire fighting and Communication facilities

   (a) Telephone
   (b) Fax
   (c) Films
   (d) Games
   3. Education and Entertainment facilities
   4. How is the market
      A. Potential
      B. General
   5. Labour from Local and other plants
      5.0 Raw materials
      5.1 Cost
   6. Transport facilities
      6.1 Rail, road, air
      6.2 Raw materials
      (a) Building
      (b) Infrastructural
   7. Recommendations
   8. Conclusions
5. Develop the material given below into a formal out-line with appropriate numbers and correct indentation. The outline is not arranged logically. Rework the out-line into its logical form.

Animals I Have Had As Pets

I. Members of the cat family
   (a) Baby lion
      1. Three days old
      2. Mother died at local zoo
   (b) House kittens
      1. Five of these

II. Members of the dog family
   (a) Two young dogs, mixed breed
   (b) A poodle
   (c) German shepherd
   (d) Other dogs

III. Canaries
   (a) Two males
   (b) Three females
   (c) Parakeet

IV. Guppies

V. Members of the reptile family
   (a) Young grass snake
   (b) Green lizard
   (c) Two snails

VI. Goldfish

6. 'The District Collector, Jhunjhunu, is concerned about the rapid increase in the number of road accidents in Pilani. The Chairman, Municipal Corporation, Pilani, has been asked to submit a report investigating the causes and suggesting measures to improve the situation.'
   Prepare an outline for the above report.

7. You, as the Collector of Bhuj District, have been asked by the Secretary, Home Department, Gujarat, to submit a report on the relief work that was under-taken after the devastating earthquake hit the area last year. The Ministry sanctioned Rs 50 crore for the relief operations in the district, which was to be spent on free distribution of grains, water, medicine, etc. Now prepare an outline keeping in mind the principles of effective outline.

8. Fantasy Garments Corporation wants to open a new garment branch in a metropolitan city for all age groups, ranging from formals to casuals. You, as the Sales Manager of the company, have to prepare a report on the set-up of the new showroom. For this report, prepare an outline which should have nine main head-ings which contain sub-headings up to second level.

9. Rajasthan has been facing severe drought conditions for the third year in succession. In order to mitigate the sufferings of the drought-affected areas, the state government is fully geared up to start drought relief operations with the objective of generating employment, maintaining cattle conservation, providing drinking water supply, etc. As Secretary to the State Minister for Relief Operations, you have been asked to write a report on the drought relief operations, to be sent to the Council of Ministers. Now prepare only the outline for this report.

10. 'Smoking among the youth in India is on the rise. Studies have consistently shown that smoking leads to various ailments. Consequently, it is a major public health concern'.
    You, as a public health professional, have been asked to suggest measures to reduce this problem after carefully studying the extent of smoking, reasons behind it, awareness about the short and long-term effects and attitudes of the youth towards this practice. Therefore, in order to conduct this study, you have decided to carry out the survey using a mail questionnaire. Formulate 15 appropriate questions to elicit relevant information from college students about the above mentioned key areas.

11. The management of Reg International, Mumbai, is greatly concerned about growing absenteeism among the employees of its various divisions. You as the Personnel Manager, HRD, have been asked to investigate the causes of absenteeism and submit a report to the Managing Director. For collecting the relevant data, prepare a mail questionnaire to be distributed among the employees of the company. Write a covering letter also. Your questionnaire should have at least twenty questions.

12. The Managing Director, Parag Textiles, Surat, wishes to study the complaints of its employees regarding the new Bonus and Incentive Scheme announced by the Board of Directors last month. As the Public Relations Officer, you have been asked to submit a report to make the policy more beneficial and effective for its employees. Assuming you have already prepared a questionnaire, write a covering letter to be sent with the questionnaire.

13. The Ministry of Health, Government of India proposes to conduct a survey to study the occupational hazards that Information Technology

(IT) professionals face. For this purpose prepare a mail questionnaire to be sent to the IT professionals working in various multi-national companies all over Mumbai.

14. National Steel Industries Ltd, Mumbai, plans to improve the existing parking facility for the various types of vehicles used by its employees. As the Public Relations Officer, you have been assigned the task of collecting the relevant information for a report to be submitted in this connection.

 Prepare a mail questionnaire to be circulated among the employees who intend to avail the parking facility.

15. Modern Institute of Technology, Tikanpur, is a new educational institute which has well-equipped labs and a huge computer centre, with all the modern computing and Internet facilities. These facilities are made available to students for 24 hours.

 The Dean, Students' Welfare Unit, of this Institute has been receiving complaints about students suffering from various kinds of aches and pains: headaches, lower back pain, especially of coccyx (tail-bone), neck pain, along with watering of eyes and weakness, i.e., debility. Preliminary reports suggest the reasons to be prolonged sitting in front of the computers, missing of meals and eating junk food at odd hours. A disturbing news has caused greater concern at the academic level: students are using the Internet facilities for non-academic purposes like 'chatting', viewing movies, listening to music, etc. All this has led to significant behavioural changes in many students not conducive to their growth, progress, and development.

 Assuming yourself to be the Chief Warden of this residential Institute, write a memo report to be submitted to the Dean, Student Welfare Division. Your report should include findings and recommendations.

16. The Dean, Community Welfare Division, wishes to know whether the BITS supermarket should stay open at night. For collecting the relevant data 300 customers were interviewed at random. The results of the survey are given below in the tabulated form:

 Table showing % response to the question: 'Do you want the store to stay open at night?'

Age Group	Yes	No	Don't Know
10–15	43.5	38.0	18.5
16–21	64.2	29.0	6.5
22–30	54.5	44.0	1.5
31–45	37.5	47.5	15.0
46–60	18.4	70.0	11.6

Now, as Research Officer, write a Letter Report to be submitted to the Dean, CWD, to enable him to take necessary decision in this regard. Invent the necessary details.

17. The Human Resource Department of Kalka Paper Industry, Mumbai, runs a two-day orientation for the new office employees every other month. At this orientation they are apprised of company history, objectives, etc., with a purpose to familiarize the participants with the industry and give them a sense of belongingness. Recently, the participants have expressed their dissatisfaction with such orientation seminars. The HRD Manager has asked you as the Personal Assistant to analyse the importance of the orientation and the presentation techniques used and also provide recommendations to make it more interesting and useful for the participants. Now write a memo report to be sent to the HRD manager.

18. The Supreme Court has recently banned the plying of diesel-run buses within a metropolitan city. According to the ruling, only CNG (Compressed Natural Gas) buses are allowed to run on the road. This has resulted in a lot of inconvenience for the public. The State Government has received a number of grievances about the inadequacy of the Metro Transport Corporation (MTC) in meeting the needs of the commuters. You, as the chairman of MTC, have been asked to study the existing situation and report the details in the form of a letter to the Transportation Ministry of the state. Write a report assuming the following details:
 - Total number of buses: 10,000
 - Number of CNG buses: 4,500
 - Number of school buses: 1,000

19. Raj Pareek Singh (RPS) University, Kareempur, wishes to introduce the internal evaluation system and has written to the Registrar, BITS, Pilani.

Write a letter report to be sent to the Chairman, Examination Committee, RPS University, containing relevant information. Your letter report should be in full block form.

20. The Research Division of National Marketing Council (NMC) conducts research designed to keep its members informed on general marketing matter. You are a Research Officer of the NMC. You have been asked by your Marketing Manager to gather information on purchasing patterns of consumers across the nation during January–March 2015. The commodities include food, tobacco & liquor, housing, clothing, medical care, recreation, education, reading, transportation, etc.

Assuming that you have completed this task, write your findings in the form of a memo report to the Marketing Manager of NMC. Invent any other necessary details.

21. The Department of Agricultural Research and Education, Government of India is currently collecting data on the amount of milk, fruit and vegetables, eggs, etc. produced in the year 2014–2015 to formulate a viable National Agriculture Policy which would take care of animal husbandry and dairying also. The data collected are given below:

Milk	78 million tones
Potato	241.5 lakh tones
Fruit & Vegetables	104 million tones
Fish	55.81 lakh tones
Eggs	31,320 million tones
Onion	47.5 lakh tones

As the Senior Scientist of this department, write a report in the form of a letter analysing the above data and giving the necessary details. Use full block form and open punctuation. Address your letter to the State Minister of Agriculture. Invent the necessary details.

22. The Managing Director of Allied Publishers Company Ltd, New Delhi 110 001, wishes to have a report on the readership of various magazines published by it. As the Public Relations Officer of this company, you have compiled the data for writing this report.

Table showing the readership in percentages

Age Group	Women Today	Financial Reporter	Sports Week	Focus
18–25	14	21	35	29
26–35	16	13	41	48
36–45	17	32	53	25
45 onwards	12	32	32	25

Write this report inventing the necessary details.

23. The Rajasthan State Council for Science and Technology (RCST) has been playing a significant role in promoting the use of science and technology for the process of development in the state, especially in the rural areas. The table given in the next column shows the percentage distribution of expenditure incurred by this Council for the year 2012–2015.

Table showing percentage distribution of expenditure.

S. No.	Items	Years		
		2012–13	2013–14	2014–15
1	Major Projects	36.7	40.6	44.5
2	Development Programmes	36.350.4	41	48.4
3	Secretariat and Travel	23.3	8.3	6.3
4	Building and Equipment	3.7	0.7	0.8
	Total	100.0	100.0	100.0

As Secretary of this Council, analyse this data and write an analytical report to be submitted to the Chairman, RCST, Jaipur. Wherever necessary use illustrations to support your analysis. Invent the necessary details.

24. Two years ago, Nisbit.com started an online trans-action portal for users to do online shopping. After two years, the Chief Executive Officer (CEO) of Nisbit.com feels concerned about the slow growth of the company. It is observed that the number of consumers switching to online shopping have not met the expected level. To find out the reasons for the poor response from the customers, the company has decided to conduct a nation-wide survey. The following table gives the finding of the survey.

Reasons	Age-Group					
	19–29		30–50		51 and above	
	M	F	M	F	M	F
Unfamiliar technology	15.1%	23.2%	36.6%	45.3%	81.2%	89.3%
Security concerns	40.5%	48.4%	55%	59.7%	4.4%	2.1%
Resistance to change	14.6%	17.4%	24.6%	29.5%	45.6%	51.7%
Lack of access to Internet	38.4%	41.1%	37.6%	44.4%	7.9%	9.6%
Inefficient delivery (unreliable)	34.4%	36.2%	31.4%	35.5%	4.8%	3.5%
Preference for traditional shopping	25.1%	58.6%	19.5%	62.3%	4.6%	38.2%
Other reasons	17.5%	24.3%	16.4%	23.6%	3.5%	4.1%

As the survey manager of the group, write a formal report to be submitted to the CEO of Nisbit.com.

25. Motivating employees for optimal performance and retaining the trained personnel in an organization is not only very important in today's competitive business world but also one of the most difficult tasks that the personnel manager faces in the organization.

A large-scale survey by opinion polls was conducted among Senior Level Managers, Mid-Level Managers and Floor Supervisors. 300 from each group participated in the survey in which the respondent indicated one most preferred motivator for himself at work. The following table gives the frequency count for responses from each group.

	Senior-level managers	Mid-level manager	Floor supervisors
Money Rewards	80	60	90
Knowledge Training	85	30	15
Social Rewards	40	65	60
Free Lunches	20	55	65
Club Privilege	50	60	10
Flexible Hours	25	30	60
	300	300	300

As the Personnel Manager, write a recommendatory report to the Vice President, Human Resource Development, of a large corporation. Invent necessary details.

26. At its last meeting, the Executive Committee of Maruti Udyog Ltd indicated an existing need for training the three levels of company managers. Hence, the Chairman has asked the Managing Director to analyse the various areas in which training is required for them. As the Managing Director, you have collected the data tabulated below:

Table Showing Training Needs by Different Levels of Management

Area	Entry-Level Managers		Middle Managers		Senior Executives	
	No.	%	No.	%	No.	%
Managing People	34	31.5	95	34.7	22	32.8
English Writing & Oral Presentation Skills	36	33.3	79	28.8	14	20.9
Accounting Computer	26	24.1	63	23.3	29	43.3
Proficiency	22	20.4	23	8.4	02	03

Analyse and interpret the given data so as to enable the Chairman to plan for a Training Programme for the three levels of Managers.

27. The Chairman, Energy Development and Conservation Council, New Delhi, has been concerned about the recurring mismatch between the demand for electricity and its supply all over the country. As the Secretary of this Council, you have been asked to study the sector-wise power consumption pattern and write a report to be submitted to the Chairman. The following table gives the data for three consecutive years.

Sector	Year		
	2012–13	2013–14	2014–15
Agriculture	70.70	79.30	85.74
Commercial	14.14	15.97	16.99
Domestic	43.34	47.92	52.54

Table showing sector-wise power distribution (in billion units):

Interpret the above data using appropriate illustrations to support your analysis.

28. Recently your college had organized a cultural festival 'Waves' where 50 teams from different places participated. Write an event report incorporating all the elements discussed in the chapter.

CHAPTER
14

Technical Proposals

OBJECTIVES

You should study this chapter to know
- ○ the definition, purpose, and types of technical proposals
- ○ the characteristics and structural elements of technical proposals
- ○ how proposals are evaluated, and thus how to write winning proposals

Introduction

Proposals are an important type of job-related writing because their acceptance can lead to significant operational improvements, new business, additional jobs, and safer working conditions. The planning, organizing, evaluating, and writing skills we have learned so far in the preceding chapters can be utilized for writing technical proposals as well. However, when preparing these longer messages, we may need to add some tasks to the familiar categories or pay special attention to tasks we are already familiar with. For example, besides analysing the purpose and the audience, the situation needs to be carefully studied to determine whether a proposal is necessary and which type would be the most appropriate.

In today's competitive business environment, it is a challenge to receive new orders, to get new customers, or to find new business partners. When a firm is considering several companies to outsource a major part of its operations to, it will evaluate the potential of these companies on the basis of their proposals to the firm, showcasing their suitability for the purpose. Hence, proposals have become a very essential way to get new orders and increase the sales of products. Proposals may include a bid on contract, or a business proposal to a government authority, another company, or organization.

Let us imagine a situation when one endeavours to set up a new laboratory in an institution and needs to seek the approval of the administration for the same. The approval may be for a requirement of a new room, purchase of equipment, or acquiring more manpower. Hence, the proposal should include the probable location of the laboratory, its functional strategies, its importance in the present infrastructure, its likely expenditures, etc. Certainly, some proposals must sell the ideas or projects they offer, but all proposals must sell the writer (or the writer's organization) as the one to do the project.

> 'There is only one way to get anybody to do anything. And that is by making the other person want to do it.'
> –Dale Carnegie

Definition and Purpose

A proposal is an offer by one party to provide a product or service to another party in exchange for money. It is usually a sales presentation seeking to persuade the reader to accept the written plan for accomplishing a task. Proposals may be written to people within an organization, to an outside company, or to the government.

In other words, *proposals are written offers to solve a technical problem or to undertake a project of practical or theoretical nature.* Consultants submit proposals to companies offering help to solve problems within a company, building contractors submit proposals to the government on some constructions for the Public Works Department, reputed advertising agencies offer to publicize a product or an idea for a company, construction companies submit proposals to governments abroad to build everything from bridges to skyscrapers—the list is endless.

> Proposals are written orders to solve a technical problem or to undertake a project of practical or the theoretical nature.

As said earlier, proposals, in general, aim to solve a problem, alter a procedure, find answers to questions, offer advice and training, or conduct research on a topic of interest to both parties.

However, proposals have varied purposes with a wide or narrow scope. Given below are a few examples to illustrate the diverse purposes of proposals:

COUNCIL OF INDUSTRIAL RESEARCH

'We agree that this is an innovative proposal. But we can't accept it as we've never seen such a proposal before.'

- To construct parking slots, buildings, bridges, highways
- To sell property, such as buildings, machines, airplanes
- To survey areas for possible water sources
- To plan and construct airport baggage conveyor systems
- To modernize the office procedures of a company
- To train international managers for work in foreign countries
- To conduct the basic research before developing an automobile factory in a foreign country
- To improve engineering performance within a company

> Proposals serve to review and improve existing products and services to meet the ever-increasing and complex requirements of today's highly competitive business environment.

Proposals serve to review and improve existing products and services to meet the ever-increasing and complex requirements of today's highly competitive business environment. They aim at providing new and sound ideas to accelerate the advancement of our society.

In the following sections, we will discuss the types, characteristic elements, style and appearance, and evaluation of technical proposals.

Types

The two basic types of proposals are sales proposals and research proposals. Both these types may be either solicited or unsolicited.

Sales proposals are also known as *business proposals*. They are sent outside the company to potential clients or customers. Sales proposals rarely duplicate one another in either structure or style. In fact, they often take quite different and creative directions like successful advertisements.

Research proposals are usually academic in nature and mostly solicited. Professors, or the institutions for which they work, may submit a proposal to obtain a grant in response to a request or announcement from the government or other agency. A research proposal may even appear in a foreign language. For example, a research proposal submitted by an academic institution in India to a multinational company in Germany may be in German. Whatever the research project, the basic content does not vary. All research proposals will contain the elements discussed later in this chapter.

Importance of Proposals

1. Proposals, like reports, are valuable records of information in an organization.
2. They act as an index of the company's growth or progress.
3. Successful proposals give financial returns to the organization.
4. They help promote various research activities that are vital for the individual, organization, or government.
5. Proposals attempt to win contracts for the company undertaking the project. Proposal writing develops certain favourable and useful skills such as communicative, persuasive, and organizational skills. It also enhances the power of estimation, judgement, and discrimination in the writer.

An organization often knows in advance those individuals and corporations that are qualified to bid on a job or help solve a problem. Hence, requests may come via mail or, in the case of the government, via newspapers.

While preparing a *solicited proposal*, the company should remember that, in all likelihood, it will have many competitors bidding for the contract. To be successful at acquiring the contract, the company will not only have to present excellent reasons to the solicitor to follow its recommendations but will also have to try to overcome the resistance from its competitors, i.e., the company's proposal should have stronger and more powerful arguments than those of the others. So, it has to meticulously follow the proposal requirements of the solicitor regarding the problem, the required solution, specific work to be done or equipment to be installed, format of the proposal, deadlines, etc.

Similarly, when a company prepares an *unsolicited proposal*, it needs to convince the reader that it understands the receiver's problem and that it is qualified to solve the problem successfully.

Characteristics

Technical proposals, whether they are sales proposals or research proposals, are a persuasive blend of information, organization, and reason. Essentially, technical proposals should

- Demonstrate to appropriate decision makers that their needs would be met with
- Be more creative than other forms of professional writing
- Permit informality and personal approach in style to some extent
- Keep in view the customer's convenience, financial gain, and prestige
- Look neat and attractive
- Include summary, background, objective, description of the problem, methodology, and cost estimate
- Anticipate any possible reasons for rejection and provide suggestions for overcoming them
- Follow meticulously the requirements of the solicitor
- Contain certain or all elements of structure according to the purpose, usefulness, and requirements of the recipient's language
- Use plain, direct, and unambiguous expressions

Whether a proposal is long or short, simple or complicated, a writer can improve the chance of securing conviction by making sure that its contents answer the following questions:

- What do we propose to do?
- How do we propose to do it?
- What evidence can we propose to use that will actually get the desired results?
- What evidence can we present to show that ours is the best way to get the desired results?
- How can we demonstrate our ability to do what we propose to do?
- What evidence must we present to show that the cost will be acceptable and, perhaps, that we can meet a satisfactory time schedule?

In providing the information called for, it will be necessary to explain what methods we propose to use, to show that we have or will obtain the resources necessary to use these methods, and to offer enough information about costs to show that our estimates are realistic.

Structure of Proposals

Major business proposals on selling a company's services, expertise, equipment, or extensive installation facilities may use the structure discussed in this section. This structure may also be adapted for a research study within an academic body. Of course, only long, comprehensive proposals require most of or all these parts. Proposals on smaller projects may use only a few. For example, information required for a grant or sales contract of ₹ 10,000,000 will undoubtedly be longer and more insightful than that for a research allowance of ₹ 25,000. Therefore, the elements desirable for a specific proposal have to be chosen carefully.

Similar to the structure of reports, all proposals have three main divisions: prefatory parts, body of proposal, and supplementary parts. Depending upon the need and existing practice, we may choose the required elements from these parts for the proposal.

I. Prefatory parts

- Title page
- Letter of transmittal
- Draft contract
- Table of contents
- List of tables/figures
- Executive summary

II. Body of proposal

- Introduction
  - Problem
  - Need
  - Background
  - Objectives or purpose
  - Scope and limitation
- Technical procedures
  - Methods and sources
  - Plan of attack
- Managerial procedures
  - Sequence of activities
  - Equipment, facilities, products
  - Personnel qualifications
- Cost estimate
- Conclusion

III. Supplementary parts

- Appendices
- References

Prefatory Parts

This segment gives the solicitor or recipient an overall idea about the proposal such as highlights and coverage. Special attention should be given to this segment, as it creates the first impression in the reader's mind. This segment can have the following parts.

Title page   The title page of a proposal is similar to that of reports. Most organizations specify the information to be included in the title page, some even provide special forms that summarize basic administrative and fiscal data. The title page should include at least the title, the name of the person or company to whom the proposal is submitted, the name of the person submitting the proposal, and the date.

Letter of transmittal   This is a cover letter that accompanies or is bound along with the proposal. Proposals submitted to government organizations may contain the letter of transmittal immediately after the title page. This cover letter includes a brief introductory, middle, and concluding paragraph. The topic and purpose are clearly mentioned in the introductory section of the letter. The middle section contains the proposal highlights and the concluding section motivates the recipient towards responding positively to the proposal.

Draft contract   A draft contract is the rough draft of the contract prepared by the proposer. When the proposal is accepted, the original or rough draft may need changes in clauses such as terms of finance, duration of the project, and delivery schedule. Thus, the contract will be finalized and signed only after the proposal has been accepted.

Table of contents   Brief proposals do not require a table of contents. But if the proposal is long then a table of contents is essential.

List of tables and figures   This list enables the reader to easily locate visual aids, if any, quickly.

Executive summary   Even brief proposals should have an executive summary. Seeking to gain a quick review, some evaluators will initially read only this summary. Hence, the executive summary should be a concise version of the detailed proposal. It should provide a brief background, telling the reader the need for taking up this project, and summarize the objectives, how they will be met, what procedures will be adopted, and also the outcome of the project. Budget figures are frequently omitted because proposal summaries or abstracts may receive wide distribution. The summary generally ends with a re-emphasis of the proposal's strengths. The length of the summary is usually between 100 and 300 words depending on the complexity of the proposal. Many proposal consultants believe that the executive summary is the most important part of a proposal. It should create a positive impact, so as to induce the reader to read more of the proposal.

> The importance of an effectively written executive summary cannot be underestimated. Many consultants believe that a project is accepted or rejected solely based on the impression created by the summary.

Body of the Proposal

The main body of a technical proposal consists of the following five sections: introduction, technical section, management section, cost estimate, and conclusion.

Introduction

Problem and need   The problem statement clearly specifies what it intends to investigate. It should elaborate the existing facilities/procedure and the shortcomings arising out of the same. It should explain why the problem exists and what benefits will come from the proposed research.

Background   This includes information such as the following:

- Previous work completed on identical or related projects
- Literature reviews on the subject, particularly the proposer's evaluation of them
- Statements showing how the proposal will build on the already completed projects and research

Purpose   The objective or purpose of the proposal should be stated clearly. It is often stated in infinitive form, for example:

- To offer the supply of forty aircraft engines to Aviation Supplier Corporation
- To provide the required training for the newly employed graduates

Scope   This part defines the boundaries of the project. For example, the proposal on a research study should clearly specify whether it will study one or more areas of a community, company, department, or a particular problem. The proposal will specify which topics will be outside its scope. The writer of the proposal has an ethical and legal obligation to clarify the limits of his/her responsibility to the client.

Limitations   This section describes the restrictions over which the proposer has no control, such as the non-availability of some classified information.

Project team/personnel   Even some short proposals include a listing of the individuals who will work on the project, including project director/coordinator. In long proposals, such information is a must including a brief résumé of each individual (educational qualifications, professional achievements, experience in the area, publications in the relevant field, etc.).

Methods and sources   The reader should be informed about the methods and sources that would be used to collect the required statistical data for the project. It may also include a discussion on the reliability of the sources from which the required information or data would be collected. For example, a proposal offering software consultancy services to some other organizations may include a discussion about the journals that have been consulted, the personnel who has been interviewed to understand the existing problem, and also the authorities with whom the legal issues have been discussed.

Technical section

Procedures   Here, a brief discussion on how the technical requirements of the reader will be met should be given. This discussion incorporates the following aspects:

Plan of attack   Here, the methodology that would be adopted to carry out the project should be presented. For instance, in a proposal offering to set up Solar Water Heating Systems in a university campus, one needs to explain each step of the process, starting from procuring materials to installation of the systems. This section can be further divided into small headings such as materials, system overview, and installation details. In general, this section presents the various solutions available for the problem and the one that has been chosen, justifying the reasons for selecting it.

'I need this proposal approved. Get the best designer in town to draw the blue prints!'

Management section

Sequence of activities   This section pertains to managing the job in question. By means of a Gantt chart or milestone chart, this section presents to the client a clear picture of the phases of activities of the project and how long each phase will take. This section not only guides the reader but also facilitates a systematic approach to the execution of the project. The charts will also show the reader how the bidder is adequately planned and prepared for the various activities.

Equipment, facilities, and products This section explains about the existing equipment, facilities, etc., and also the additional facilities that may be needed to carry out the project. It may also elaborate upon the infrastructure by listing all the available equipment, products, facilities, etc. In addition, a detailed list can be included mentioning all those things that are necessary for the job.

Human resources This section presents the details of the human resources requirement to complete the project. It also elaborates the technical expertise the different personnel have to accomplish the project. The team organization can also be elaborated upon here, mentioning the names of the chief coordinator, co-investigators, etc. The company profile can also be included in this section. The proposer also provides the human resources requirement, which should include a detailed list of number of persons and also the type of expertise they should possess.

Budget/cost estimate

The budget or cost section is mandatory for all proposals. This provides a breakdown of all estimated costs for the project. It should include such items as materials and supplies, salaries, travel, duplicating, consumable items, etc. Some budget sections may be in tabular form or even in the form of visual aid. It is customary to include a budget justification section, in paragraph form, stating the various items of expenses the project would incur and also the potential sources of funding for the project. This would be a further rationale for the financial figures. The recipient will appreciate the bidder's acknowledging responsibility for potential cost overruns and funding shortfalls and the thorough preparation in presenting this estimate.

Conclusion

This last section provides a final opportunity for the provider to re-emphasize and persuade the recipient that they have all the resources in terms of material, expertise, and enthusiasm to accomplish the project. No new ideas should be added here and this section should be very brief, maybe one paragraph.

Supplementary Parts

Appendices, as in formal reports, are optional in proposals as well. Visuals (maps or graphs) and some pertinent letters of support and endorsement can be added. But when in doubt it is better to leave out appendices.

References give the list of sources that have been used or quoted in the proposal. References are usually a part of research proposals that require documentation.

Style and Appearance

All techniques and principles that are applicable to technical communication and report writing are equally applicable to technical proposals, whether they are for research grants or for a sales contract.

It must be borne in mind that the physical appearance of the proposal makes an important non-verbal impression. As proposals are evaluated immediately in terms of general appearance, neatness, specific appearance of the table of contents, list of figures, title page, consistency of style, completeness, and professionalism, it is very essential to spend a considerable amount of time in refining these aspects. Each item must be checked and rechecked. It would also help to adopt appropriate means of visual persuasion. Company logos are often found on each page. Colour and visual aids are used to add effectiveness. For example, a marketing executive's major, unsolicited, successful proposal can have on its cover an accurate sketch of the prospective customer's buildings. The proposal should be attractively bound and protected by a plastic cover.

Evaluation of Proposals

It might seem that in a chapter on writing proposals, a section on evaluation would be inappropriate, but generally writers produce better products if they understand how their work will be judged.

By giving a scale of values to the following set of questions, the reader can make a point comparison between competing proposals:

1. Understanding of purposes, objectives, and tasks—thirty points.

   (a) Does the bidder demonstrate clear understanding and acceptance of the requirements presented in the RFP (request for proposal)?
   (b) Are the tasks outlined in the proposal clear and well defined?
   (c) Are there important omissions in the specified tasks?
   On a scale of thirty points please assign a rating to this area.

2. Technical quality of methods proposed—thirty points.

   (a) Are sufficient time and resources specified to accomplish the quality outlined in the proposal?
   (b) Does the proposer emphasize quality as an important criterion when presenting methods?
   (c) Will the quality of the proposed methods be monitored throughout the contract period?
   (d) Is this monitoring sufficient to ensure quality?
   On a scale of thirty points please assign a rating to this area.

3. Quality of management plan and planning—ten points.

   (a) Has a management plan been designed to ensure receipt of materials at certain specified times?
   (b) Does the proposal clearly identify working relationships within the contractor's staff and with this agency's staff?
   (c) Is sufficient technical management assigned to the task to ensure production and quality of output?
   (d) Will sufficient information be available to this agency to permit analysis of cost and effectiveness?
   On a scale of ten points please assign a rating to this area.

4. Qualification of staff—twenty points.

   (a) Have individuals to whom the task is assigned had prior experience in the required technical areas?
   (b) Have key personnel been assigned to the project for a substantial time?
   (c) Have project directors and those assigned management roles been in similar management positions before?
   (d) Is there sufficient depth in the staff to provide backup and overload capabilities?
   On a scale of twenty points please assign a rating to this area.

5. Corporate capability and experience—ten points.

   (a) Has the organization had previous experience in planning and managing efforts of this type?
   (b) Has the organization previously managed projects of this size and complexity?
   (c) Is the organization of sufficient size and stability to undertake the responsibility called for?
   (d) Is there any 'track record' of performance available, indicating consistent meeting of schedules with quality output within fiscal limits, or the inverse?
   On a scale of ten points please assign a rating to this area.

Although only one of the preceding questions deals directly with expenses, costs are a primary consideration. In commercial proposals, the first consideration is cost, followed by performance, reliability, economy of operation, and early delivery of the product.

Beyond these considerations are many others, but one of the most important is the tone of the proposal. Those addressed are extremely concerned that the responsible officials of the proposing organization are genuinely interested in doing the work and are committed to providing complete satisfaction, even beyond normal guarantees. In other words, reviewers are likely to reject a proposal if they believe that they would have to put up with questionable practices—or with indifference.

Many proposals turn out to be unsuccessful because of the following reasons:

- Questionable project design
- Inadequate explanation of the research
- Lack of experience of the investigator

Other major reasons include vague experimental purpose and poorly prepared knowledge of the literature. The important point to note is that most of these reasons derive from the presentation of the material, that is, from how the proposals were written rather than from the nature of the research. In other words, if the investigators had prepared their proposals more carefully, they might have been successful, and in the world of research and grants as well as in business, a successful proposal often means the difference between working and looking for another job.

Exhibits 14.1 and 14.2 will give you a fair idea of how to write well-structured, persuasive proposals.

EXHIBIT 14.1   Sample Proposal

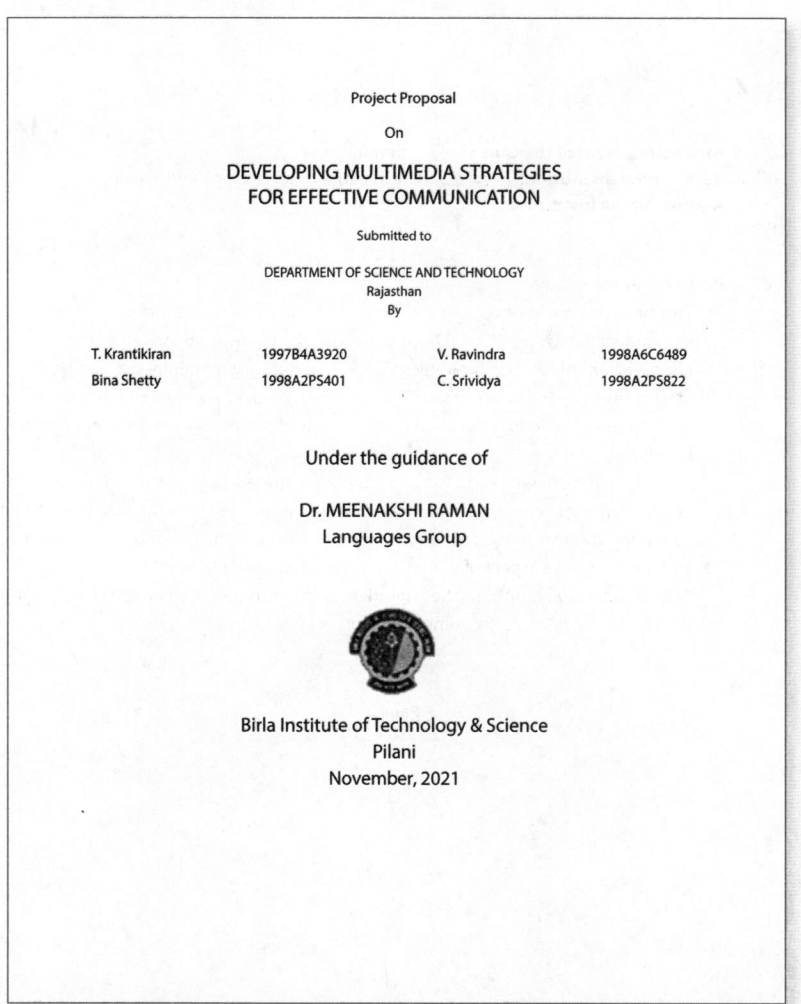

Project Proposal

On

**DEVELOPING MULTIMEDIA STRATEGIES
FOR EFFECTIVE COMMUNICATION**

Submitted to

DEPARTMENT OF SCIENCE AND TECHNOLOGY
Rajasthan
By

| T. Krantikiran | 1997B4A3920 | V. Ravindra | 1998A6C6489 |
| Bina Shetty | 1998A2PS401 | C. Srividya | 1998A2PS822 |

Under the guidance of

Dr. MEENAKSHI RAMAN
Languages Group

Birla Institute of Technology & Science
Pilani
November, 2021

(Contd)

STUDENT PROJECT PROGRAMME

1. Title of the project:	Developing Multimedia Strategies for Effective Communication

2. Name of the students:

	Name	ID No.
(i)	T. Krantikiran	1997B4A3920
(ii)	V. Ravindra	1998A6C6489
(iii)	Bina Shetty	1998A2PS401
(iv)	C. Srividya	1998A2PS822

3. Class/Year of the students: (i) 4th year.
(ii), (iii), (iv) 3rd year.

4. a. Name of the Project Guide Dr. Meenaskhi Raman, Assistant Professor

b. Experience of the Guide: Teaching at various levels for the past 14 years; taken up projects in multimedia courseware development

c. Address of the Guide: Dr. Meenakshi Raman
Languages Group
Faculty Division I
BITS, Pilani-333031
Telephone:
Office: 45073-Extn. 307
Residence: 42238/44736

d. Broad Area/Field of the Guide: Communication.

5. a. Name of the institution: Birla Institute of Technology and Science

b. Address of the institution: BITS, Pilani
Rajasthan - 333031
Ph. 45073 - 307

6. a. Project Summary

This project aims at developing through multimedia certain strategies for effective communication, which is vital for the growth of an individual as well as the society. Effective communication strategies, if developed through multimedia, can be easily grasped even by the lower strata of the society. Moreover, these strategies may prove to be indispensable for education institutions and professional organizations of any kind.

This multimedia package would incorporate strategies for both verbal and non-verbal communication. The main focus would be on the significant aspects of oral communication such as body language, voice modulation and audience awareness, and the illustrative aspect of written communication namely, graphic aids.

The package would not only be user-friendly but also be informative, thus enabling the users to develop better communication skills.

Page 2

(Contd)

b. Technical details of the project

Since effective communication is indispensable for the progress of any society, the project will focus on developing strategies for the same. Though the Internet and other audio-visual aids like audio and video cassettes, etc. throw some light on techniques for developing language skills, they do not deal in detail with the various aspects of communication.

Even though there are many aspects of communication, the project would limit itself only to certain aspects of oral and written communication. The project would require the following:

Software Requirements

- Multimedia development kit
- Adobe Photoshop

Hardware Requirements

- A personal computer with Pentium II processor
- Multimedia kit (speakers, sound care)
- 32 MB RAM
- Internet connection (preferable for research and analysis of the existing products and to download latest software to compress audio and video files, e.g. MP3 format)
- Recent publications (literature) on multimedia

7. Introduction of the project
a & b. Definition and origin of the proposal

The development of any society largely depends on the interaction among its people. This interaction is essential for the ideas, facts, feelings, and courses of action to be transmitted and interchanged. But without adequate communication skills, it would be difficult for the people to interact effectively.

In a country like ours where illiteracy is still prevalent among the lower strata of many societies, oral and visual communication can greatly help the flow of interaction in a society. To achieve effectiveness in communication, people need to follow certain strategies. There is no doubt that these strategies can be developed in various ways.

We feel that multimedia is the most effective of all and hence designing effective communication strategies through this medium would be of great help to the students, professionals, and laymen. For example, strategies for effectively using the various means of body language such as posture, gesture, eye contact, etc. would certainly prove to be of immense help for all these categories of people. They can use each of these means of non-verbal communication according to the situation in which they communicate. For instance, people working in an organization may have to deliver many short or long presentations or participate in meetings, etc. during their professional career; students may have to give several presentations such as seminars and talks, or participate in group discussions and interviews during their academic career; laymen need to communicate their ideas, decisions, etc. to others. Hence developing effective communication strategies is vital for the overall progress of any society.

c. Objective

This project aims at developing multimedia strategies for effective communication (both verbal and non-verbal). It would mainly focus on:

1. All aspects of body language, voice modulation, and audience awareness pertaining to non-verbal communication

(Contd)

2. Graphic aids, which are vital for both oral and written communication

d. Work plan

For effective scheduling of time among students and for convenience, the project will be split into two broad phases of four months each. In the first phase, the package development will focus on the written form of verbal communication, and certain aspects of non-verbal communication such as personal appearance and posture, gestures. The various patterns of communication within an organization will also be dealt with.

In the second phase, the oral aspect of verbal communication, facial expression, eye contact, and space distancing of non-verbal communication will be developed. Management of information within the organization, audio-visual aids on business correspondence, reports, group discussion, meetings, seminars, and conferences will also be dealt with.

e. Methodology

- Literature survey
- Collection of materials
- Scripting
- Developing strategies
- Designing strategies through multimedia

f. & g. Organization of work element and time schedule

Phase I	4 months (approximately)
Phase II	4 months (approximately)
	as per work plan

8. Details of facilities to be provided by the institution

- Library
- Computer hardware
- Software for the use of multimedia
- Recording
- Internet access

9. Budget estimate:

I. Minor Equipment

1.	Consumables	₹ 8,000.00
2.	Report writing	₹ 1,000.00
3.	Contingency & other costs	₹ 1,000.00
	Total	₹ 10,000.00

10. Utilization of the outcome of the project

The multimedia package developed in this project will be informative as well as user-friendly. It will not only create an awareness among the public about the various aspects of effective communication, but also enable them to modify and develop their communication strategies. This in turn will prove to be beneficial for the progress of the society at large.

Page 4

EXHIBIT 14.2   Sample Proposal (with covering letter)

Multi-modal Gymnasium
Varanasi-221004

January 21, 2022

Mr Anuj Sharma
Chairman
Diesel Locomotive Works
Varanasi-221004

Re: Project proposal for setting up of a multi-modal gymnasium in DLW

Dear sir

The attached document, 'Proposal for Setting Up of a Multi-modal Gymnasium in DLW Institute', outlines our project for a modern gym. Reduced man-days and associated costs due to medical problems of the DLW staff and officers has been a long-standing issue. Our proposal aims to suggest a remedy for these problems. The project is also expected to satisfy the long-pending demand of DLW staff for setting up of a gymnasium with multifarious facilities.

This proposal provides you with an overview of the proposed plan, an outline of the work plan along with the cost estimate, and the suggested plan of action for utilization.

This proposal also explores the alternative facilities provided and the utility of each.

The authenticity of the proposal is supported by the fact that many leading organizations in the world including Intel, IBM, GE, TATA, and others have implemented this concept successfully. If you have any questions or concern about our proposal, please feel free to contact me over my mobile 9830038796 or by e-mail at anirudh@vsnl.com.

Yours truly

Anirudh Gautam
Dy Chief Personnel Officer
DLW

Enclosure: proposal for multi-modal gymnasium

(Contd)

PROJECT PROPOSAL

ON

SETTING UP OF A MULTI-MODAL GYMNASIUM IN DLW

SUBMITTED TO

CHAIRMAN
DLW

BY

ANIRUDH GAUTAM
DY CHIEF PERSONNEL OFFICER/G

DIESEL LOCOMOTIVE WORKS
VARANASI 221004 (UP)

January 2022

(Contd)

STAFF WELFARE PROGRAM
Draft Contract

Project Title	SETTING UP OF A MULTI-MODAL GYMNASIUM IN DLW INSTITUTE
Name & designation of proposer Officer	Anirudh Gautam, Dy Chief Personnel
Postal address of the proposer	DLW, Manuadih, Varanasi
Name of the institute in which the gymnasium is proposed to be set up	North DLW Institute
Time required for commencement of the project on receipt of approval	6 months
Duration of the project	6 months
Amount of money required	One-time Cost: Rs 9 million approximately
Recurring Expenses: ₹ 2.2 million	Expected Annual Income: ₹ 2.0 million

(Contd)

EXECUTIVE SUMMARY

This proposal is about setting up of a multi-modal gymnasium in DLW. Last year DLW Hospital registered about 150 heart ailment cases. This year, the figure has risen to 200. Similarly, there has been phenomenal increase in the high blood pressure, depression, and anxiety cases. In addition, other stress-related medical problems have shown a rising trend, notably that of the cardio-vascular systems, digestive, and the nervous systems. Total cost incurred by DLW due to lost working days and also due to the cost of medical treatment was calculated to be ₹ 57 lakh for last year alone. This year, the figure is expected to be at least double that of the previous year.

In order to arrest these alarming trends through preventive means, it is proposed to set up a multi-modal gymnasium at DLW for use by staff and officers, and their families. Contrary to popular belief, a multi-modal gymnasium offers varied health and fitness programmes, ranging from iron-pumping machines to Yoga therapy and Chinese acupuncture. This technical proposal enumerates the suggested outline of the proposed gymnasium, conventional and non-conventional programmes proposed to be offered, and the consequent benefits due to the same. The proposal also brings out the estimated time schedules for completion and the cost likely to be incurred.

The outcome of this project is envisaged in terms of reduction in lost working days and associated costs due to medical problems of DLW staff and officers. Also, the project is expected to fulfil a long-pending demand of DLW staff for setting up of a multifarious gymnasium in DLW.

Page 3

(Contd)

TECHNICAL DETAILS OF THE PROJECT

A conventional gymnasium conjures images of weight benches, trestles, parallel bars, balance beams, tread mills, and weight machines. This was true about gymnasiums about 20 years ago. The modern gymnasium offers a wide range of fitness regimens, ranging from the conventional ones to alternative therapies based on traditional health-care systems. Gold's Gym, a world leader in health and fitness facilities, offers a wide variety of programmes, including injury prevention and care, nutrition and supplementation, weight loss and gain, and anti-aging and senior health, amongst others. The traditional 'dhyan' yoga, acupuncture, 'pranayam', reiki, and other programmes are all offered under one roof.

The current proposal for setting up a multi-modal gymnasium in DLW visualizes provision of a wide variety of health and fitness programmes as mentioned above. It is proposed to set up modern cardio-vascular fitness machines such as treadmills, stair-climbers, elliptical gliders, upright and recumbent bicycles, and rowing machines. Concurrently, setting up of an ambient Yoga Centre with the help of local expertise is also proposed. Upgrading the existing badminton, squash, tennis, and swimming facilities are also planned.

(Contd)

BACKGROUND

In the year 1962, when DLW was set up with American collaboration, the stadium, the golf course, the indoor badminton courts, the squash courts, the tennis courts, and the basketball courts were constructed with the aim of making the fitness facilities available to the DLW employees. It can be said with some pride that DLW employees and their children have excelled in a number of sports and some of them have even found place in the national teams. Availability of adequate and wide variety of sporting facilities has been primarily responsible for a healthy atmosphere in DLW as the number of lost working days due to sickness have been low compared to other production units of Indian Railways and also IR as a whole.

The recent years have, however, seen a rise in the working pressures as DLW has strived to compete with the global market. There have been demands on DLW system to bring out new designs of locomotives in less cycle times and at reduced costs. The competition from Chinese and other Asian suppliers have had a telling influence on DLW's operating ratios. Amidst the rumours of possible privatization and a reducing budget from the Railway Board, DLW has not only been able to survive but has made a place for itself in the Mid-east, South East, and African markets. Exports to countries like Bangladesh, Tanzania, Jordan, Sri Lanka, Vietnam, and Malaysia have been successfully executed.

Uncertainty and diversity of production have had a detrimental effect on the mental and physical health of its employees. The number of cases of cardiovascular diseases has almost doubled in the last three years. The number of lost working days due to sickness has also significantly increased in the past few years. There has also been a general increase in grievance levels of the employees with regard to their future, especially when they compare themselves with other government departments, which are still insulated from the market economies. The Staff Welfare Committee during its last meeting with the Chairman, DLW, had recommended certain steps to alleviate the troubles of employees. Setting up of a multifarious gymnasium figured as one of the recommendations. On this basis the Chairman had asked the Personnel Department to put up a proposal for setting up of a multifarious gymnasium in DLW.

(Contd)

STATEMENT OF THE PROBLEM

Figure 1 shows a year-wise break-up of sickness cases registered by DLW hospital.

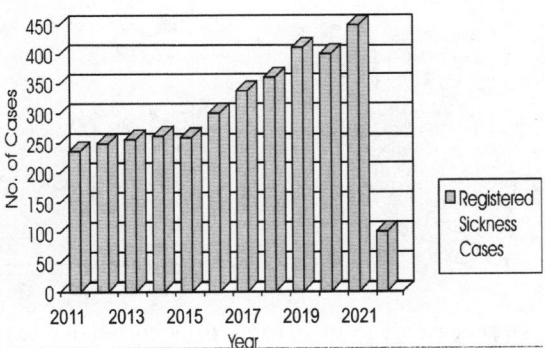

Figure 1   Year-wise break-up of registered sick cases in DLW hospital

As can be seen, there has been a gradual increase in the number of registered cases over the past ten years. Year 2020 shows a slight reversal in the number of cases, mainly because of a large number of retirements during that year. The above figure only refers to the in-patients department cases. If the out-patients reporting is also added, then the problem assumes larger proportions. The trend, however, remains the same.

Given in Figure 2 is the break-up of the cases in 2011.

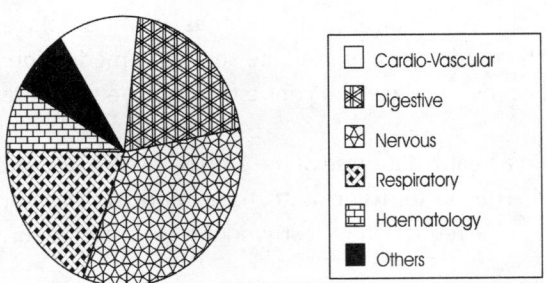

Figure 2   Break-up of registered cases in Year 2011

Figure 3 shows the working days lost in the last year due to medical problems and the associated total costs. This is compared with the projections for the year 2002.

(Contd)

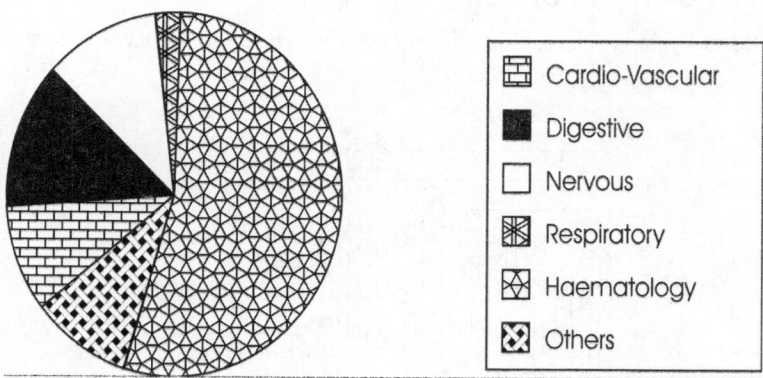

Figure 3   Working days lost and total cost incurred due to medical problems

As can be seen, the associated total costs on account of sickness are projected to double in the current year. The actual expenditure due to lost working days and due to medical treatment is expected to touch ₹ 120 million in the current year. The share of cardiovascular diseases has increased significantly. More significantly, this has had an adverse effect on the morale of the employees.

Objectives

The main objective of setting up the gymnasium is to ensure fitness for the employees through sports, exercises, healthy food habits, relaxed life style, and meditation. A multi-modal gymnasium will act as a counselling centre for employees with tailor-made fitness programmes.

It is also proposed to maintain a health database of the employees in association with the DLW central hospital. The bottom line is to have more satisfied and healthy employees with consequent reduction in lost working days and medical expenditure.

(Contd)

WORK PLAN

Phases

For scheduling purposes it is proposed to split up the project into three phases. In the first phase, it is envisaged to set up the cardio-fitness centre. The second phase is aimed at establishing the Cybex circuit weight-training area. In the final phase, the meditation hall and upgrading of the tennis courts and the swimming pool are planned.

Coordinator

As a first step, a full-time coordinator needs to be selected. The minimum qualifications and the work experience of the Gymnasium Coordinator have to be determined and the emoluments which can be offered have to be decided. The coordinator shall be responsible for looking after the management of assets worth ₹ 10 million and shall also be responsible for effective management of the gymnasium. Therefore, selection criteria are required to be approved by the Chairman.

Location

It is proposed to locate the cardio-centre and the weight centre of the gymnasium in the North Institute of DLW. The Yoga and meditation centre is intended to be put up at the Officers' Club. The location of the courts and the swimming pool remains the same but the skirting area around these is planned to be concreted and tiled.

Area

The cardio-centre and the weight-training centre require an area of about 20,000 square feet. The area has already been surveyed and the vacant stretch in front of the basketball court in the North Institute premises can be used for construction of the building. For the Yoga centre, the space near the Officers' Club is proposed for building the main hall and the annexe. The area in front of the swimming pool is planned to be utilized for construction of wash rooms.

Types of Equipment

The cardio-centre is envisaged with the following equipment:
  a. Treadmills

(Contd)

    b. Stair Climbers
    c. Elliptical gliders
    d. Upright & recumbent bicycles
    e. Rowing machines

    The weight-training centre is proposed to be equipped with the following equipment:

    a. Weight benches
    b. Wall bars
    c. Parallel bars
    d. Incline boards
    e. Balance beams
    f. Trestles
    g. Weight machine centre

    The Yoga centre needs no specific equipment, but requires a hall with proper ventilation. There are plans to have a tie-up with the Art of Living Foundation for meditation courses.

Human Resource

It is estimated that a skeletal staff of about five shall be required for proper administration. For this purpose no additional staff is planned to be recruited, rather volunteers from the existing class 'C' and 'D' categories shall be screened.

Charges

For DLW staff and officers, the charges shall be deducted from the salary at a flat rate of ₹ 500/- per month. For external members the fee shall be ₹ 2000/- per month.

Timings

On Saturday and Sunday the gymnasium is proposed to remain open from 7:30 hrs to 17:30 hrs. On working days the timings need to be decided in consultation with the staff council and the officers' association. However a timing of 6:30 to 8:30 in the morning and 16:00 to 19:00 in the evening appears to be convenient.

(Contd)

Methodology

For civil construction works, it is first proposed to draw up the detailed layouts in association with DLW's civil engineering wing. The specifications of the buildings also need to be firmed up. Thereafter, potential supplier assessment of reputed civil contractors will be done after floating limited tenders. Electrical works shall be clubbed with the civil works.

For purchase of the equipment, it is intended to buy these on single tender basis from Gold's Gym who are the leading manufacturers of gymnasium equipment.

Organization of work elements and time schedule

Given below in Table 1 is the list of activities and the expected durations for each. The detailed Gantt chart shall be worked out after the approval of the proposal. The expected completion time after paralleling of activities has been worked out to be about six months from the date of commencement of work.

TABLE 1: List of Activities and Expected Durations

ACTIVITY	DURATION
Getting approval for the technical proposal	10 days
Discussion with staff council & officers' association	5 days
Nomination of core group	3 days
ACTIVITY	DURATION
Briefing of the core group	5 days
Budget approval	15 days
Forming of specification for civil work	10 days
Forming of specification for equipment	10 days
Freezing criteria for co-ordinator selection	5 days
Civil contractor survey	15 days
Discussions with Gold's Gym regarding equipment	10 days
Calling of volunteers from staff	10 days
Tender for civil works	30 days
Single tender for equipment	25 days
Selection of coordinator	30 days
Selection of other staff	25 days
Completion of civil works	90 days
Installation of equipment	90 days
Tie-ups for Yoga centre	30 days
Suggested plan of action for utilization	

(Contd)

There is a requirement to increase fitness and health awareness amongst DLW's employees. DLW's internal cable TV shall be used to promote the gymnasium. Schools and colleges shall be given sets of fliers for distribution. It is proposed to organize weight-lifting and body building competitions for children as well as for adults to attract interest of DLW's employees.

In association with the DLW hospital, dedicational and physical therapies shall be worked out for some volunteers. This shall give an impetus to popularizing the concept of fitness. It is expected that the full capacity of the gymnasium will be on the lines similar to that of the DLW golf course.

It is proposed to have separate timings for family members. This shall ensure active participation by ladies and children. This experiment has already been successfully tried out with the swimming pool.

It is expected that a few success stories in terms of improved fitness levels among the staff shall impart momentum to the whole concept of gymnasium.

Budget Estimate

Given below in Table 2 is an estimate of cost likely to be incurred in setting up the gym facilities. The recurring costs shall be additional to this estimate and have been worked out separately. Also indicated are the expected earnings/support annually.

TABLE 2: Expected Expenditure and Earnings for the Gymnasium

S no.	Category of Expenditure	Cost in ₹ million
	One-time cost	9.0
1	Civil Construction	2.5
2	Electrical Works	0.5
3	Equipment	3.0
4	Selection process	0.5
5	Core team expenses	0.5
6	Travel expenses	0.5
7	Misc expenses	1.5
	Recurring Expenses/Year	2.2
1	Salaries and wages	0.5
2	Maintenance	0.7
3	Contingencies	1.0

(Contd)

	Expected Income/Year	2.0
1	DLW members	0.5
2	Non-DLW members	1.5

It is anticipated that with the setting up of systematic fitness facilities such as a gymnasium, the overall health levels of the employees and their families shall improve. This is supported by the fact that many leading organizations in the world including Intel, IBM, GE, TATA, and others have successfully tried and implemented this concept.

The Chairman is therefore requested to accord approval to the setting up of a multifarious gymnasium at DLW.

Please refer to the Online Resource Centre for more samples of technical proposal.

SUMMARY

Technical proposals are sales offers to solve a problem. In writing such proposals, academicians offer their domain expertise in solving technical problems; business professionals offer their know-how to take up some major or minor projects in order to provide consultancy, training, or innovative solutions to certain problems that arise in their organizations or other companies.

Technical proposals are written in formal and persuasive language and include standard structural elements. An understanding of the purpose, structure, characteristics, and style of technical proposals will enable us to prepare winning proposals.

EXERCISES

1. The Technical Writing Department of Bell Educational Supplies and Technologies (BEST) needs new computer equipment. Currently, the department has outdated hardware, outdated word processing software, an outdated printer, and limited graphic capabilities.

   Because of these problems, the company's user manuals, reports, and sales brochures are being poorly reviewed by customers. Further, BEST has no website for product advertisement and/or company recognition. All these factors have led to a decline in profits.

   As Technical Writing Department Manager, you have consulted with your five staff members to correct these problems. As a team, you have decided that the company needs to purchase the following new equipment:
   - Six new personal computers
   - Two laser printers
   - Word processing software
   - Graphics software
   - Scanner

   Draft a technical proposal to the CEO of BEST company for the purchase and installation of the equipment. Invent all the necessary details.

2. As a fresh graduate in engineering you have decided to establish a manufacturing unit in your hometown. For this purpose, you have decided to avail yourself of the liberalized loan facility under the self-employment scheme. Therefore, you wish to submit an unsolicited technical proposal for the manufacture of an item of your choice seeking loan from the State Industrial Development Corporation (SIDC), Jaipur.

   Now draft this proposal to be sent to the Director, SIDC, Jaipur, inventing the necessary details.

3. Assume that as Project Development Officer you have been asked by the National Institute of Computer Education, Futura Building, 24/377 Peetampura, New Delhi-110031, to set up a Communication Technology Centre at Pilani for training professionals in the use of latest technological aids for face-to-face and distance communication.

   Inventing all the required details, write in proper format this technical proposal which is to be submitted to the Director of the Institute.

4. The Government of Odisha has invited proposals from Non-Governmental Organizations (NGOs) for setting up small-scale industries in the areas of electrical goods, motor spare parts, oil-crushing units, cotton textiles, wool manufacturing, etc., for the rehabilitation of the flood-affected people of the state.

   Choose an industry of your choice and write a technical proposal to be submitted to the Director of Industries, Government of Orissa, assuming yourself to be the Managing Director of a public limited company. Invent the necessary details.

5. Excel Computer Manufacturing (ECM), a multinational company, wants to improve the existing parking facilities for the four-wheeler and two-wheeler vehicles of its staff.

   As the Personnel Manager, ECM, draft a proposal to be sent to the Secretary, Board of Directors, of your company for improving the parking facilities. Your proposal should include the following issues:

Area availability, lighting, provision of stands and roofs, arrangement of security, issue of identity cards, fixing the rent, etc.

6. The Government of India has recently launched a company to produce personal computers on a large scale. As Finance Manager of this company draft a proposal containing all the required details like space, space-conditioning and dust-proofing, hardware, peripherals, consumables (like disks, tapes, ribbons, floppies, etc.), networking, power supply, technical and other staff, ancillary support systems, etc. This proposal is to be submitted to the Secretary, Department of Company Affairs, for approval.

7. Assume that your organization has been spending about ₹ 500,000 per annum on getting its various documents such as routine forms, brochures, reports, proposals, manuals, etc., printed by outside agencies. Now, the organization has decided to have its own publishing unit to cater to all the above printing needs. As the Office Manager, draft a proposal for the creation of this unit for the consideration of the Board of Directors of your organization.

8. As the District Health Officer, Vidisha, Madhya Pradesh, you have decided to set up a Rehabilitation-cum-Health-Care Centre in the district, especially for the benefit of the rural population of about 200,000. The objective of the rehabilitation programme is to help persons with total or partial disability and to enable them to live with human dignity. The health-care programme mainly aims at carrying out immunization programme effectively in addition to developing awareness about the need for nutrition to fight the menace of malnutrition and the related diseases among children. The proposal is to be submitted to the Director of Health Services, Madhya Pradesh, Bhopal. Draft this proposal.

9. The Government of Rajasthan has earmarked ₹ 100 million for drought relief work, for the worst-affected district of Jhunjhunu in May 2021. This money has to be spent on the following:
   • Provision of work to labourers on afforestation, road construction, digging wells, etc.
   • Distribution of seed, fertilizers, and fodder at subsidized rates
   • Supply of drinking water
   • Supply of essential commodities through fair price shops
   • Free distribution of medicines
   • Supply of free lunch to school children

   As Collector, Jhunjhunu, draft a proposal to be sent to the Secretary, Home Department, Government of Rajasthan, Jaipur, containing the details of how you propose to proceed with the relief work under the various categories mentioned above.

10. The Northill University, Guwahati, wishes to revise and revamp its curriculum in business. As the Research Associate in the Bureau of Business Research, Shillong, you have been asked to study the data collected from business executives and business leaders regarding the strengths and weaknesses of the graduates' knowledge of business and business skills. The table given below contains the tabulated responses of 1000 respondents.

Respondents' Views on the Importance of Subject Areas in Business (in per cent):

Courses \ Ratings	Important	Very important	Unimportant
Accounting	88.2	11.8	0.0
Business Policy	81.5	14.0	4.5
Business Communication	91.7	8.3	0.0
Management	68.5	22.7	8.8
Management Information System	59.8	32.8	7.4
Marketing	80.7	19.3	0.0
Organizational Behaviour	62.0	33.7	4.3
Personnel Management	11.8	37.7	50.5
Strategic Planning	39.6	51.7	8.7

As the Research Associate of this Bureau, draft a technical proposal to be sent to the Registrar, Northill University, Guwahati. Invent the necessary details and use the statistics given in the table.

CHAPTER 15
Formal Letters, Memos, and Email

OBJECTIVES

You should read this chapter to know
- ○ the various written modes of communication
- ○ the seven C's of written communication
- ○ the purpose, significance, and structure of memos, letters, and emails and how they differ
- ○ how to write various kinds of business letters
- ○ the classification of memos
- ○ the advantages and limitations of email
- ○ about email etiquette, effectiveness, and security

Introduction

Having studied and understood the basics of writing skills through the preceding chapters in this part of the book, let us now look at their applications in the various modes of formal written communication. These forms are used extensively in the daily communication activities of any organization. Formal written communication includes business interactions in the form of memoranda (memos), letters, emails, reports, manuals, bulletins, etc. In present day workplaces, the transfer of information on paper has been substantially replaced by electronic communication. However, the value of well-written messages remains unchanged.

Effective letter writing requires a strong command over language, knowledge of the various popular formats, coherent thoughts, and a good choice of words.

Any document has to be designed with respect to the target audience. It is essential to be precise and concise. Often the reader may be interested only in the key information being conveyed rather than the details. Hence, avoid peripheral details and directly address the issue.

In some instances, a conversational style may be preferred over a formal style. However, while very formal writing may alienate the readers, an extremely informal approach may appear very casual and unprofessional. Therefore, it is important to strike a balance in the writing style, keeping in mind the audience being addressed. One should always be aware of the purpose of writing and never compromise on the content. Often written documents serve as the first point of contact, and hence are crucial in creating the first impression.

In this chapter, we will learn the characteristics and principles of formal correspondence, as also understand the various formats and types through numerous examples.

Formats of Written Correspondence

Organizations usually have a set format for memos and letters, and hence each organization has its own established pattern of written communication. However, the current approach of communication through emails has become more informal. The traditional formats of letters will be discussed later in the chapter.

There are similarities as well as differences in the structure of letters, memos, and emails. A memo consists of a To, From, Subject, and Date approach, which makes inter-office communication easier. Since it has recommendation towards the end, it quickly gives the action plan. Letters are longer and more formal than emails and memos. While memos and emails can be as short as two lines, the length of emails should not exceed what can fit one screen. Further, a letter is generally addressed to one person, while memos and emails address a group of individuals. Also, emails can have additional attachments such as pictures, scanned images, documents, and audio/video clips.

Of the three formats we are discussing here, letters are the most formal. Before selecting the mode of communication, one has to consider several factors, including the purpose of writing, the role of the receiver, the urgency of the information being conveyed, and the expected impact.

Types of Messages

A message can be written for different purposes. It can be congratulatory, for cancellation of a ticket, for accepting an invitation, for breaking bad news, or informatory. We will discuss the four broad types of messages here: general message, positive news, bad news, and negative message.

General message

Any message being conveyed has to be formulated very clearly. It is always better to have a direct approach of communicating the information rather than using buffer statements or beating around the bush. For example, 'All the employees of the institute are invited to gather in the community hall for high tea and an entertainment programme on the eve of New Year 2022'. A general message can be anything from birthday celebration, apology message, congratulatory message, etc.

Positive news

Positive news is mainly written to reassure or convey constructive information. It portrays the positive aspect of an issue, and helps in playing down the negative elements. Such communication helps develop a good relationship between the sender and the recipient. The following guidelines will help organize the elements of positive news:

- Summarize the main points of the news
- Provide details, clarification, and background
- Present negative elements, if any, as positively as possible, between the positive elements
- Highlight the benefits
- Be courteous and use goodwill ending

A positive message can be a transmittal, confirmation, summary, or clarification. Like for a general message, while drafting a positive message, it will serve well to be aware of the purpose, the recipient, the contents to be included, the benefit to the readers, and the response anticipated.

Bad news

Though inevitable, we all try to avoid breaking bad news. Since it must be done, a few things one can do to lighten the effect are as follows. Begin a negative message with a buffer statement. Then state the bad news along with proof and always conclude on an upbeat and positive note, may be including a possible solution to the situation. A neutral opening, followed with by a brief reminder of the positive aspects of the relationship in the past might constitute an encouraging beginning. Then address the bad experiences and the problems faced, the possible reasons, if appropriate, and end by expressing desire to continue the good relationship.

See the example below:

> Dear Mr Grover,
>
> Our organization has used your taxi service for a very long time and had a very satisfactory experience in the past. We have not faced any problems earlier. However, this time when our employee, Mr Ashok Behl, who was to be picked up from the airport, came out of the terminal, the driver was nowhere to be seen. When Mr Behl tried to contact the driver, his phone was switched off. We then called your office, and the driver came after 45 minutes and was not at all apologetic. Our employee was put through great inconvenience because of this delay.
>
> Since we have been your customers for the last 10 years, we do not expect poor services from you. We wanted to bring this to your attention so that you can personally find out the reason behind such reluctant attitude.
>
> We hope that this problem can be quickly addressed to ensure that our future dealings with you are hassle free.
>
> Regards,
>
> Sameer

Use neutral terms as far as possible when conveying a message. Avoid severe words such as terrible, bad, regret, unfortunate, and cannot. Always allow the recipient an opportunity to express their views so that the communication does not end abruptly.

Negative message

It is never too difficult to deliver positive or neutral messages, but a negative message needs to be considered well and worded carefully. Although it is a difficult task, the sooner it is delivered, the better. One may have to adopt a direct or an indirect approach depending on the situation. For example, if a researcher has applied for funds for a project to an agency and the agency is not granting the fund, the agency is expected to inform the candidate about the same at the earliest and in clear words. Often, conveying a negative message to a known person needs to be handled much more delicately and carefully than conveying the same to a stranger. The following points might help drafting a negative message:

- Approach the issue directly rather than covering up.
- Be gentle in delivery. Use of passive voice may sometimes work better. Be tactful and do not dwell too long on the negative information.
- Provide adequate reasons without sounding apologetic.
- Conclude on a positive note.
- Be sincere in approach.
- Allow the recipients an opportunity to express their views. Do not end the communication abruptly.

Letter Writing

It is important for professionals to be educated on the functional importance of effective writing. In letter writing, precision and clarity of meaning are extremely important. Time is a valuable resource, and precise communication helps save the time of the writer as well as of the recipient. People who value time and communication are bound to gain out of their effective communication skills.

Further, the reader creates an image of the writer or the organization through their letters. A good letter should make for effortless reading. It should be clear and concise, with short sentences and simple words. It should keep to the facts and be easy to read and understand.

The following sections will take us through the various characteristics and components of professional letters.

> 'Letter writing is the only device for combining solitude with good company.'   *–Lord Byron*

The Seven Cs of Letter Writing

To write an effective letter, one needs to understand the purpose of writing and then draft the letter, focussing on the reader's perspective. Formal letters should be clear, courteous, firm, and as friendly as the topic allows. The best letters have a conversational tone and read as if the reader is being spoken to. The following are the seven Cs of letter writing:

- Clarity
- Conciseness
- Correctness
- Courtesy
- Cordiality
- Conviction
- Completeness

While writing a letter, we usually try to convince our readers or derive a positive reaction from them. The readers will respond quickly only if the meaning is crystal clear and conveyed well. We need not attempt to present ourselves or our organization as perfect and flawless. Instead, by taking a stance and accepting responsibility, we may create a more reliable impression.

Significance

Any organization will have to correspond in writing with its customers, branches, suppliers, bankers, and other vendors with whom it has or would like to have a lasting professional relationship.

There have been enormous changes in communication patterns in the past decade. However, traditional formal letters still retain their importance for the following reasons:

- Just as personal letters help maintain personal relationship with friends and relatives, formal letters assist in sustaining relationships with other organizations, clients, and vendors.
- They are the appropriate forms of communication when the information to be conveyed is complex.
- They serve as permanent records and are a valuable repository of information, which can be referred to in future.
- They help reach out to a large and geographically diverse audience economically.

Purpose

Although each individual letter has a unique message, the ultimate purpose of writing any formal letter is to sustain existing business relationships or to create and establish new business relationships. For example, when we apply for a job, we offer our services to the prospective employer. Similarly, when the purchase manager of a company writes a letter complaining about a damaged consignment, his/her aim is not only to bring the flaw to the notice of the recipient but also to request for corrective action. Formal letters therefore are written for varied purposes as mentioned below:

- To inform
- To enquire
- To request
- To complain
- To sell a product, service, or scheme
- To congratulate
- To order
- To collect dues
- To make an adjustment
- To apply for a job

Structure

Unlike personal letters, formal letters have a distinct structure and layout. One needs to be familiar with not only the different elements or parts of a letter but also their positioning in the letter and the purpose for which they are included in the letter. Several of these elements appear in all letters, while others appear only when desirable or appropriate.

Standard elements

Heading　Also known as the letterhead, the heading shows the organization's name, full address, and telephone numbers. If a separate letterhead is not provided, the heading includes the sender's address. Letterheads have the potential to create a favourable impression and hence need to be designed with thought and imagination. The colour and quality of paper, the size of the letterhead, the type of fonts that are used, and the spacing are factors to be carefully considered. Many companies even seek the help of advertising agencies to design their letterheads. A sample letterhead is shown here:

Whenever a letterhead paper is not used, the sender's address is either aligned with the left margin or centre-aligned, depending on the layout of the letter. For example, the same heading given above can be displayed as follows:

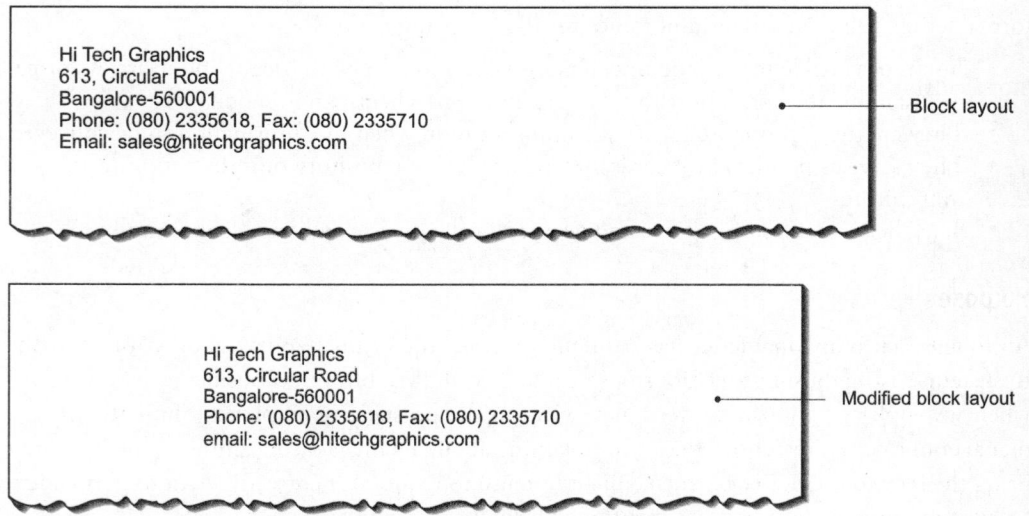

Dateline　This component refers to the date on which the letter was written. It includes the date, month, and year. It can be represented in either of the following two ways:

3 February 2022 or February 3, 2022

The former represents the British style and the latter is in American style. Some companies may use their in-house style. However, the British style is gradually giving way to the American style throughout the corporate world. For international correspondence, the accepted format for the recipient's region has to be checked first. Ordinals (e.g., 3rd, 7th, etc.) are normally not used to indicate datelines. The month should preferably be spelt out to avoid confusion with the date.

Inside address   This part of the letter identifies the recipient of the letter and is separated from the date by at least one blank line. The amount of space separating the inside address from the date may also be adjusted to suit the length of the letter. A courtesy title should precede the recipient's name. Table C1 in the Online Resource Centre shows the appropriate titles that can be used.

It is always best to address the letter to a specific person. The name of the person can usually be obtained by checking the organization's website or by telephoning the organization. If we do not know the name of a specific person, it is acceptable to address the letter to the department or to a job title.

Immediately following the addressee's name and title, separate lines should contain the name of the company, the street address or post office box number, and the city and state or province with proper postal code or zip code. When writing internationally, the addressee's country should follow in capital letters and occupy the last line of the inside address. Here are a few samples of inside addresses:

Ms Christina George	*The General Manager*
Corporate Communications Inc.	*Lion Industries Limited*
3456 Grant	*No. 3-A, East Patel Nagar*
Chicago, IL 60611	*Manasarovar*
USA	*Mumbai-400001*
Messers Lal Chand and Sons	*Professor (Ms) Gayathri Devi*
56, Nehru Marg	*Department of Chemistry*
Greater Kailash	*Indian Institute of Pharmacy*
New Delhi-110002	*Patel Colony*
	Secunderabad-500017

Salutation   Always try to address the letter to a person by name rather than by title. If a specific name is not available, a salutation such as the following may be chosen:

Dear Committee Member	Dear Students
Dear Colleagues	To All Sales Reps
To Whom It May Concern	Dear Sir or Madam
Dear Madam or Sir	*Dear Purchasing Agent*

If your letter is addressed to a company with men and women employees, the proper salutation is:

Gentlemen and Ladies	*Ladies and Gentlemen*

If your letter is addressed to an organization of only women or only men, use:

Dear Sirs	*Dear Ladies*
Dear Mesdames	*Gentlemen*
Ladies	

If we do not know the gender of the person to whom we are writing, one of the following representations may be used:

Dear Mallika Pande	*Dear M. Pande*

The way we address a recipient is governed by our relationship with that person. Whether we use the first or the last name and whether we drop the Dear makes the letter either more formal or less formal:

Dear Mr Gupta: *(most formal)*

Dear Ashok:

Ashok,   *(very informal)*

Current day corporate letters/emails use 'Hi' or 'Hello' to address people. Use of the first name to address someone is a common practice in the US and UK. How a person signs off is usually a safe indicator of how to address them. Surnames are seldom used. If a person signs off a letter as 'Thanks and regards, Sir Jeffrey Gallus', we should be addressing that person as 'Dear Sir Jeffrey Gallus'. However, if the same person signs off as Jeff, 'Hi Jeff' would be a more appropriate way of addressing him.

Message   This is the main content of the letter and usually occupies the maximum amount of space. It should be single-spaced, with a blank line separating it from the preceding and the following parts of the letter. In addition, each paragraph of the message should also be separated by a blank line.

The main text of the letter can be formatted to make it convenient for the reader to gather information quickly and easily. For example, bulleted lists, italics, and bold fonts can be used to organize the content and emphasize where required. However, while doing so one must adhere to the style of writing practised within the organization.

Complimentary close   This element is a single word or phrase, separated from the message by a blank line. Here is a list of expressions that can be used for complimentary closing:

Sincerely, Cordially, Truly, Faithfully, Faithfully yours, Cordially yours, Truly yours, Sincerely yours, Yours sincerely, Yours truly, Yours obediently, Yours faithfully

Of these expressions, the single-word expressions are more contemporary and may suit almost any formal letter. The choice of close provides a way to create just the desired tone. Closings such as, *Sincerely yours, Truly yours, Very truly yours*, etc. seem outdated. Sincerely and Cordially are widely accepted closings. *Thanks and regards, Warm regards, Best regards, Cheers!, Best*, etc. are the more common closings used currently.

Signature block   The complimentary closing line is followed by the signature block, which includes the writer's signature, name, and title. Every letter must end with a signature to give authenticity to the information contained in it. An unsigned letter is of no consequence. The signature block is placed four lines below the complimentary close. It includes the sender's name and title. If your name might leave the reader in doubt about your gender, you may include a title in the signature block as shown in the sample given below:

Cordially,

Ms Santosh Singh

Senior Executive

The letterhead indicates that the writer represents the organization. However, if the letter is on plain paper or runs onto a second page, we should emphasize that we are writing legally for the company. The accepted way of doing so is to place the company's name in capital letters a double space below the complimentary close and include the sender's name and title four lines below that.

Sincerely,

SHAREWELL INDUSTRIES

Mr Atul Chauhan
President

If an organization has delegated the authority of signing letters to an executive by the Power of Attorney, that executive will add per pro. or pp. (indicating that the letter is signed on someone else's behalf) just before the name of the organization and sign below it as follows:

Cordially,
Per Pro. Sharewell Industries
Lakshmi Deshpande

Additional elements

Formal letters differ considerably from each other in their subject matter, the identifying information they need (such as addressee notation, attention line, subject line, or reference line), and also the format they adopt. The following elements may be used in any combination, depending upon the requirements of the particular letter, but generally in the order given here:

- Addressee notation
- Subject line
- Enclosure notation
- Mailing notation

- Attention line
- Reference initials
- Copy notation
- Postscript

Addressee notation　This notation generally appears a double space above the inside address, in all capital letters. PERSONAL, CONFIDENTIAL, PLEASE FORWARD, and THROUGH PROPER CHANNEL are examples of such notations that are used in letters that have a restricted readership or that must be handled in a special way.

Attention line　An attention line is used when the inside address does not include the name of an individual. It can be used to draw the attention of a particular person or a particular department in an organization so as to ensure a quick and prompt action in response to the letter. The attention line may be placed two spaces below the inside address. It is generally given in bold as well as capital letters, and is included in the following manner:

ATTENTION: DR SATISH YADAV, PRODUCTION UNIT
ATTENTION: PERSONNEL MANAGER

Subject line　This element lets the recipient know at a glance what the letter is about; it also indicates where to file the letter for future reference. It usually appears below the salutation. But sometimes it is placed above the salutation and below the attention line:

Dear Mr Gupta:
SUBJECT: INFORMATION REGARDING LAST WEEK'S INSPECTION
or
ATTENTION: PERSONNEL MANAGER
SUBJECT: INFORMATION REGARDING LAST WEEK'S INSPECTION

Dear Mr Gupta:

The subject line is also generally given in bold as well as capital letters.

Reference initials   Often, one person may dictate or write the letter and another may produce it. On such occasions, reference initials are included to show who helped prepare the letter. Reference initials appear two spaces below the last line of the signature block. While the writer's name appears in the signature block, only the initials of the preparer are necessary. If only the department's name appears in the signature block, both sets of initials should appear, usually in one of the following forms:

Ksm/rk, Ksm:rk, KSM:RK

The first set of initials is the writer's and the second set is the helper's. At times, the letter may be written, signed, and prepared by different persons. In such cases, at least the file copy of a letter should bear all the three sets of initials (KSM/AS/rk: signer, writer, preparer). When people key in their own letters, reference initials are not included. With the increased use of electronic mails, the use of reference initials has become obsolete.

Reference line   Formal letters often carry a reference line, which is used for sequential correspondence with the recipient. The reference line consists of an alphanumeric reference number, which uniquely identifies the letter. An official reply to such a letter usually quotes this reference number, as follows: 'With reference to your letter, Ref. no. ABCD/03/07, …' or 'Further to your letter, Ref. no. ABCD/03/07, dated 10 January 2022…'. Reference numbers are also used to keep a record of letters sent or received. There can be several methods of inserting reference numbers in formal letters. Some examples are as follows:

By serial number: This method uses a running number for all letters generated.

By department and serial number: This method uses the originating department's initials followed by a serial number.

By project ID and serial number: This method includes the project ID, which could be initials unique to the name of the project, followed by the date of the letter or the serial number.

There are no standard methods for inserting reference numbers except probably in government letters. In other companies, the method to be used is normally decided internally by the concerned department. Generally, the reference line appears below the dateline.

Enclosure notation   This notation appears at the bottom of the letter, one or two lines below the reference initials. Some common forms are:

Enclosure: Draft of proposal

Encl.: Draft of proposal

Enclosures: 1. Report (10 pages)

2. Photographs (2)

3. List of participants

Copy notation   This is an optional component. It indicates who is receiving a courtesy copy (cc). Some companies indicate copies made on a photocopier (pc), or they simply use copy (c). Recipients are listed in the order of rank if they hold different ranks or in alphabetical order if they hold equal ranks. This part follows reference initials or enclosure notations:

Cc: Charles Mathew

Pc: Leela Sampson

Copy to Ben Adams

C: Rahul Bhatia

In addition to the name of an individual, copy notation may include any combination of that person's courtesy title, position, department, company, and complete address, along with notations about any enclosures being sent with the copies.

Cc: Charles Mathew, with the list of absentees and with a request to look into the matter.

At times, copies are sent to benefit readers other than the person who receives the original letter or courtesy copy, without the knowledge of these recipients. In that case, the notation *bc*, *bcc*, or *bpc* (*blind copy*, *blind courtesy copy*, or *blind photocopy*) is placed where the copy notation would normally appear, but only on the blind copy, not on the original.

Mailing notation   This is placed either at the bottom of the letter after reference initials or enclosure notations, or at the top of the letter above the inside address on the left-hand side. Mailing notations such as BY REGISTERED POST, BY COURIER, BY SPEED POST, etc. will generally appear in capital letters to catch the attention. In addition, the same notation will also appear on the envelope.

Postscript   Letters may also bear postscripts, i.e., afterthoughts to the letter, to the messages that require emphasis, or personal notes. A postscript is usually the last item on any letter and may be preceded by *P.S.*, *PS*, or nothing at all. It can also be shown as a second afterthought with the notation *P.P.S.*, meaning *post postscript*.

As far as possible, try to avoid using postscripts as they convey an impression of poor planning. However, they can be used in sales letters, not as an afterthought but as a punch line to remind the reader of a benefit of taking advantage of the offer. Exhibit 15.1 shows both standard and additional elements.

Layout

Suitable and correct layout enhances the overall effectiveness of any letter. The layout helps to arrange all the elements of a formal letter in an organized manner. Although the basic parts of a letter have remained the same for centuries, the layouts have changed to quite an extent. Sometimes an organization adopts a certain format as its policy; sometimes the individual letter writer is allowed to choose the format most appropriate for a given letter or to settle on a personal preference. In general, two major letter layouts are widely used.

- Block layout (complete block layout)
- Modified block layout

Business or official letters use either mixed or open punctuation. In the former style, which is most popularly used, a colon is used with the salutation (informal letters use a comma here) and a comma is used with the complimentary close. In the latter style, no punctuation is used with the salutation and the complimentary close. Some organizations prefer to omit both the salutation and complimentary close. However, this format is rarely used.

Block layout

The block layout, also known as the complete block layout (Exhibit 15.2), is extremely popular as it makes the letter look attractive, elegant, and efficient. The main characteristic of this layout is that all elements except the letterhead heading are aligned to the left margin.

Modified block layout

The modified block format (Exhibit 15.3) differs from the block format in the positioning of certain elements: the heading is centre aligned whereas the dateline, complimentary close, and signature block are right-aligned.

Although organizations seem to prefer the full block format, the modified block is also acceptable. The modified block's appearance is often considered to be more balanced and traditional.

EXHIBIT 15.1 Formal letter with standard and additional elements

Residents' Welfare Association
Nehru Colony, Jawahar Nagar, Coimbatore

Date ——————→ January 12, 2022

Mailing Notation ———→ **By Speed Post**
Address Notation ———→ Personal
Inside address ————→ Water Works Department
Moorthy Complex
Bapu Nagar
Coimbatore

Attention line ————→ **ATTENTION**: Mr Rohan Kumar, In Charge, Water Supply
Salutation ————→ Dear Mr Rohan Kumar,

Subject line ————→ **SUBJECT**: NO WATER SUPPLY ON 5 JANUARY 2022
I am extremely sorry to inform you that our colony did not have any water supply on the 5th of this month and we had to face acute water shortage because of this

I contacted your office on 5th January at 10 a.m. and the person at the receiving end informed us that the water supply has been stopped for certain areas because of some problems in pumping. But he ensured that the supply would resume by 5 p.m. To our dismay, we could not receive water on that day. The complaint letter signed by all the residents of this colony is enclosed. Please look into the matter and see to it that such lacuna does not occur at least in future.

Complimentary close ——→ Regards
Sincerely,

Signature block ————→ *Manokaran*

Manokaran, President
Reference Initials ————→ gk

Enclosure line ————→ Enclosure (1)
Copy line ————→ Copy: Chairman, Water Board
Post script ————→ PS: Please instruct your office to give prior intimation in case of not supplying water on a particular day.

345, Nehru Colony, Jawahar Nagar, Coimbatore -641011
Phone: 9817580324

EXHIBIT 15.2   Block layout

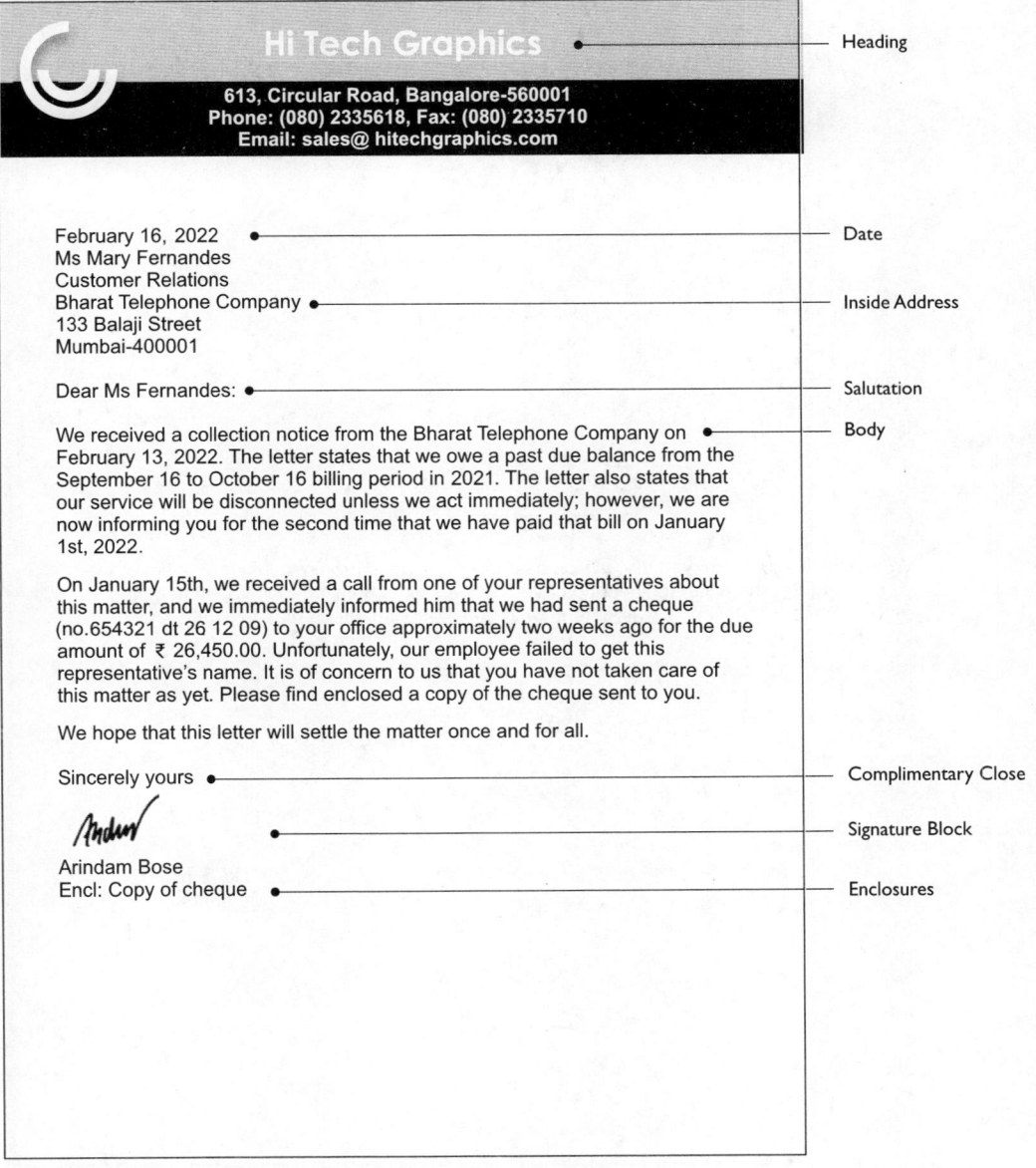

Hi Tech Graphics •——————————— Heading

613, Circular Road, Bangalore-560001
Phone: (080) 2335618, Fax: (080) 2335710
Email: sales@ hitechgraphics.com

February 16, 2022 •——————————————— Date
Ms Mary Fernandes
Customer Relations
Bharat Telephone Company •————————————— Inside Address
133 Balaji Street
Mumbai-400001

Dear Ms Fernandes: •——————————————— Salutation

We received a collection notice from the Bharat Telephone Company on •———— Body
February 13, 2022. The letter states that we owe a past due balance from the
September 16 to October 16 billing period in 2021. The letter also states that
our service will be disconnected unless we act immediately; however, we are
now informing you for the second time that we have paid that bill on January
1st, 2022.

On January 15th, we received a call from one of your representatives about
this matter, and we immediately informed him that we had sent a cheque
(no.654321 dt 26 12 09) to your office approximately two weeks ago for the due
amount of ₹ 26,450.00. Unfortunately, our employee failed to get this
representative's name. It is of concern to us that you have not taken care of
this matter as yet. Please find enclosed a copy of the cheque sent to you.

We hope that this letter will settle the matter once and for all.

Sincerely yours •——————————————— Complimentary Close

Arindam

•——————————————————— Signature Block

Arindam Bose
Encl: Copy of cheque •————————————— Enclosures

EXHIBIT 15.3   Modified block layout

International Association for Teachers of Chemical Engineering (IATCE)

August 20, 2021

Ref: Your letter M/124/VC dated 25 July 2021

The Vice Chancellor
Nehru University
262, Lake View
Bangalore 560 012

Dear Prof. Pathak,

SUB: ACCEPTANCE OF YOUR NOMINATIONS

We would like to inform you that IATCE has accepted the following faculty members of the Department of Chemical Engineering of your university as nominated in your letter cited to represent the Energy and Environment Wing of IATCE with effect from 1 November 2021:
Prof. Rajendra Prasad
Prof. Amit Deshpande

Please find attached their membership cards.

IATCE looks forward to their active participation in all endeavours related to its Energy and Environment Wing.

Sincerely yours,

Vicky

Vicky Sowerby
Membership Officer
Encl: Two cards

14589 FM 1878 W. Pmb # 140, Houston, TX 86075
Ph: 95874621357

Principles

Read the following message contained in a business letter.

> Dear Sir,
>
> It is so unfortunate for me to write this complaint letter to you. I wonder why you had been so careless in sending me five sets of the books on 'Technical Communication' by Raghav Mimani. Don't you know I had asked for 'Business Communication' by the same author? Are you out of your senses? Don't you know I have my comprehensive exam of business communication tomorrow? What if I fail? You will be responsible for it. I am fuming with anger. I will never ask you for anything in future if you keep doing this. Just check out your records and give my money back immediately.
>
> Your angry customer,
> Shantanu

How do you find this letter? If you are the recipient of this letter, how would you receive it? Will you react positively to the complaint? Most probably, you will not feel like responding. Even though you will respond to the complaint, you may also decide to end your relationship with this customer.

As we know, formal letters are written mainly to create, establish, or sustain business relationships. A good relationship is based on respect and courtesy. Hence, when trying to establish good relationships through formal messages, remember the following principles of formal letter writing:

> In the midst of joy do not promise to give a man anything, in the midst of great anger do not answer a man's letter.

- Use the 'you' attitude
- Be clear and concise
- Be correct and complete
- Emphasize the positive
- Be courteous and considerate

'You' attitude

Focus on the recipient's needs, purposes, or interests instead of our own. Even if we have to talk about ourselves in a formal letter a great deal, we should do so in a way that relates our concerns to those of the recipient. This recipient-oriented style is often called the 'you' attitude. It does not mean using the word 'you' more often, but making the recipient the main focus of the letter. Even an unpleasant situation can be changed by using the 'you' attitude.

At the simplest level, one can adopt the 'you' attitude by replacing the terms that refer to oneself and the company with terms that refer to the audience. In other words, use more of you and yours instead of I, me, mine, we, us, our, and ours. Table 15.1 gives some examples.

Table 15.1   Example of sentences with 'you' attitude

We/matter-of-fact attitude	'You' attitude
We are glad we can now send our smartwatches.	You would be glad to know that your smartwatches are ready for sale.
We are happy to receive your request for the automatic locks.	Thank you for your request for the automatic locks.
We regret that the goods did not reach the buyers in time.	We regret that you could not receive your goods in time.
We are pleased to announce our new insurance scheme.	Now you can avail our new insurance scheme.
We offer the printer cartridges in three colours: Black, red, and blue.	You may choose your printer cartridges from three colours: black, red, and blue.

However, there are always exceptions to rules. On some occasions, it may be better to avoid using you. For instance, when someone makes a mistake, we may want to minimize ill feeling by referring to the mistake impersonally rather than pointing it out directly. 'We have a problem' would work better than 'you caused a problem'. Look at the examples given in Table 15.2. The more impersonal, matter-of-fact tone shows greater sensitivity and avoids creating unnecessary hard feelings in the recipient.

Table 15.2 Examples of when not to use 'you' attitude

'You' attitude	Impersonal attitude
You should never use this type of paper for photocopiers.	This type of paper does not work well in photocopiers.
You have not connected the wires properly.	The wires have not been connected properly.
You have not yet sent copies of the 2021–2022 Bulletin.	Copies of the 2021–2022 Bulletin are yet to reach us.
You are not supposed to keep the lights switched on when you leave your hostel room.	Lights should be switched off when students leave their hostel rooms.
Your indifferent attitude has caused this problem.	A little more care could have avoided this problem.
You failed to respond to the letter on time.	The letter was not responded to on time.

Although the *you* attitude brings in effectiveness in letters, some organizations may have a tradition of avoiding references to *you* and *I* in their technical or formal documents. Instead they may follow a unique style and restrict the use of personal pronouns in these documents.

Clarity and conciseness

Read the following two versions of the same message included in a formal letter and see which one conveys the information clearly and succinctly:

Version 1:   I am extremely sorry to have to point out to you that we do not have these brands in stock at the present moment of time.

Version 2:   These brands are presently out of stock.

The second version is more direct and concise not only because of the use of fewer words but also due to the absence of redundant expressions that camouflage the meaning.

Conciseness means saying what needs to be said in as few words as possible—a required skill in technical communication. While many writers can get their meanings across to readers, only skilled ones can do so effectively in a few words.

The message must be clear and specific. A cluttered sentence might not only make the message difficult to understand but also be interpreted in more than one way. By being unambiguous, we can achieve conciseness as well as clarity. Lack of clarity and conciseness is often because of the following reasons:

- Long, involved sentences
- Sentences revealing over-enthusiasm
- Verbosity or wordiness
- Redundancy or use of low information content (LIC) words

 Please refer to the Online Resource Centre for examples of how to avoid these pitfalls. Further, Table 15.3 gives examples of some common phrases that can easily be replaced with short, more concise words or descriptions.

Table 15.3 Concise versions of long phrases

Phrase	Concise version		
along the lines of	like	in order to	to
as a matter of fact	in fact	in spite of the fact that	although/ though
at all times	always	in the event that	if
at the present time	now/currently	in the final analysis	finally
at this point in time	now/currently	in the nature of	like
because of the fact that	because/since	in the neighbourhood of	about
by means of	by	until such time as	until
due to the fact that	because	in the near future	soon
for the purpose of	for	in accordance with	according to
have the ability to	can/be able to	enclosed please find	enclosed is

Correctness and Completeness

If an enquiry to a company about the mode of payment for equipment purchased from them receives a reply such as this: 'Send your payment by demand draft as soon as possible', the customer will obviously be annoyed because the message

- does not give the complete details about the payment, i.e., the name on which the draft has to be taken and also whether it should be account payee
- does not specify the deadline for sending the draft.

Even if the formal letters are concise and clear, unless they are specific and complete in all respects, they may not fulfil the desired objective.

The term correctness refers to accuracy or precision, and the term completeness refers to thoroughness or giving all the required details. To be correct and complete, one has to understand the purpose of the letter and convey it clearly. These qualities can be achieved in letter writing by following certain guidelines:

- Use evaluative and factual words/phrases rather than abstract and general expressions.
- Use unambiguous words/phrases.
- Proofread the message for accuracy of spelling and grammar before sending it.
- Check whether all queries have been answered and all details provided.

In Table 15.4 given below, the italicized words in the original version refer to abstract/general, ambiguous, and grammatically incorrect words/phrases. They are changed to factual, unambiguous, and grammatically correct ones in the revised version.

Table 15.4 Examples of sentences revised for correctness and completeness

Original	Revised
We need a *large* office space.	We need at least *10,000 square feet*.
A number of customers filed their returns today.	*Ten* customers filed their returns today.
On July 1, the company will *close its doors*.	On July 1, the company will *close down*.
This *antiquated merchandising* strategy is ineffectual in contemporary business operations.	This *old sales* strategy will not work in today's business.
If you have any questions, please *feel free to* contact us.	If you have any questions, please *contact* us.
We *are submitting* a cheque *for the amount of* ₹ 20,000.	We *are sending* a cheque *for* ₹ 20,000.

The examples given in Table 15.5 show how abstract phrases can be substituted with more factual ones in order to make writing more correct and concrete.

Table 15.5 Examples of factual phrases replacing abstract ones

Abstract	Factual/evaluative
leading company	first among 2000 companies
substantial discounts	discount of 20%
light in weight	feather light/lightweight
as soon as possible	latest by/on or before 2nd August
sold a good number of books	sold 25,300 books
publications	books/magazines/manuals/bulletins/brochures
huge area	7000 square feet
many storeys	four floors
the majority	65%
excellent attendance record	100% attendance record

Positive approach

A message can be framed in several ways. Much of the difference lies in the meaning of the words as meant by the writer and understood by the reader. Consider, for example, the following words of Ernest Hemingway, a renowned American novelist:

> Never think that war, no matter how necessary nor how justified, is not a crime.

If you need to understand the author's opinion of war, you must shift the negative terms never and not to positive. In doing so, if you omit one, you may arrive at the intended meaning which is:

> All war is crime.

Positive words are always best to achieve your goal. Most information, even bad news, conveyed through official letters has some redeeming feature. However, the audience should be made aware of this feature. Emphasize the positive side of the message rather than the negative side. Words that reflect a positive attitude are more acceptable to readers. In addition, the positive side of a message will also help in offering criticism or advice without offending the receiver. Compare the differing results obtained from the negative and positive versions of each of the messages given in Table 15.6.

Table 15.6 Negative and positive versions of messages

Negative	Positive
We *never* exchange damaged goods.	You *may* exchange the purchased goods provided they are in good condition.
We regret to inform you that we *may not* be able to grant you request for credit.	For the time being we *can* serve you on cash basis only.
It is *impossible* to repair your car in two days.	Your car *will be ready* by Thursday.
We *cannot* send you the gowns until you tell us what colour and size you want.	*Please check* your size and colour preferences in the enclosed order form and send to us so that you receive the gowns.
You should have known that the contact lenses *cannot* be washed with tissue, for it is clearly mentioned in the instructions.	The instructions explain why the contact lenses *should* be washed only with the particular solution.

Also avoid words with negative connotations, i.e., try to frame messages with words that will not offend or hurt the readers. Replace words with the unpleasant connotations with milder terms or euphemisms (see Table 15.7).

Table 15.7 Mild replacements for offending words/phrases

Unpleasant words/phrases	Mild words/phrases
elderly	senior citizen
second-hand cars	resale cars
cheap goods	inexpensive goods
toilet paper	bathroom tissue
high-calorie food	high-energy food
two of them died....	two of them breathed their last...
you're being fired	you're being let go
global warming	global climate change
servant	domestic help
dull	slow learner
poor people	lower income group
What's the problem?	What's the issue?
This contains artificial flavour.	This contains 95% natural flavour.
disabled	differently abled
bizarre	unusual

Courtesy and consideration

Given below is a message received by a candidate a week after submitting an application for a job:

Your application for the post of Purchase Manager cannot be considered by our company.

Such curt and discourteous responses are not very pleasant to the receiver. It is not possible for the company to employ every applicant, but being courteous while rejecting an applicant would convey the message more amicably, and sometimes even help retain the opportunity to offer employment to a rejected candidate at a later point, if so desired. The same message framed as mentioned below would have been more positive in its approach, leaving the lines of communication open.

Thank you very much for your application dated _____ for the post of _____ in our company.

We are sorry that we are not able to offer you this position at present. However, we have noted down your credentials and have filed your application for future use. Whenever an opportunity arises in future, we shall consider your application.

We appreciate your interest in working with us. Our best wishes to you.

Being polite to the readers helps earn their respect and sustain the relationship with them. A courteously written correspondence shows not only the writer's friendly concern but also consideration for the reader's needs and feelings. This helps forge a stronger bond between the two individuals.

> 'Courtesy is like oil; it reduces friction.'

When we meet somebody face to face, courtesy and consideration can be expressed through non-verbal means. Similarly, the following guidelines will help convey courteousness and consideration to the recipient while writing a letter:

- Use a conversational tone (be natural)
- Avoid dogmatism (do not preach)
- Avoid anger (be patient)

Table 15.8 gives an idea of how courtesy and consideration bring in a positive and desirable change in the style of an official letter. Sometimes, we may have to offer more explanation while being courteous, and in this process we may end up with more words. However, although being concise is important, cutting down on words is not the main objective in writing a letter.

Table 15.8 Examples of courteous and considerate writing

Original	Revised
Your indifferent attitude has caused a great loss.	Had you been a bit more careful, we could have avoided this 20% loss.
We cannot grant you the leave you have asked for in your application of 25 July 2014.	As we have to complete the project by the end of this month, you may take leave after 30 July 2014. But sorry, at present it may not be possible for us to grant you leave.
If you do not respond to this letter also, we do not have any option other than lodging legal proceedings against your company.	We hope that you will certainly respond to this letter at least so that we can avoid taking any legal action against your company.

The Online Resource Centre provides more phrases that express courtesy and consideration in formal writing.

Planning a Letter

Writing is serious business as it creates permanent records. As effective letters can accomplish their objectives effortlessly, they need to be properly planned and prepared. We need to determine the structure of the letter before we start writing. It is important to remember that by taking time to plan our letters, we save time in writing and proofreading them. The meticulous planning of formal letters involves the following four steps:

- Being clear with the purpose
- Knowing the reader
- Understanding the message to be included
- Deciding whether a response is required

Clarity of purpose

We need to be clear in our mind about the main goal to be accomplished through our letters. Various purposes for which formal letters are written have been discussed earlier. For instance, if we write a letter to tell potential customers about our product, our purpose will not only be to inform them but also to persuade them to buy our product.

Audience awareness

Before writing a letter we should ask ourselves, 'Who am I writing this letter to?' Are we writing it to a person known to us or to a new person? This kind of question will enable us to decide the degree of formality of our letters. Identifying our readers is an important step in our planning process.

Deciding the content

Depending on the purpose and the audience, we can decide the details to be included in the letter. As already mentioned earlier, the seven C's of effective letter-writing should be kept in mind while planning the content of our letters. For instance, when we write a letter to an author to appreciate her book, we may need to start with a brief introduction of ourselves and then move on to congratulate the author for writing an excellent book. Then few salient points of the book that had interested us can be mentioned. Finally, we can conclude the letter by congratulating the author once again.

Understanding the need for response

While planning our letters, we should be clear whether we need a reply from our readers and if so, we should ask: What kind of reply is it? Is it an action to be taken, a policy to be implemented, an instruction to be followed or some details to be sent? These questions will enable us to decide the supporting details to be included in our letters.

Business Letters

Business communications, to a large extent, depend on well-written letters. We become part of an organization by sending to them our job application along with résumé. Once selected, we write them an acceptance letter. After becoming an integral part of the organization, many business letters such as instruction, complaint, sales, tender, and collection letters are written on a daily basis. Some of the important business letters are discussed in this section along with samples.

Depending on their purpose, business letters can be classified into various types, as shown in Table 15.9. The table also lists the purposes of these letters. Among these, order, enquiry and claim, and adjustment letters are mostly informative. They either ask for information or provide information for further action. On the other hand, credit, collection, sales, fundraising, and job application types of letters are persuasive, as they attempt to persuade or motivate readers toward a desired action. In the following sections, a few routine types of business letters are discussed.

Table 15.9 Business letter types and their objectives

Types	Objectives
Credit	To enquire or request for loan
Collection	To collect past due accounts
Enquiry	To enquire the status of something; price lists; catalogues
Order placement	To place an order for products or services
Claim	To lodge a complaint and ask for a remedy
Adjustment	To inform how the complaint would be taken care of
Sales	To sell a product or service
Fundraising	To request the readers to donate money or time
Job application	To apply for an employment
Covering letter for job application	To introduce the applicant
Thank you/follow-up	To express courtesy/get updated information
Acceptance/rejection	To take up/refuse an offer or assignment
Resignation	To give up a job without creating an unpleasant situation
Persuasive	To convince someone

Credit Letters

Credit means that we pay later for what we buy now. Goods are sold on credit to increase profit, as it encourages people to buy more. It does have a few disadvantages such as an increased workload in business in order to keep an extra record of collecting payments. It is also risky as the debts can increase inordinately. However, it also helps a retailer to stock goods and make the payment after the sale is over. This acts as an impetus to business. The inconveniences of separate payments on each purchase are also reduced. The different types of credit letters include (a) requests for credit, (b) status enquiries, and (c) replies to status enquiries.

Letters requesting for credit

There are two types of letters in this category—request for enhancing credit limit and request for enhancing the credit period. Exhibit 15.4 is a sample of a letter written to seek credit limit enhancement.

Refer to the Online Resource Centre for a sample of a letter requesting enhancement of credit period.

Status enquiry letters

When clients apply for credit, the finance company needs to check upon their credit worthiness to ensure that it does not incur losses owing to bad debts (credit that is not paid back). While making enquiries about the financial status of any individual, a polite tone, but formal in approach, must be used, keeping the following points in mind:

- Seek general information about the applicant
- Assure that the information provided will be kept confidential
- Get the employer's opinion on the application for the grant of credit within the requested limit.

Some companies use a proforma to make status enquiries, instead of writing letters. A self-addressed, postage paid envelope is usually enclosed to encourage timely replies. When the requested information is received, ensure to acknowledge by writing a suitable letter. Letters should be marked confidential and addressed to the same person who has replied or to any other senior officer in the same department (see Exhibit 15.5).

Replies to status enquiries

When the reply to the status enquiry about the credit of an individual or firm is positive, the reply is simple and faces no difficulty in writing. However, if the case is uncertain, care needs to be taken while responding. The reply should convey the message without being rude. Mark the letter confidential and keep in mind the following points (see Exhibits 15.6 and C2 in the Online Resource Centre):

- Provide an honest opinion about the facts
- Reflect that the information supplied by you may be found useful
- Remind that the information provided by you must be kept confidential and you are not responsible for distortions

Collection Letters

Collection letters can be written by either the individual or the organization when goods are taken on credit, and the payment due is to be collected (see Exhibits 15.7 and 15.8). Sometimes companies offer products on credit basis to create goodwill among their clients. Collection letters are written to communicate with customers for collecting payment. The letters act as a reminder for return of the credits that have become overdue. The desired result is an increase in the company revenues. These letters begin as pleasant, friendly reminders, and become increasingly stern and demanding in case of unsatisfactory responses.

EXHIBIT 15.4   Request for enhancing credit limit (block layout)

Fashion Flash

23 September 2021

The Manager
State Bank of India
Anna Nagar Branch
Anna Nagr
Chennai-600045

Dear Sir:

SUB: REQUEST FOR INCREASING CREDIT LIMIT

Fashion Flash has been a loyal customer of your bank since its inception. We carry out all our financial transactions solely through your bank.

As you may be aware, our brand has gained popularity over the years. Our business has expanded from ₹ 40 lakhs per annum to ₹ 3 crore per annum during 2019–2021 thereby increasing our economic activities. Hence, in order to meet this challenge, we request you to kindly extend your credit limit to Fashion Flash from the existing amount of ₹ 30 lakhs to 50 lakhs.

We eagerly await your positive reply.

Yours truly

Prakash Jain
(MD, Fashion Flash)

260, Modern Plaza, Focus Point, Chennai-600031
Phone: (044) 2445577, Fax: (044) 24455791, email:design@fashionflash.com.

EXHIBIT 15.5 Letter seeking status enquiry (modified block layout)

Oracle Global Finance Ltd.

20 October 2021

Manager HRD
Alfred Industries Ltd.
30/2nd Cross, Adyar Road
Chennai-60

Dear Sir,

SUB: CREDIT-WORTHINESS STATUS ENQUIRY

We are a vehicle and white goods global finance company having worldwide headquarters in Washington, DC and India headquarters in Delhi. We have business operations with a financing portfolio of $2000 billion as on 31.3.2021 out of which $100 billion is accounted for by our India operations. We have been established in India since 2018 and enjoy a market share of 10% and 7% in vehicle and white goods financing, respectively.

Mr Lokesh Jain, employed with your organization as Deputy Manager, has approached us for a vehicle finance of ₹ 10 lakhs. It would be highly appreciated if the following information is made available to us to assess his credit-worthiness and take a decision on his application. In addition, kindly inform us whether the application can be considered.

Please be assured that the information shared with us shall be kept confidential and will be used only for the purpose of processing his loan application.

(a) His present emoluments and take-home salary
(b) Designation
(c) Date of joining
(d) Present age and remaining service period with you
(e) His conduct as an employee
(f) Outstanding loan to be paid to you, if any; Date of receiving the loan and the deadline to repay the same
(g) Mode of loan recovery: Can it be from the salary if you receive a proper mandate from the applicant?

We assure you again that the information provided by you about your employee shall be kept confidential and used exclusively for the stated purpose. An early reply from your office in this regard will help us to act upon his application quickly.

Thank you and regards.
Yours faithfully,

Manager (Retail Credit)

345, Indraprastha Enclave, Pandu Nagar, New Delhi-110023
Ph: 011-25211605

EXHIBIT 15.6   Careful drafting of a reply to a credit status enquiry (modified block layout)

Alfred Industries Ltd.

02 November 2021

Manager (Retail Credit) Oracle Global Finance Ltd.
345, Indraprastha Enclave, Pandu Nagar
New Delhi-110023

Dear Sir

SUB: LOAN APPLICATION OF LOKESH JAIN

We thank you for your letter Cs//Y389 dated 20.10.2021 with regard to the captioned subject. In this connection, please be advised that Mr Jain aged about 40 years has been working with us for the last eight years and is designated as Deputy Manager (Engineering).

He is one of our most sincere and reliable employees. His present take home salary is ₹ 30,000 per month and as on date he does not have any outstanding loans with us. The company superannuates its technical employees at the age of 60 years.

As per the policy of the company, no third-party installment payment is entertained even upon the appropriate mandate from the employees, except premiums to be paid towards LIC policies bought by employees.

We hope this information serves your purpose.

Regards
Yours faithfully

Alfred Industries Ltd
(Manager HRD)

30/2nd Cross, Adyar road, Chennai-600050
Ph: 9437216200

EXHIBIT 15.7   Initial collection letter (block layout)

Money on Demand

January 28, 2022

REF: ACCOUNT 2130
Mr Dhyanchand
150 Everest Building,
Carol Bagh, New Delhi

Dear Mr Dhyanchand:

This is a friendly reminder that your account number 2130 has an overdue payment of ₹ 45,000.

We would appreciate your payment of this amount within ten days of receipt of this letter. If this letter and your payment have crossed, we thank you and apologize for any inconvenience caused.

If you want to discuss your account, please call us Monday through Saturday, 9 am to 5 pm at 011-6787657. You can also contact us through email at accountstaus@moneyondeman.com

Sincerely

Sharma

Devesh Sharma
Accounts Receivable

Pusa Road, New Delhi – 110025, Ph: 01125974661
Website: www.moneyondemand.com

EXHIBIT 15.8 Stern collection letter

Money on Demand

February 10, 2022

REF: ACCOUNT 2130
OUR REMINDER DATED January 7, 2022

Mr Dhyanchand
150 Everest Building,
Karol Bagh, New Delhi 110076

Re: Account 2130

Dear Mr Dhyanchand:

This is the final notice regarding your past due payment of Rs. 45,000 linked to your account number 2130.

You have repeatedly ignored our written requests for payment of the above-noted invoice and you have not contacted us with any explanation.

Unless we receive the due payment in full by the end of the business day, February 28, 2022, we will have to take the unpleasant step of turning your account over to a professional collection agency. We urge you to give this matter your full attention now, before it is too late, and send your payment to us immediately.

If you want to discuss your account, please call us Monday through Saturday, 9 am to 5 pm at 011-6787657. You can also contact us through e-mail at accountstatus@moneyondemand.com

Sincerely,

(Devesh Sharma)
Accounts Receivable

Pusa Road, 205 South Patel Nagar, New Delhi 110025
www.moneyondemand.com, Phone: 011-6787657

Letters of Enquiry

An enquiry letter is useful when we need information, advice, names, or directions. However, avoid asking for too much information or for information that could easily be obtained in some other way, for example by a quick search on the Web.

Solicited and unsolicited enquiry letters Solicited letters of enquiry (Exhibit 15.9) are written when a business or agency advertises its products or services. For example, if a software manufacturer advertises a

EXHIBIT 15.9   Solicited letter of enquiry (complete block layout)

Institute of Business Management & Research

15 February 2022

Mr Suresh Mennon
General Manager (Business Development)
LG Electronics Ltd.
Plot No. E- 456, Mohan Industrial Estate
Okhla, Mathura Road
New Delhi- 110038

Dear Mr Mennon

I am writing this letter to find out more about the newly launched LCD projector by your company. We happened to see your product in operation at a recently held international conference at Pune and subsequently read a couple of your advertisements in Times of India, New Delhi.

We have gone through your website and other sources to get detailed information on the product but have not been able to find the exact information we are looking for.

We are a large technical institute operating in Gurgaon with 800 students in various professional courses. To make our classroom lecture delivery effective, we are in the process of installing LCD projectors in each class.

For evaluation and assessment of this product, we would like to seek your help in finding answers to the following questions:
1. What are the limitations of your product?
2. How long it can be used on a continuous basis?
3. Can the colour combination be changed on the spot?
4. What is the life time of the product?

We would appreciate technical and objective answers to these questions, which will help us come to a decision with respect to this product.

I look forward to an early response.

Yours Faithfully

Dr Rameshwar Kumar
Dean – Academics

Sector – 56, Gurgaon
Ph: 0124 – 41424344, Fax: 0124 – 43464749
www.ibmr.edu.in Email: deanacad@ibmr.edu.in

new package developed and it cannot be inspected locally, we can write a solicited letter to that manufacturer asking specific questions. If information on a technical subject is required, an enquiry letter to a company involved in that subject may provide the data. In fact, the company may be able to offer much more help than expected, provided the enquiry letter is written effectively and clearly.

A letter of enquiry is unsolicited (Exhibit 15.10) if the recipient has not prompted the enquiry. For example, if we read an article by an expert, we may have further questions or want more information. We seek help from these people in a slightly different form of enquiry letter. Unsolicited enquiry letters must be constructed more carefully, because recipients of unsolicited letters of enquiry are not ordinarily prepared to handle such enquiries. Exhibits 15.9 and 15.10 are samples of letters of enquiry (solicited and unsolicited).

Another sample of a solicited letter of enquiry is given in the Online Resource Centre.

EXHIBIT 15.10   Unsolicited letter of enquiry (complete block layout)

249 Ashok Bhawan
BITS, Pilani-333031
7 February 2022

The Placement Officer
Lokesh Technology Solutions

12, Barakhambha Road
New Delhi 110 005

Dear Sir

I, Anil Nath, am a 4th year student pursuing BE Mechanical at the Birla Institute of Technology and Science (BITS), Pilani, which is Asia's premier engineering institute.

My areas of interest are Power Plant Engineering, Prime Movers & Fluid Machines, Design of Machine Elements, Production Techniques.

I look forward to working in a research group environment, so that I can contribute actively to the field of my interest. Joining your research group would be immensely beneficial to me. I am confident that it will give a positive synergy to our mutual interests.

I request you to kindly enlighten me about the job openings in Generation Next Technology for a Mechanical Engineering Graduate.

I assure my full commitment and sincerity in handling any task that will be assigned to me. Please find my curriculum vitae enclosed with this letter for your kind reference.

I hope to receive a favourable reply soon.

Thank You

Yours faithfully
Anil Nath
Mobile: 9887706110
Email: anilnath.bits@gmail.com

Enclosure: CV

Practice 1

Write a letter of enquiry for leather purses, bags, and other goods, asking for all details regarding variety, quality, colour, price list, etc. for a shop you are opening in a developing node of your city. Use the modified block form for your letter.

Reply to enquiry letters

Letters of enquiry must be replied to promptly to maintain healthy and pleasant business relationships. If the enquiry is from an established customer, appreciate their effort. If the enquiry is from a prospective customer, let your reply reflect that you are glad to receive the enquiry and express the hope that this new relationship will last long. The reply should address all the points referred to in the letter of enquiry clearly and concisely. The terms of payment must be stated, if applicable, in the reply, so that there is no room for confusion. Keep the reply brief and to the point (see Exhibit 15.11). Some more sample replies to enquiry letters are given in the Online Resource Centre.

Letters of quotation

A quotation is similar to a letter of enquiry or is a promise to supply goods on the given terms. It is not obligatory to buy goods for which a quotation is requested and suppliers never quote the prices for goods that they cannot supply. Therefore the quotation must be reasonable.

While requesting a quotation, the buyer must also enquire about the additional charges of transportation and insurance, as it becomes difficult to include them later. Exhibit 15.12 is an example of a typical quotation letter. Refer to the Online Resource Centre for tender letters, which are another type of quotation letters, with more examples.

Order Placement Letters

An order is a request for something to be made, supplied, or served. Some companies have their own printed forms for placing orders. These forms are pre-numbered for easy reference. The printed details ensure that no important information is missed out. However, small-scale companies do not use printed forms and place orders by writing letters. An order placement letter should be written very clearly and accurately, including the complete description of the goods required, quantities, price, catalogue number, delivery requirements, and the terms of payment as agreed by both the parties. Exhibit 15.13 is a sample of an order placement letter. Refer to the Online Resource Centre for more samples.

Claim Letters

A claim (a formal complaint) or an adjustment request (a claim settlement) is made when a company's product or service is not satisfactory. Although a complaint could be lodged over telephone, we may like to document the complaint, as the written word always has a greater value than the spoken word, and serves as a permanent record for future reference.

While we may be understandably upset or dissatisfied, the person (say the Sales Manager of a firm) who is reading the letter may not have anything to do with the origin of the problem. Hence, a courteous, clear, concise explanation will impress him much more favourably than an abusive, angry letter. Remember that the objectives of writing a claim letter are:

- To bring the mistake/fault to the notice of the supplying company
- To rectify the mistake either by repair or replacement

EXHIBIT 15.11   Reply to enquiry letter

LOKESH TECHNOLOGY SOLUTIONS
Where innovation meets vision

20 February 2022

Mr Anil Nath
249 Ashok Bhawan
BITS, Pilani 333031

Dear Anil:

Thank you for your interest in a career with Generation Next Technology.

To submit an online application and CV or resumé, please follow the simple steps
given below. The LOKESH TECHNOLOGY SOLUTIONS online application process is
our sole method of ensuring a prompt and professional review of your background and
qualifications.
1. Log on to http://www.lts.com/careers/.
2. Select "How do I get a job at LTS?"
3. Select "Apply for a position."

You can apply for the following positions.

Field engineers
Field engineers work on land and sea, in arctic cold, and in desert heat to deliver
services to our customers. While challenging, it is one of the most rewarding and
respected jobs in the industry. Required: bachelor's or master's degree in engineering
or applied science.

Research & Development Scientists and Engineers
Research & Development scientists and engineers create, design, and develop
Schlumberger's industry-leading equipment, technology, and software worldwide.
Required: master's or doctorate degree in engineering or applied science.

Manufacturing, Supply Chain and Logistics Professionals and Engineers
These professionals and engineers specialize in the manufacture and deployment of
our industry-leading equipment and technology around the world. Required: bachelor's
or master's degree in engineering, manufacturing, or supply chain and logistics.
A local recruiter will review your application and contact you as soon as possible to
inform you about any suitable positions currently available within LTS.

Best regards,

(Rajesh Gupta)

The LTS Recruiting Team

12 Barakhamba Road, New Delhi
Phone: 011-23327584, www.lts.com

EXHIBIT 15.12   Quotation letter (complete block layout)

SURBHI GRANITES

3 February 2022
The Chief Distributor
Housing Materials
Ajmer

Dear Sir

This letter is in connection with the telephonic conversation held with Mr. Rushi.
We are expanding our present facilities for block cutting and polishing. The present facilities
are overburdened; so we are looking for adding more capacity to cater to the growing demand
and rising customer expectations.
Please send us the most competitive prices and service terms for the following items.

S.No.	Item	Grade	Quantity	Total
1	Diamond Cutter	7, 8, 15	15	each grade 45
2	Polishing Stones	A, B16, C	30 each grade	90
3	Chain saw	94, 95	1 each	2

I look forward to a prompt reply from your end.

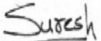

Suresh Jain
(Managing Director)
Surbhi Granites
Corporate office: Opp. Goods Station, Station Road, Jaipur
Ph: 0141-2431586

5 Amba Bari, Jaipur-302001
Ph: 0141-2431586

EXHIBIT 15.13   Order placement letter

 Star Publishers

25 January 2022

Mr Dinesh
UNICORN
36 Chandpole Bazaar
Jaipur 302 001

Dear Sir

This letter is in response to the conversation held with you over phone and subsequent mails. We are opening our office in a prime location of the city. We would like to place an order with you for stationery items for our office. Please ensure the following items are of the highest quality and latest trend:

ITEMS	QUANTITY	CATALOGUE NO.
Camel office	files 100 pieces	52
Steindler pens	25 pieces of each category mentioned	54, 55, 57, 58
Decorative accessories	2 pieces of each category mentioned	59, 62

Payment will be made by DD once we receive the goods at our office.

Yours Truly

Hitesh Jain
HR Manager
Star Publishers

1 Akbar Road, New Delhi 110 002, Ph: 011-43544357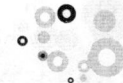

A complaint letter can be written for any of the following reasons:

- Billing errors
- Damaged goods
- Wrong goods
- Wrong quantity
- Unsatisfactory quality
- Goods not matching the sample
- Difference in agreed prices
- Late delivery
- Non-delivery
- Poor service

When we make a complaint, we should remember:

- To write as soon as the mistake is found
- To base the complaint only on facts
- To think patiently and not to assume that the supplier is obviously to be blamed
- To avoid rudeness

Here are some guidelines to remember while writing a claim letter (see Exhibit 15.14).

1. Provide a reference point, namely, consignment number/invoice number, date, and items ordered either in the beginning of the message or in the subject line.
2. Explain the problem clearly and give specific details.
3. Briefly explain the inconvenience/difficult situation being faced because of the problem.
4. Appeal to the supplier's reputation and sense of responsibility and fairness.
5. State clearly as to what action is required or what adjustment is expected from the supplier.
6. Mention the relevant documents being enclosed with the letter and also your availability for receiving the corrected consignment.

Companies usually accept the customers' explanation of what is wrong. So, ethically it is important to be entirely honest while filing claims. Also be ready to support the claim with sales receipts, letters, catalogue descriptions, or invoices. Always send copies of the documents, retaining the originals. Refer to the Online Resource Centre for another sample claim letter.

 Refer to the Online Resource Centre to learn how to write adjustment letters, which are written in reply to claim letters.

Sales Letters

Sales letters are the most cost-effective and time-efficient means of marketing products or services. They are also a form of advertising. However, unlike press and television advertising, which are meant for everybody, sales letters are targeted at selected types of customers. The primary objective of any sales letter is to convert its readers into potential customers.

Before drafting a sales letter (Exhibit 15.15), one must gain a thorough understanding of the product or service. If it is a product, we must be aware of the following details:

- Appearance
- Working
- Price
- Mode of delivery
- Manufacturing
- Packaging
- Discount offers

If it is a service such as an organization offering consultancy, we must understand the following:

- People involved
- Details of jobs undertaken
- Terms and conditions
- Duration
- Changes

Besides having a thorough understanding of the product or service, one must also be aware of the readers' requirements and their need for our product or service. Gather as much information as possible about their status (academic/economic/financial), age, interests, emotional concerns, nationality, and culture.

EXHIBIT 15.14   Claim letter (block layout)

STEELCO FURNITURES

Our Reference: TT/472

Your reference: MA/32

08 March 2022

APEX MATTRESSES LTD
Daryagunj
Vijay Nagar

Dear Sir:

Thank you for promptly delivering 200 mattresses in response to our order no. TT/472 on 3 March 2022. However, we are sorry to state that upon checking of the consignment, we found that 65 of the 200 mattresses sent are badly soiled and faded.

I had placed this order for your high-quality, affordable products, based on the recommendation by one of my friends. We propose to sell these mattresses at a winter sale in our area, five days from now. Advertisements for the same have already been made through the local media.

The mattresses received at our end must have been affected during transportation by water seepage owing to the recent rains. Therefore, I request you to replace these 65 damaged mattresses immediately. I have very little time left to arrange them from elsewhere. The damaged products will be sent back to you within two days at your cost by transport on receipt of the confirmations about replacement.

Kindly send us the replacement within two days of receipt of this letter and oblige. We thank you in advance for your cooperation.

Yours sincerely

Madhurima Gupta
Sales Manager

50, Lenin Sarani, Kolkata-700031
Ph: 044-48975612

EXHIBIT 15.15   Sales letter (complete block layout)

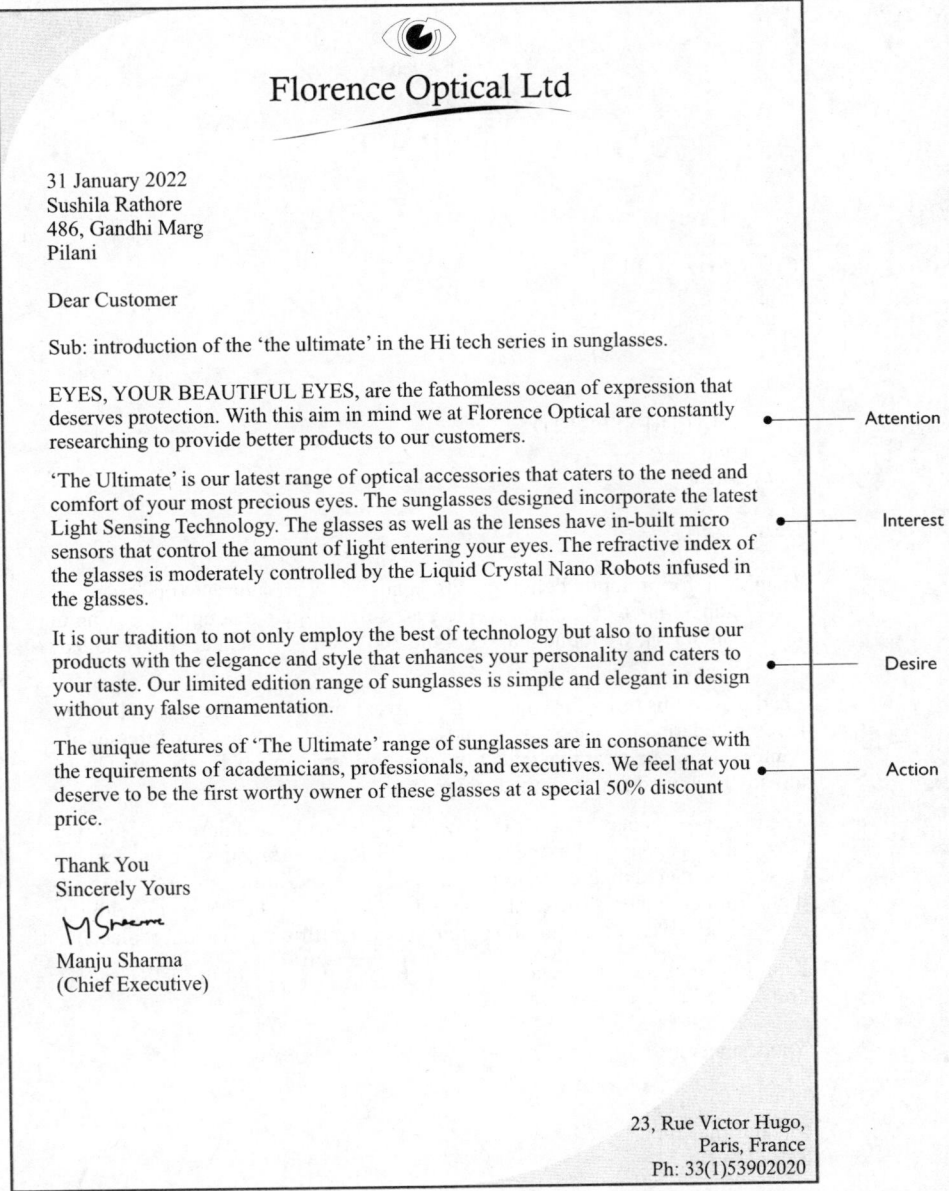

With the product or service and our prospects in mind, we may then start drafting the sales letter. As already mentioned, sales letters are effective when they sound like successful advertisements. In order to make sales letters effective, we should ensure that they accomplish the following objectives, through the AIDA (attention, interest, desire, action) approach:

- Catch the reader's attention (A)
- Arouse the reader's interest (I)
- Create a desire in the reader (D)
- Motivate the reader to action (A)

Accordingly, sales letters can have three or four paragraphs as follows:

1. Attention (introductory paragraph)
2. Interest and desire (discussion paragraph)
3. Action (closing paragraph)

When using the AIDA approach in sales letters, begin the letter with an attention-catching device, generate interest by describing the features of the product, increase the desire for the product by highlighting the benefits that are most appealing to the reader, and close by motivating the reader to act upon the letter.

Catching attention

The introductory paragraph of a sales letter tells the readers why we are writing and what we are writing about. Some unique strategies may help gain the attention of the intended readers of the sales letter. Gaining attention thus would interest them to read further. The following devices can be used to grab the attention of the reader, depending on the product or service involved:

- Question
- Anecdote
- Central selling point
- Quotation
- Statistics
- Appeal

Question or a series of questions   This method will make the reader expect an answer. For example, the following questions can be used by a publisher to lead a reader to become interested in the online resources accompanying a book:

> How can I develop my interpersonal skills? How am I going to perform in my interview? Will my performance satisfy my prospective employer? Do these questions bother you? The online resources accompanying our book has the answers.

Quotation   A quotation gives credibility to the product or service, and hence, readers would like to know what the product is. See the following example aimed at technical writers:

> 'Reading makes a full man; conference a ready man; and writing an exact man'—Francis Bacon
>
> If you want to become exact in your writing approach, use 'Write Exact', our new office communication service.

Anecdote   Begin the letter with a brief dramatic story. The following example illustrates the attention-catching opening of a sales letter written to advertise a photocopier:

> I was so upset over what had happened that morning. As soon as I entered my boss's room, he yelled at me and I was puzzled. But alas! When I saw the photocopies of the survey report in his hand I had no words to say. They were so poorly produced that I could not make out anything. At that same moment I decided to replace the existing office photocopier with a CLEARIMAGE photocopier, which I came across in a client's firm.

Statistics   Provide the statistics of a survey result to the target audience, e.g., a non-government organization persuading people to pledge eye donation:

> Please do not ever think that you are alone. You will be one among those people who have agreed to donate their eyes after their death. A recent survey conducted by 'Drishti' says that among the 1000 people who have signed for this noble cause, 20% are above 60 years of age, 30% are in the age group of 50–60 and the remaining 50% belong to your age group.

Central selling point   Emphasize the main benefit of the product to the reader. See the following example about air conditioners:

> You are sure to like such a gift that gives you a cooler and more comfortable home, free from dust, and also a saving of over 20 per cent in electricity.

Appeal Begin the letter with an appeal to the reader's emotions and values when offering the services of tax consultants.

There is nothing worse than paying taxes when you do not have to do so.

Building interest

After catching the reader's attention by using any one or a combination of the devices discussed earlier, one should move on to build an interest in the reader about the product or service. At this stage, highlight the product's key selling point. For example, if we wish to sell a tourist package offer, mention how relatively inexpensive it is, how many places it would cover, and what concession it gives for children.

Although all these are attractive features, we should focus on that very important feature that most concerns our prospects and that which distinguishes our product. Hence, concentrate on the central selling point and build the audience's interest by highlighting this point. Make it stand out by using additional effects such as using different fonts, some designs, or a unique style of writing. For a better understanding of how to build interest among readers, go through the following extract taken from a sales letter:

Introducing the Sanyo PLC-XU35.XU30.SU30 series of ultra-portable projectors with revolutionary Media Card Imager (MCI) unit. You no longer have to carry along your laptop as well as your projector to make a presentation. You can convert your computer presentations or any screen data into images and save the images onto the MCI flash card with up to 1 GB data storage capacity.

Just set up the projector, slide in the flash card into the MCI unit attached to the projector and project the images just like you would do in a normal presentation.

Despite being compact and lightweight, these new Sanyo projectors have a full range of input/output capabilities to be compatible with all latest computer and video sources taking them to the top of their class. Along with proprietary Digital Realized Video Scaling technology and Digital Realized Interpolation technology, these projectors give video and data images of highest quality. They have such a stylish design that they impress your audience even before you begin your presentation.

Increasing desire

This is an extension of the previous section. Here, the main benefit of using the product is mentioned repeatedly by elaborating and expanding. Use of persuasive words and also some pictures would help in convincing the audience that the product is worth what we are claiming. This main benefit will enthral the audience, motivate them to read further, and lead them towards the last step of the sales letter.

The reader's desire towards the product may be further enhanced by using the 'you' attitude, action terms (strong verbs such as 'supports' instead of 'is designed to support'), colourful verbs and adjectives ('you will be delighted to see your floor sparkle like effulgent diamonds'), and by talking about price. Support the claim by describing how the product has benefited others. It includes testimonials (praise from satisfied customers) or statistics related to the product's performance. Guarantees of exchange or return privileges may be included in this part of the letter, indicating that we have faith in our product.

Try to anticipate every question the audience may like to ask and answer them in the letter. This part of a sales letter can be written effectively as shown in the following excerpt:

With research having shown that students grasp and learn much more when aided by a projector, your classrooms definitely deserve the latest Sanyo projectors to create that extra edge that will translate into a big difference and revolutionize teaching practices in your institute.

Motivating action

The sales letter has now grabbed the attention of the audience, created an interest in them, and increased the desire to buy. Having accomplished these three objectives, the next stage is to motivate the reader to act as soon as possible. Although the ultimate objective is to persuade the reader to buy the product, the letter must at least motivate the reader to ask for a demonstration, a sample, or at least some more information about the product.

The stronger the effort to sell, the stronger will be the drive for action. For example, 'Order your washing machine today—while it is on your mind' might be more effective than 'Please send us your order today.' Before including these words, tell the reader how to go about placing the order. A few samples are given below:

> Fill in your preferences on the enclosed stamped and addressed order form and drop it in the mail box today!
>
> Call the number 011-25525526 right now and we'll deliver your washing machine in no time. Order it now, while you are thinking of Washwell 2010.
>
> *Just visit our website to place your order for…*

You can also persuade readers by informing them about a trial offer, free sample offer, limited time or limited quantity offer, and festival or seasonal offers.

> Mail the enclosed card today and get ready for the festival of lights.
>
> Order today be the first to own this prestigious coffee maker.

The sales letter can be concluded by re-emphasizing the basic appeal or the central selling point. This emphasis would enable the reader to recall the benefits of the product.

Postscript

Although postscript (PS) appears as an afterthought in other letters, it can be used in sales letters as a part of its design. It can be used to emphasize the central appeal, to motivate the reader to act, to suggest that the reader pass along the sales message, or to invite the reader's attention to other enclosures. See the following samples for the effective use of PS.

> PS   *Know someone whose birthday is approaching? Our digital photo frame makes a distinctive gift.*
>
> PS   *Remember! If you feel that Washwell 2021 is not for you, we'll give you every paisa of your money back. We are confident that Washwell 2021 will become one of your household essentials.*

Instruction Letters

Instruction letters serve the purpose of instructing, but in the form of a letter. These letters are used so that the receiver can keep a record of the instruction conveyed. These letters use a forceful and authoritative tone. The clarity of thought that comes through these letters helps in getting the job done efficiently.

While writing these letters, information of similar nature should be presented together. The instructions should be arranged in the order of the most important to the least so that the reader can act accordingly. If the letter conveys only one instruction, then write the purpose of the instruction, giving the details and the action expected. For instance, see the instruction letter given in Exhibit 15.16 written to the instructors for smooth conduct of examinations.

The Online Resource Centre includes another sample instruction letter. Refer to the Online Resource Centre also to learn how to write pessuasive letters.

EXHIBIT 15.16 Sample instruction letter

25/01/2021

Dear Prof. S. Sharma,

All staff members are requested to take care of the following points during comprehensive examination with full alertness.

A. Question Paper Production:-

A1. The question papers must be word-processed.

A2. Be present in the reprography while copies of the question paper are being made.

A3. Collect Original question paper, rota master, etc.

A4. Erase the word processed file of the question paper from the hard disc and keep the soft copy with you in secondary storage form.

A5. Confirm that the Rubber Blanket in the Rota machine is cleaned after the work is over and also collect the Master Copy of the question paper as well as the copies of question paper not properly produced (if any).

A6. Do not bring question papers outside the reprography room without keeping them inside envelopes.

A7. Question paper & other material related to it must be kept under lock & key.

B. Invigilation:

Make the following announcements for students necessarily before the start of the exam :

1 Do not carry your mobile phones along with you. If you have one with you, please keep the same in switched-off mode on the invigilator's table.

2 In case of a closed book test, deposit any unwanted slip / paper or any other materials, which may be of direct / indirect help in the exam.

3 Please keep your id cards on your desk during the exam.

4 Do not involve yourself in any kind of direct / indirect copying or cheating attempt.

B1. Check the identity of the student by his/ her identity card/ library card.

B2. Take attendance on the prescribed form and tally the total by counting the students present. Draw a rough sketch showing how the attendance was taken.

B3. Do not leave the examination hall unattended at any time.

B4. Be vigilant and keep moving most of the time.

B5. During invigilation, no other work should be done, for example, correction of answer scripts, reading a book etc.

B6. Keep only one door open for exit during the last half an hour. Further, at least one of the invigilators should be present at the exit so that no student leaves the room without submitting the answer book.

B7. Name, ID No and Section No/ Name of the Instructor on the cover page of the answer book(s) must be checked before the student leaves the hall.

B8. A focused effort has been made to keep those people free, who had supplied the necessary information in this regard. But, still in case of any ambiguity regarding invigilation duties assigned to you, please contact Dr. Rajesh Prasad Mishra, In-charge, Test scheduling at phone no. 8456.

I expect cooperation from all of you for the smooth functioning of the division.

DEAN

INSTRUCTION DIVISION

Cover Letters

A covering letter serves the purpose of creating the necessary background to any submission. It also indicates the origin of the submission by specifying the authorization for a study or project. Any document in the form of a proposal, a questionnaire, a résumé, or a report should be accompanied with a covering letter. The receiver decides if he/she should read the accompanied document right then or later based on the cover letter. Thus, the covering letter offers a first impression to the reader, and must be written with care.

A common example of a covering letter is that accompanying a résumé, together forming a job application. This type of covering letter should complement the résumé. Since the employer is inundated with a lot of letters and résumés, the cover letter should be written in such a way that it impresses the selection panel.

Covering letters must be planned and executed well, ensuring that they are to-the-point and free of typographical and grammatical flaws. However, these days the importance of covering letters is reducing; sometimes a very brief letter informing the recipient that the document has been submitted suffices.

Writing the Cover Letter

While writing the cover letter for any document, the following points must be taken into account:

- The purpose of the document
- The highlights
- The benefits drawn
- The expected response
- A courtesy close

Exhibit 15.17 shows a general cover letter.
Refer to the Online Resource Centre for some useful cover letter openers.

Academic and Business Cover Letters

An academic cover letter accompanies the documents written for an academic job, such as a proposals, questionnaire, reports, or job in a university (Exhibits 15.18 and 15.19), and business cover letters (refer to the Online Resource Centre for a sample business cover letter) accompany the documents related to the industry, such as job description, salary statements, reports, etc. The document submission is incomplete without the cover letters.

At the basic level, both the letters serve the same purpose, but they differ in their content. One cannot write the same cover letter for academic and business purposes. In an academic cover letter, the focus is on education, research, publication, and teaching. Write in simple and grammatically correct English. While writing a business cover letter, one needs to take additional care of the format, structure, font, tone, etc.

Remember the following points:

- Give complete information
- Be precise
- Be polite

Cover Letters accompanying Résumés

The résumé is always accompanied by a cover letter. If the cover letter is able to catch the employers' attention, there are fair chances to get shortlisted for the interview. The following points will be helpful while drafting a cover letter to accompany a résumé:

- Catch immediate attention of the reader
- Give the reasons for writing. If this is in response to an ad or a recommendation
- Underscore the key item of the résumé, by including the qualities that you possess, but which are not mentioned in the résumé
- Close the letter on a positive note, expressing enthusiasm
- Highlight your abilities by stressing on the accomplishments

General considerations

- Be careful about the name and the address
- Know the head of department
- Apply without delay

EXHIBIT 15.17   General cover letter

Kastroy India Pvt. Ltd

29 September, 2021
The General Manager (Projects)
Punj – Loyd Ltd.
Gurgaon

Sub: Submission of Bid Documents

Dear Sir

We are pleased to submit a tender bid in response to your offer letter No. PL/Prj /2021/ 4367 dated 08 September, 2021.

You will be glad to know that we have infrastructural giants spread across three continents. We have started our operations in India five years back and our annual turnover is ₹ 5,000 Crores, which comes from the construction of roads and bridges, commercial buildings, gas pipes lines, refineries, etc.

A complete set of bid document is enclosed for your perusal and scrutiny. It contains technical and financial details and a comprehensive write up on the company. We feel sure that you will find our bid highly competitive.

Kindly contact us for any further information and clarification. We assure you that if we win the bid, we will deliver high-quality on time and within budget.

Sincerely Yours

C.K. Naidu
General Manager (Commercial)

Encl:
   I. Technical Specifications
   II. Commercial Specification
   iii. Company Brochure
   IV. Company Balance Sheet of last 3 years
   V. Details of Projects Completed during Last 3 years.

Vatika Towers, Golf Course Road, Gurgaon
Ph: 09845670011

Tips for Writing Cover Letter

- A cover letter serves as a sales representative for the individual.
- The cover letter should complement the résumé
- Be specific; avoid a general objective. Know the pulse of the industry.
- If already working, then do not demand the expected salary, but a make reference of the current salary.

- The cover letter on email is often shorter than the traditional cover letter.
- Aim for flawless writing by checking the spelling and grammar.

EXHIBIT 15.18 Cover letter for travel grant request (block layout)

13 September, 2021

The Director
DST
New Delhi

Subject: Request for International travel grant

Dear Sir

Please find attached my application for travel grant for presenting my paper entitled 'Enhancing employability through the application of technology' in the *International Conference on Technology and Management* at IMT Dubai.

I look forward to a positive response.

Regards
Poonam Singh
Research Scholar

Cover letters for résumés are addressed to the individual and not the organization, and hence knowing the name of the person being addressed creates a better impression. Ensure that the spelling of the name is correct. Let the letter be specific and brief.

- Keep the letter short
- Express enthusiasm and interest for the positive
- Highlight your strengths
- Organize the letter in three parts, indicating reference, special qualities and availability at convenient time.
- Start with the central selling point—education or experience—which is going to benefit the company
- Make your keenness for the specific job obvious

Please refer to Exhibits 15.20 and 15.21 for some sample cover letters accompanying résumés. More samples of cover letter are given in the Online Resource Centre.

Refer to the Online Resource Centre also to learn about official letters such as demi-official letters, government letters, and letters to authorities, as well as other formal letters such as letters to the editor, permission letters, thank you/follow-up letters, acceptance/rejection letters, and resignation letters.

EXHIBIT 15.19   Cover letter for establishment of a language lab (block layout)

28 October, 2014

Ramya Jayesh
Coordinator (Marketing)
Orell Techno Systems (India) Pvt. Ltd
Bangalore

The Head of the Department
Languages Group
BITS, Pilani 333031

Dear Madam

SUB: Oréll Digital Language Lab (ODLL)

Oréll Digital Language Lab or ODLL, in a nutshell, is a professionally engineered, comprehensive and end-user friendlylanguage learning software brought to you by Oréll Technosystems (India) Pvt Ltd, a dynamic I.T solutions provider focused on delivering cutting-edge solutions primarily for educational institutions.

We are forwarding herewith our most competitive quote tailored to your specifics along with ODLL product details and optimalsystem requirements for your immediate perusal.

You may please refer to the rates in IIT Delhi quotation.

Some of our more recent clients:

- I I T, Guwahati
- I M T, Ghaziabad
- Indian Army, Dehradun
- Shrinathji Institute of Technology & Engineering, Udaipur
- Biyani Girl's College, Jaipur
- BITS-Pilani (Birla Institute of Science & Technology)
- UV College, Ganpat University, Gujarat.
- Amity International Schools, Delhi
- Institute of Computer and Communication Technology, Anand.
- Shankersinh Vaghela Bapu Institute of Technology, Gandhinagar.
- Truba College of Science & Technology – 2 LABS, Bhopal.
- Sanjay Ghodawat Institute of Technology, Kolhapur.
- Sanjeevan Engineering & Technology Institute, Kolhapur.
- Miraj College of Engineering, Miraj. Learners Academy, Kota. (Rajasthan)
- Chinar Public School, Alwar. (Rajasthan)

Kindly visit us at www.orell.in for a detailed review of our range of products, services and clientele/testimonials.

Should you require any further details/clarifications, it would only be our pleasure to oblige at once. Thank you once again and looking forward to a mutually rewarding and long-term association.

Yours truly

Ramya Jayesh
Co-ordinator, Marketing

Memos

There are four important channels through which information flows within the various sections of an organization. Employees can communicate face-to-face, over telephone, through email, or through an inter-office memorandum, which, in short, is called a memo. Memos (or memoranda) are written by everyone from junior executives and engineers to Chief Executive Officers. When you think of a memo, what do you think of? It could be a small piece of paper with a letterhead that says something like:

EXHIBIT 15.20   Cover letter accompanying a résumé 1

249, Ashok Bhawan
BITS, Pilani
Rajasthan – 333031

25 November 2021

The Manager
Cisco Private Limited
5th cross, Sebastian Road
Hyderabad
India – 530016

Dear Sir

I was very interested to see your advertisement for a Software Engineer in *The Hindu* (20 November 2021). I have been seeking just such an opportunity as this, and I think my background and your requirements may be a good match. I am very much interested in working as a Software Engineer in your esteemed organization. I enclose my résumé as a first step in exploring the possibilities of employment with Cisco Private Limited.

I have worked as a project trainee in Satyam Computer Services Limited for the past 6 months. I was involved in developing a graphical user interface for Metadata Management System. So I have hands-on experience in Java Swing, Java Security, JDBC, and Oracle.

As a Software Engineer in your organization, I assure you that I will work hard for the improvement of your company. Furthermore, I work well with others.

I would appreciate your keeping this enquiry confidential. Thank you for your consideration.

Yours faithfully

(Gopinath M.C.)

Enclosure: Résumé

'From the desk of ...' or 'Don't forget ...' or 'Reminders ...'

The message itself may be very simple—something like:

'Buy more paper clips' or 'Meet with President at 2:30' or 'We are running out of storage space'.

While these memos are informative or persuasive, and may serve their simple purposes, more complex memos are often needed in an office setting. However, even though business memos may be more formal and complicated, the intention in writing one is still the same—to formally communicate within the organization and keep a record of this communication. These are brief written communications circulated within an organization. They not only facilitate communication about various operations, but also play an important role in arriving at some quick decisions. For example, the production manager of a fabrics company can decide which type of fabric needs to be produced more on the basis of information provided in a memo from the marketing manager.

Memos also help solve problems either by informing the reader about new information, such as policy changes, price increases, etc., or by persuading the reader to take an action, such as attend a meeting, use less paper, or change a current production procedure.

Inter-office memos enable the flow of information in all the three types of organizational communication, namely vertical, horizontal, and diagonal. The President of an organization can send a memo to the Vice

EXHIBIT 15.21   Cover letter accompanying a résumé 2

311 Nelson Street
West Lake Circle, Jaipur

March 6, 2022

Ms Vibha Acharya
Engineer
ACE Monitoring and Analysis, Inc.
P.O. Box 233, Mumbai

Dear Ms Acharya

Dr Samuel Johnson, a consultant to your firm and my Organizational Management professor, has informed me that 'ACE Monitoring and Analysis' is looking for someone with excellent communications skills, organizational experience, and leadership background to train for a management position. I believe that my enclosed resume will demonstrate that I have the characteristics and experience you seek. In addition, I would like to mention that my work experience last summer makes me a particularly strong candidate for the position.

As a promoter for Sansui Training at the 2021 Singapore Show, I discussed Sansui's products with marketers and sales personnel from around the world. I also researched and wrote reports on new product development and compiled information on industry trends. The knowledge of the mass communication industry I gained from this position helped me analyse how Sansui products can meet the needs of regular and prospective clients, and the valuable experience I gained in promotion, sales, and marketing would help me use that information effectively.

I would welcome the opportunity to discuss these and other qualifications with you. If you are interested, please contact me on my cell number 98292 27400. I look forward to meeting with you to discuss the ways my skills may best serve 'Aerosol Monitoring and Analysis'.

Sincerely,

(Mohan Gokhle)

Enclosure: Resume

President (vertical), a Vice President of one division can send a memo to the Vice President of another division (horizontal), or the President can send a memo to an employee directly without going through the hierarchical set-up inside the organization (diagonal).

In short, memos help in bridging the communication gap among the various sections of any organization and also serve as permanent record of information.

Classification and Purpose

No other kind of written communication reaches so many people at so many levels as does a memo in an organization. The larger the organization and the more levels of authority it has, the more inefficient phone calls and face-to-face discussions become. A memo is a good way to reach many people at once. Of course email is nowadays used in most organizations to convey information in whatever written form you may choose—memo, circular, or notice.

A memo is important not only because of its frequency of use and the wide range of subject matter that can be presented in memo form, but also because it represents a component of interpersonal communication skills within a work environment. To write effective memos that will contribute to efficient functioning within the organization, one has to keep in mind the purpose of writing the memo and the readers' interest. One also needs to take care of the organization of information, completeness and tone while writing an effective memo.

Depending on their purpose, memos can be classified into three major categories:

- Documentary
- Congratulatory
- Disciplinary

Documentary memos

As the name suggests, these memos are mainly used for conveying information, such as memos written to a subordinate to remind, to announce, to give instructions, to explain a policy or procedure, to a peer or superior to make a request or routine recommendation, or to confirm an agreement. For instance, a memo explaining the new method of maintaining medical records of employees in an organization, requesting the head of another division to provide additional manpower for shifting some huge machines, providing some suggestions for improving the existing billing system—all fall under this category. Short reports also can be submitted in the form of documentary memos. Such reports are called memo reports. These are discussed in Chapter 13.

Congratulatory memos

Memos are also used to give credit to employees of an organization for the outstanding work they have accomplished. It is appropriate for the Vice Chancellor of a university to send a memo to the faculty members congratulating their outstanding contribution to the field of research. Similarly, employees can also send their compliments in the form of a memo to their officers, for the awards or achievements that the latter may have earned.

Uses of Memos

1. To request for action or information. This allows one to have a written record of the request. As compared with an oral request, this type of written request is more difficult for the audience to forget or ignore.
2. To explain to the reader something that is not understood. The purpose in this case is to clarify something to the reader.
3. To announce or to give formal notice to readers, publicly informing them about new procedures, new products, or anything that needs to be publicly known.
4. To confirm the details of a meeting, conversation, or telephone call. This would enable one to have a written record of decisions or agreements that were made.
5. To suggest solutions to business problems, to offer one's services or those of the department, or to bring up new ideas or methods of doing things.
6. To report the details of a project at regular intervals as a way of helping the organization keep track of progress and problems.

Disciplinary memos

When employees violate the rules or breach the code of conduct in an organization, they will be served either with a severe warning or any other punishment as decided by the management. The memo conveying this action is known as a **disciplinary memo**. For instance, a memo may be issued to an officer who has accepted a bribe from one of the customers.

Structure and Layout

Standard memos are divided into five main segments to organize information and to help achieve the writer's purpose. However, depending on the requirements, we may need to add two more segments, one for attachments and the other for distribution of copies.

- Heading
- Discussion
- Signature
- Distribution (optional)

- Opening
- Closing
- Necessary Attachments (optional)

Organizations generally provide printed memo forms to their various divisions, which contain all the segments mentioned above. A sample template is given in Exhibit 15.22.

Heading

The heading segment follows this general format:
- Name of the organization and address (Printed Letterhead)
- Date: (Complete and current date)
  To: (Designation of the recipient)
  From: (Designation of the sender)
  Subject: (What the memo is about, highlighted in some way)

Since memos are used for communication within the organization, it is enough if the designations of the sender and the recipient are mentioned against To and From in the layout.

Almost every recipient reads the subject line, which gives a clear idea of the topic discussed in the memo. The subject line, usually typed in capitals, communicates to the reader(s), the purpose of the memo. One-word subject lines do not communicate effectively, as in the following flawed subject line.

Subject: SUPERVISORS

Such a subject line gives a vague idea about the contents of the memo, but lacks focus. A better subject line for this would be:

Subject: SALARY INCREASE FOR SUPERVISORS

A few more samples of subject lines are given below:

- PERMISSION TO CHANGE PROCEDURE
- REQUEST FOR FOUR MACHINES
- DETAILS OF TRAINING PROGRAMME
- TERMINATION OF SERVICES
- ARRANGEMENTS FOR THE CONFERENCE

You may find that the topic and the main focus are connected by a preposition in all these examples. Such a combination works well in all subject lines and clarifies the actual subject matter of the memo to the reader.

EXHIBIT 15.22　Memo template

National Steel Industries Ltd
12, Gandhi Marg, New Delhi 110002

Inter-office Memorandum

DATE:

TO:

FROM:

SUBJECT:

_____
_____ _____ (Opening)
_____
_____
_____
_____
_____
_____ (Discussion)
_____
_____
_____
_____
_____
_____ (Discussion)
_____
_____ (Closing)

Signature

Attachments:

Distribution:

Opening

The purpose of a memo is usually found in the opening paragraphs and is presented in three parts: the context and problem, the specific assignment or task, and the purpose of the memo.

The context is the event, circumstance, or background of the problem being resolved or the topic handled in the memo. The first paragraph establishes the background. State the problem or simply the opening of a sentence, such as, 'In our effort to reduce the absenteeism in our Division ....' Include only what your reader needs, but be sure it is clear.

In the task statement, the steps taken to help resolve the problem must be mentioned. If the action was requested, the task may be indicated by a sentence opening such as, 'You asked that I look at ....' To explain our intentions, we might say, 'To determine the best method of controlling the percentage of absenteeism, I took recourse to three methods ....'

Finally, the purpose statement of a memo gives the reason for writing it and forecasts what is in the rest of the memo. Make sure that this statement is forthright and explains to the reader exactly what is in store. For example, we might say: 'This memo presents a description of the current situation, some proposed alternatives, and my recommendations.' If one intends using headings for the different memo segments, the major headings can be referred to in the forecast statement to provide a better guide for the reader.

Some guidelines for the memo's opening segment:

- Include only as much information as is needed by the decision makers in the context, but be convincing in establishing that a real problem exists. Do not ramble on with insignificant details.
- If one has trouble putting the task into words, consider whether you are clear in the mind about the situation. More planning might be required before writing the memo.
- Ensure that the forecast statement divides the subject into topics most significant to the decision maker.

To summarize the opening segment, the memo should start with one or two clear sentences informing the reader of the need and purpose of the communication. For instance, the introductory paragraph of a memo from the Manager of the Training Division of a company to the Vice President of that company may contain the following few lines:

> As directed by you in your memo dated 2 March 2022, I analysed the possibilities of offering a three-week training programme to our supervisors. I am submitting my views on organizing this programme in the lines that follow:

Generally, when we write a memo requesting somebody to provide something, the memo may be very short, and hence need not contain a separate introductory paragraph. In this case, the introduction and discussion would be combined as shown below:

> As directed by our President, we are trying to complete the work by tonight. To accelerate the pace of work, I request you to spare two computers for tonight only.

Discussion

The discussion segment is the part where we develop the arguments that support our ideas. For example, if a memo is being written to a superior who has asked for an analysis of the feasibility of offering some new services to employees, the details of the analysis can be explained in this section. If one has to direct a subordinate to conduct a survey on the effectiveness of the new machines introduced in the division, the specific details regarding the aspects that need to be examined can be mentioned in this segment. Since very few readers read every line of the memos they receive, keep the communication brief.

The following two examples will show you how the discussion segments of a memo appear:

Example 1

I personally went to the reprography section of our institute and found out that the photocopier is not effective because of the poor quality of stationery used. The paper used is very thin and hence the impressions of one side fall on the other.

Example 2

Our committee examined the case and the details are given below:

1. Adequate quotations were not received for the purchase of the two machines. We found out that there are five dealers for the sale of these machines in our locality.
2. The machines were not properly checked as soon as they were received. They were sent to the production division directly.
3. The Purchase Manager does not have adequate explanations for this casual action.

Closing

After the reader has absorbed all of the information, close with a courteous ending that states the actions expected from the reader. Always consider how the reader will benefit from the desired actions and how those actions can be made easier. For example, we might say, 'I will be glad to discuss this recommendation with you during our Tuesday trip to Delhi and follow through on any decisions you make'.

A memo can end with some complimentary remarks or directive statements. While a complimentary close motivates the readers and makes them feel happy, a directive close tells them what exactly is to be expected or what they have to do next. Here are examples of these two types of closing statements.

Complimentary Close

- If our results continue to improve at this rate, we will attract more students during the coming years. Congratulations!
- Please accept my compliments for introducing this new computing system in your Division.
- There is no doubt that your conscientious efforts would help us accomplish our task without any difficulty. Keep it up!

Directive Close

- I would like to resolve the issue only after hearing from you. Hence, kindly inform me before 25 March 2022.
- To complete your analysis in time, our Finance Manager would provide the necessary data tomorrow, 25 March 2022. Please bring along with you the registration details of the newly acquired land.

Hints on the Discussion Segment

1. Begin with the information that is most important; i.e., start with the key findings or recommendations.
2. Follow the inverted pyramid pattern of communication. Start with the most general information and move to the specific or supporting facts.
3. Try to make the text more reader-friendly by applying boldface type, headings, columns, and graphics.
4. For easy reading, list the important points or details rather than writing in paragraphs when possible.
5. Be careful to make lists parallel in grammatical form.

Necessary attachments

Make sure all findings are documented to provide detailed information whenever necessary. This can be achieved by attaching lists, graphs, tables, etc. at the end of the memo. Be sure to refer to the attachments in the memo and add a notation about what is attached below the closing, like this:

- Attached: Director's approval letter
- Attached: Several Complaints about Product, January–June 2021
- Attached: List of absentees on 17 July 2021

Distribution

This last segment is used to mention the designations of those people to whom a copy of the memo has been sent. As already said, this segment is not mandatory in a memo. The short form of complimentary copy, that is, Cc, can also be used instead of the word *distribution*:

Distribution:
Assistant Manager, Operations
Supervisor, Manufacturing

Distribution:
All Associate Professors

Cc: Personnel Manager with a request to circulate among the employees
        Budget Officer
        Assistant Manager, Finance

Style

Regarding the style of memos, the organization and the individual's relationship with the readers suggest the degree of formality or informality that should be adopted in a memo. In some companies, a formal style is expected; in others, a handwritten note or informality in style is the rule.

Useful Tips to Prepare Memos

1. Use the standard format or the one prescribed by the organization.
2. Include all the necessary segments.
3. State clearly the context and purpose in the opening segment.
4. Keep in mind your relationship with the recipient to choose the degree of formality.
5. Maintain a positive tone.
6. State in the closing segment what action is expected from the recipient.
7. Use features like highlighting, bold face, etc. to draw attention.
8. Keep the memo short.

Some bosses—those who believe in the importance of upholding status distinctions—want memos to sound formal and distant in a way that is appropriate while communicating to a superior. Others—those who have an open and participative approach to managing—would prefer the use of first person, contractions, and even sentence fragments to create an informal and conversational style. While writing a memo, therefore, one may choose the style that suits the organizational culture, but keep in mind that a friendly tone and courtesy are always required to suggest an association with the reader.

Generally, the tone is kept neutral or positive, but one may occasionally have to issue complaints or reprimands in memo form. Use caution in negative situations, and be aware of the effect of the correspondence. If the communication is spiteful, blunt, or too coldly formal, it might annoy the recipient.

Flowery language, excessively technical jargon, or complicated syntax will make one sound pompous. Therefore, one should aim to sound cordial, straightforward, and lucid. Develop a relaxed and conversational style without being too chatty. Projecting an image of consideration creates a greater chance of being viewed as knowledgeable and competent in carrying out the professional responsibilities.

Ensure that a memo is as short and to-the-point as possible. Whether the news is good, bad, or neutral, address the issue in the opening segment. If the memo is lengthy, provide an indication of its organization in the opening segment. Exhibits 15.23 and 15.24 illustrate a memo. The Online Resource Centre contains more samples.

EXHIBIT 15.23   Sample memo 1

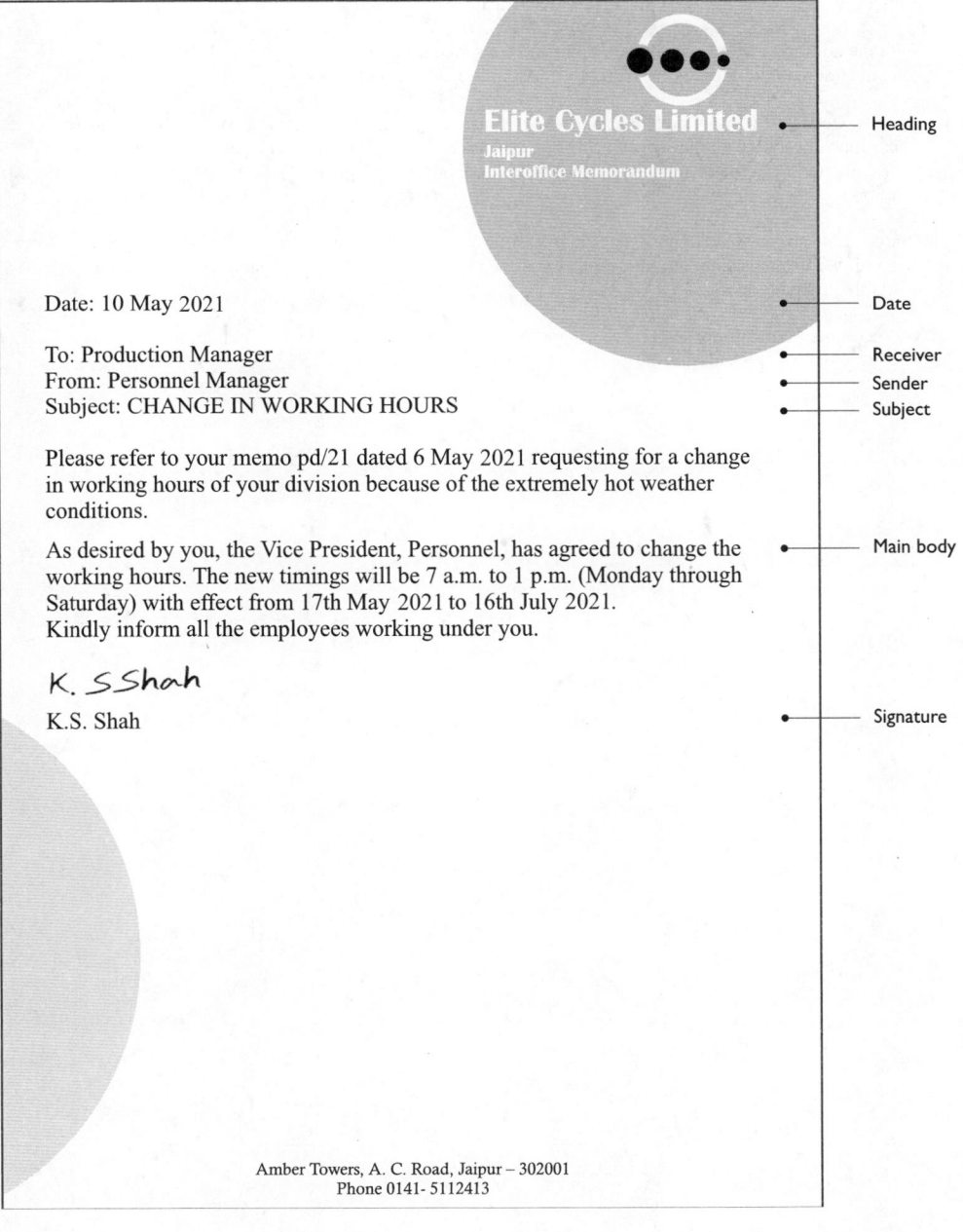

Elite Cycles Limited
Jaipur
Interoffice Memorandum

— Heading

Date: 10 May 2021 — Date

To: Production Manager — Receiver
From: Personnel Manager — Sender
Subject: CHANGE IN WORKING HOURS — Subject

Please refer to your memo pd/21 dated 6 May 2021 requesting for a change in working hours of your division because of the extremely hot weather conditions.

As desired by you, the Vice President, Personnel, has agreed to change the working hours. The new timings will be 7 a.m. to 1 p.m. (Monday through Saturday) with effect from 17th May 2021 to 16th July 2021.
Kindly inform all the employees working under you. — Main body

K. S Shah
K.S. Shah — Signature

Amber Towers, A. C. Road, Jaipur – 302001
Phone 0141- 5112413

EXHIBIT 15.24   Sample memo 2

National Institute of Technology
Worli, Mumbai

Inter-office Memorandum

Date: 25 August 2021
To: Dean, Educational Hardware Division
From: Manager, Reprography Unit
Subject: PURCHASE OF THREE PHOTOCOPIERS

With the addition of four new departments and consequent increase in the number of both faculty and students, the volume of work in the reprography unit has considerably gone up.

The two CopyFast photocopiers that we have at present are no longer adequate to meet the demands. These machines were bought seven years ago and have become obsolete. Moreover, they break down frequently and need major repairs. This year alone we have spent ₹ 25000/- to keep them in working condition.

Our estimate shows that we now require at least three more photocopiers to cope with the increasing demand. We have also studied the features and the cost of various brands of photocopiers currently available in the market.

We recommend the purchase of three ImageX5 photocopiers from Singhania Imaging Ltd., Mumbai. The price list is enclosed.

I request your approval for the purchase of these three photocopiers.

Saif Ali

Saif Ali

Emails

The phenomenally rapid growth of the Internet and its widespread use in business has changed the way in which organizations communicate. All organizations have Internet access, and most individuals have a personal email address. Many companies are promoting the use of email for most—if not all—in-house correspondence, and a great deal of communication with outside organizations also relies on email.

Email stands for electronic mail. These are digital messages that can be sent through an Internet connection. Email offers some tremendous advantages. It is fast—a message can be sent to as many people as necessary instantaneously. It is inexpensive, as it saves paper and is promoted in most organizations as a green initiative. It is convenient and saves time. When dealing with external agencies, especially important clients, it is the most unobtrusive mode of communication, as the recipients can read it at their own leisure and pace, and respond after due reflection.

When email technology was introduced, it brought in a completely different world in terms of writing. For one thing, it was very informal, as personal emails did not require strict attention to grammatical rules. Page design did not exist. People have adopted the informality of email that had become a standard feature of the genre. However, as emails gradually replaced office letters and memos of all kinds, a certain amount of formality has been introduced. Governments and ministries now use emails for their official communication.

Email also has its own vocabulary of acronyms. Some of the most common ones are listed in this chapter. Email combines most of the best—and a few of the worst—characteristics of the more well known methods of communicating, including postal mail, telephone, and fax. You can do several things with email that are difficult or impossible with any other form of communication. While email is an efficient way to communicate, it is also subject to limitations. In particular, we must realize when email is appropriate and when it is not. For example, when writing about emotionally charged subjects (or when the person we are writing to may become angry, defensive, or otherwise upset about a subject, or when we are angry), a face-to-face conversation is generally more appropriate than email. Similarly, discussing confidential matters or very complex issues is often better done in other ways than through email.

Advantages and Limitations

Advantages

- It is possible to communicate quickly with anyone through the Internet. Email usually reaches its destination in a span of minutes or seconds.
- It is easy to send messages to more than one recipient simultaneously by just typing in several email addresses. It also allows one to maintain mailing lists on the computer, which allows quick distribution to many people. It results in avoiding repetition or reproduction of text.
- Email can be used to access vast pools of information stored on the Internet.
- Thousands of email messages can be saved and stored, and one can search message files electronically.
- All or part of an email can be pasted into other computer documents.
- Some email services allow access to a printer-friendly version of email messages.
- Most email systems have a reply button that enables one to include all or part of the original message when replying. This feature speeds up replying to messages, as people spend a lot of time establishing a context for their reply in a phone call or a letter.
- Drawings, sounds, video clips, and other computer files can be attached to an email.
- Letters, notes, files, data, or reports can all be sent using emails.
- One need not worry about interrupting someone when sending an email. The email is sent and delivered by a computer system communicating with the Internet. Although it is put into someone's mailbox, the recipient is not interrupted by the arrival of email.

- The received emails can be dealt with at a convenient time in leisure. Also, it does not have to be written or sent only at a time when the recipient will be available. This is known as asynchronous communication.
- Emails are not anonymous—each message carries the return address of the sender—but it is possible to write to anyone with an email address. All the messages appear the same to the person who gets the email. The messages are generally judged on the basis of their content, not their source.
- As in the case of postal mail, emails can be marked with high, medium, or low priority. The email can also be marked for receiver's action.
- The cost of sending an email is independent of the distance, and in many cases, the cost does not even depend on the size of the message. Most Internet access charges are based on the number of hours one uses the Internet per month, or a flat monthly fee.

> 'What a wonderful thing is mail, capable of conveying across continents a warm human hand-clasp.'
> –Author Unknown

Limitations

- Email is editable. Email communication is subject to security issues. It is therefore insecure.
- Email is anonymous. The identity of a message's author can be completely masked or lost in just two generations of the message. It might be impossible to be certain as to where an email originated. Without an identifiable source, any claim based on it cannot be validated.
- Email cannot be retracted. Once the 'Send' button is pressed, there is no bringing it back.
- Email is not necessarily private. Since messages are passed from one system to another, and sometimes through several systems or networks, there are many opportunities for someone to intercept or read email. Many types of computer systems have built-in protections to stop users from reading others' email, but it is still possible for a system administrator to read the email on a system or for someone to bypass the security of a computer system.
- Some email systems can send or receive text files only. Although we can send and receive images, programs, files produced by word processing programs, or multimedia messages, some recipients may not be able to properly view the message.
- It is possible to forge email. This is not common, but it is possible to forge the address of the sender. We may want to take steps to confirm the source of some emails that we receive.
- We can receive too much or unwanted email, just like other types of junk mail. Junk email is called spam. One may have to take active steps to delete the junk mail received and try to stop it from being sent in the first place.
- We may not know about the person with whom we are communicating. The communication is often all in text and it is possible for us to get an incorrect impression of the person sending us email. Also, some people misrepresent themselves. One must be aware of phishing, which can be used for identity threat. Phishing is the process of attempting to access sensitive information such as usernames and credit card details by pretending to be trustworthy entity.

Style, Structure, and Content

Email messages are equivalent to brief informal memos used to communicate information or ask questions. Email messages must be kept *brief*, ideally, under 200 words. In some circumstances, such as writing an email report, longer messages will be required. However, in most cases, short messages are likely to get the point across more clearly and are also more likely to get answered. Few people are interested in reading messages much longer than what fits on their computer screens. In fact, some people do not read lengthy messages or stop reading after the first few hundred words.

Also exercise caution in terms of the tone used. For example, a message from a junior employee that addresses the president of a major company using an informal tone in an email note to a colleague would be inappropriate. In other words, the audience and purpose when writing email should be considered as carefully as when writing formal letters.

Also, just because the medium is electronic, do not assume the messages being sent are short-lived. Many people archive their email and system administrators can retrieve long-deleted messages. An ill-conceived email may still be available in the archives or deleted mail.

Poor spelling and grammar in email messages could lead some readers to question the writer's competence. Although problems with spelling and grammar are generally ignored in forums such as the various *Internet Newsgroups* (indeed, commenting on these sort of issues is generally considered bad Internet manners), they are generally frowned upon in schools and industry. Sending an email memo filled with spelling errors to an instructor, client, or supervisor is ill-advised. Always take the time to consider the effect that the tone, style, grammar, and spelling of a mail may have on the recipient.

As far as the structure of email is concerned, there are three parts: header, message, signature. The header identifies the sender, receiver, people who receive a copy of the mail, the date on which the mail is sent, and the subject. The message includes the content and the signature block contains the name of the sender. The senders can also add their address and phone numbers to the signature block.

Emoticons and acronyms

Emoticons, or *emotional icons*, are used to compensate for the inability to convey voice inflections, facial expressions, and bodily gestures in written communication. Some emoticons are better known as *smileys*, as they are faces showing different expressions. As they are read from a monitor screen, emails tend to be a cold and emotionless medium. It is also very easy to offend people without even knowing that we have. Sarcasm, even meant in good heart, is usually misinterpreted. That is why emoticons were developed. They allow you to make it clear when what you are saying is not meant to be taken seriously. Those with a dry sense of humour, for example, can use ;-) which is a wink and a grin.

Some commonly used emoticons and email acronyms are listed in Table 15.10. While there are no standard definitions for the following emoticons, we have supplied their most usual meanings. Refer to the Online Resource Centre for common email jargon.

Table 15.10 Common emoticons and acronyms

Emotion	Expression	Emotion	Expression
:) or :-) 😊	Expresses happiness, sarcasm, or joke	:Q or :-Q	Expresses confusion
:(or :-(😦	Expresses unhappiness	:@ or :-@ or 😠	Expresses shock or screaming
:] or :-] or :D or :-D	Expresses jovial happiness	😐 :-s	Worried
:[or :-[	Expresses despondent unhappiness	😄 :))	Laughing
:I or :-I	Expresses indifference	😴 I-)	Sleep
:/ or :\ or :-/ or :-\	Indicates undecided, confused, or skeptical	😷 :-&	Sick
:S or :-S	Expresses incoherence or loss of words	😡 X-(	Angry
😶 I-(	Not talking	💀 8-X	Skull

(Contd)

Emotion	Expression	Emotion	Expression
(:|	Tired	=:)	Alien 1
:-?	Thinking	>-)	Alien 2
=D>	Applause	:-L	Frustrated
%%-	Good luck	<):)	Cowboy
**==	Flag	[-o<	Praying
~o)	Coffee	:)>-	Peace
*-:)	Idea	O:)	Angel

Acronyms	Expansion	Acronyms	Expansion
ASAP	as soon as possible	AAMOF	as a matter of fact
BBFN	bye bye for now	BFN	bye for now
BTW	by the way	BYKT	but you knew that
CMIIW	correct me if I'm wrong	EOL	end of lecture
FAQ	frequently asked question(s)	FITB	fill in the blank
FWIW	for what it's worth	FYI	for your information
HTH	hope this helps	IAC	in any case
IAE	in any event	IMCO	in my considered opinion
IMHO	in my humble opinion	IMNSHO	in my not so humble opinion
IMO	in my opinion	IOW	in other words
LOL	lots of luck or laughing out loud	MGB	may God bless
MHOTY	my hat's off to you	NRN	no reply necessary
OIC	oh, I see	OTOH	on the other hand
ROF	rolling on the floor	ROFL	rolling on the floor laughing
RSN	real soon now	SITD	still in the dark
TC	take care	TIA	thanks in advance
TIC	tongue in cheek	TTYL	talk to you later
TYVM	thank you very much	WYSIWYG	what you see is what you get
<G>	Grinning	<J>	Joking
<L>	Laughing	<S>	Smiling
<Y>	Yawning		

Email Etiquette

There are many etiquette guides and many different etiquette rules. Some rules will differ according to the nature of the business and the corporate culture. Here, we list what we consider as the 24 most important email etiquette rules that apply to nearly all formal communication situations.

Answer swiftly

People send an email because they wish to receive a quick response. If they did not want a quick response, they would send a letter or a fax. Therefore, each email should be replied to within at least 24 hours, and preferably within the same working day. If the email is complicated, just reply saying that the email has been received and that you will get back to them. This will put the reader's mind at rest and usually they will then be very patient.

Do not overuse reply all

Use Reply All only if you really need your message to be seen by each person who received the original message.

Use templates for frequently used responses

If you often tend to receive the same queries, such as directions to your office or how to subscribe to your newsletter, save your replies as response templates and paste these into your message when you need them. You can save your templates in a Word document, or use pre-formatted emails. Even better is a tool such as *ReplyMate* for Outlook (allows you to use 10 templates for free).

Use proper structure and layout

Since reading from a screen is more difficult than reading from paper, the structure and layout are very important for email messages. Use short paragraphs and blank lines between each paragraph. When making points, number them or mark each point as separate to keep the overview.

Identify yourself and the topic

Where possible, identify yourself on the From: line using your full name rather than just email address. For example, use 'Amit Kumar Saxena' aks@vu.edu.in rather than just aks@vu.edu.in. Recipients are more likely to respond if they can easily identify the sender. In addition, knowing whom a message is from helps the recipient put the message in context.

At the end of the message, include an alternative way to be contacted (i.e. phone number, FAX, postal address) along with the name. This information can be provided in a signature field that can be turned off for more personal emails. Providing contact information is especially important when asking for an answer that is likely to be quite complex. Often, less time is required to explain something complex over the phone or in person than to type out the message.

The information in the subject line should be meaningful to the recipient as well as the sender. For instance, when sending an email to a company requesting information about a product, it is better to mention the actual name of the product, e.g., 'Product A information', than to just say 'Product information' or the company's name in the subject.

Answer all questions, and pre-empt further questions

An email reply must answer all questions, and pre-empt further questions. If all the questions in the original email are not answered, it will likely bring further emails regarding the unanswered questions, which will not only waste the time of the sender and the recipient but also cause considerable frustration. Moreover, if one is able to pre-empt relevant questions, the reader will be grateful and impressed with the sender's efficiency and thoughtfulness. Imagine for instance that an off-campus student sends you (assume that you are a professor of Electronics Engineering) an email asking some doubts on a lesson. Instead of just explaining the answer to the student's problem, if you mention some other sources that he/she can refer to for further understanding, the student will definitely appreciate this extra information.

Be concise and to the point

Do not make an email longer than it needs to be. Remember that reading an email is harder than reading printed communications and a long email can be very discouraging to read.

Use proper spelling, grammar, and punctuation

As in all forms of written communication, this is not only important—because improper spelling, grammar, and punctuation give a bad impression of the individual or the company—but also essential for conveying the message properly. For example, using u, r, and ur for *you*, *are*, and *your* respectively is inappropriate for formal messages. Mails with no full stops or commas are difficult to read and can sometimes even change the meaning of the text. If your program has a spell checking option, why not use it?

Do not write in CAPITALS

IF YOU WRITE IN CAPITALS IT SEEMS AS IF YOU ARE SHOUTING. This can be highly annoying and might trigger an unwanted response in the form of a flame mail. Therefore, try not to send any email text in capitals.

Avoid long sentences

Try to keep the sentences to a maximum of 15–20 words. Email is meant to be a quick medium and requires a kind of writing different from letters.

Use active instead of passive voice

Try to use the active voice of a verb wherever possible. For instance, 'We will process your order today' sounds better than 'Your order will be processed today'. The first sounds more personal, whereas the latter, especially when used frequently, sounds unnecessarily formal.

Keep your language gender-neutral

It is important to be gender-sensitive. Avoid using discriminatory language such as: 'The user should add a signature by configuring his email program'. Apart from using he/she, you can also use the neutral gender: 'The user should add a signature by configuring the email program'.

Maintain coherence

When replying to an email, include the original mail in the reply, Click 'Reply', instead of 'New Mail'. Some people opine that the previous message must be removed since this has already been sent and is therefore unnecessary. However, if a person receives several emails, it is difficult to remember each individual email. This means that a 'threadless email' will not provide enough information and the recipient may have to spend a frustratingly long time to find out the context of the email in order to deal with it. Leaving the thread might take a fraction longer in download time, but it will save the recipient much more time and frustration in looking for the related emails in their inbox.

Do not overuse the high priority option

We all know the story of the boy who cried wolf. Overuse of the high-priority option will make it lose its function when really needed. Moreover, even if a message has high priority, it will come across as slightly aggressive if it is flagged as 'high priority'.

Do not attach unnecessary files

Large attachments can annoy readers and even bring down their email system. Wherever possible, try to compress attachments and only send attachments when they are productive. Moreover, one should have a good virus scanner in place to prevent the readers from receiving documents containing viruses.

Re-read the email before you send it

A lot of people do not bother to re-read an email before they send it out, as can be seen from the many spelling and grammatical mistakes contained in emails. Besides, reading the email from the recipients' perspective will help frame a more effective message and avoid misunderstandings and inappropriate comments.

Take care with abbreviations and emoticons

In business emails, try not to use abbreviations such as BTW and LOL. The recipient might not be aware of the meanings of the abbreviations, and in business emails these are generally not appropriate. The same goes for emoticons. It is advisable to avoid using any entities that the recipient might not be familiar with.

Be careful with formatting

Remember that when an email is formatted, the sender might not be able to view the formatting, or might see fonts that are different from the ones intended. When using colours, use a colour that is easy to read on the background. One also needs to be aware of the fact that there are some accessibility norms that do not allow the use of certain colours, keeping in mind colour-blind people.

Take care with rich text and HTML messages

When sending an email in rich text or HTML format, be aware that the receiver might be able to receive only plain text emails. If this is the case, the recipient will receive the message as a .txt attachment. Most email clients, however, including Microsoft Outlook, are able to receive HTML and rich text messages.

Do not use email to discuss confidential matters

Sending an email is like sending a postcard. Do not send confidential information by email. Moreover, never make any vilifying or discriminating comments in formal emails, even if they are meant to be jokes.

Avoid using URGENT and IMPORTANT

Even more so than the high-priority option, try to avoid the use of words such as 'Urgent' and 'Important' in an email or subject line. Use this only if it is a really, really urgent or important message.

Use the Bcc: field or do a mail merge

When sending an email, some people place all the email addresses in the To: field. There are two drawbacks to this practice: (1) the recipient knows that the same message has been sent to a large number of recipients, and (2) someone else's email address is being publicized without their permission. One way to get round this is to place all addresses in the Bcc: field. However, if the To: field appears blank, it might look like spamming to all the recipients. Instead, the list containing the email addresses of all recipients could be included in the To: field, or even better, with Microsoft Outlook and Word, it is possible to mail merge and create one message for each recipient. A mail merge also allows the use of fields in the message so that each recipient can be addressed personally. For more information on how to do a Word mail merge, consult the Help feature in MS Outlook.

Using the cc field

Try not to use the Cc field unless the recipient in the Cc field knows why they are receiving a copy of the message. This will depend on the situation. Only the recipients in the To field are supposed to act on the message. The Cc field is used to keep others informed about the project, e.g. the manager or a co-worker.

The Bcc is used when a copy of the mail is to be sent to other recipients without the knowledge of the main recipient. For example, it is used in official correspondence such as appraisals being done by first-level managers. The first-level manager provides some feedback to his/her subordinate over email, but includes the second-level manager in Bcc, just for his/her information. The subordinate doesn't get to see that the email is also copied to the second-level manager. This is mostly used for providing sensitive feedback that is supposed to be private.

> 'I consider it a good rule for letter-writing to leave unmentioned what the recipient already knows, and instead tell him something new.'
> –Sigmund Freud

Do not reply to spam

Replying to spam or unsubscribing confirms that the email address is 'live'. Confirming this will only generate even more spam. Therefore, just hit the delete button or use email software to remove spam automatically.

Effectiveness and Security

Email is, of course, a form of written communication, but it is different from traditional written communication. Email is not bound by the physical limitations of a page of paper, it can be transmitted and received very quickly, and a single message can be sent to a group of thousands of people as easily as it can be sent to one or two people. Since email is written communication and it is not done on paper, we have to do what we can to make it easy to read and comprehend. Because messages are sent electronically, it is possible to get a response in a matter of minutes or seconds. When we are communicating, trading comments separated by only a few seconds, email is similar to spoken communication. It tends to get informal and personal, and that is probably just the way we want to be during quick exchanges with another person. On the other hand, when we are communicating using email, we cannot display our facial expressions or gestures or express intonation the same way we would when speaking.

Some Tips for Email Effectiveness

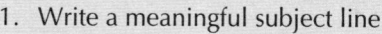

1. Write a meaningful subject line.
2. Keep the message focussed and readable.
3. Use attachments sparingly.
4. Identify yourself clearly.
5. Be kind. Do not flame.
6. Proofread.
7. Do not assume privacy.
8. Distinguish between formal and informal situations.
9. Respond promptly.
10. Show respect and restraint.

There is no substitute for a well-thought-out and well-expressed message. There is also no excuse for misspellings or grammatical errors in professional or business communications. To make an email most effective, it has to be written clearly, take into account the fact that people will likely be reading the message on a computer screen, and take full advantage of the medium itself.

However, one must be aware of the ease with which email messages can be sent to unintended readers and of the possibility that old email messages will come back to haunt the sender. The following guidelines are worth following:

- Ensure that personal messages are sent to the correct individual and not to a mailing list. Many email users have been embarrassed by accidentally sending personal or confidential messages to an entire mailing list or organization. Before sending an email, one of the last things that should be done is check that it is being sent to the intended recipient(s).
- Do not send confidential or personal material via email unless it is encrypted because most email systems are insecure. We suggest that any email message sent should be treated as a public document simply because email servers typically maintain copies of email and many people archive the email they receive. An email message sent several years ago could turn up when least expected.

Although you may be familiar with sending and receiving email messages, the Online Resource Centre provides some sample email messages.

Now that we are aware of the three forms of written communication, namely letters, memos, and emails, Table 15.11 concludes the chapter with a comparison of the various characteristics of these very important forms of communication.

Table 15.11 Characteristics of memos, letters, and email

Characteristic	Letters	Memos	Email
Structure	Contains several elements starting from heading to distribution list	Contains lesser elements than letters and does not include solutation and complimentary close	Contains most of the elements which a letter has but addresses are the email addresses or email ids
Layout	Elements are arranged like any of the layouts discussed earlier in the chapter	Heading elements are aligned with left margin vertically	Given in prescribed format by the email service providers
Purpose	To persuade and to share information	To share information, to direct to recommend, or to congratulate	To convey short routine messages quickly
Audience	Generally low-tech and lay readers, such as vendors and clients	Generally high-tech or low-tech, mostly business colleagues	Generally multiple readers with various levels of knowledge; could include instructors, company supervisors, and subordinates as well as family and friends
Degree of formality	More formal than memos because of external destination	Generally informal because of internal circulation	Degree of formality is less than both in memos and in letters and follows a conversational style
Short forms, abbreviations,	To be avoided	To be avoided unless very common among the employees	To be judiciously used
Circulation	External	Internal	Both internal and external

SUMMARY

Writing letters, memos, and emails is an integral part of all academic and work environments and cannot be avoided. To create and maintain a cordial relationship with people one has to deal with, one should develop the art of writing. While good command over language is a must, one must also know the structure of the various forms of written communication. Effective writing possesses the seven Cs of writing, that is, it should be clear, concise, correct, courteous, conversational, convincing, and complete. As a rule, any letter must to be responded to at the earliest.

The purpose of writing a letter is to sustain a relationship or build a new one. Business letters are of different types, such as credit and collection, enquiry, order placement, complaint and adjustment, instruction letters, and persuasive letters. The principles of business writing will also enable one to write impressive job applications and résumés. More common today is the online submission of a résumé, accompanied with a covering email. Besides these types, there are official and personal letters.

Memos are a very common medium of communication within an organization. The purpose is to inform, persuade, or reprimand. They are classified as documentary, congratulatory, and disciplinary. Most organizations have their own prescribed format for memos; however, the standard format can be followed otherwise. Memos use a more informal language than business letters.

Emails have come to be used widely throughout the world today. It is one of the most convenient ways to contact people in any part of the world for business or personal purposes. Communication through emails has its advantages and disadvantages. Emails also have their own set of vocabulary, jargon, and etiquette. It is prudent to be aware of the email etiquette. Learning how to use emails effectively will definitely go a long way in helping one to develop good communication skills. It is also worthwhile knowing the security issues involved, so that we are able to make the best use of this wonderful mode of communication.

A good idea of these three major forms of written communication will help us to use the most effective mode in any given situation, and help us achieve our goals effectively and efficiently.

EXERCISES

1. Answer briefly the following questions:
   (a) How are letters, memos, and emails different from each other? Do they also have any similarity? Explain.
   (b) Emails are very fast means of communication, but have some drawbacks. Discuss.
   (c) Discuss the important points to be considered while drafting a cover letter to accompany a résumé.
   (d) Discuss email etiquette.
   (e) You have submitted a proposal to University Grants Commission (UGC) on 'Enhancing education in rural area through technology'. Write a letter urging action. Invent the necessary details.

2. Recall or refer to the principles of writing business letters discussed in this chapter and point out the weakness in the following sentences taken from business letters. Then rewrite them so as to make them effective.

   (a) I beg to inform you that owing to the nationwide shortage of packing material we are not in a position to comply with your request.
   (b) We do not find any reason why you are asking for the replacement of computers, which you have mentioned in your letter reached you in damaged condition.
   (c) As instructed, we will bill you for this amount. We are sending the goods today by Green Grass Couriers.
   (d) Your misunderstanding of our June 7 letter caused you to make this mistake.
   (e) Even though you were late in paying the bill, we did not disallow the discount.
   (f) In replying to your esteemed favour of the 5th, I submit under separate cover the report requested by you.
   (g) This is to strongly protest the inappropriate behaviour of your sales manager when I

requested her to kindly permit me to meet you because I wanted to tell you about the external agencies which are creating some problems.

(h) With reference to your request for the supply of 10 Kg of wheat flour to be used on the auspicious occasion of your house-warming ceremony, we are sorry to say that we may not be able to make any commitment at the present moment of time.

(i) You should be aware of the fact that there is no better dealer for Greeting Cards than us not only in this whole city but also in the other 3 metropolis.

(j) Though I have carefully gone through the report prepared and submitted by you to us in your letter of 15th June, I regret to say that owing to the reason that I am extremely busy, I am not in a position to take any action for the proper implementation of recommendations offered by you with great consideration.

(k) I hereby acknowledge the receipt of your letter and beg to tell you that as per the rules, it is not required to submit your request for loan to the branch manager.

3. Rewrite these opening sentences of claim and adjustment letters:

(a) We are sorry that we cannot accept the return of the TV that you bought on August 5.

(b) You are certainly being unfair to us when you insinuate that we tried to put something over on you when you bought a defective lamp.

(c) We cannot understand how your records could have been broken as you claim in your letter of March 10.

(d) We are surprised to learn that you are already having trouble with your Cine movie projector.

(e) In answer to your letter expressing dissatisfaction with your dictating machine, I wish to state that we stand behind anything we sell.

4. Revise these statements granting or refusing adjustments:

(a) Although we are not at fault, we are willing to accept return of the lampshade.

(b) It is simply impossible for us to grant your request. Everyone gets the same fair treatment at Gordon's.

(c) Since the delay in delivery was not our fault, we cannot accept the responsibility for your loss.

(d) We shall be willing to exchange the machine in order to retain you as a valued customer.

(e) We will make this concession to you, even though it is much more than should be expected under the circumstances.

5. Revise these sales letter beginnings to give them more reader appeal. You may use the method of gaining attention in the opening sentence.

(a) We are conducting an intensive sales campaign to get the public to know about the Book Club.

(b) November 5 was a night of darkness for millions of people in the East. There was a power failure that lasted for several hours.

(c) Can't you remember the fun you had at Lake Joy last summer?

(d) The quality of the paper you use will affect your reader's reaction to your message.

(e) We believe that the Current Review is a winner.

6. Change these negative introductions to make them affirmative:

(a) You don't want to waste your money when you buy tyres.

(b) Srickem was developed to prevent your plastic floor tiles from buckling and curling. You will never have to worry about unsightly playroom floors again.

(c) Don't you remember last December 26? Were you prepared to cope with the record snowfall? Were you not huffing and puffing with your snow shovel while your neighbours guided their snow blowers along their walks?

7. Revise these sentences to eliminate dangling phrases:

(a) Before rejecting these designs, we suggest that you compare them with other plans.

(b) Relying on his ability to react quickly in emergencies, the car picked up speed.

(c) Already filled with students, the visitors could find no place in the auditorium.

(d) Referring to your letter of March 13, your complaint was ill-advised.

(e) Having sent the incorrect invoices to you, we assure you that we will adjust it at once.

8. Rewrite these sentences to eliminate all double negatives:

    (a) Didn't you hear nothing from the personnel manager about your promotions.

    (b) The problems had arisen so unexpectedly that scarcely nobody in the office knew what to do.

    (c) Plan your vacation trip now. Don't plan to go nowhere this summer.

    (d) That cannot be done no longer by any member of the tax department.

9. Assuming that you have the requisite credentials, draft Job Application Letters in response to the following advertisements:

    (a) Wanted a Plant Manager (Operations) at our new factory in Gaziabad, UP Engineering Graduates with minimum 5 years experience in manufacturing industries as Plant Managers can apply. Salary is negotiable. Apply with particulars to Box 650, The Hindu, Chennai-600004 latest by 30th July.

    (b) If you are a recent post-graduate in science and interested in research, apply for the post of Junior Research Associate in our R&D Division. You should hold a post-graduate degree in mathematics, physics, chemistry, or biology. If selected you will be given a Research Assistantship of ₹ 8000/- p.m. and you will have the opportunity to work in excellent laboratories. Apply within 15 days to Manager Research, R&D Division, Wipro Industrues, Bangalore–560 012.

    (c) Safe Insurance Company seeks personable, college-trained person to manage office of five employees. People skills and good communication ability a must. Knowledge of office procedures and Word processing essential. Send application within 10 days to Human Resource Office, P.O. Box 719, Kanpur.

    (d) Healthy Foods Ltd, a fast-growing manufacturer in the food-processing industry, has openings in its training program. Only highly motivated, dynamic, and result-oriented people with excellent communication skills need apply. Opportunities for advancement to management positions based on performance. Applicants must demonstrate a professional image and possess skills in working with people. Computer literacy required. Apply to Personnel Manager, P.O. Box 520, Bangalore.

10. Answer as directed.

    (a) As the Manager of Sangam Hotel, New Delhi, write a claim letter to the General Manager of Bharat Potteries, Aligarh Road, Bhavanipur, telling him that most of the contents of the chinaware which you had ordered from their firm have reached you in a damaged condition. Demand replacement or suitable compensation. Invent necessary details.

    (b) As the Purchase Manager of Satyam Computers, 9 Naidu Road, Hyderabad-500007, you had ordered two dozen Personal Computers from Hindustan Computers Limited (HCL), 140 M.G. Road, Bangalore-500001. When the consignment arrived, you found some of the pieces in the damaged condition.

        (i) Write a *complaint letter* to the Sales Manager of the company asking for repair, replacement, or compensation.

        (ii) As the Sales Manager of HCL, draft a *suitable reply*.

        Use full block format in both the letters.

    (c) As the Purchase Officer of a Company, write a complaint letter to Uniflex Ltd, New Delhi, pointing out the damage which was discovered after checking the consignment containing Compact Discs sent to you by the supplier. Invent the necessary details.

    (d) Playing the role of Senior Sales Manager, Apex Ltd, draft a suitable reply to this claim letter.

        Use full block letter format for the letters.

    (e) Ajay purchased a VCR on August 13 from Smiley's TV Town, Mumbai. The VCR came with a 'ninety-day warranty against all defects' and a guarantee for 'in-home free repairs and labour'. On October 30, the VCR showed a horizontal line across the screen when she replayed tapes. Ajay called the store manager, Vikas Mallik, and explained the problem. Mallik said the horizontal lines were caused by a dirty head and told Ajay to bring the VCR in for cleaning. He also told Ajay that he would be charged for this service since dirty VCR heads were basic wear and, therefore, not covered by the warranty. Ajay was angered by this response from the store manager and decided to write a letter of complaint. Based on the information provided, write Ajay's letter of complaint.

(f) As the Sales Director of Fitness Plus Centre, Bangalore, draft a sales letter to Business Professisons selling them on your 3 Wellness Packages: 1. The 3-day Fitness Weekend 2. The 7-day Total Fitness Program 3. The Individualized Corporate Well-being Program.

(g) Assume that you are the Marketing Manager for a professional hockey team. At present, you are concerned about season-ticket sales for the coming season. They are well below sales for previous years and hence you plan to do something about it.

Draft a sales letter to those 500 people who have bought season ticket last year but did not this year.

(h) Realizing the need of packing services in Faridabad, Elite Professional Packers have recently started their services in the city. You, as the publicity manager of this company have the onus of promoting this service.

Draft a sales letter to be sent to the prospective users.

11. Draft a memo as directed in each of the following:

(a) Various universities in India offer their academic programmes through distance learning mode of education. A large number of junior officers of a company have sought permission to improve their educational qualifications through these programmes. As the Manager of Personnel Department of this company write a memo to be sent to all Junior Officers spelling out (a) the company's policy in this regard and (b) the deadline for submitting their applications. You can also assure them that a decision would soon be taken and communicated to the applicants.

(b) Workwell Industries Limited has observed that a sizeable number of employees take leave on certain occasions such as mega music event, test cricket matches, and international film festivals. You as the Vice President, Personnel, feel that this hampers the smooth functioning of the industry and tells upon the efficiency of the organization. Appealing to the Divisional Heads of your organization to take suitable action to prevent such mass absenteeism, draft a memo and mark a copy of the same to your Managaing Director.

(c) The General Manager of Comfort Home Appliances Ltd, Kolkata, is worried about the wastage of stationery in almost all sections of the company. Draft a memo to be signed by him and sent to all Sectional Heads, asking them to identify the reasons behind such wastage and also advise the employees under their control to restrain from such wastage.

(d) Elite Industries Ltd, Gurgaon, has decided to replace the furniture in its Marketing and Human Resource Development Divisions. As the Office Manager, write a memo to the Purchase Officer to identify the items of furniture to be replaced, identify the supplier, and submit a memo containing all the related details for administrative approval. Mark a copy of this memo to the Finance Manager.

(e) The Clean Food Products Ltd, Kanpur has decided to provide Internet surfing facilities to all its employees. As the Managing Director of this company, draft a memo to be circulated among all the Divisional Heads containing the following details:

The location, number of computers to be made available, timings, and procedure for lodging complaints if any.

12. Recently you read in the editorial column about honour killing practised in the conservative Indian society. Write a letter to an editor drawing attention to the stigma in society.

13. Through the medium of letter to the editor, you want to share your feelings about the award winning film 'Slumdog Millionaire'. Write a letter to the editor reflecting the fact that this film does not only deal with the down trodden urchins but also want to generate awareness about the imbalances in the country.

14. You are the coordinator of the English drama club, and you want to stage 'Othello' by

Shakespeare this Saturday. Write to the Welfare Division to book the auditorium and get the arrangements done for light and sound system.

15. Write a letter of declining the invitation to the course corrdinator of National Academy of Defense Production who has invited you to deliver the guest lecture on 'Interview Skills'. Invent the necessary details.

16. Email
   (a) Should email replace the communication forms such as memos and letters? Explain your answer.
   (b) Imagine yourself to be the instructor of a course in which 75 students have registe red. Draft an email to be sent to all these students asking them to select a topic of their choice and prepare for a professional presentation of 10 minutes duration.
   (c) As the Sales Manager of a company, draft a reply which has to be sent in the form of an email to three customers who have complained about your product. Invent necessary details.
   (d) Assume that you are the Managing Director of a company dealing with electronic equipment. Inform all your employees about the new community hall which the company is going to inaugurate next week. Highlight the important features of both the hall and the inauguration.
   (e) As the Personnel Manager of a multinational firm, draft an email to be sent to those candidates who were not selected in the interview conducted few days before. Take care not to be courteous and sympathetic while conveying the negative message.
   (f) What are the various email service providers that you have come across? Compare the structure and layout of the email facility offered by them.

17. The following formal letters are ineffective. Rewrite each of them keeping the principles of business letter writing in mind:

(i)
Spotless Washing Machines Pvt.Ltd.
10, Browning Street, Kanpur, U.P.
Website: swmpl.org Phone: 0123-44455566

23 January 2022

Mr. Vipul Kumar
35, Race Course Road
Kanpur
U.P.

Dear Customer:

Sub: Invalid cheque sent by you

Your cheque sent by you for the balance payment of the washing machine purchased by you a few months ago from us as you had wanted more efficiency in your washing and also wanted to have more choices in terms of color, capacity, model, etc., as compared to the earlier one you had been using for quite sometime has bounced back from the bank today.

It is your duty to check your bank balance befor sending the balance payment to us! Now we have to wait for your cheque again for some time.

As you know, we are one of the leading dealers in Washing Machines in Kanpur and we have different types of clients. Some pay in time; some pay on time; but so far nobody has sent a bouncing cheque! Of course, you are one among those who pay the amout in time but what's the use?
Here's your cheque and we expect you to do the needful at your earliest without fail.
Ashish Kumar
Sales Manager
Enclosure: Your cheque

(ii)
11 March, 2022
Avy Trading Corporation
Daryagunj, Vijay Nagar,
New Delhi-110005
Telephon: 011-2345678
Mr Ravi Malhotra

Sales manager, Reva Computers

Ajmer Road, Jaipur-302006

Dear Mr Malhotra

We saw your advertisment in the *The Times of India* about one of your important products. The advertisment caught or attention because we are interested in this product. In fact, we want to equip our corporate office with modern facilities and we would like to buy this product.

However, we cannot send the purchase unless we know more about this product. Send us more information about the product as soon as possible. We want to know many things, which include product product specification and special features of this model of the product, details of discount for bulk purchase, an estimate for the cost of the product, and details regarding terms of business and delivery dates.

Respond to this letter as early as possible.
Yours Sincerely
Amit Jain

(iii)
24, Malaya Road
Vellore, India
Commercial Art & Design
Subject: Freelance Graphic Designers for advertising and marketing Campaign Enclosure: Sample copy of graphic Design and Campaigns
Onoff Pvt. Ltd.
Connaught Place
New Delhi, India

Our Reference: MA / 1230

Dear Business owners,

Designing a great marketing piece is a lot like baking a cake. Anyone can throw some ingredients together and plunk them in the oven. But to take a mouth watering, threelayer chocolate cake you need all the right ingredients in just right proporitions and the knowledge to put them together correctly.

Our firm provides freelance graphic designers are highly qualified and trained professionals we are proud to boast that we have a long list of clients. Do you dread the thought of a new advertising campaign for your business? Does the thought of creating a new sales brochure overwhelm you?

Call us today at 09888880000 and let's finalize the deal otherwise you will have to pay a heavy penalty for it. Thanks for your time.

It's time you should obliged by our services and gives us maximum profits. Great commercial art and graphic design business is just a phone call away.
Director 'Sharma'
Your sincerely
10 March 2022

CHAPTER 16
Research Papers, Thesis, and Technical Descriptions

OBJECTIVES

You should read this chapter to know

○ the characteristics and components of a research paper and the steps involved in writing one

○ guidelines for creating effective technical descriptions

○ components, structure, and style of writing a thesis

Introduction

Research is an essential ingredient of all fields of study as well as all professions. Whether an academician or a working professional, one has to be involved in research to become better equipped in the chosen field. Research can be performed individually on one's own initiative or as a member of a research project team.

Any systematic investigation towards increasing the sum of knowledge can be termed as research. Sometimes one may undertake research work that is a replica of some previous study in order to test the reported findings or the relevance of such findings under different circumstances. At other times, one may undertake research to make decisions regarding a new development or to refine or qualify the findings of earlier studies. Research may also be based on a meticulous search of material in journals, books, or other publications, or carefully designed experiments. However, every piece of research must make an original contribution to existing data and knowledge, irrespective of the method of enquiry.

At the end of a research, once books and magazines have been referred to, field trips have been completed, and experiments have been carried out in the laboratory, there remains the task of organizing the results and publishing them. The research would hardly be of any value unless the findings are disseminated to others working on or interested in the same sphere of activity or knowledge. Now, how does one go about this? The findings can be presented in a professional gathering, or if one wishes to reach a wider audience, the work can be published in a journal. In either case, the facts and opinions have to be organized so as to present their meaning as clearly and descriptively as possible. Such an *organized analysis of a subject* written mainly to record and disseminate information or knowledge, or to present a point of view on a selected topic, is known as a *research paper*.

It is a documented prose work incorporating the results or findings of an original work. It may also be called a scientific paper, investigative paper, or library paper. In fact, it is a long essay often supported by relevant references from suitable sources. It gives a concise account of the work performed, the materials used, the methods adopted, and the results arrived at.

A research paper must be the first disclosure of a new research finding, presenting information that would enable peers (people associated with the field of study) to (a) assess observations, (b) repeat experiments, and (c) evaluate intellectual processes. It is obvious that the knowledge required from research should be quickly publicized to avoid wasteful duplication of work (by another researcher) and to establish the researcher's claim to the priority of discovery. Further, when the findings of a research are published in the form of a research paper, the efforts acquire a permanent value. However, it is worth noting that all research journals of repute control the number of papers published by a rigorous referring system to ensure the originality and quality of contribution.

The other important technical document, which is an indispensable part of any technical project, is technical description. Technical descriptions are an important part of any technical document, be it a letter, a thesis, or a report. We have learnt so far that whenever we correspond on technical matters, our intention must come across very clearly through our communication. The section on technical descriptions in this chapter explains how large amounts of information about any unfamiliar or novel object or process can be presented in an understandable and useful manner.

Research Paper

In its style, structure, and approach, a research paper closely resembles a formal report. Hence, the characteristics are also similar, except for a few differences. While a research paper is written mainly to disseminate new knowledge acquired through research, a report is written to facilitate decision-making or problem resolution.

The audience of a research paper might choose not to read a research paper, but a report will definitely be read by the target audience. The incentive of a research paper may be one's professional advancement, whereas a report always arises out of a specific need.

Characteristics

Given below are the chief characteristics of a research paper:

1. A research paper is the most important form of expository discourse. It may be written on any topic or subject—scientific, technical, social, cultural, etc., but the treatment is scholarly in nature.
2. It is highly stylized and contains a high concentration of certain writing techniques such as definition, classification, interpretation, abstraction, and description.
3. It is objective in nature and the presentation of information is accurate, concise, direct, and unambiguous.
4. Generally, it contains almost all the formal elements that a technical report includes.
5. Most research papers are characterized by the use of visual aids, and scientific, technical, or specialized vocabulary.

'Guys we have to justify the existence of this analog receiver department. We need some fresh research.'

6. Every research paper is a unified composition arising out of the study of a particular subject, assembling the relevant data, and organizing and analysing the same.
7. A research paper is a documented prose work. All important analyses have to be supported by adequate evidence. In short, documentation is essential for all research papers.

Components

A research paper is a piece of written communication organized to meet the needs of a standard, valid publication. It is therefore highly structured, with distinctive and clearly evident component parts, which are listed as follows:

- Title
- Authors, affiliations, and addresses
- Abstract
- Introduction
- Materials and methods
- Results

- Discussion
- Conclusions
- Acknowledgements
- List of symbols
- References or bibliography

 When a research paper is published, it is properly formatted according to the style used by the journal publishing the article. The journal editor provides inputs to the authors on the various standard elements and format required by the journal. The author then submits the article in the required format. Refer to the Online Resource Centre for an exhibit of a journal article, showing the components listed above.

Title

The title of a research paper may be defined as the fewest possible words that adequately describe the contents of the paper. It ought to be well-studied and should give a definite and concise indication of the rest of the paper.

In preparing the title for a paper, remember that the title will be read by thousands of people. Many people will read the title either in the original journal or in one of the secondary (abstracting and indexing) publications. Therefore, choose all the words in the title with great care and ensure that their association with one another is meaningful. Remember that the indexing and abstracting services depend heavily on the accuracy of the title. Also, an improperly titled paper may be virtually lost and may never reach its intended audience. An effective title

- Is a clear indication of the contents of the paper
- Is neither too short nor too long
- Contains specific and not general terms
- Is built on careful syntax
- Is a label and not a sentence
- Avoids the use of common abbreviations, specific notations, and reference numbers

Table 16.1 gives examples to illustrate how to prepare appropriate titles.

Table 16.1 Examples of titles for a research paper

Poorly-worded title	Well-worded title
• March2 serves as an automated validation tool (a sentence; not specific)	• March2: An Automated Validation Tool for the APOLLO2 Code
• Using Parallel Computers for Numerical Studies in the Atmosphere, Ocean Interaction (careless syntax, general)	• Numerical Studies on the Interaction between Atmosphere and Ocean Using Different Kinds of Parallel Computers
• Evaluation of the Measures to Adjust the Increase of Outpatients in MIS (ambiguous abbreviation)	• Evaluation of the Measures to Adjust the Increase of Outpatients in Medical Insurance System

Names, affiliations, and addresses of authors

The full names of the authors and their designations are mentioned just below the title of the research paper. For instance, if the author ABC is a Professor of Mechanical Engineering at XYZ University, the name and designation can be written as follows:

ABC
Professor, Department of Chemical Engineering
XYZ University, Address

An address serves two purposes: it serves to identify the author and it also supplies the author's mailing address. The mailing address is necessary for many reasons, the most common one being to enable the readers to get reprints if required. Unless the writer of a research paper wishes to publish anonymously, it is obligatory to provide a full name and a full address. The authors' names should be spelt and given in the same way in all their publications. Departure from this causes confusion at the time of accumulating information.

In the case of a multi-author paper, if different authors of the same paper have different addresses, the addresses are linked to the names using superscripted indicators. The addresses are placed either directly below the names of the authors or in the form of footnotes. The author to whom correspondence concerning the paper is to be addressed is indicated by an asterisk (*) and the footnote, 'Author for correspondence.'

Only those authors who have actively contributed to the overall design and execution of the experiments are listed. The authors are listed in the order of importance to the experiments, the first author being acknowledged as the senior author and primary progenitor of the work being reported. However, there is no uniformity regarding the order in which the names of authors should be given; they are sometimes listed alphabetically. Lack of uniformity in this respect sometimes causes problems and confusion at retrieval and retrospective research stages. There is an obvious need for consensus in this regard at the international level.

Abstract

With increased importance acquired by secondary services, particularly abstracting periodicals, the abstract of a research paper has assumed special significance. It has two main functions:

1. To enable readers identify the basic content of a document quickly and accurately in order to determine its relevance to their interests and thus to decide whether they need to read the document in its entirety, and
2. To meet the requirement of abstracting journals.

There are two types of abstracts: *informative* and *indicative*. Normally, a research paper should have an informative abstract that gives information about the purpose of the study, newly observed facts, conclusions of an experiment or argument, and, if possible, the essential parts of any new theory, treatment, apparatus, technique, etc. Sometimes, the abstract is read instead of reading the full paper. It should, therefore, be self-contained with regard to the new information being presented in the paper. The other common type of abstract is an indicative or *descriptive* abstract. This is more suitable for long, descriptive papers. An indicative abstract indicates the contents of the paper and the scope of the work carried out without giving much information about the results and conclusions.

The characteristics of an abstract are as follows:

1. It is as concise as possible and does not exceed 3 per cent of the total length of the paper.
2. It is self-contained.
3. It does not contain any bibliography, figure, or table references.
4. It does not contain any unfamiliar abbreviations and acronyms.
5. It is generally written after the paper is prepared.

The steps involved in preparing an abstract are mentioned below:

1. Read the introductory paragraph of the study to identify the objective.
2. Scan the summary and conclusions at the end to note down the main findings of the study.
3. Read through the text for information on methodology adopted, new data, and any other vital information.
4. Prepare a draft arranging the various items in the following order: objective, new methodology or equipment used, data of fundamental value, and major conclusion and/or correlations derived.
5. Modify and trim the abstract to the required size.

Refer to the exhibit in the Online Resource Centre of a journal article. The opening paragraph is the abstract of the article.

Introduction

The purpose of an introduction is to supply sufficient background information so as to allow the reader to understand and evaluate the results of the study. It may, therefore, become necessary to refer to work performed earlier only in strict relevance to the above purpose. Sometimes it is necessary to outline the author's earlier attempts to solve the problem along with citations to relevant literature. It is, however, redundant to attempt a complete historical survey of the earlier work. Very often, it is possible to cite a single reference to an important recent review article instead of giving a long list of references; all of them might have been referred to in the review article. The following are guidelines for writing a good introduction:

- First, present the nature and scope of the problem investigated.
- Review the pertinent literature to orient the reader.
- State the method of investigation and, if necessary, the reasons for the choice of a particular method.
- State the principal results of the investigation and also the principal conclusions suggested by the results.

Materials and methods

The main purpose of this section is to describe (and if necessary defend) the experimental design, experimental technique, or theoretical derivation, and then provide enough details so that a competent worker can repeat the experiments.

If a well-known technique or approach is used, it is enough to cite the relevant literature reference where the description is available. If the original source is difficult to understand, the method must be described more in detail than just citing a reference. In cases where the technique or approach adopted involves some modification over the earlier technique or approach, give only a detailed description of the modification.

For materials, mention relevant specifications. Describe the experiments performed, the ranges covered, the new equipment used, etc. in sufficient detail. Also include quantities and even physical properties of the reagents used besides providing the technical specifications. This section usually has subheadings.

Results

This section forms the core of the paper—the data. There are three ways of presenting the data: (a) text, (b) tabular form, and (c) illustration form. A particular set of data should be given only in one of these forms. Avoid duplication as far as possible.

Choice of data   While selecting data for inclusion in a paper, avoid the two extremes in this respect. One extreme is the tendency to shift almost the entire data present in the laboratory notebook to the paper. At the other extreme is the situation where data is so insufficient that the reader cannot understand the logic of the conclusions drawn. Give essential data that form the basis for the major conclusions emerging from the study. Let the rest of the data (supporting data) stay in the laboratory notebook. Extensive data for future reference should be deposited in a central data repository, to be made available on demand. Insert a footnote to this effect in the paper at a suitable place.

Choice of form of data presentation   Present only simple and descriptive data in the text. Highlight important data in tables or graphs. This facilitates comprehension and makes descriptions and comparisons vivid. Information given in tables or figures should supplement, not duplicate, the information provided in the text.

As to the choice between tabular and illustrative forms of presentation, the broad criterion is that for values (requiring a high degree of exactness), a tabular presentation is preferable. However, illustrations are preferred to highlight trends. Avoid presentation of trivial data in tables or illustrations, as it leads to distraction.

Presentation of data in tabular form   Tables should be self-sufficient, capable of conveying a message independently. Each table should have a self-explanatory title.

All tables should be numbered consecutively in Arabic numerals and should be referred to in the text by numbers and not by terms such as 'above', 'below', 'preceding', or 'following'.

Column headings should be brief. For units of measurement, standard abbreviations should be used and these should be placed below the column headings. Avoid structural formulae and other such material that cannot be readily typeset inside tables.

Do not make the tables complex by including too many items and too many details in them. Rather than making a cumbersome single table, complex data should be divided and presented in two or more tables. It is also quite often possible to simplify tables by taking out common data and putting them as running matter below the title as headnotes. Footnotes can also be used judiciously to reduce the number of columns.

In the case of big tables, which can neither be shortened nor be split up, try to avoid printing on a bigger sheet and folding the same. After some years, through ageing of paper and constant use in the library, the paper is likely to crack at the fold. In such cases, the table may get detached and misplaced. It is preferable to continue the table on the next page. Of course, the title and the column headings have to be repeated on the second page, but this does not add much to the overall cost.

Presentation of data in illustration form   Each illustration adds to the cost of production. Therefore, include only absolutely essential figures. For example, a figure showing a linear relationship between two parameters can be safely dropped; a mere statement to this effect in the text would suffice. Infrared, ultraviolet, and Nuclear Magnetic Resonance (NMR) Spectra and Differential Thermal Analysis (DTA) curves need not be included; it is enough to give the significant numerical data in the text. Restrict the number of structural formulae to the bare minimum.

Considerable economy in cost and space can be achieved in a number of ways, such as combining several simple graphs into a composite illustration, particularly when either both the parameters or only one parameter is common, and making judicious alterations in scales to reduce the size of the illustrations.

Research papers are usually set in double-column setting. For economy of space, efforts should be made to accommodate as many illustrations as possible in single-column width; page width size should be chosen only in exceptional cases. See the journal articles exhibited in the Online Resource Centre for examples of single-column figures.

Line drawings should be made with Indian ink on white drawing board, cellophane sheet, or tracing cloth. Along with the original drawings, one or two sets of blueprints or photocopies should be supplied to the editor for use at the refereeing stage and other operations prior to the final printing of the paper.

All illustrations (graphs, photographs, flow diagrams, circuit diagrams, etc. taken together) should be assigned consecutive numbers in Arabic numerals in the order in which they are referred to in the text. All figures (graphs, structures, formulae, schemes, charts, etc.) should be referred to in the text as 'Fig. 1', 'Scheme 1', 'Chart 1', etc. and not by expressions such as 'below', 'above', 'preceding', or 'following'.

Captions and legends should be simple, but self-explanatory   As in the case of tables, crowding of data in illustrations is likely to cause difficulty for the reader to comprehend correlations and should be avoided. Also avoid including too many explanatory notes and other details inside illustrations.

While getting illustrations made, pay due attention to the reduction they are going to undergo while printing. The size of letters, numbers, dots, etc. should be such that on reduction they become neither illegible nor too big. Ensure that the lettering of all the illustrations is such that on reduction it closely matches the text font size. For best results, the optimum size of an original illustration is two to three times the reduced size. This affords enough reduction to conceal drafting imperfections. When explanations for symbols used inside illustrations are given in captions, use only such symbols that are available with the printer.

For presenting statistical data, various types of illustrations can be used: graphs, histograms (bar diagrams), pie charts, etc. When there is continuity of variation between two parameters (e.g., temperature vs pressure), use graphs. When data are taken over periodic intervals (e.g., census taken every ten years), use a bar diagram (histogram). When the purpose in making an illustration is to show relative proportions of components of an entity (e.g., percentages of different minerals occurring in a country), it is preferable to use a pie chart.

Photographs (half-tone illustrations)   Submit photographs that show important features prominently and clearly. The background should be unobtrusive, free from distracting objects, and uniform in tone, yet in good contrast with the principal objects of interest. White glossy prints, one and a half to three times larger than the reproduction desired, give best results. Prints on matt paper are not suitable for reproduction. In the case of instruments and equipment, preference should be given to line drawings over photographs.

By suitably marking unwanted portions of a photograph, the size can be shortened and the reduction kept to the minimum. In the case of photomicrographs and electron micrographs, magnification should be mentioned along with the caption (e.g., ×1500). Size reduction of such illustrations should be avoided as far as possible. If, for an unavoidable reason, minor reduction becomes necessary, this change should be taken into account while giving magnification. Another way to indicate magnification is to insert a scale in the illustration itself at a suitable place.

Discussion

The main functions of this section are to interpret data and to highlight the significant features of the data and the possible causes of these features. It should also mention the limitations, if any, of the data and point out any sources of error.

Avoid the tendency to repeat the description of data in this section. What is obvious from the tables or figures need not be described in the text again. This section should interpret the data depicted in the figures and tables.

Conclusions

Conclusions should stem directly from the data presented and no extra material should be introduced. When there are significant findings, conclusions are a necessary part of the paper. The major function of conclusions is to make recommendations based on the results of the study. If no recommendation emerges, this section can be avoided.

In such cases where the study has led to clear-cut findings, it is preferable to give the conclusions in the form of a series of numbered points.

Acknowledgements

In the acknowledgements section, two possible ingredients require consideration. First, any significant technical help received from any individual, whether in the laboratory where the work was performed or elsewhere, should be acknowledged. Also acknowledge the source of special equipment, other materials, etc. Second, acknowledge any outside financial assistance such as grants, contracts, or fellowships in this section.

This section does not include scientific details, but is equally important as it conveys courtesy and gratitude for all the help received for the completion of the project.

List of symbols

Standard abbreviations can be used. It is a good practice to give the full version followed by the abbreviation within parentheses at the first occurrence. Thereafter, only the abbreviated form may be used throughout.

One- or two-letter symbols may be used to represent physical quantities, units, or chemical elements. Use only standard symbols. All symbols other than those that are in common use must be explained in the 'nomenclature' section.

Names of units and their abbreviations should conform to standard practices. The most appropriate system is the International System (SI) of Units.

References

The main purpose in citing references to the work of earlier researchers is to enable the reader to consult the original source. Therefore, unless the references are complete in respect of all bibliographic details, the readers will face immense difficulty in locating the original sources. Only such references should be cited as have been actually consulted.

Thesis

A thesis is a long research report. The report concerns a problem or series of problems in a particular area of research. It describes what was known about it previously, the progress made by the current work in solving it, an interpretation of the results, and where or how further progress in the field can be made. A thesis is not an answer to an assignment question.

One important difference is that the reader of an assignment is usually the one who has set it. He/she already knows the answer (or one of the answers), not to mention the background, the literature, the assumptions and theories, and the strengths and weaknesses of them. The readers of a thesis do not know what the 'answer' is. If the thesis is for the degree of Doctor of Philosophy (PhD), the university requires that it make an original contribution to human knowledge: the research performed must discover something hitherto unknown.

> The examiners reading and assessing a thesis will be experts in the general field of the research, but on the exact topic of the thesis, only the researcher is the world expert.

Obviously, the examiners reading and assessing a thesis will be experts in the general field of the research, but on the exact topic of the thesis, only the researcher is the world expert. A thesis should be written in such a way that the topic is clear to a reader who has not spent years thinking about it.

The thesis will also be used as a scientific report and consulted by future workers in the laboratory who will want to know, in detail, the work performed. Theses are occasionally consulted by people from other institutions, and the library sends electronic versions if requested. More commonly, theses are now stored in an entirely digital form, as .pdf files on a server at the university. The advantage is that the thesis can be consulted easily by researchers around the world.

It is often helpful to have someone other than ourselves read some sections of the thesis,

'Your thesis on "How to water plants"? Interesting.'

particularly the introduction and conclusion chapters. It may also be appropriate to ask other members of staff to read some sections of the thesis that they may find relevant or of interest, as they may be able to make valuable contributions. In either case, provide the revised versions, so that they do not waste time correcting the grammar, spelling, poor construction, or presentation.

The sections that follow provide a guide to thesis writing: on the problems of getting started, getting organized, dividing the huge task into less formidable pieces, and working on those pieces. The discussion is divided into the following sections:

- Outline
- Organization
- Style
- Presentation

- Timetable
- Iteration
- Structure

Thesis writing is a long, difficult task. Fortunately, it will seem less daunting once a couple of chapters have been completed. Towards the end, it might even become an enjoyable task—an enjoyment based on satisfaction of achievement. Like many tasks, thesis writing usually seems worst before beginning; so let us look at how to make a start.

Outline

First, make up a thesis outline: several pages containing chapter headings, subheadings, some figure titles (to indicate which results go where), and perhaps some other notes and comments. Once a list of chapters has been prepared and, under each chapter heading, a reasonably complete list of things to be reported or explained has been included, it provides the momentum required to drive the task further. With that done, the aim of writing a thesis is no longer a daunting goal, but something simpler. The new aim is just to write a paragraph or section about each subheading. It helps to start with an easy one; this makes one familiar with the habit of writing and builds self-confidence. Often the Materials and Methods chapter is the easiest to write—it comprises the work carried out, written carefully, formally, and in a logical order.

How to make an outline of a chapter? Assemble all the figures that will be used in the chapter and put them in the order that explains them best, especially if one were to explain to someone what each of them meant. One might as well rehearse explaining it to someone else—after all there will probably be several seminars to be presented based on the thesis work. Once the most logical order of figures has been established, note down the keywords of the explanation. These keywords provide a skeleton for much of the chapter's outline.

Once an outline is ready, discuss it with a supervisor or a guide. This step is important: the guide will have useful suggestions, but it also serves notice that a steady flow of chapter drafts can be expected that will make high-priority demands on his/her time. Once, with the supervisor's approval, a logical structure has been decided, the guide will need a copy of this outline for reference when reading the chapters.

Organization

It is encouraging and helpful to start a filing system. Create a word-processor file for each chapter and one for the references. Include notes as well as text in these files. One might think of something interesting or relevant to another chapter while working on one chapter. In that case, promptly place a note to do so in the file for that particular chapter. The more such notes accumulate, the easier it will be to develop the content for that chapter.

> 'My thoughts, such as they are, come crowding in so fast upon me that my only difficulty is to choose or reject.'
>
> –Dryden

Keep a backup of these files and do so every day at least. One should also have a physical filing system—a collection of folders with chapter numbers on them. This will make one feel good about getting started and also help clean up the desk. The files will contain not just the plots of results and pages of calculations, but all sorts of old notes, references, calibration curves, suppliers' addresses, specifications, speculations, letters from colleagues, etc., which might suddenly strike as relevant to one chapter or the other. Place these items in the respective folders. As bits and pieces of text are written, place the hard copy, figures, etc. in these folders as well. Consider making a copy of the lab book. This has another purpose beyond security: usually, the lab book stays in the lab, but a copy might be required for your own future use. Further, scientific ethics require lab books and original data to be preserved for at least ten years, and saving a copy would help if the data are unfortunately lost.

While getting organized, make sure to also take care of any university paperwork. Examiners have to be nominated and they have to agree to serve. Various forms are required by the department and by the university administration. Make sure that the rate-limiting step is the production of the thesis, and not some minor bureaucratic problem.

Timetable

Consult the supervisor and make a timetable for writing the thesis—a list of dates on which the first and second drafts of each chapter will be submitted to the supervisor. This structures the time and provides intermediate targets. An aim 'to have the whole thing done by (some distant date)' can be deceptive and it also leads to procrastination. Once we have committed to the supervisor on delivering a first draft of Chapter 3 on Wednesday, it helps focus our attention.

The timetable may be represented as a chart with items that can be checked off as and when they are completed. This is particularly useful towards the end of the thesis when there will be quite a few loose ends here and there.

Iteration

Beginning to write is often the most difficult part. However, it is very important to write *something* every time we sit down to write, even if it is just a set of notes or a few paragraphs of text that we might never show to anyone else. It would be nice if clear, precise prose leapt easily from the keyboard, but it usually does not. Most of us find it easier, however, to improve something that is already written than to produce text from nothing. So put down a draft (a rough draft) for your own purpose, and then clean it up before presenting it to the advisor. Word processors are wonderful in this regard—the first draft need not be started at the beginning; it is also all right to leave gaps with notes to ourselves, and then clean it all up later.

The supervisor might expect to read each chapter in draft form. He/she will then return it with suggestions and comments. *Do not be upset if a chapter—especially the first one written—returns covered in red ink.* The adviser will want the thesis to be as good as possible, because his/her reputation, as well as yours, is at stake. Scientific writing is a difficult art, and it takes a while to learn. As a consequence, there will be several ways in which the first draft can be improved. So adopt a positive attitude towards criticism—each comment suggests a way of making the thesis more perfect.

As we write the thesis, our scientific writing is almost certain to improve. Even for native speakers of English, who write very well in other styles, one notices an enormous improvement in the first drafts from the first to the last chapter written. The process of writing the thesis is like a course in scientific writing, and in that sense, each chapter is like an assignment in which we are taught, but not assessed. Remember, only the final draft is assessed—the more comments the supervisor adds to the first or the second draft, the better the final outcome.

Before submitting a draft, run a spell check so that the supervisor does not waste time on the typos. Take special care to avoid any characteristic grammatical failings one may have.

Style

The text must be clear. Good grammar and thoughtful writing will make the thesis easier to read and understand. Scientific writing has to be formal. Native English speakers should remember that scientific English is an international language. Slang and informal writing will be harder for a non-native speaker to understand.

As we have learnt in earlier chapters, short, simple phrases and words are often better than long ones. On the other hand, there will be times when complicated sentences will be required because the idea is complex. Some lengthy technical words will also be necessary in many theses. Do not sacrifice accuracy for the sake of brevity.

Sometimes, it is easier to present information and arguments as a series of numbered points rather than as one or more long and awkward paragraphs. A list of points is usually easier to write. However, be careful not to overuse this presentation as a thesis must be a connected, convincing argument, and not just a list of facts and observations.

One important stylistic choice is between active and passive voice. Active voice ('I measured the frequency...') is simpler, and often clearer. Passive voice ('The frequency was measured...') might lead to ungrammatical or awkward sentences. Be especially wary of dangling participles when using passive voice. People generally avoid active voice in a thesis because of two reasons: (a) to adhere to the trend as many theses are written in the passive voice, and (b) the use of 'I' is considered to be immodest. However, there is no harm in using the first person singular when reporting work that we did ourself.

Presentation

There is no need for a thesis to be a masterpiece of desktop publishing. Time can be more productively spent improving the content rather than the appearance.

In many cases, a reasonably neat diagram can be drawn by hand faster than with a graphics package, and it can be scanned if an electronic version is required. Either is usually satisfactory. A black-and-white, moderate-resolution scan of a hand-drawn sketch will be bigger than a line drawing generated on a graphics package, but not huge. While discussing the size of files, it should be mentioned that photographs look attractive, but take up a lot of memory. There is also another important factor. A photograph pays attention to the camera angle, the focus, etc., while a schematic diagram focusses on the components that need to be depicted and the way in which the components of the system interact with each other. Hence, the numerically small information content of the line drawing may bear much more useful information than a photograph.

Another note about figures and photographs—do not save ordinary photographs or other illustrations as bitmaps in the digital version of a thesis, because these take up a lot of memory and are, therefore, very slow to transfer. Nearly all graphic packages allow them to be saved in compressed formats such as .jpg or .gif files. Further, space and time can be conserved by reducing the number of colours. In vector graphics (as used for drawings), shades of grey are often produced by black and white pixels, so one-bit colour is adequate.

In general, students spend too much time on diagrams—time that could have been spent on examining the arguments, making the explanations clearer, thinking more about the significance, and checking for errors in the algebra.

In general, students spend too much time on diagrams—time that could have been spent on examining the arguments, making the explanations clearer, thinking more about the significance, and checking for errors in the algebra.

There is no strong correlation between the length of a thesis and its quality. Readers will not appreciate large amounts of vague or unnecessary text.

Structure

The list of contents provided in this section is appropriate for most theses. In some cases, one or two of them may be irrelevant. Results and Discussion are usually combined in several theses. Select the content that would add the most value to the thesis. Then make a list, in point form, of what will go in each chapter. Try to make this rather detailed, so that a list of points is derived that corresponds to subsections or even to the paragraphs of the thesis. At this stage, think hard about the logic of the presentation. Making a plan of each chapter and section before we begin to write will probably make the content clearer and easier to read. It will also be easier to write.

> Think hard about the logic of the presentation. Making a plan of each chapter and section before we begin writing will probably make the content clearer and easier to read. It will also be easier to write.

Title page

This may vary among institutions; for example, Title/Author/A thesis submitted for the degree of Doctor of Philosophy in the Faculty of Science/The University of Rajasthan/Date. Exhibit 16.1 shows a sample title page for a thesis.

Declaration/certificate

Check the wording required by the institution, and whether there is a standard form. Many universities require something like the following: 'This is to certify that the thesis on the topic ... submitted by ... embodies his/her original work supervised by me (signature/name/date).'

Acknowledgements

Most thesis authors include a page of thanks to those who have helped them in the scientific work, and also indirectly by providing educational resources, funds, advice, emotional support, etc. *If any of the work is collaborative, it should be mentioned clearly as to who did which sections.*

Table of contents

If the introduction starts on page 1, the earlier pages, such as Certificate and Acknowledgements, should take Roman numerals for page numbers. It helps to have the subheadings of each chapter, as well as the chapter titles. Remember that the thesis may be used as a reference in the laboratory; so, a detailed Table of Contents helps to find things easily.

Abstract

Of the entire thesis, this part will be the most widely published and read because it will be published in compilations of thesis abstracts. It is best written towards the end, but not at the very last minute because several drafts might be required before the final copy is ready. It should be a summary of the thesis—a concise description of the issue(s) addressed, the method used to resolve it/them, the results, and conclusions. An abstract must be selfcontained. Usually, it does not contain references. When a reference is necessary, the

EXHIBIT 16.1   A sample title page for a thesis

**ENVISIONING THE UNBORN: ART, ANATOMY,
AND THE PRINTING PRESS IN THE EARLY MODERN ERA**

by

ARJUN BHATT

Professor Mallika Singh, Advisor

A Thesis Submitted in Partial Fulfilment of The Requirements
for the Degree of Master of

Arts in Sociology

at

Delhi University

May 2019

---

relevant details from that reference should be included in the text of the abstract itself. In any case, restrict its length to 2 to 5 per cent of the thesis.

Introduction

The Introduction should describe the topic and its significance. State the problem(s) as plainly as possible. Remember that having worked on this project for a few years gives you the kind of familiarity with the topic that others might not possess. Hence, try to step back mentally and take a broader view of the problem.

Do not overestimate the readers' familiarity with the topic, especially in the introduction. The thesis will be read by researchers in the general area, but not all of them need be specialists in that particular topic. It may help to imagine such a person—we could probably think of a researcher whom we might have met at a conference for our subject, but who was working in a different area. He/she is intelligent, has the same general background, but knows little of the literature or tricks that apply to this particular topic.

> The introduction should be interesting. If the reader loses interest here, then it is unlikely to revive his/her interest in the further sections.

The introduction should be interesting. If the reader loses interest here, then it is unlikely to revive his/her interest in the further sections. Reading several thesis introductions available in the library will familiarize one with the general style for writing an introduction.

This section might go through several drafts to make it read well and logical, while keeping it short. For this section, it is a good idea to ask someone who is not a specialist to read it and comment on the following

> 'I wish he would explain his explanation.'
> –Byron, on Coleridge

lines: Is it an adequate introduction? Is it easy to follow? The introduction must be written—or at least major revisions carried out—towards the end of the thesis writing because this section indicates the progress of the thesis, which may become clearer during the writing.

Literature review

Where did the problem come from? What is already known about this problem? What other methods have been tried to solve it?

A literature review proves that the thesis discusses something of importance and interest. Ideally, most of the work in the chosen field of interest would already have been carried out by other researchers. We are only trying to improve the sense of writing about a specific subject. The review is a concise but comprehensive summary of different articles to establish the significance of the topic. To be able to balance one's view and opinions on the topic, it is very important to keep up with the literature right from the beginning of the study, and note down any important papers over the years. A summary of these papers serves as a good starting point for the review.

How many papers to include and how relevant do they have to be before being included is a matter of judgement. About a hundred is a reasonable figure, but it will depend on the field. It gives an opportunity to demonstrate the fact that the thesis author is the world expert on the (narrow) topic. See the following sample paragraph from a literature review about gender and alcohol:

> Much of the research examining alcohol effects has focused on males. Yet, epidemiological and clinical data have shown notable sex differences in alcohol use and propensity for abuse and dependence. For instance, women have shorter intervals between the onset of drinking and the emergence of problem drinking than men (Greenfield, 2002). Women also differ from men in their sensitivity to a number of acute and chronic consequences of ethanol (e.g. NIAAA, 2004; Fillmore and Weafer, 2004). Sex differences in drinking behaviors often become more pronounced during adolescence and may be associated, in part, with puberty-related increases in gonadal hormones [see Witt (2007)].

> 'Research is to see what everybody else has seen, and to think what nobody else has thought'
> –*Albert Szent-Gyorgyi*

(*Source*: Vetter-O'Hagen, Elena Varlinskaya, and Linda Spear, 'Sex Differences in Ethanol Intake and Sensitivity to Aversive Effects during Adolescence and Adulthood', Alcohol and Alcoholism, vol. 44, no. 6, 2009.)

Middle chapters

In some theses, the middle chapters are the journal articles of which the student is the major author. There are several disadvantages to this format.

One is that a thesis is both allowed and expected to have more details than a journal article. For journal articles, one usually has to reduce the number of figures. In many cases, all of the interesting and relevant data can be included in the thesis, and not just those that appeared in the journal. The degree of experimental details is usually greater in a thesis. Often, a researcher requests a thesis in order to obtain more details about how a study was performed.

Another disadvantage is that journal articles may have some common material in the Introduction and the Materials and Methods sections.

The exact structure in the middle chapters will vary among theses. In some theses, it is necessary to establish some theory to describe the experimental techniques, then to report what was done on several different problems or different stages of the problem, and then finally to present a model or a new theory based on the new work. For such a thesis, the chapter headings might be: Theory, Materials and Methods, [first problem], [second problem], [third problem], [proposed theory/model], and then Conclusions.

For other theses, it might be appropriate to discuss different techniques in different chapters rather than to have a single Materials and Methods chapter.

Here are some comments on the elements Materials and Methods, Theory, and Results and Discussion, which may or may not be separate sections in a thesis.

Materials and methods   This section varies enormously from thesis to thesis and may be absent in theoretical theses. It should be possible for a competent researcher to reproduce exactly the same experiment with the same or similar results by following the description provided in this section. There is a good chance that this test will be conducted; another researcher might at some point want to perform a similar experiment either with the same gear or with a new set-up in a foreign country. Therefore, this section must be written keeping in mind these possibilities.

In some theses, particularly multidisciplinary or developmental ones, there may be more than one such chapter. In this case, the different disciplines should be indicated in the chapter titles.

Theory   Irrespective of the category the thesis belongs to, one chapter should be included that discusses the basic theory on which the thesis is built. For example, if the thesis is on a management topic, this section should be included to explain the basic concepts and theories involved in the detailed research.

Results and discussion   The results and discussion are very often combined in theses. This is sensible because of the length of a thesis: there may be several chapters of results, and, if we wait till they are all presented before beginning the discussion, the reader may lose continuity and have difficulty in comprehending. The division of the Results and Discussion material into chapters is usually best done according to subject matter.

Make sure that the conditions under which each set of results was obtained is described precisely. What was held constant? What were the other relevant parameters? Also make sure that appropriate statistical analyses have been used. Where applicable, show measurement and standard errors on graphs. Use appropriate statistical tests.

In most cases, a discussion is needed to interpret the results and define their significance. What do they mean? How do they fit into the existing body of knowledge? Are they consistent with current theories? Do they give new insights? Do they suggest new theories or mechanisms?

One must try to distance themselves from the usual perspective and look at the work from a broader perspective. We should not just ask ourselves what it means in terms of the orthodoxy of our own research group, but also try to understand how other people in the field might perceive it. Does it have any implications that do not relate to the questions that we set out to answer?

Conclusions and suggestions for further work

As mentioned, the abstract should include a brief account of the conclusions. However, the actual conclusion section is much longer than this account in the abstract, and allows one to be more explicit and more careful with the prescribed criteria or conditions. It might be helpful to note down the conclusions in point form.

It is often the case with scientific investigations that more questions than answers are produced. Does the work suggest any interesting further avenues? Are there ways in which the work could be improved by future workers? What are the practical implications of the work? Usually, the chapter on further work should be reasonably short—perhaps a few pages. As with the introduction, it is a good idea to ask someone who is not a specialist to read this section and comment.

References

It is tempting to omit the titles of the articles cited, and the university allows this, but think of all the times when one would have seen a reference in a paper and gone to look it up only to find that it was not helpful after all. Include all the important sources consulted, used, or quoted in the thesis.

Appendices

If there is material that should be included in the thesis but which would break the flow or make it laborious for the reader, include it as an appendix. Some topics that are typically included in appendices include important and original computer programs, data files that are too large to be represented simply in the results chapters, and pictures or diagrams of results that are not important enough to keep in the main text.

Table 16.2 recapitulates the similarities and differences pertaining to the three forms of technical documents discussed in this chapter.

Table 16.2 Comparison of research papers, dissertations, and theses

	Research paper	Dissertation	Thesis
Objective	Presenting an organized analysis of research work in a journal specific to the field of research.	Presenting an organized analysis of research work as a part of completing an assigned task spanning over a few months/one term/one semester.	Presenting an organized analysis of research work as the culmination of doctoral research spanning over two or more years.
Length	Three to ten pages	Thirty to fifty pages	Longest among the three
Style	Formal and objective	Formal and objective	Formal and objective
Evaluation	Evaluated by the editor of the journal	Evaluated by the guide	Evaluated by the guide and other external examiners
Structure	Does not contain a cover page. Only the title is given and there is no separate title page	Contains a cover page and a title page	Contains a cover page and a title page

Technical Description

Technical description is a key part of any descriptive technical document because it defines objects and/or processes. Basically, a technical description divides a complex item or topic into more manageable components. It defines, describes, and illustrates the various elements contained within the whole—whether an object, process, or concept. As an example, one might break down something as complex as a human body into its various components, such as the nervous, circulatory, and digestive systems, and muscles, organs, and tissues. This process (called division and classification) helps to manage the overwhelming amount of detail inherent in complex tasks such as the description of the body. It is simply breaking down large amounts of information into more manageable segments.

In many cases, the division process continues through many levels of detail; for example, one component of the digestive system might be the stomach, which is further divided into valves and secretive pores, and other more minute elements that make up that particular organ. The process of classification groups similar items on the basis of function or location or material in order to put them into some category relevant to the descriptive task. The general guideline for writing technical descriptions is to generate iterations (cycles) of *naming, defining, describing,* and *illustrating*. The technical description usually ends with a description of *one complete cycle of operation* of the object or process.

Guidelines for Writing Good Descriptions

The following are guidelines for writing good technical descriptions with respect to organization, content, and structure.

Organization

Overview   Begin with a brief overview that reveals the object's

  (a) Overall framework, arrangement, or shape
  (b) Purpose or function

Parts   Divide the object into parts and describe each part

  (a) In sufficient detail so that reader is able to use, make, or draw it
  (b) In a way that reveals its role and its relation to other parts

Order   Organize the part descriptions in one of the following orders:

  (a) Spatial order (top to bottom, outside to inside)
  (b) Priority order (most to least important)
  (c) Chronological order (order of [dis]assembly)

Content

Specifics

  (a) Include relevant specific features (such as size, shape, colour, material, and technical names)
  (b) Omit irrelevant background, confusing details, and needless words.

Comparison   Compare features or parts with other things already familiar.

Contrast   Contrast properties with the properties of others to reveal their significance.

Structure

Format   Clarify the text with

  (a) *Heads* Identify topics with clear, nested section headings.
  (b) *Lists* Itemize related features with indenting and marks.
  (c) *Figures* Integrate figures and text with labels and references.

Verbal cues   Guide the reader through the instructions with

  (a) *Parallelism* Use parallel words and phrases for parallel ideas.
  (b) *Proleptics* Use verbal links (*also*, *but*, *however*, etc.) to signal how the description fits together.

Checklist

As you reread and revise your instructions, watch out for problems such as the following:

- Make sure you provide real instructions—explanations of how to build, operate, or repair something.
- Write a good introduction—in it, indicate the exact procedure to be explained and provide an overview of contents.
- Make sure that you use the various types of lists wherever appropriate. In particular, use numbered vertical lists for sequential steps.
- Use headings to mark off all the main sections and subheadings for subsections. (Remember that no heading 'Introduction' is needed between the title and the first paragraph. Remember not to use first-level headings in this assignment; start with the second level.)
- Use special notices as appropriate.
- Make sure you use proper format for all headings, lists, special notices, and graphics.
- Use graphics to illustrate any key actions or objects.

- Provide additional supplementary explanation of the steps as necessary.
- Remember to create a section listing equipment and supplies, if necessary.
- Include strong sections of definition, description, or both, as necessary, using the guidelines provided on content, organization, and format.
- Share the draft with the people of similar aptitude and knowledge for whom the instructions are made.

Writing Technical Descriptions

The main steps involved in writing technical descriptions include naming, definition (assigning meaning to objects), description (highlighting certain aspects of the object), and illustration (description using graphic elements rather than words). These four activities are explained as follows:

Naming

This most basic activity, naming, is how we understand the world around us. Without names, we would have to refer to objects, people, or places with sounds and gestures: 'uhh, that big thing over there' or 'you know, what's his name, the little fat guy with the blue hair'. This might be acceptable to some extent in a face-to-face communication, but it is impossible in written communication. We describe the world around us by naming what we see. Hence, start with naming the objects and processes that are being described.

Definition

A technical description begins with a definition (formal or extended) of the object or process to be described and a general breakdown of the components that will be detailed. The introductory paragraphs (or pages, in some cases) provide the reader with general information of the content that follows. Usually, the definition is followed by a list of the components and a brief note on the detailed description of each. In the case of the digestive system, for example:

First, the definition

> The digestive system is essentially a tube passing through the body from the mouth to the anus. It is responsible for the ingestion and processing of food into useable energy, which is taken up by the body's cells, and non-useable waste products, which are eliminated.

Then, the list of components

> The digestive system consists of the mouth, oesophagus, stomach, small and large intestines, the colon, and the anus.

Finally, an outline of the detailed description that follows

> Each of these components is described in terms of its function and cellular makeup in the subsections that follow.

A *definition* fixes an object, concept, or process within some system of knowing. Definitions appear in one of three forms: *formal, informal,* and *extended* definition. A formal definition will contain a *term* (that which is to be defined), a *category* to which the term belongs, and a set of *differentia* (a set of words that separate the term from those elements within the category in which the term is located). An informal definition appears in brackets in a sentence to help clarify a concept (for example, one was used in the definition of *term* above). An extended definition is a form of technical description itself and may run to tens, hundreds, or even thousands of pages in length. The formal definition is the most common form and is the one given in dictionaries. As a rule, a definition cannot use the term itself in the category or the differentia. For example, we cannot say that a Calico cat is a cat that is calico in colouration—this is referred to as a circular definition.

If you have ever played 'twenty questions' (the game in which the players try to guess the identity of an object in the mind of another player by asking questions to which the answer can only be yes or no), you will have a good grasp of categories. To start a game like this one, a player will ask a question such as 'Is the object manufactured' or 'Is the object natural'. It would be useless to ask 'Is it a comb' right away since you only have twenty questions that you can ask. The strategy to playing successfully is to narrow the category to which the object belongs sufficiently so that naming specific objects becomes a possibility.

Asking whether an object is big or small provides no useful information because the terms big and small are relative—one quickly learns to relate the size to something specific—like a Volkswagen or a Canadian dime. For example, a float-plane is an object that is larger than a Volkswagen, but you would still have a difficult time determining the name of the object from that category.

Functional categories are also useful in this case; such as 'What does it do?'. Questions such as 'Is it a form of transportation?' or 'Is it a tool?' might help. To guess that the object is a form of transportation gets us somewhere closer to a correct guess, in the case of the float-plane, but so are cars, trains, and ships. As it is obvious, narrowing the category is critical to winning this game—guess air transport, guess fixed-wing air transport, guess fixed-wing air transport that is able to take off or land on water and you have a working definition of a float-plane.

Description

Description is the process of making an object, idea, or process known to someone who is unfamiliar with it (it is very much like a definition in that respect). A description will use words and illustrations to outline the shape, the material, the purpose or function, and the relationship of one object, idea, or process to other objects, ideas, and/or processes. A description attempts to make the unknown familiar; therefore, it occasionally uses the light of the familiar to illuminate the darkness of the unfamiliar. In this quest, it is common for descriptions to use analogies, metaphors, or similes to get an idea across (for example, the metaphors of light and darkness in the preceding sentence were used to carry the meaning of the process of gaining knowledge). Description also relies upon strategies of organization such as division and classification, comparison and contrast.

Division and classification is the process of breaking down complex systems into more manageable components and then grouping the components together based on some over-riding determinant such as spatial relationship, functional or genetic similarity, chronological relationship, or a host of other bases upon which a classification can take place. The result is usually informative or analytical in nature rather than comparative. When the programming takes place for solving an algorithm, the programme is made of subprogrammes which are called within the main programme to get the final output of the algorithm.

Comparison and contrast is familiar to most people who have ever shopped for something that is made by many suppliers, or that comes in a variety of models—like a car or a computer. The compare and contrast method depends on a fixed set of criteria (such as cost, practicality, efficiency, or options available) in order to analyse the choices presented to the user. The criteria are applied to each of the cars or computers, for example, and a comparison of the results usually points to a better car or computer. Compare and contrast analysis is result oriented, and sometimes persuasive, rather than being strictly informative.

Illustrations

It has been said that a picture is worth a thousand words; however, in technical writing, it must be understood that a picture does not *replace* words; rather it *enhances the meaning* of the words. A prudent technical writer uses graphics to his/her advantage to show an overall view of the object or process and to illustrate each of the sub-divisions into which the object or process is divided.

Graphics are very useful as aids to transmit meaning, especially when language is a barrier to understanding, but they are limited to describing something abstract. Consider the following: How would you describe the beautiful maple leaf to someone who has never seen one? Or a hockey stick?

Graphics are ideal for representing things that are complex. However, they do not define themselves; we have to label and describe each part of the graphic based on what we want to convey. Consider this example: Is the definition of a hockey stick different for someone who has never seen one compared to someone who uses one professionally? The answer is yes. The hockey player defines the stick in a much different way because he/she has a different need—the player defines the stick in terms that are more refined: blade width, amount of curve allowed, shaft flexibility, and a host of other regulations imposed by the governing body of whichever league he/she plays in. The person who has never seen a hockey stick will most likely want to know what it looks like and how it is used.

SUMMARY

Research papers may be classified under the advanced forms of technical writing, generally taken up by academicians and researchers. This formal form of writing possess certain specific characteristics and include some essential components in its structure. It is also necessary to follow certain steps in writing them. Knowledge of all these features will enable one to write this important document with effectiveness in order to disseminate research-related information to their readers. The importance of technical descriptions is also growing fast with the development of technologically advanced products. A technical description is required to understand the details of an object or a process, and to be able to identify the problem if something goes wrong in the functioning. Hence, one needs to keep in mind organization, content, structure, classification, and level of detailing in order to provide a clear and effortless description that is easy for the reader to grasp.

A thesis is, in fact, a long research report. It takes approximately three or more years to complete a thesis. Students who are interested in research must know the essential components, structure, and style of a thesis.

EXERCISES

1. Read the following statements and say whether you agree or disagree with each of them. Then write the justification/explanation for your point of view.
   (a) There is no significant difference between a research paper and an article.
   (b) The title of a research paper should be self-explanatory.
   (c) The language of a research paper should be like the language of a technical proposal/report.
   (d) The objective of writing a research paper is the same as that of a technical report.
   (e) There is no defined order of writing different elements of a research paper.
   (f) Acknowledgements should be placed at the beginning of a research paper.
   (g) Informative Abstract and an Executive Summary are one and the same type of synopsis used in research papers.
   (h) A research paper contains only References, no Bibliography.
   (i) The 'Introduction' to the research paper only mentions the earlier relevant work.
   (j) The writer of a research paper should use only standard symbols, abbreviations, and nomenclature.

2. What are the points you need to bear in mind while preparing the title of a research paper?

3. Write the technical description of word processor following the guidelines discussed in the chapter. Also give illustrations to enhance the description.

4. Select an object from your environment and write the technical description either individually or in the team. You can describe any tool, mechanism, or a piece of equipment. You should have the exact and precise description of these to write the technical description.

5. Find examples of technical description and technical process from text books, journals, professional magazines and the Internet. Analyse them and see how effectively they are written. If you find any flaw, rewrite them after applying the techniques discussed in the chapter.

6. Keeping in mind the areas of your research interest, write down any five topics on which you would like to pursue your PhD research.

PART IV

Review of Grammar

Chapter 17: Grammar and Vocabulary Development

CHAPTER
17

Grammar and Vocabulary Development

OBJECTIVES

You should study the chapter to know the
- ○ history and origin of words
- ○ significance of dictionaries and thesauri for vocabulary enrichment
- ○ different ways of word formation
- ○ what confusables, homonyms, homophones, and eponyms are
- ○ phrasal words and the phrasal verb pattern
- ○ how to use the following important parts of speech: nouns, adjectives, adverbs, prepositions, and infinitives
- ○ the role of gerunds and infinitives
- ○ subject–verb agreement in sentences
- ○ tenses as well as active and passive voice
- ○ articles and conditional sentences

Introduction

As we have learnt throughout this book, communication is a process of sending and receiving intricate and complicated messages in commonly accepted language. The persons taking part in a conversation should be competent in both written and spoken communication. To hone these skills, the training pertaining to listening, speaking, reading, and writing, covered in the preceding parts of this book, has become necessary. However, the basis of all these forms of communication is sound vocabulary and grammar.

Vocabulary is important as it aids expressions and communication with its size usually linked with reading comprehension ability. Linguistic vocabulary is synonymous with thinking vocabulary and people are judged by others based on their vocabulary.

In the same manner, it is mandatory to understand the rules of grammar in order to formulate correct sentences.

In this chapter, we will learn the basics of English vocabulary and grammar. We will start with vocabulary building and then touch upon the important grammatical elements of the language, explaining these with several examples.

A Brief History of Words

The basis of modern English is old English or Anglo-Saxon, a Germanic language spoken by the invaders of the British Isles in the fifth century AD.

"The grammar is faulty and spelling ridiculous; otherwise it is an impressive résumé.

Over the centuries, English has been influenced by the languages of many other people—Scandinavian, Norman (French), and Latin.

As the language has grown and developed, words have undergone transformations in their spelling and meaning. The pronunciation of old English words differs considerably from that of their modern equivalents. The long vowels (wherever the vowel sound is prolonged, e.g., /a:/, /I:/, /u:/) in particular have undergone considerable modification. For example, the old English 'stan' is the same as the modern English 'stone', 'heafod' is now 'head', and 'sawol' is 'soul'. Look at the following list of old English words and their modern versions.

Old English	Modern English	Old English	Modern English
halig	holy	bat	boat
gan	go	fot	foot
ban	bone	cene	keen
rap	rope	fyr	fire
hlaf	loaf	riht	right

Over a period of time, words stopped being used in a particular context. Instead, they would crop up in a new context. This resulted in a change of meaning. For example, the word manuscript originally meant 'that which is written by hand' (from the Latin *manu*, 'by hand', and *scriptus*, 'written'), but today it refers also to material written on a typewriter or computer. Another example is the word 'place'. Earlier, it only referred to an open square in a village or town. Today, it covers a considerably wider range: you speak of a place in a book, and you can put someone in his/her place. Some more examples are given below.

Word	Original meaning
citizen	city-dweller
fellow	associate
box	boxwood
junk	old rope
bead	prayer

There is another category of words where the meaning has become restricted. 'Wife' originally meant 'woman' but is now restricted to a particular relation.

Word	Original meaning
deer	animal
meat	food
lesson	reading
liquor	liquid

Closely connected with expansion and restriction of meaning is the change whereby a concrete word has taken an abstract meaning, or an abstract term has become more concrete. An appropriate example is the word 'humour', which in medieval medicine was thought to be one of the four fluids responsible for one's health and disposition. This word now refers to 'temperament' and 'mood', thus shifting from a concrete to an abstract meaning. The abstract Latin word *fructus* (derived from the verb *fruor*, 'to enjoy') originally meant enjoyment, but even in classical Latin, we find it frequently used in the sense of 'profit', 'income,' 'fruit' (as in 'the fruits of my labour'). Later, it became restricted to the enjoyment of tangible objects such as apples, oranges, grapes, and pears.

Through the ages, words have risen or fallen on the social scale by a process called pejoration and amelioration. In the first instance, words that had a favourable implication may 'degenerate' in meaning. For example, take the words boor, churl, silly, and foolish. Originally, these words had the following meanings:

Word	Original meaning
boor	peasant
churl	freeman
silly	soul
foolish	blessed fool

What a transformation! In modern English, they have acquired such negative connotations that they are virtually banned from polite conversation.

The opposite of this process, that is, amelioration and elevation of meaning also happens. Take some commonly used adjectives, such as fond, nice, and naughty. Fond was used by Shakespeare in the sense of foolish. Nice (origin *nescius*, meaning ignorant) meant stupid, but later acquired a pleasant connotation, such as the expression 'a nice way', or a nice person. Naughty literally meant 'naught-like', that is, worthless, and it meant only that. Later it was used in the sense of wicked, but nowadays it has weakened to a mild term of reproach. Words keep getting added to this list, and therefore, you should constantly attempt to expand your vocabulary.

Let us now look at some of the more recent trends that the language has undergone. The widespread use of mobile phones and short messaging service (SMS) has spawned an entirely new 'language' whose purpose is to send quick messages. These messages are not based on any rules of grammar and are created by the user as he/she sees fit. Look at the following messages:

- Hw r u?
- Hi wassup? Hw hve u been?
- Ive been bz al dis while, need 2 catch up asap!
- Hey I g2g now. Brb
- Recd yr msg. Get bk 2 u soon.
- M in ofc bk to bizness

As the purpose is to send and comprehend quickly, these messages are usually short and have telegraphic language. They are very specific to the individual who is involved in the communication.

Slang is another category of language, very informal in nature. It is often used by members of a particular group, schoolboy slang, for instance, or 'campus lingo'. Words such as *sams, fundoo, sploosh, junta, zook,* and *cracked* are examples of campus lingo in India. A layperson might be completely mystified hearing a student saying, 'He went to the ski to have sams with a dayski' (meaning 'He went to the skylab to have samosas with a day scholar'). But the members of the student's group would understand this very clearly. Words and phrases belonging to this genre are subject to fad. They come and go with equal rapidity and what was in vogue yesterday marks you as out of touch if you use it today.

Slang does not belong only to particular groups. Most people use slang, at times, to convey humour, or sound forceful or vivid. You must be familiar with some of the following:

Slang	Meaning	Slang	Meaning
a blue moon	rare phenomenon	in a wax	angry
peg	drink of brandy and water	tin	money
never say die	don't despair	on the top	moving
mind your eye	be careful	all serene	all right
mug up	get by heart	no go	of no use
I feel very fit	well		

A fairly recent development in India is the widespread use of 'Hinglish', or the combination of Hindi and English spoken by a wide cross section of people. Many Indians, who are more comfortable speaking in their own language, find it convenient to speak this hybrid language, switching from one to the other effortlessly. While purists may complain, the popularity of Hinglish can be gauged by the fact that it is now used even by reputed English-language newspapers in India.

While Hinglish is a modern phenomenon, it would be worth remembering that many words from native Indian languages have long been assimilated into English and can be found in virtually all English dictionaries. For example, pyjama, jungle, bungalow, thug.

Hinglish word	Meaning
desi	local, of Indian origin
phirangi	foreign
junta	people in general
masti	fun
bindaas	carefree, unorthodox

Throughout history, language has continued to evolve and change in many directions. Users of the language should be familiar with words, their usage, and meaning in today's context. Since words are the primary symbols of communication, you must build an extensive vocabulary. This will enable you to become fluent writers and speakers.

Vocabulary, or the words one knows, can be broadly divided into two categories: active and passive. Words that fall into your active vocabulary are those that you know well and feel confident about using in your speech and writing. Our passive vocabulary consists of words that we have heard or read and understand, but do not feel too confident about using. To increase your active vocabulary, you must move words from your passive to your active vocabulary by consciously using them. A dictionary will help you here greatly. You can also increase your passive vocabulary by reading widely or watching and listening to more serious programmes on television and radio.

Learning Words by Etymology

If you learn a word by understanding its etymology (origin and history), then you can learn a string of words at a time. Larger dictionaries can provide you with this information. For example, animus means mind and it appears in words like:

Pusillanimous—A person who lacks courage

Equanimity—When a person shows evenness of temper, especially under stress

Magnanimous—A large-hearted person

An even more systematic way to improve your vocabulary is to fit every new word you come across in a pattern:

- Its grammatical class and usage
- Its etymology (or origin)
- Which other words it combines with

- Its exact meaning and how it relates to words with similar meanings
- Social situations where it can or cannot be used

Using the Dictionary and Thesaurus

A dictionary is an invaluable tool for building up your vocabulary. Often you come across a word whose meaning you guess by its context. But you could miss out its exact meaning by a narrow margin or by a mile. This is where a dictionary comes in handy. For example:

The institute had eight special officers patrolling (not paroling) during the ragging period.

Proper study methods will ensure (not induce) good grades in college.

A dictionary will help you to use words and phrases more exactly. It will also help you find a range of possible words for a particular situation. A dictionary lists the words of a language in alphabetical order and gives information about their meaning, their grammar, spelling, usage, pronunciation, their history, and so on. The degree of completeness of this information depends upon the size and purpose of the dictionary.

In a thesaurus, the words are arranged in groups that have similar meanings. It can be arranged alphabetically or thematically. The largest thesaurus in the world is the *Historical Thesaurus of the Oxford English Dictionary.* Two of the most commonly used thesauri are

- *Roget's Thesaurus of English Words and Phrases*
- *The Oxford American Thesaurus of Current English*

Online dictionaries are also available now where you can search for the word you require. Some well-known dictionary websites are

- www.dictionary.com
- www.yourdictionary.com
- www.onelook.com

Some well-known dictionaries are

- *Oxford Advanced Learner's Dictionary*
- *Concise Oxford English Dictionary*
- *Webster's New World Dictionary*
- *The American Heritage Dictionary of the English Language*

- *Merriam-Webster*
- *The Chambers Dictionary*
- *Longman Dictionary of Contemporary English*

Thesaurus

The name 'thesaurus' comes from a classical Greek word meaning 'treasury' or 'store' and that is exactly what it is: a treasure-house of words. It was invented in the mid-nineteenth century by Peter Mark Roget who published, in 1852, his Thesaurus of *English Words and Phrases*, classified and arranged so as to facilitate the expression of idea and assist in literary composition.

Changing Words from One Form to Another

There are various methods of word formation such as blending, compounding, coinage, borrowing, and clipping. It is only in the process of derivation where you find change in the forms of the words. Indeed derivation is accomplished by means of a large number of small 'bits' of the English language. These small 'bits' are called affixes, such as un-, mis-, pre-, -ful, and -less. Affixes can be divided into two categories—prefixes and suffixes. If affixes are added at the beginning of a word (e.g., un-) these are called prefixes. Added to the end of the word (e.g., -ish) these are called suffixes.

Derivations are used to make new words in the language and are often used to make words of a different grammatical category from the stem. For example, the addition of the derivational morpheme '-ness' changes the adjective 'good' to the noun 'goodness'. Similarly, the noun 'care' becomes the adjectives 'careful' or 'careless' via the derivational morpheme -ful or -less. The Online Resource Centre contains tables showing how derivational affixes change the grammatical category of words, and change words from one part of speech to another.

Word Formation: Prefixes and Suffixes

New words can be formed with the help of affixes, which include both prefixes and suffixes. By using these, new words can be formed by making some modification in the root word. For example, if the root word is 'regulate', you can change it to *deregulation* by adding two affixes—*de-* (prefix) and *-tion* (suffix). By using prefixes and suffixes, the meaning of the word changes. As another example, if you consider the word 'easy', you can see how the prefix *un-* changes the word to *uneasy* and the suffix *-ness* further changes it to *uneasiness*. After becoming aware of this pattern of word formation, you can easily learn new words, thus enhancing your vocabulary.

The prefix need not have any meaning in isolation but, in some cases, it has a meaning. For example, the prefix *man-* in *manmade, manhole, manhandle,* and *manhood* has a meaning. Man can also be used as a suffix as in clergyman, fireman, etc. Though the prefix *dis-* has a meaning *not/the opposite of* in isolation, it does not act as a prefix in words such as *distribute, distinguish, disturb,* and *district*. But when used as a prefix in words like disappointment, it changes the meaning of the word a great deal. *Disinterested, disapprove,* and *disadvantage* are some other examples.

Let us now see some examples of suffixes that change the word class. The suffix *-able* added to the verbs *read, work,* and *suit* changes them into the adjectives *readable, workable,* and *suitable*. Likewise, the suffix *-ly* added to the adjectives *slow, personal,* and *quick* converts them into the adverbs *slowly, personally,* and quickly; the suffix *-ion* added to the verbs *infect, expedite,* and *complete* changes them into the nouns *infection, expedition,* and *completion*.

Many words used in the English language today are not originally English and are borrowed from other languages. Therefore, it would be useful to know the etymology of the suffix and prefix. For example, the Latin word *septa* means 'seven' and when we say *septuagenarian*, we mean someone in his seventies (in terms of age).

A large number of English words have Latin or Greek origins. It is interesting to note that the three words, affixes, suffixes, and prefixes, themselves are also different combinations of the same concept with its base as *-fix*. Some examples of prefixes and suffixes are provided in the Online Resource Centre.

Synonyms and Antonyms

In any kind of writing, the choice of words is very important. If you use the same word several times, it becomes monotonous. To break this monotony, it is a good idea to use synonyms. It is believed that no two words have exactly the same meaning; however, they can be interchangeably used. A word can be substituted with its synonym only if it is the same part of speech. You must have complete knowledge of the meaning and categorization of the word. A few examples given below show the same category of synonyms:

- *entertainment* and *amusement* (noun)
- *broaden* and *widen* (verb)
- *slowly* and *leisurely* (adverb)
- *enticing* and *tempting* (adjective)

Synonyms

A synonym is word or expression that has the same or nearly the same meaning as another in the same language. Words that are synonyms are said to be *synonymous*. If you talk about a *long time* or an *extended* time, long and extended become synonyms. In the figurative sense, two words are often said to be synonymous if they have the same connotation. For example, *dark* is synonymous to *gloomy* as *merry* is to *happiness*.

> The word good has nineteen synonyms.

Synonyms can be any part of speech (e.g., nouns, verbs, adjectives, adverbs, or prepositions), as long as both members of the pair are the same part of speech. More examples of English synonyms are given below:

- *Beaker* and *receptacle* (noun)
- Opening and *aperture* (noun)
- *Wash* and *clean* (verb)
- *Beautiful* and *attractive* (adjective)
- *Saturnine* and *woebegone* (adjective)
- Quickly and *rapidly* (adverb)
- *In* and *into* (preposition)

Synonyms are defined with respect to certain senses of words; for instance, *pupil* as the 'aperture in the iris of the eye' is not synonymous with *student*. Similarly, the word *expired* meaning 'having lost validity' (as in medicines) does not necessarily mean *death*. The word usage is very important. Even words similar in meaning like *close* and *shut* may have slightly different nuances. 'Closing a shop' implies that the shop is no longer operational and no one can do the business. But 'shutting the shop' means the shop is being made secure so that nothing can be taken out.

It is an accepted fact that no synonyms have exactly the same meaning because of several reasons such as etymology, orthography, phonic qualities, ambiguous meanings, and usage. Only in some situations, the synonyms identical in meaning can be interchangeably used for practical purposes. Different words that are similar in meaning usually differ for a reason; for example, *feline* is more formal than *cat*. *Long* and *extended* are only synonyms in one usage and not in others, such as a *long arm* and an *extended arm*; in isolation, *long* and *extended* are very different in meaning.

Synonyms are also a source of euphemisms. When you want to say 'he died', you will use the expression 'he passed away'. A thesaurus offers a

'Kailash, I'm Shankar and this is Shiva.'

listing of similar or related words; these are often, but not always, synonyms. A close synonym of *strong* is *muscular*, but it places much more emphasis on physical strength. By contrast, *stalwart* and *staunch* are synonyms that emphasize more abstract aspects of this meaning of strength. Synonyms can sometimes be group of words, as in *gain* and *get one's hands on*.

Even if a word has innumerable synonyms, it is not necessary that the word in a given context can be substituted by a synonym. The following example makes it clear:

Example *Alleviate*

Synonyms: abate, allay, assuage, mitigate, relieve, temper

To *alleviate* is to make something easier to endure (alleviate the pain following surgery); *allay* is often used interchangeably, but it also means to put the rest, to quiet or calm (to allay their suspicions). *Assuage* and *allay* both suggest the calming or satisfying of a desire or appetite, but *assuage* implies a more complete or permanent satisfaction (we allay our hunger by nibbling *hors d'oeuvres,* but a huge dinner assuages our appetite). To *relieve* implies reducing the misery or discomfort to the point where something is hearable (relieve the monotony of the cross-country car trip) and *mitigate,* which comes from a Latin word meaning 'to soften', usually means to lessen in force or intensity (mitigate the storm's impact). *Abate* suggests a progressive lessening in degree or intensity (her fever was abating). To *temper* is to soften or moderate (to temper justice with mercy), but it can also mean the exact opposite—to harden or toughen something (tempering steel; a body tempered by lifting weights).

For more examples, refer to the Online Resource Centre.

Antonyms

Antonyms are the word pairs that are opposite in meaning such as dark and light, tall and short, and acme and abyss. Words may have different antonyms depending on the meaning. Both *long* and *tall* are antonyms of *short*. There are different kinds of opposites as discussed below.

Absolute   Such antonyms have no comparison or with no middle ground, e.g. true and false. Logically, a statement is either true or false; it cannot be slightly true or rather false.

Gradable   These are the opposites with gradations of meaning, e.g. hot and cold; it makes sense to say something is rather hot or very cold. There are many words to represent intermediate stages of hot and cold, such as tepid, cool, and warm. So, hot and cold are at opposite ends of a continuum, rather than being absolutes.

Relational   The pairs such as buy and sell, pedagogue and protégé, where the relationship between the two objects is described, fall in this category.

Auto   These are the words that mean opposite in different contexts such as wind up (close, start), fast (quick movement, fixed firmly), stay (at a specific place, postpone), out (visible, invisible), and cleave (to split, to adhere).

Languages often have ways of creating antonyms as an easy extension of lexicon. Examples are the English prefixes *dis-* and *un-*. Respect is the antonym of *disrespect* and *comfortable* is of *uncomfortable*. The Online Resource Centre gives a list of antonyms.

Idioms

An idiom is a phrase whose meaning is difficult or sometimes impossible to guess by looking at the meanings of the individual words it contains. For example, the phrase '(being) in the same boat' has a literal meaning that is easy to understand, but it also has a common idiomatic meaning. Consider the following sentence:

I found the job difficult at first. But we were all in the same boat; we were all learning.

Here, (being) *in the same* boat means 'to be in the same difficult or unfortunate situation'. Some idioms are imaginative expressions such as proverbs and sayings:

Too many cooks spoil the broth.

The expression means that if too many people are involved in something, it will not be well done. If the expression is well known, part of it may be left out:

Well I knew everything would go wrong— it's the usual story of too many cooks!

Other idioms are short expressions that are used for a particular purpose:

> Hang in there! (used to encourage somebody in a difficult situation)
>
> Get lost! (a rude way of saying 'go away!')

Many idioms, however, are not vivid in this way. They are considered as idioms because their form is fixed:

- for certain
- in any case

 The Online Resource Centre provides a listing of popularly used idioms.

Confusables

The term 'confusable' or 'confusible' is a semi-technical term for one of two or more words that are commonly or easily confused with one another, e.g., *luxuriant* with *luxurious*; *they're* with *there* and *their*. The British lexicographer Adrian Room (*Dictionary of Confusing Words and Meanings*, 1985) separates *confusibles* or 'lookalikes' such as *dominating* and *domineering* from *distinguishables* or 'meanalikes' such as *faun* and *satyr*. At least seven factors contribute to confusion:

1. *Homophony*, in which words have the same sound but different spellings and meanings: slay, sleigh.
2. *Homography*, in which words have the same spelling, but different sounds and meanings: wind moving air, wind to turn or twist.
3. *Shared elements*: *mitigate* and *militate* share the same number of syllables, the same stress pattern, and the same opening and closing syllables.
4. *Transposable* or *exchangeable elements*: *cavalry* and *Calvary*, *form* and *from*, *accept* and *except*. Factors 3 and 4 become more potent still when words have similar meanings and uses: *affect* and *effect*.
5. *Words mistaken for phrases* or *vice versa*: *already* and *all ready*.
6. *Semantic proximity*: *baroque* and *rococo*, *nadir* and *zenith*. In this case, confusion may be encouraged by different but related applications of the same terms by different people: *acronym* and *initialism*, *subconscious* and *unconscious*. Some words are very different in meaning but sometimes displace one another because of close association: *acid* and *alkali*, *defendant* and *plaintiff*.
7. *Uncertainty arising from different uses in different varieties of English*: *biscuit* and *cookie* in British English and American English.

Look at the following example illustrating the wrong usage of words:

> A boy of twelve is in intensive care in hospital after a group of teenagers doused him in *inflammatory* liquid and then threw a lit match at him.

Here the writer meant *inflammable*, capable of being set on fire, not *inflammatory*, tending to stir up trouble.

 The Online Resource Centre provides lists of frequently confused homophones, as well as several words that are commonly confused, with a summary of their main meanings.

One-word Substitutes

Often we have words on the tip of our tongue, and we all know how hard it is sometimes to remember the exact word appropriate to a certain situation or object. The following is a list of some such words. The entries are listed under the key concept. The list is arranged alphabetically with respect to the key concepts.

Act

Act on another's behalf: **represent**

Abroad

Sell abroad: **export**

Absence

Employee's statement that his/her absence was due to illness: **self-certification**

Acupressure

Spot where pressure is applied: **acupoint**

Aircraft

Aircraft's main body: **fuselage**

Belt

Belt attachment to hold a sword: **frog**

Beyond

Beyond what is necessary: **supererogatory**

Broadcast

Unscheduled broadcast of important news: **newsflash**

Clean

Sterilize food by gamma radiation: **irradiate**

Computer hardware

Hardcopy describing the uses and design of hardware: **documentation**

Concrete

Concrete layer under a building: **oversite, raft**

Direction

Designed to operate in a particular direction: **directional**

Disease

Indication of a disease: **symptom, stigma**

Preventing disease: **prophylactic**

Enjoy

Enjoy without hurrying: **savour**

Enlarge

Enlarge by internal stretching: **distend**

Exchange

Mutual exchange: **reciprocity**

False

Writings whose authorship is falsely assigned: **pseudepigrapha**

Fee

Fee paid to recompense someone doing 'unpaid' work: **honorarium**

Filter

Filtration process for blood of a patient suffering kidney failure: **dialysis**

Glacier

Material deposited by a glacier: **moraine**

Gravity

Fall controlled by gravity only: **free fall**

Grid

Grid for data display: **matrix**

Heat

Heat and then cool slowly as a toughening process: **anneal**

Decompose by exposure to extreme heat: **pyrolyse**

Hope

Denoting a hope unlikely to be fulfilled: **pious, vain**

Identity

Item habitually seen with a person or thing, and serving to identify them: **attribute**

Information

Alter information to make it more acceptable: **launder**

Internet

Proper way to use the Internet: **netiquette**

Joint

Flat ring sealing a joint: **gasket, washer**

Lathe

Person working wood on a lathe: **turner**

Lens

Counterbalanced camera platform that can be raised or turned: **crane**

Light

Use of light for medical treatment: **phototherapy**

Map

Ratio between actual and mapped distances: **scale**

Meaning

Study of meaning: **semantics**

Medication

Denoting the application of medication through skin: **transcutaneous, transdermal**

Night

Night blindness: **nyctalopia**

Notice

Accidental failure to notice: **oversight**

Number

Fixed or permitted number: **quota**

Office

Period of holding office: **tenure**

Origin

Study of the origin of names: **onomastics**

Paper

Role of paper for continuous printing: **web**

Plastic

Overlay with plastic: **laminate**

Quality

Assessed by quality: **qualitative**

Radar

Metal foil released to evade radar detection: **chaff, window**

Record

Collection of documents or records: **archive, registry**

Remove

Remove from a dangerous place: **evacuate**

Satellite

Satellite's communication link with earth: **downlink**

Scent

Perforated container for sweet-smelling substances: **pomander**

Sea

Mapping of seas: **hydrography**

Telephone

Direct telephone line for emergency use: **hotline**

Temperature

Excessively low body temperature: **hypothermia**

Theory

Supporter of a theory: advocate, exponent, proponent, **protagonist**

University

Former university student: **alumnus**

Valve

Externally operated valve regulating flow through a pipe: **stopcock**

Vehicle

Weighing apparatus for vehicles: **weighbridge**

Vibration

Strong vibration: **judder**

Wall

Recess in a wall: alcove, niche

Water

Irrational fear of water: hydrophobia

Wind

Revolving pointer showing wind direction: weathercock, weathervane

X-rays

Screen for direct viewing of X-ray images: fluoroscope

Yield

Policy of yielding to the demands of a potential aggressor: appeasement

Zigzag

Zigzag course taken by a vessel sailing into the wind: tacking, traverse

Homonyms

A homonym is a word that has both the same pronunciation and the same spelling as another, but is etymologically unrelated to it. The following are a few examples of homonyms:

- *bill* (statement of charges): *bill* (beak)
- *fair* (just): *fair* (sale, entertainment)
- *pole* (long slender rounded piece of wood or metal): *pole* (each of the two points in the celestial sphere about which the stars appear to revolve)
- *pulse* (throbbing): *pulse* (edible seeds)
- *row* (noun, a line): *row* (verb, propel boat)
- *soil* (earth): *soil* (make dirty)

Traditionally, homonyms of this type are treated as separate words and given distinct dictionary entries (e.g., 'pole 1' and 'pole 2') whereas more closely related meanings are treated as offshoots of the same word, which historically speaking they are (so 'Each of the two terminals of an electric cell or battery etc.' comes under 'pole 2').

Popularly, homonyms may or may not include pairs whose two words have the same meaning but do not belong to the same grammatical category (e.g. red, noun, and adjective).

 Loosely, *homonym* is sometimes used for a word that has either the same sound or the same spelling as another (but not both). The Online Resource Centre provides a list of common homonyms.

Homophones

A homophone is a word that is pronounced the same as another. This term is usually used for partial homonyms, which are distinguished by both meaning and spelling. Examples of homophones are

feat: feet, no: know, none: nun, stare: stair

Some English pairs are homophones in some accents but not in others, for example:

saw: sore, pore: pour, wine: whine

The occurrence of homophones is largely a matter of historical chance, in which words with distinct meanings come to coincide phonologically: *byre*—a cowshed, *buyer*—one who buys. Words may be homophones in one variety of English but not another: *father/farther* and *for/four* are homophonous in received pronunciation (the standard form of British English pronunciation, based on educated speech in southern England, widely accepted as a standard elsewhere), but not in American English and Scottish English; wails/Wales are general homophones; wails/Wales/whales are homophones for many, but not in Irish English and Scottish English. Whether/whither are homophones in Scotland, but not whether/weather, which are homophones in England. The Online Resource Centre provides a basic list of homophones.

Eponyms

An eponym is a person or thing, or the name of a person or thing, after whom something is named, such as a building, an institution, an organization, a machine, a product, or a process. The following are a few examples of eponyms:

Alzheimer's disease	from Alois Alzheimer (1864–1915), German neurologist
Braille	from Louis Braille (1809–52), French inventor
diesel	from Rudolf Diesel (1858–1913), German engineer
mackintosh	from Charles Macintosh (1766–1843) (with a change of spelling)
Morse code	from S. F. B. Morse (1791–1872), American inventor
sandwich	from the 4th Earl of Sandwich (1718–92)
stetson	from John B. Stetson (1830–1906), US hat manufacturer

An eponym is also the person whose name is used as follows: The Roman emperor Constantine, who gave his name to Constantinople.

The adjective eponymous is used in the following way:

- *Beowulf* is the *eponymous* hero *of the Old English poem* of that name.
- *Emma* is the *eponymous* heroine of the novel *Emma* by Jane Austen.
- *Robinson Crusoe* is the *eponymous* hero of *The Life and Strange and Surprising Adventures of Robinson Crusoe* by Daniel Defoe.

The process of eponymy results in several forms:

- Simple eponyms such as *atlas*, which became popular after the sixteenth century Flemish cartographer Gerardus Mercator put the figure of the titan Atlas on the cover of a book of maps.
- Compounds and attributive constructions such as *loganberry*, after the nineteenth century US lawyer James H. Logan, and Turing machine after the twentieth century British mathematician Alan Turing.
- Possessives such as *Parkinson's Law*, after the twentieth century British economist C. Northcote Parkinson, and the *Islets of Langerhans*, after the nineteenth century German pathologist Paul Langerhans.
- Derivatives such as *bowdlerize* and *gardenia*, after the eighteenth century English expurgator of Shakespeare, Thomas Bowdler, and the nineteenth century Scottish-American physician Alexander Garden.
- Clippings such as *dunce* from the middle name and first element of the last name of the learned thirteenth century Scottish friar and theologian John Duns Scotus, whose rivals called him a fool.
- Blends such as *gerrymander*, after the US politician Elbridge Gerry (born 1744), whose redrawn map of the voting districts of Massachusetts in 1812 was said to look like a salamander, and was then declared a gerrymander. The word became a verb soon after.

Phrasal Verbs

A phrasal verb is a short phrase made up of a verb and one or two prepositions or adverb. Each phrasal verb has its unique meaning(s), which is different from the meaning(s) of the verb itself. For example, the meanings of 'take', 'take in', and 'take up on' are totally different. In other words, phrasal verbs are verbs with particles. A common example is the verb 'to fix up' as in 'He fixed up the car'. The word 'up' here is a particle, not a preposition, because its position can be changed: 'He fixed the car up.' These movements of the particle 'up' quickly distinguishes it from the preposition 'up'.

Phrasal verbs are different from verbs with helpers. The particle that follows the verb changes the meaning of the phrasal verb in idiomatic ways. Some particles can be separated from the verb so that a noun and pronoun can be inserted, and some particles cannot be separated from the verb. In addition, some phrases are intransitive, meaning they cannot take a direct object.

- Separable 'add up' (meaning: to add)
  She added up the total on her calculator. (Correct)
  She added it up on her calculator. (Correct)
- Inseparable 'get around' (meaning: to evade)
  She always gets around the rules. (Correct)
  She always gets the rules around. (This construction makes no sense in English.) (Incorrect)
- Intransitive 'catch on' (meaning: to understand)
  After I explained the math problem, she began to catch on. (Correct)
  She began to catch on the math problem. (Catch on cannot take a direct object in this meaning.) (Incorrect)
  She began to catch on to the math problem. (the word 'to' makes the math problem an indirect object, which is acceptable in this meaning.) (Correct)

Phrasal Verb Patterns

A phrasal verb is a combination of either prepositions or adverbs, or both. It may also combine with one or more pronouns or nouns.

Particle verbs

Phrasal verbs with adverbs resembling a preposition are sometimes called particle verbs. These are related to separable verbs. Basically, there are two main patterns in this: transitive and intransitive.

- An intransitive particle verb does not have any object. For example:
  When he entered the room I *looked up.*
- A transitive particle verb has a nominal object in addition to the adverb. Here one thing should be taken care of—that if the object is an ordinary noun, it can usually come on either side of the adverb, though very long noun phrases tend to come after the adverb, as in the third example given here:
  Switch off the light.

  Switch the light off.

  Switch off the lights in the corridor outside the room.

However, with some transitive particle verbs, the noun object must come after the adverb. In these cases 'inseparable' phrasal verbs are involved. For example:

  The hot balloon gave off the fumes. (not The hot balloon gave fumes off.)

Still there are transitive particle verbs that require the object to precede the adverb. For example:

  They let the lady through. (not They let through the lady.)

In all the transitive particle verbs, if the object is a pronoun, it must precede the adverb. For example:

  Switch it off. (not Switch off it.)

  They let her through. (not They let through her.)

Prepositional verbs

Phrasal verbs with a preposition are called as prepositional verbs. These verbs are always followed by their nominal object and are different from transitive particle verbs because the object still follows the preposition if it is a pronoun. For example:

  We *look after* our children.

  We *look after* them. (not look them after)

When a verb has its own object, it usually precedes the preposition. For example:

He *helped* Namita *to* an extra portion of apples.

He *helped* her *to* some. (with pronouns)

Prepositional verbs with two prepositions are possible. For example:

Sahil *talked to* his mother *about* his new bike.

Phrasal prepositional verbs

A phrasal verb can be combination of both an adverb and a preposition at the same time. the verb itself can have a direct object. For example:

The novice driver *got off* to a flying start. (no direct object)

Onlookers *put the accident down to* the driver's loss of concentration. (direct object)

The Online Resource Centre lists a few common phrasal verbs. The Online Resource Centre also provides a list of technical vocabulary, common errors in usage, commonly misspelt words, and British words along with their American equivalents.

Some Pointers to Good Word Usage

1. *Be positive* We all have a 'passive vocabulary'—words we know but never use. There are also words we have read or have a vague idea about. Try shifting these words into the vocabulary that you are familiar with and use.

2. *Dictionary and Thesaurus* These are invaluable tools. Use a dictionary to check the meanings of words. Make sure you can use it easily and confidently. A thesaurus will give you all the options to a word, so that you can choose the best one for the occasion.

3. *A wide choice* Remember, English gives you several choices of words for a single meaning. When you write, do not opt for the first word that comes to mind. Make an effort to think of other words. Use a thesaurus to help you in your work.

4. *Who is your audience?* When speaking or writing, always keep your audience in mind. Communication is a two-way street. Use words that are in keeping with your audience's level of language skills and their knowledge of the subject.

5. *Social setting* Do you have an informal or formal relationship with your audience? Words that suit one social setting could be disastrous in another and ruin your communication, however accurate your language may be. Choose your style of speaking or writing according to the occasion.

6. *Jargon* Unless your target audience is in the know, jargon tends to put off people. Avoid it.

7. *New words* Just as words fall into disuse and vanish from the language, new words are constantly being added. Some words will last and become widely used; others will fade away. While it is interesting to try out new words, be wary of trying to sound 'fashionable' all the time, or you will end up looking ridiculous.

8. *Sentence construction* Learn how words work in sentences before you can start developing your vocabulary. This will greatly benefit you, especially in writing.

9. *Understand word structure* Many words have common 'building blocks'—try to figure out how these work. Also try to understand how common prefixes and suffixes are used in constructing new words.

10. *Enjoy the language!* English is an amazingly creative, flexible language, capable of assimilating various influences and expressing subtleties of thought. As Thomas Carlyle said, 'Be not the slave of words'. Do not let words scare you—rather, get a hold on them and enjoy using them in your speaking and writing.

Nouns

Noun is defined as a word used for a name of a person, place, thing, or idea. In English, it is one of the eight parts of speech. It is further classified as common, countable/uncountable, concrete/abstract, collective, proper, gerund, and compound.

1. The common noun names general things such as city, country, chair, car, dog, etc. Common nouns are general names and are not capitalized unless they begin the sentence or are part of a title.

2. The nouns that refer to things which can be counted are countable nouns. They are easy to recognize. For example, we can count pens, chairs, people, dogs, cats, etc. They can be singular or plural. When they are singular, we must use a/an/the/my/this. For example,

   I want an apple.

   or

   Where is my pen?

   When countable nouns are plural, we can use them alone. For example,
   I want apples.

   or

   Pens are lying on the table.

3. Uncountable nouns refer to items, concepts, etc. that cannot be divided into separate elements. These are further classified as concrete and abstract nouns. Examples of concrete nouns are iron, rice, and furniture. You can experience this group of nouns with your five senses: you *see* them, *hear* them, *smell* them, *taste* them, and *feel* them. For example,

   Ramya licked the ice cream.

   Here *ice cream* is a concrete noun. We can *see* the pink colour. We can *taste* the vanilla flavour. We can *feel* our tongue growing numb from the cold. Any noun that one can experience with at least one of the five senses is a concrete noun. Whereas, we cannot experience abstract nouns in the same way as concrete nouns. For example, concentration, homework, and freedom. We cannot see the colour of concentration, we can neither taste it, nor hear it.

4. Nouns that refer to group of people or things are collective nouns. Collective nouns are different names given to collections or groups, be they birds, animals, or things. For example, committee, crowd, army, navy, conglomeration, and family.

5. Nouns that refer to specific people, organizations, or places are proper nouns. A proper noun has two distinctive features: (a) it will name a specific item, usually a one-of-a-kind and (b) it will begin with a capital letter no matter where it occurs in a sentence. For example, France, America, Kolkata, Ashok, and Indian Space Research Organization.

6. Nouns that are formed from a verb by adding 'ing' are called gerund nouns. It can follow a preposition, adjective, or most often a verb. For example,

   I love *dancing*.
   I love *participating* in drawing competitions.

7. Nouns that are made up of two or more words are called compound nouns. These nouns are explained in detail in the following section.

Compound Nouns

Compound nouns are formed by nouns modified by other adjectives or nouns. For example, toothpaste is a compound noun formed by two nouns. Black bird has black as an adjective and bird as noun. In both these examples, the first word modifies or describes the second word, telling us what kind of object or person it is, or what its purpose is. And the second part identifies the object or person in question. Other examples are as follows:

Weekly magazine	adjective + noun
Swimming pool	verb + noun
Underground	preposition + noun
Haircut	noun + verb
Hanger-on	noun + preposition
Dry washing	adjective + verb
Input	preposition + verb

A compound noun can be formed by just putting the word side by side, e.g., bedroom, or putting a hyphen, such as check-in, and also by separating them with a single space, such as full moon.

Noun Phrases

There are five types of phrases, which are named after the part of speech that is the head of the phrase. They are noun phrase, verb phrase, adjective phrase, adverbial phrase, and prepositional phrase. The noun phrase is the nucleus of every sentence. A noun phrase can have infinite length as well as any number of other phrases, e.g., adverb, adjective, and noun within its syntactical structure.

> A noun phrase has as its head a noun, pronoun, nominal adjective, or a numeral.

A noun phrase has as its head a noun, pronoun, nominal adjective, or a numeral. For example, 'flowers', 'they', and 'the flowers' are noun phrases, but 'flower' is just a noun, as is indicated in the following sentences (the noun phrases are italicized):

Question:	Does he like *flowers*?
Answer:	Yes, he likes *them*.
Question:	Do you like *the flowers over there*?
Answer:	Yes, *they are* beautiful.
Question:	Do you like *the flowers that bloom in my garden*?
Answer:	Yes, I like *them*. (*Note*: Here, 'them' refers to 'the flowers', not 'flowers'.)
Question:	Do you also like *the pink coloured flower*?
Answer:	Yes, I like it.

Some noun phrases consist of only one word like the noun *flowers* (in the first question) and the pronouns *it*, *them*, and *they*. But most noun phrases have more than one word.

A noun phrase comprises a noun (obviously) and any associated modifier. For example,

- The tortuous and meandering *road*
- A black *car*
- Any related *report*

Different types of modifiers of a noun are given below:

Determiners:	The scientist assembled *the* machine.
Possessives:	She brought the *machine's* parts.
Adjectives:	The *cumbersome* machines are kept underground.

| Prepositions: | The pedestrians *over* the bridge … |
| Clauses: | The conference *we went to* … |

The noun phrase formed by modifiers can be in different forms and combinations. For example, the brilliant and assiduous scientist (*adjectival*), the boundary following the edge of the college premises (*participial*), the first woman to walk in space (*infinitive*), the interview that he had given the day before (*modifying clause*), and the motel next to the supermarket over the bridge (*prepositional*).

The possible functions of a noun phrase are given below:

The subject:	*The magnificent aircraft* soared upwards.
The object:	I saw *that movie about buccaneers*.
The complement:	She dreams of being *an engineer*.
Possessive:	*My husband's* grandfather was a professor.
The object of a preposition:	He jumped over *the wall*.

If you look closely at any sentence, you will find many noun phrases that contribute to the length and profundity of the sentence. For this reason, it is necessary to know noun phrases and their structure.

> You can identify a noun phrase by observing all the words that are contiguous with the noun itself.

You can identify a noun phrase by observing all the words that are contiguous with the noun itself. But sometimes, the noun phrase can be broken up to become discontinuous. You will also find that a part of the noun phrase is placed at the end of the sentence (usually, modifying phrases—participial or prepositional) so that it receives more attention. See the sentence given below:

Recently, innumerable rallies have been reported **involving the hoi polloi to have turned into imbroglios**.

The entire noun phrase can be put together: 'That *innumerable rallies involving the hoi polloi have turned into imbroglios* have been reported recently.' However, shifting the modifying phrases, the bold italicized part of the phrase, to the end of the sentence puts additional emphasis on that part. Some more examples are given below.

A rumour spread among the students that the exams are being postponed till June. (instead of *'A rumour that the exams are being postponed till June spread among the students.'*)

The time had come to *start working on the project seriously and to involve more people for making it successful*. (instead of *'The time to start working on the project seriously and to involve more people for making it successful had come.'*)

Sometimes, you will find the noun phrase as a long compound phrase. Such a noun phrase is called a *stacked* or *packed* noun phrase. Many a time, one noun modifies another as in head student, story book, and water bodies. But by creating a long thread of attributive nouns or modifiers, we sometimes make things difficult, as shown in the following sentence:

Students who develop webpages would be familiar with what is now known as the uniform resource locator protocol problem.

The difficulty here is to understand what is modifying what. Also, as you keep expecting the string to end, the energy of the sentence dwindles into a series of false endings. Such phrases are a particular temptation in technical writing. You can improve the sentence by identifying the extended compound noun phrase and then removing the last noun series and converting one of the modifying nouns into a prepositional phrase as shown below:

The problem with the protocol of uniform resource locators is now recognized by students who develop webpages.

So, in this situation, making a sentence longer is probably an advantage as it simplifies the complex original sentence.

Gerunds

Any verb with the *-ing* form used as a subject of a verb and acting like a verb-noun is called a *gerund*. Read this sentence:

Meditating is her favourite kind of relaxation.

Meditating is formed by adding *-ing* to the verb *meditate*. Some more examples of gerund are given below:

Poaching is a prohibited activity in India.

In the above sentences, the gerund, like a noun, is the subject of a verb and, like a verb, also takes an object; hence, it indicates that the gerund also has the force of a verb.

I love singing classical songs.

In this sentence, the gerund, like a noun, is the object of a verb and, like a verb, also takes an object, thus indicating that it also has the force of a verb.

> Gerund is the form of the verb that ends in '-ing' and has the force of a noun and a verb.

Finally, the gerund, like a noun, can be governed by a preposition and, like a verb, also takes an object as in the following sentence:

He has a penchant for listening to melodious songs.

Compound gerund forms are formed by keeping a past participle after the gerunds of *have* and *be*. Some sentences of compound gerund forms are given below:

Having worked for ten hours, he felt sleepy.

She was angry about *not having been invited*.

He loves *being appreciated*.

The gerund of a transitive verb has the following forms:

Active	*Passive*
Present: steering	Present: being steered
Perfect: having steered	Perfect: having been steered

As both the gerund and the present participle end in the *-ing* form, be careful to discriminate between them. A *verbal noun* is the gerund that has the force of a noun and a verb, whereas, the *verbal adjective* has the present participle that has the force of an adjective and a verb.

Examples of gerund:

Smoking is bad for health.

The child was tired of *swimming*.

We were stopped from *watching* the match.

Examples of participle:

He is *smoking* too much these days.

He was *swimming* when I saw him.

She was *watching* the match when her phone rang.

In the compound nouns *dancing-shoes, frying-pan,* and *writing-table,* dancing, frying, and writing are gerunds. They mean 'shoes for dancing', 'a pan for frying', and 'a table for writing', respectively.

Remember to use the possessive case of nouns and pronouns before gerunds, as in the following sentences:

They celebrated at *his* being promoted.   (not *him*)

I heard of *their* having shifted to another town.   (not *them*)

We have no trust in *his* keeping his promise.

Why do you insist on *her* being present?

All plans depend on *Sona's* passing the entrance examination.

The mishap was due to the *pilot's* ignoring the signals.

> Always use the possessive case of nouns and pronouns before gerunds

Uses of Gerunds

A gerund being a verb-noun may be used as the following:

- Subject of a verb:
  Smoking is bad for health.
  Reading is his favourite pastime.
- Object of a transitive verb:
  Stop *talking*.
  Girls love making dolls.
  I love *reading* serious stories.

- Object of a preposition:
  I am tired of *writing*.
  He is fond of *singing*.
  He was put behind bars for *telling* a lie.
- Complement of a verb:
  Doing is *learning*.
  What is most detested is *drinking*.

Infinitives

An infinitive is a kind of noun with some features of the verb, especially that of taking an object (when the verb is transitive) and adverbial qualifiers. In short, the infinitive is a verb-noun. Read the following sentences:

She never *finds* fault with me. (finite verb)

She never tries to *find* fault with me. (verb infinitive)

In the first sentence, the word *finds* has *she* for its subject; hence, the verb *find* is restricted by person and number. You call it a finite verb (all verbs in the indicative, subjunctive, and imperative moods are finite, as they are restricted by the person and number of their subject). In the second sentence, *to find* merely names the action denoted by the verb *find* and is used without mentioning any subject. It is thus not limited by person and number as a verb that has a subject and is, therefore, called the *verb infinitive* or simply the *infinitive*. Read the following sentences:

> Though the word 'to' is frequently used with the infinitive, it is not an essential part or sign of it.

To *err* is human. (Here the infinitive, like a noun, is the subject of the verb is.)

Girls love to *dance*. (The infinitive here, like a noun, is the object of the verb love.)

To *respect* our elders is our moral responsibility. (Though the infinitive, like a noun, is the subject of the verb is, it, like a verb, also takes an object.)

She refused to *obey* the rules. (In this sentence, the infinitive, like a noun, is the object of the verb refused and, like a verb, also takes an object.)

Many people desire to *make* wealth quickly. (The infinitive, like a noun, is the object of the verb desire, but, like a verb, also takes an object and is modified by an adverb.)

Though the word *to* is frequently used with the infinitive, it is not an essential part or sign of it. Thus, after certain verbs (*bid, let, make, need, dare, see, hear,* etc.), we use the infinitive without *to* as shown in the following sentences:

Bid him *goodbye.*

Let them *play.*

I will not let you *sleep.*

Make him *do* his homework.

I need not *go* there.

You need not *wait* for him.

You dare not *watch* it.

He saw me *do* it.

I heard her *shout.*

The infinitive without *to* is also used after the modal auxiliary verbs *shall, will, should, would, may, might, must, can,* and *could.*

You *shall* come. (You will be compelled to come.)

I *will* study. (I am determined to study.)

He *may* leave. (He is at liberty to leave; He is permitted to leave.)

You must attend the class. (You are ordered to attend the class.)

I *can* dance. (I am able to dance.)

The infinitive without *to* can also be used after *had better, had rather, would rather, sooner than,* and *rather than.* See the following sentences:

She *had better* come seek permission.

He *had rather* write than read.

I *would rather* live than die.

Uses of Infinitives

The infinitive, with or without adjuncts, may be used like a noun as discussed below:

- As the subject of a verb:
  To point mistakes is easy.
  To practise regularly is important.
- As the object of a transitive verb:
  I do not mean to hurt.
  It's beginning *to rain.*
- As the complement of a verb:
  Her most cherished dream is *to write.*
  Her habit is *to walk daily.*

- As the object of the preposition:
  I had no alternative but *to experiment.*
  The company had no choice except *to file* for bankruptcy.
- As an objective complement:
  Mother wanted him *to study.*

When an infinitive is used as a noun, it is called the *simple infinitive.* The infinitive is also used as follows:

> When an infinitive is used as a noun, it is called the 'simple infinitive'.

- To qualify a verb, usually to express purpose:
  He called *to see* the programme. (for the purpose of seeing the programme)
  We breathe *to live.* (purpose)
  I came *to see* her off. (purpose)
  It upset him *to find* that a theft had taken place in the house. (cause)

- To qualify a noun:
  This is not the time *to cook.*
  She will have cause *to regret.*
  He is a person *to be adulated.*
- To qualify a sentence:
  To share my feelings, I often remember you.
  He was horrified *to hear* the news.

- To qualify an adjective:
  Mangoes are good *to eat*.
  That horse is hard *to catch*.
  The girls are anxious *to dance*.
  My mother is too ill *to do* any household chores.

In these uses, the infinitive is called the *gerundial* or *qualifying infinitive*. In the first two groups of sentences, the gerundial infinitive does the work of an adverb; in the third group, it does the work of an adjective; and in the fourth group, it is used absolutely.

An infinitive may be active or passive. When active, it may have a present and a perfect form, may merely name the act, or may represent progressive or continued action.

Active

Present: to write　　　　　　　　　　　Perfect: to have written

Present continuous: to be writing　　　Perfect continuous: to have been writing

When passive, the infinitive has a present and a perfect form.

Passive

Present: to be written　　　　　　　　Perfect: to have been written

You will see that the infinitive and the gerund are similar in being used as nouns, while still having the power that a verb has of governing another noun or pronoun in the objective case. Therefore, gerund is the form of the verb that ends in *-ing* and has the force of a noun and a verb.

Both the gerund and the infinitive have the same uses of a noun and a verb. Thus, you may find in many sentences either of them being used without any special difference in meaning, as shown below:

His brother started *playing* the guitar when he was in second standard.
His brother started *to play* the guitar when he was in second standard.

To earn is better than *to spend*.

Earning is better than *spending*.

Subject–verb Agreement

The 'subject' should agree with the 'verb' in number and person. A sentence that has a singular subject is accompanied by a singular verb. On the other hand, a sentence that has a plural subject should have a plural verb. The complex subject of the sentence is followed by a verb that agrees with the main noun in the subject.

He plays in the playground.　　　(singular subject)
They play in the playground.　　　(plural subject)

There are some nouns that can be treated as both singular and plural forms.
The government has (have) announced its (their) new employment scheme.

Other words that can have either singular or plural verbs are school, class, department, team, university, press, public, crowd, firm, committee, community, family, generation, electorate, group, jury, orchestra, and the names of specific organizations such as the Reserve Bank of India, Infosys, Maruti, and BBC.

> You use a singular verb if you see an institution or organization as a whole unit and plural verb if you see it as a collection of individuals.

You use a singular verb if you see an institution or organization as a whole unit and plural verb if you see it as a collection of individuals.

There is not much difference in meaning, although in formal writing, use of singular verb is more common. But in some contexts, a plural form of the verb is needed. See the following sentence:

The jury disagree about the guilt of the accused.

However, in the following sentence, a singular form is preferred and you would say

The orchestra is about to play.

This cannot be ' The orchestra are …' as you are referring to the orchestra as a unit, and not as the individuals comprising it.

There are some nouns that are plural and take a plural verb, such as premises, particulars, belongings, clothes, goods, earnings, surroundings, stairs, riches, savings, congratulations, and thanks.

Congratulations are due to you on your grand success.
His *belongings are* kept in the locker for safety.

Plural verbs are used with the nouns such as police, people, and staff. Some nouns always end in -s and look as if they are plural, but when we use them as the subject of a sentence, they have a singular verb as shown in the following sentence:

The *news* about the kidney racket is very disturbing.

Other words that end in -*s* and take a singular verb are means (method or money); academic disciplines, e.g., mathematics, physics, statistics, linguistics, economics, phonetics, and politics; sports such as athletics and gymnastics; and diseases such as diabetes, measles, and rabies. However, let us compare the following:

In academic disciplines:

Politics is the favourite of many students in this university.

Statistics was always the most feared subject for students.

Economics has only recently been added to the course list.

But in general use:

What *are* your politics?   (political beliefs)

The statistics *are* not indicative of this fact.   (information shown in numbers)

The economics of the project *are* not very encouraging.   (the finances)

When a subject is made up of two or more items joined by *either … or* or *neither … nor*, we use a singular verb if the last item is singular (although a plural *v*erb is sometimes used in informal English) and a plural verb if the last item is plural.

Either the classroom or the auditorium *is* a good place to hold the lecture.

Neither the secretary nor his representatives *are* to join the club.

If the last item is singular and the previous item plural, you can use either a singular or a plural verb.

Either the teachers or the principal *is/are* to blame for the problem in the school.

Let us now run through some important rules.

Use the singular verb in the following cases:

(a) With words such as *any of, none of, the majority of, a lot of, plenty of, all (of) some (of)* and an unaccountable noun

(b) With *every* and *each* (normally used with a singular noun):

Every room in this guest house *has* an attached kitchen.

Each of the boys *plays* well.

(c) With *everyone, everybody*, and *everything*, (every one is two words when the meaning is each one).

Every one in the audience *is* considered for inclusion in the workshop.

Use the plural verb in the following cases:

With *a/the majority of, a number of, a lot of, plenty of, all (of), and some (of)* and plural noun, we use a plural verb. But if we say *the number of*, we use a singular verb.

The number of tigers in the country is decreasing.

but

A number of reports have been questioned.

A lot of lectures are planned.

Plenty of showrooms now accept credit cards.

Use either the singular or the plural verb in the following cases:

With *any of, each of, either of, neither of,* and *none of*, usually a singular verb is used. The plural verb is used in informal style.

I doubt if *any* of them *knows* where the illegal arms are hidden.

Neither of the popular Indian games *has* (have) got international recognition so far.

Some phrases with a plural form are thought of as singular and, therefore, take a singular verb. These include phrases referring to measurements, amounts, and quantities as shown in the following examples:

Three quarters of a ton *is* too much.

The three hundred rupees I earned *was* kept in the locker.

When a subject has two or more items joined by *and*, a plural verb is used. However, phrases connected by *and* can also be followed by singular verbs if you think of them as making up a single item, as in the following sentences:

Choco pie and ice cream *is* Amrit's favourite at the moment.

Research and development *involves* myriad of activities.

After per cent (also percent or %), we use singular verb.

An inflation of 10 per cent per annum *makes* a big difference in any economy.

A 75 per cent likelihood of winning *is* worth the effort.

Tenses

There are three main tenses—present, past, and future. The tense of a verb reflects the time of an action or event. Read these three sentences:

> The etymology of the word 'tense' is from the Latin word 'tempus', which means 'time'.

(a) I sing a song to entertain you.

(b) I sang a song in the competition.

(c) I shall sing a song tomorrow for Gandhi Jayanti.

In sentence (a), the verb *sing* refers to the present time. In sentence (b), the verb *sang* refers to the past time, and in sentence (c), the verb *shall sing* refers to the future time.

A verb that refers to the present time is said to be in the present tense, e.g., I eat, I swim, and I dance. A verb that refers to the past time is said to be in the past tense, e.g., I ate, I swam, and I danced. A verb that refers to the future time is said to be in the future tense, e.g., I shall eat, I shall swim, and I shall dance.

Sometimes, a past tense may refer to the present time and a present tense may express future time, as in the following sentence:

I wish I knew swimming.

These sentences are in the past tense but refer to the present time. The following sentence is in the present tense but refers to the future tense:

Let us wait till he comes.

Present Tense

Simple present tense

The simple present tense is used in the following cases.

1. To express a habitual action:

   She exercises every day.

   I sleep at ten o'clock every day.

   My dog keeps the house safe.

2. To express general truths:

   The earth revolves around the sun.

   Medicines are bitter.

   Self-confidence boosts morale.

   There are no real secrets to success.

3. In exclamatory sentences, beginning with *here* and *there* to express what is actually taking place in the present:

   There goes the rocket!

   Here she comes!

4. In vivid narrative, as substitute for the simple past, and in broadcasting and commentaries on sporting events, instead of the present continuous tense, to talk about the actions and events:

   Sachin now advances forward and plays the fast ball.

   Anuj rushes home immediately.

5. To express a future event that is part of a fixed timetable or fixed programme:

   The next train is at 6 tomorrow evening.

   The circus show begins at 10 o'clock.

   The flight leaves at 5.

   When does the shop open?

6. To introduce quotations:

   Mark Twain says, 'The right word may be effective, but no word was ever as effective as a rightly timed pause.'

Present continuous tense

The present continuous tense is used in the following cases.

1. To describe an ongoing action while speaking:

   He is reading.   (now)

   The girls are riding their cycles.

2. To describe a temporary action that may not be actually happening at the time of speaking:

   I am reading *Gone with the Wind*.   (But I am not reading it at this moment.)

3. To mention an action that has already been planned/arranged to take place in the near future:

   I am going to Kolkata tonight.

   My cousin is leaving tomorrow.

It has been pointed out before that the simple present is used for a habitual action. However, when the reference is to a particularly obstinate habit, something which persists in spite of advice or warning, we use the present continuous with an adverb such as always, continually, and constantly.

> My pet is naughty; she is always crying out for food.

The following verbs, on account of their meaning, are not normally used in the continuous form.

1. Verbs of perception, e.g., see, hear, smell, taste, feel, notice, and recognize.
2. Verbs of appearance, e.g., appear, look, and seem.
3. Verbs of emotion, e.g., want, wish, desire, like, love, hate, hope, refuse, and prefer.
4. Verbs of thinking, e.g., think, suppose, believe, agree, consider, trust, remember, forget, know, understand, imagine, mean, and mind.
5. Verbs indicating possession, e.g., own, possess, belong, contain, consist, be (except when used in the passive).

Some examples having such verbs are given below:

Wrong	Right
This honey is tasting sweet.	This honey tastes sweet.
I am thinking you are right.	I think you are right.
Were you hearing that?	Did you hear that?
He is having a luxurious car.	He has a luxurious car.

However, the verbs listed above can be used in the continuous tense with a change of meaning as in the following sentences:

> I am feeling great this morning.   (feel = feel physically)
>
> She's seeing her friend this evening.   (see = visit)
>
> I am looking at the strange figure.   (look = stare at)

Present perfect tense

The present perfect tense is used in the following cases.

1. To indicate completed activities in the immediate past (with just):
   > She has just entered.
   >
   > It has just turned blue.
2. To express past actions whose time is not given and not definite:
   > Have you read *The Children of a Lesser God*?
   >
   > I have never seen him in a pensive mood.
   >
   > Mr Singh has been to Mumbai.
3. To describe past events when you think more of their effect in the present than of the action itself:
   > Anuj has eaten all the bananas.   (There aren't any left for you.)
   >
   > She has hurt my sentiments.   (I am upset.)
   >
   > I have completed my work.   (Now I am free.)
4. To denote an action begun sometime in the past and continuing up to the present moment (often with since and for phrases):
   > She has known him for many years.
   >
   > My teacher has been ill since last week.
   >
   > He has warned her every day for the past week—but to no effect.
   >
   > I have not seen Sameer for several weeks.

5. To express habitual or continued actions:

> We have lived here for twenty years.
>
> He has worn contact lenses all his life.

The following adverbs or adverbial phrases can also be used with the present perfect (apart from those mentioned above): *never, ever* (in questions only), *so far, till now, yet* (in negatives and questions), *already, today, this week, this month,* etc.

The present perfect tense is never used with adverbs of past time. You should not say, for example, 'He has gone to Kolkata yesterday.' In such cases, the simple past should be used: 'He went to Kolkata yesterday.'

> The present perfect tense is never used with adverbs of past time.

Present perfect continuous tense

The present perfect continuous is used for an action that began at some time in the past and is still going on. It is used with since and for to denote point and period of time, respectively:

> She has been studying for five hours.   (She is still studying.)
>
> They have been building the house for several months.
>
> They have been dancing since 4 o'clock.

This tense is also sometimes used for an action already finished. In such cases, the continuity of the activity is emphasized as an explanation of something.

> 'Why are your shoes so torn?''I have been working in the garden.'

Past Tense

Simple past tense

The simple past is used to indicate an action completed in the past. It often occurs with adverbs or adverbial phrases of the past time.

> I danced enthusiastically yesterday.
>
> She received compliments after the speech.
>
> He left for the USA last month.

Sometimes, this tense is used without an adverb of time. In such cases, the time may be either implied or indicated by the context.

> I read Akbar and Birbal's tales during the holidays.
>
> I did not concentrate on my studies.
>
> The girls defeated the boys in the cricket match.

The simple past is also used for past habits:

> I chanted hymns for many hours every day
>
> She always carried her sunglasses.

Past continuous tense

The past continuous is used to denote an action going on at some time in the past. The time of the action may or may not be indicated.

She was watching television all night.
It was getting late.
The rock fell on him while he was sitting.
When I saw him, he was playing badminton.

As in the last two examples above, the past continuous and simple past are used together when a new action happened in the middle of a longer action. The simple past is used for the new action. This tense is also used with *always, continually,* etc. for persistent habits in the past.

She was always upset.

Past perfect tense

The past perfect describes an action completed before a certain moment in the past and before the happening of another event in the past.

When I arrived at the party, she had already gone home.

The actions of the past should indicate the chronology. The simple past is used to indicate the later event and the past perfect is used to indicate the earlier event, for example, in the following sentence, the earlier event of the bus having started, is expressed using the past perfect tense.

When I reached the bus stop, the bus had started.   (So, I could not get into the bus.)
If I had received your letter, I would have replied.
I had finished the work before he arrived.

Past perfect continuous tense

The past perfect continuous is used for an action that began before a certain point in the past and continued up to that time, having linked to the period or point of time in the past.

At that time, he had been writing a book for two months.
When the visitors came to the school in the year 2000, Mr Singh had already been teaching there for two years.

Future Tense

In English language, the future tense is referred to in two ways—the simple future (the 'going to' form) and the simple present.

Simple future tense

The simple future tense is used to refer to the things that you cannot control. It expresses the future as a fact.

I shall be forty on 13th July.
It will be Valentine's Day next week.
You will know your exam results in July.

We also use this tense to talk about what we think or believe will happen in the future:

I am sure India will win the cricket match.
I think I will stand first in the obstacle race.
I believe Mayank will be the president.

As in these sentences, this tense is often used with the verb of thinking such as *I think, I am sure, I expect, I believe,* and *probably.* We can also use this tense when we decide to do something at the time of speaking:

It is very foggy, I will take my car.
'She is in a meeting right now.''Alright, I will wait.'

We will discuss below the use of 'going to', 'about to', and 'be to' forms in simple future tense.

Going to   Remember that if an action is already decided upon and preparations have been made, you should use the 'going to' form, not the simple future tense. The simple future tense is used for an instant decision. The 'going to' form (be going to + base of the verb) is used in the following cases.

- When it is decided to do something before taking any action about it:

  'Have you finalized the dates for the conference?' 'Yes, I am going to the vice-chancellor.'
  'Why do you want to leave the job?' 'I am going to join another organization.'
- To talk about what seems likely or certain, when there is something in the present that tells us about the future; to make predictions:

  It is going to be a bright day. Look at the sun.
  The house is in a dilapidated condition. It is going to fall.
  She is going to have a tough time.
- To express an action that is on the point of happening:

  Let us get into the class. It is going to start soon.
  Look! The balloon is going to burst.

About to   The 'be about to + base' form is used to indicate the immediate future:

Let us get into the class. It is about to start.
Do not go to sleep. We are about to have dinner.

Be to   We use 'be to + base' form to talk about official plans and arrangements:

I am to visit your institute next week.
The panel discussion is to talk on 'brain drain' tomorrow.

'Be to' is used in a formal style, often in news reports. 'Be' is usually left out in headlines, e.g., 'President to visit the Tsunami-hit areas'.

Use of present tense in simple future

Simple present   The simple present tense is used for official programmes and timetables:

The institute reopens on 1st August.
The class starts at 10 and finishes at 12 noon.
When does the next flight for Delhi leave?

The simple present is often used for future time in clauses with *if, unless, when, while, before, after, until, by the time, and as soon as*. The simple future tense is not used in the following cases:

I will not get into the pool if the water is cold.
Can I see the film before I leave?
Let us relax till he completes his writing.
Please call me up as soon as you get the information.

Present continuous   The present continuous tense is also used for simple future when we refer to something that we have planned to do in the future:

I am going for the conference day after tomorrow.
We are partying out today evening.
Mr Salim Khan is leaving this afternoon.

The present continuous (not the simple present) should be used for personal arrangements.

Future continuous tense

The future continuous tense is used to discuss the actions that will be in progress at a time in the future.

I suppose she will be singing when we enter the auditorium.

This time tomorrow, I will be delivering my lecture in your institute.

This tense is also used to mention the actions in the future that are already planned or are expected to happen in the normal course of things as shown in the following sentences:

She will be staying in the room till the evening.

The president will be talking to us tomorrow.

The milkman will be coming soon.

Future perfect tense

The future perfect tense is used to talk about the actions that will be completed by or before a certain future time.

I shall have finished my writing by then.

You will have left before he comes to see me.

By the end of the day, I will have worked in this place for six hours.

Future perfect continuous tense

The future perfect continuous tense is used for the actions that will be in progress over a period/point of time and will end in the future. This tense is not very common in use.

By next year, we shall have been living here for one year.

I'll have been training in this institute for fifteen months by next September.

Active	Simple	Continuous	Perfect	Perfect continuous
Present	I read.	I am readintg.	I have read.	I have been reading.
Past	I read.	I was reading.	I had read.	I had been reading.
Future	I will read.	I will be reading.	I will have read.	I will have been reading.

Active and Passive Voice

In technical and scientific writing, it is advisable for the writers to opt for passive without agents to achieve objectivity. Therefore, the personal pronouns *I* and *we* are avoided so that impersonal style can be achieved, as in 'This solution was, therefore, found to be far more suitable' and 'The observations were recorded to be put forward at appropriate time'.

The impersonal passive is preferred for two reasons—it deletes the subject of an intransitive verb and a dummy is used in the construction of the sentence. This dummy has neither thematic nor referential content. For example, there is a dummy word in the sentence: '*There* are two paintings'.

The simple rule of converting active voice into passive is by changing the subject into a direct object and the direct object into a subject. In most of the cases, the indirect object is a person, which you call personal passive, such as 'the children' in the following sentence:

Active voice:	The father explained the exercise to *the children*.
Passive voice:	*The children* were explained the exercise by the father.

When you replace the pronoun for 'the children', you observe that this pronoun is in its subject form. This is the reason it is called personal passive.

Active voice:	The father explained *them* the exercise.
Passive voice:	*They* were explained the exercise.

Many a time, the agent is removed in passive sentences as 'by the father' in the above sentence. Contrarily, impersonal passives can be defined as passives of reporting verbs. Verbs that refer to saying, thinking, knowing, and believing are mostly followed by the to-infinitive form of passive.

> Impersonal passive can be defined as passives of reporting verbs.

Active:	We all believe that she is in India.
Impersonal passive:	She is believed to be in India.

There are four infinitive forms that can be used depending on the tense in the active sentence:

- Simple: to do
- Continuous: to be doing
- Perfect: to have done
- Perfect continuous: to have been doing

The phrase *It is said* ... is an impersonal passive construction. Passive constructions are popularly used in news. Some more examples of impersonal passive voice are given below:

Active:	They say he works arduously.
Impersonal passive:	He is said to work arduously.
Active:	They say she sang well.
Impersonal passive:	She is said to have sung well.
Active:	They said she had scored well.
Impersonal passive:	She was said to have scored well.
Active:	They think he has been having a complex.
Impersonal passive:	He is thought to have been having a complex.
Active:	People say that children are afraid of the dark.
Impersonal passive:	It is said that children are afraid of the dark.
Active:	We can generate heat for welding in many ways.
Impersonal passive:	Heat can be generated for welding in many ways.
Active:	We pass an electric current across the electrodes.
Impersonal passive:	An electric current is passed across the electrodes.

The following examples show two variations in impersonal passive.

- People *believe* that Chinese **is** the most widely spoken language.
  - *It is believed* that Chinese **is** the most widely spoken language.
  - Chinese *is believed* **to be** the most widely spoken language.
- Police have reported that it was the mischievous man who **caused** the accident.
  - It *has been reported* by the police that it was the mischievous man who **caused** the accident.
  - The mischievous man *has been reported* by the police **to have caused** the accident.
- A journalist *reports* that they **are leaving** Chennai tomorrow night.
  - *It is reported* by a journalist that they **are leaving** Chennai tomorrow night.
  - They *are reported* **to be leaving** Chennai tomorrow night.
- Their parents *thought* that the teenagers **were dancing** at the disco.
  - *It was thought* that the teenagers **were dancing** at the disco.
  - The teenagers *were thought* **to be dancing** at the disco.

In scientific writing, active verbs are preferred over passive verbs. Active verbs render the writing more reader friendly and clear, whereas passive verbs only tend to make it more complex and unmanageable.

Conditional Sentences

Conditional clauses or 'if' clauses is another name given to conditional sentences. As the main clause, these are used to express that an action (without if) can only take place after fulfilling a certain condition (in the clause with *if*). The three types of conditional sentences are listed below:

Type	Condition
I	Condition possible to fulfil
II	Condition in theory possible to fulfil
III	Condition not possible to fulfil (too late)

Form:

Type	'If' clause	Main clause
I	Simple present	will-future (or modal + infinitive)
II	Simple past	would + infinitive*
III	Past perfact	would + have + past participle *

*You can substitute could or might for would (should, may, will).

Examples

Type		Example
I	Positive	If I *swim*, I *will cross* the river.
	Negative	If I *swim*, I *won't sink* in the river.
		If I *don't swim*, I *will sink* in the river.
II	Positive	If I *swam*, I *would* cross the river.
	Negative	If I *swam*, I *wouldn't sink* in the river.
		If I *didn't swim*, I *would sink* in the river.
III	Positive	If he *had swum*, he *would have crossed* the river.
	Negative	If he *had swum*, he *wouldn't have sunk* in the river.
		If he *hadn't swum*, he *would have sunk* in the river.

In the following examples, observe the 'if' clause:

1. My uncle has lost his glasses. He thinks they may be at the neighbour's house.

   Uncle: I think I left my glasses at your house. Have you seen them?

   Neighbour: No, but I'll have a look when I get home. If I find it, I'll tell you.

   In this example, the neighbour thinks that there is a real possibility that he will find the glasses. So, he says, 'If I find…, I'll…'.

2. The situation given below is slightly different.

   Raj says, 'If I found an urchin in the city, I'd inform the police.'

   Here Raj is not thinking about a real possibility; he is imagining the situation and does not expect to find an urchin in the city. So, he says, 'If I found…, I'd (= I would)…' (*not* 'if I find…, I'll…').

When you imagine something like this, you use if + past (if I found/if you were/if we didn't, etc.). But the meaning is not past. See the following examples:

What would you do if you won the first prize? (We don't really expect this to happen.)

I don't really want to go to their conference, but I probably will go. They'd be offended if I didn't go.

Snigdha has decided not to apply for the teaching job. She is not really qualified for it, so she probably wouldn't get it if she applied.

Normally, *would* is not used in the 'if' part of the sentence:

She would be very scared if she faced the lion. (not 'if she would face')

If I didn't attend the dinner, they'd be annoyed. (not 'if I wouldn't go')

But it is possible to say 'if...would' when you ask somebody to do something:

I would be obliged if you would include my name in the list of participants. (As a part of formal letter)

'Shall I move the papers?' 'Yes, please, if you would.'

In the other part of the sentence (not the 'if' part), we use would ('d)/wouldn't:

If you ate more apples, you'd (= you would) probably feel healthier.

Would you mind if I used your computer?

Could and *might* can replace *would* here.

If you consume more milk, you might feel healthier. (= It is possible that you would feel healthier.)

If it stopped raining, we could go to the institute. (= We would be able to go.)

Look at the following examples:

She wants to go home but she can't because she doesn't remember the way. She says, 'If I knew the way, I would go home.'

This tells us that she does not know the way. She is imagining the situation. The real situation is that she does not know the way. In this situation also, if + past (if I knew/if you were/if we didn't, etc.) is used. However, the meaning is present, not past.

> After 'if' and 'wish', you can use 'were instead of 'was'.

I would take more sessions if I had more time. (But I don't have much time.)

If you were in my position, what would you do?

It's a pity you can't give me medicines. It would be helpful if you could.

We use the past tense in the same way after wish (I wish I knew/I wish you were, etc.). We use wish to say that we regret something that is not as we would like it to be:

I wish I knew driving.   (I don't know it and I regret this.)

Do you ever wish you could swim?

It rains a lot here. I wish it didn't rain so often.

It's very lonely here. I wish there weren't so many vacant houses. (But there are a lot of vacant houses.)

I wish I didn't have to travel in this weather.   (But I have to travel.)

After 'if' and 'wish', you can use *were* instead of *was* (if I were/I wish it were, etc.). So, you can say:

If I were the president, I wouldn't address that gathering. (or 'If I was the president ... .')

I'd go out if it weren't cold. (or '... if it wasn't cold.')

I wish it were possible.

We do not normally use *would* in the 'if' part of the sentence or after 'wish':

If I were a millionaire, I would have a chartered plane. (not 'If I would be millionaire ...')

I wish you had something to eat.   (not 'I wish you would have ...')

Sometimes, *wish ... would* is possible as in 'I wish you would speak'. Note that *could* sometimes means 'would be able to' and sometimes 'was/were able to':

You could get a loan more easily.   (you could get = you would be able to get)

Observe the following sentences:

> Last week, Reena was in town for a performance. Leela didn't know this, so she didn't go to see her. They met two days ago. Leela said, 'If I had known you were in town, I would have come to see you.'

Here, the real situation was that Leela did not know Reena was in town.

When you are talking about the past, you use *if* + *had* ('d)... (If had known/been/done, etc.):

> If I'd seen you, of course, I would have exchanged greetings. (But I didn't see you.)
> I would have gone to the cinema last Sunday if I hadn't been so busy. (But I was busy.)
> If I had been with the child, I wouldn't have allowed him to dive into the deep water. (But I wasn't with him.)
> The view was panoramic. If I'd had a brush and colours, I would have painted some pictures. (But I didn't have a brush and colours.)

Compare:

> I'm not thirsty. If I was thirsty, I would drink something. (now)
> I wasn't thirsty. If I had been thirsty, I would have drunk something. (past)

Do not use *would* in the 'if' part of the sentence. Use *would* in the other part of the sentence:

> If I had seen you, I would have said hello.   (not 'if I would have seen you')

Note that 'd can be *would* or *had*:

> If I'd seen you, (I'd seen = I had seen)
> I'd have said hello. (I'd have said = I would have said)

Use *had* (done) in the same way after wish. I wish something had happened = I am sorry that it didn't happen:

> I wish I'd known that Sneha was in the garden. I would have gone to meet her.
> (But I didn't know.)
> I feel uneasy. I wish I hadn't taken so many medicines. (I took many medicines.)
> Do you wish you had studied science instead of languages? (You didn't study science.)
> The weather was hot while we were away. I wish it had been colder.

Do not use *would have*... after wish in the following types of sentences:

> I wish it had been colder. (not 'I wish it would have been colder'.)

Compare *would* (do) and *would have* (done):

> If I had gone for extra classes last evening, I would be satiated now. (I am not satiated now—present)
> If I had gone to the conglomeration last night, I would have met lots of friends. (I didn't meet lots of friends—past)

Compare *would have*, *could have*, and *might have*:

If the climate hadn't been so bad,	we would have eaten out.
	we could have eaten out.
	(= we would have been able to party out.)
	we might have gone out.
	(= perhaps we would have gone out.)

We use *would* ('d) when we imagine a situation or action:

> It would be nice to have a day off but we can't afford it.
>
> I'm not going to discuss it yet with him. I'm not prepared and I wouldn't broach the topic.

We use *would have* (done) when we imagine situations or actions in the past:

> They supported me a lot. I don't know what I would have done without their support.
>
> I didn't go to bed. I wasn't tired, so I wouldn't have slept.

Compare *will* ('ll) and *would* ('d):

> I'll stay a bit longer because I've got time.
>
> I'd stay a bit longer but I really have to go now. (So I can't stay longer.)

Sometimes, *would/wouldn't* is the past of *will/won't*. Compare:

'If the sales hadn't been so low, I would have raised your salary.'

Present	Past
• Ranjan: I'll write to you on Monday.	• Ranjan said he'd write to me on Monday.
• Anita: I promise I won't hesitate.	• Anita promised that she wouldn't hesitate.
• Leela: Damn! The car won't start.	• Leela was angry because the car wouldn't start.

Observe the following situation:

It is raining. Mayank wants to go out, but not in the rain. He says:

> I wish it would stop raining.

This means that Mayank is complaining about the rain and wants it to stop.

We use *I wish… would…* when we want something to happen or somebody to do something. Mayank is unhappy with the situation. Some more examples of similar situations are given below:

> We use 'would have' (done) when we imagine situations or actions in the past.

> The door bell has been constantly ringing for two minutes. I wish somebody would open the door.
>
> I wish you would do something instead of just sleeping and wasting time.

You can use *I wish…wouldn't* to complain about the things people do repeatedly:

> I wish you wouldn't keep interfering in my business.

We use *I wish…would…* for actions and changes, not situations. Compare:

> I wish Sangeeta would come.   (= I want her to come.)
>
> I wish Sangeeta were (or was) here now.   (not 'I wish Sangeeta would be…')
>
> I wish somebody would buy me a diamond necklace.
>
> I wish I had a diamond necklace.   (not 'I wish I would have…')

You can also use *would* when you talk about the things that happened regularly in the past:

> When we were young, we lived by the riverside. In summer, if the weather was hot, we would all get up early and go for a swim. (= We did this regularly.)
>
> Whenever she was happy, she would shop a lot.

In the last sentence, besides the above meaning, *would* is similar to *used to*.

> Whenever she was happy, she used to shop a lot.

Adjectives and Degrees of Comparison

An adjective is a word that qualifies or describes a noun. It can also give the number by quantifying. It gives extra meaning to the noun by adding something to it.

Adjectives are used attributively and predicatively. When an adjective is used with the noun as an epithet or attribute, it is said to be used attributively. On the other hand, when an adjective is used along with the verb and forms part of the predicate, it is said to have been used predicatively.

> An adjective gives extra meaning to the noun by adding something to it.

> The *beautiful* girl was declared Miss India. (attributive)
> The girl is *beautiful*. (predicative)

In the first sentence, the adjective *beautiful* is used along with the noun *girl* as an epithet or attribute, whereas in the second sentence, it is used along with the verb *is* and forms part of the predicate. Some adjectives can be used only predicatively:

> She is *scared* of lizards.
> He is quite *fine*.

Types of Adjectives

Various categories of adjectives are discussed here.

Adjective of quality (descriptive adjective)

An adjective of quality shows the quality of a person or thing:

> Pilani is a *small* town.
> He is an *honest* person.
> The *stupid old* fool tried to run in the marathon.

This type of adjective will answer the question 'of what kind'. Even proper adjectives are classified as *adjectives of quality*. They are formed from proper nouns such as French fries, French beans, Hindi grammar, and Turkish tulip.

Adjective of quantity

An adjective of quantity tells us how much of a thing is intended. When an adjective answers the question 'how much', it is an adjective of quantity. However, these adjectives do not indicate the quantity in definite terms.

> I recited *some* poems.
> She has done *enough* studies.
> Children have *no idea* about the risks involved in adventure sports.
> She took *great* care of her grandchildren.
> The *entire* sum of money was spent on buying the new cottage.

Adjective of number (numerical adjective)

An adjective of number quantifies persons or things. It also indicates the order of appearance and answers the question 'how many'.

> *Few* men dare to combat a lion.
> There are *no* good articles in the journal.
> There are *several* students waiting to register for this course.
> *Most* boys opt for flying as their hobby.
> Wednesday is the *third* day of the week.
> The cow has *one* tail.

There are three types of numerical adjectives—indefinite, definite, and distributive.

Indefinite numerical adjective An indefinite numerical adjective gives a vague idea of the numbers and not definite. Such adjectives are *many, several, sundry, few, some, certain,* and so on.

> A *few* boys were absent in the class.

Definite numerical adjective   A definite numerical adjective includes the cardinal numbers such as *one, two, three,* and *five*. It also includes ordinals such as *first, second,* and *third*.

> I did not take the *first* bus.

Distributive numerical adjectives   These adjectives refer to each one of a number.

> *Each* girl has done well in the exam.
> We expect *every* student to appear for the interview.
> *Each* word of yours is an assumption.
> *Either* side of the footpath is laden with rows of flowers.
> *Neither* report on his assassination reflects the true picture.

Adjective of demonstration

An adjective of demonstration gives the answer to the question 'which'. *This* and *that* are used with singular nouns and *these* and *those* with plural nouns.

> *This* is the book I was looking for.
> *Those* boys helped in organizing the convention.

Interrogative adjective

'Wh' words such as *what, which, whose,* and *when* fall under the category of interrogative adjectives. *What* is used in a general sense while *which* is used in a selective sense.

> *What* car has she opted for?
> *Which* way is the Birla museum?
> *Whose* book is this?

Comparison of Adjectives

There are three degrees of comparisons.
- Positive degree: She is *smart*.
- Comparative degree: But her sister is *smarter*.
- Superlative degree: Their brother, however, is the *smartest*.

The first kind of adjective is in simple form and shows positive degree. It denotes the mere existence of some quality. It is used in isolation when no comparison is needed. But when there is comparison between two things, the comparative degree of an adjective denotes a higher degree of quality than the positive.

> She is *more intelligent* than her friend.
> Which of those two stories is *better* for publication?
> Today is *hotter* than yesterday.

The superlative degree of an adjective is used to reflect the best in the group.

> She is the *most beautiful* girl in the class.

Formation of comparative and superlative degrees

Most adjectives of one syllable, and some of more than one, form the comparative by adding *-er* and the superlative by adding *-est* to the positive.

Positive	Comparative	Superlative
smart	smarter	smartest
big	bigger	biggest
faint	fainter	faintest
high	higher	highest
near	nearer	nearest
kind	kinder	kindest
quiet	quieter	quietest
great	greater	greatest

When the positive ends in 'e', only -r and -st are added.

nice	nicer	nicest
cute	cuter	cutest
blue	bluer	bluest
large	larger	largest
able	abler	ablest
brave	braver	bravest
grave	graver	gravest

When the positive ends in 'y', preceded by a consonant, the 'y' is changed into 'i' before adding -er and -est.

crazy	crazier	craziest
ugly	uglier	ugliest
lazy	lazier	laziest
heavy	heavier	heaviest
merry	merrier	merriest
healthy	healthier	healthiest

When the positive is a word of one syllable and ends in a single consonant, preceded by a vowel, this consonant is doubled before adding -er and -est.

Positive	Comparative	Superlative
big	bigger	biggest
fat	fatter	fattest
red	redder	reddest
sad	sadder	saddest
hot	hotter	hottest
thin	thinner	thinnest

Adjectives of more than two syllables, and many of those with two, form the comparative by using the adverb *more* with the positive and the superlative by using the adverb *most* with the positive.

amazing	more amazing	most amazing
beautiful	more beautiful	most beautiful
arduous	more arduous	most arduous
timorous	more timorous	most timorous
courageous	more courageous	most courageous
erudite	more erudite	most erudite
proper	more proper	most proper

Remember not to use -er when qualities of the same person are compared. If you have to say that Seema is more courageous than Reema, you say 'Seema is braver than Reema', but if you have to say that Seema's courage is greater than her compassion, then say 'Seema is more brave than compassionate'.

When two objects are compared with each other, the latter term of comparison must exclude the former:

The dictionary is more useful than any other book.

Irregular comparison

Some adjectives are compared irregularly, that is, their comparative and superlative degrees are not made from their positive degrees.

Positive	Comparative	Superlative	Positive	Comparative	Superlative
bad, evil, ill	worse	worst	late	later, latter	latest, last
much	more	most	good, well	better	best
many	more	most	old	older, elder	oldest, eldest

(Contd)

Positive	Comparative	Superlative	Positive	Comparative	Superlative
high	higher	highest	(out)	outer	outermost
far	farther	farthest	(in)	inner	innermost
fore	former	foremost, first	(up)	upper	uppermost
(fore)	further	furthest			

(*Note:* The words in parentheses are not adjectives but adverbs.)

Adverbs

An adverb is a word that adds more information about a place, time, manner, cause, or degree to a verb, an adjective, a phrase, or another adverb. A typical adverb may be recognized by the -ly suffix that has been attached to an adjective such as beautifully, quickly, slowly, nicely, humbly, and mortally. We can better understand it with the help of the adjacent table.

Adjective	Adverb (Adjective + -ly)
Dangerous	Dangerously
Careful	Carefully
Nice	Nicely
Ease	Easily
Horrible	Horribly

Indeed all words ending with -ly are not adverbs. Sometimes, the adverbs and adjectives are the same word. It may be useful to contrast the adjectives and adverbs in the following pairs of sentences. The adjective is in the first sentence of each pair.

Adverbs and Adjectives with the Same Form

I caught an *early* train.
I arrived home *early*.

He aimed at the *higher* target.
He aimed *higher* next time.

He walked a *straight* line.
He went *straight* to his friend.

It's a *hard* job we have to do.
We tried *hard* to convince them.

The suffix -ly is not an inflectional suffix, as for example, the -s plural ending on nouns and the -ed past tense ending on verbs; instead, it is a derivational suffix. We should also note that the words which end in -ly are adjectives not adverbs, the -ly in this case deriving adjectives from nouns such as friendly, kingly, cowardly, manly, fatherly, worldly. Consequently, we have adjectives not adverbs in: *She is a friendly person, What a cowardly act!,* and *Here is some fatherly advice.*

Adverbs with Two Forms

Many English adverb forms without -ly look like adjectives; they clearly distribute like adverbs. Many English adverbs lack the -ly ending for historical reason (they are flat adverbs) but that ending is sometimes extended to them. Consequently, some adverbs can exist in two forms, one with the -ly ending and one without, as in *cheap* and *cheaply* and *slow* and *slowly*, which are merely alternate adverb forms. However, some users of English may resist using words like *cheap* and *slow* as adverbs when alternates such as *cheaply* and *slowly* exist. The following adverbs have two forms:

She bought them *cheap*.
She bought them *cheaply*.
Do it up *tight*!
Do it up *tightly*!

Hold it *close* to you.
Hold it *closely* to you.

Adverb Forms

Just like some adjectives, a few adverbs also have comparative and superlative forms such as badly, worse, worst, well, better, best, etc. The comparative and superlatives in most adverbs require use of more and most as in *more commonly* and *most successfully.* See the following examples:

> Score *faster* if you wish to win.
> If you read *faster,* I will award you an 'A' grade.
> He left *earlier* than expected.
> They drive *much slower* now.

Adverbs as Intensifiers

Adverbs often function as intensifiers giving greater or lesser emphasis. These intensifiers have three different functions: emphasizers, amplifiers, and down toners.

Emphasizers

> I *really* love him.
> She *literally* enjoys dancing.

Amplifiers

> The boss *completely* rejected the report.
> I am *absolutely* fine.
> They *heartily* welcomed the members.

Down toners

> She is *sort* of happy go round.
> We can improve it *to some extent.*
> The boss *mildly* refused the proposal.

Kinds of Adverb

There are many different kinds of adverbs explained in the following.

Adverbs of manner

An adverb of manner explains how an action has taken place. They are usually placed after the main verb or after the object.

> He works *well.* (after the main verb)
> Ram runs *quickly.*
> Sangeeta speaks *softly.*
> Gita laughed *loudly* to attract her attention.
> He plays the flute *beautifully.* (after the object)
> He ate the chocolate cake *greedily.*

Adverbs of time

An adverb of time explains when an action has taken place.

> She *regularly* visits France. (frequency)
> Mohan *never* drinks milk. (before the main verb)
> Students must *always* work hard. (after the auxiliary 'must')
> My father went to the principal's house *yesterday.*
> *Later* Mohan prepared tea. (before the subject)

Adverbs of place

Adverbs of place tell us where an action has taken place. These adverbs are mostly used after the main verb or after the object.

There comes the train. (before the main verb)

Books were spread *everywhere*. (after the main verb)

Sita looked up *towards* the sky.

Mohan went *outside*.

They buried the dead body *nearby*. (after the object)

Suresh threw the book *outside*.

Adverbs of degree

Adverbs of degree explain the intensity of an action, an adjective, or another adverb.

The juice was *extremely* cold. (before the adjective)

Sita is *too* tired to go out tonight.

Seema is *just* leaving the house. (before the main verb)

She has *almost* finished.

He *rarely* perceived any change in her behaviour.

Relative adverbs

Relative adverbs can be used to join sentences or clauses such as where, when, and why.

This is the place *where* we lost the money.

I remember the month *when* I joined this institute.

People were shouting *when* we entered the hall.

Mohan disclosed *why* he has stolen the money.

Adverbs of frequency

An adverb of frequency expresses how often an action is or was done.

He is *always* in time for meal.

She exercises in the morning *everyday*.

He *often* smiles while speaking.

They *sometimes* stay up all night.

Conjuctions

A conjunction is a word that joins together two words, phrases, or clauses.

Cup *and* saucer make a complete set.

My love for skiing *and* the mountains brought me to the Alps.

They are poor *but* they are honest.

> An adjective gives extra meaning to the noun by adding something to it.

Conjunctions make sentences more compact by joining them as in the following sentence:

Usha and Geeta are very beautiful.

This is better than saying 'Usha is very beautiful. Geeta is very beautiful.' Here are some more examples:

Up *and* down he goes.

I will come on Monday *or* Tuesday.

Jerry is small *but* strong.

You must be careful to distinguish *relative pronouns* and *relative adverbs* from conjunctions because they also bring together parts of a sentence.

Some conjunctions are used in pairs and are called *correlative conjunctions* or *correlatives*. A few correlatives are given below:

Either … or	*Either* attend the classes *or* go home.
Neither … nor	He is *neither* intelligent *nor* arduous.
Both … and	She is *both* pretty and intelligent.
Whether … or	I am in a dilemma about *whether* to stay *or* to go.
Not only … but also	She is *not only* a dancer *but also* a singer.

Some compound expressions are used as conjunctions and are called *compound conjunctions*, for example, *as if, in order that, even if, on condition that, so that, as soon as, as though, provided that, and in as much as*.

Conjunction Classes

The two main classes of conjunctions are coordinating and subordinating conjuctions.

Coordinating conjunctions

When a sentence contains two independent statements or two statements of equal rank or importance, *coordinating conjunction* is used. This class includes *and, but, for, or, nor, also, either… or, neither… nor*, and so on. They are of the following four types.

Cumulative   It adds one statement to another:
  We worked hard *and* we achieved the goal.

Adversative   It expresses the contrast between two statements:
  He is weak *but* enthusiastic.
  She would participate *only* if she is invited.
  I am irked, *yet* I am quiet.

Disjunctive   It expresses the choice between two alternatives:
  She must take the medicines *or* she will die.
  Either he is crazy *or* he is a fanatic.
  Run fast, *else* you will not win.

Illative   It expresses an inference:
  Somebody surely came in, *for* I heard footsteps.
  All preemptive measures must have been neglected, for the malaria spread rapidly.

Subordinating conjunctions

A conjunction introducing a dependent or subordinate clause is called a *subordinating conjunction*. A subordinating conjunction joins a clause to another on which it depends for its full meaning. These are *while, where, when, as, unless, before, till, although, though, that, if, because*, and *after*. Such conjunctions are further classified into seven types according to their meanings.

Time
  My grandfather died *before* I was born.
  No institute can flourish *till* it is under the tutelage of an efficient leader.
  Many events have been organized *since* you came.
  I returned to the office *after* he came.

Cause or reason

My energy is like the energy of a five-year-old *because* my mind is active.

Since you want, it shall be done.

As she was not there, I talked to her mother.

He may join, *as* he is eager.

Purpose

We eat apples, *so* we remain happy.

He helped me, *lest* I should fail.

Result

I was so enervated *that* I could scarcely smile.

Condition

The problem cannot be solved *unless* you identify it.

I will come to the party *if* you come.

Concession

I didn't get the job *though* I had all the necessary qualifications.

A picture is a picture, *even if* there is nothing aesthetic in it.

Comparison

He is taller *than* his brother.

The following are some examples of the use of conjunctions:

Is this right *or* wrong?

It is better late *than* never.

You can do it *if* you try.

You will only do it *when* you try.

Conjunctions Used in Adverbial Phrases and Clauses

Conjunctions used in adverbial clauses are considered as subordinating clauses. And these clauses when used with adverbial clauses, in a sentence, acts like an adverb in that sentence. For example,

Wherever he goes, he leaves the mark of his churlish behaviour. (The adverbial clause *wherever he goes* modifies the verb *leaves*.)

A subordinate conjunction may work as an, adverb, adjective, or noun in complex sentences. An adverbial clause beginning with a subordinating conjunction makes the clause subordinate or dependent.

Common subordinate conjunctions are as follows:

although	in order (that)	until
after	insofar as	unless
as	in that	when
as far as	lest	wherever
as though	no matter how	while
as if	now that	whenever
as soon as	once	why
because	provided (that)	where
even if	since	whether
before	so that	

(Contd)

even though	supposing (that)
how	though
if	till
in as much as	than
in case (that)	that

Like a single word adverb, an adverbial clause describes a verb and answers the questions such as where, why, how, when, or to what degree. For example,

When it rains, the flowers bloom.

Here *when* is the subordinating clause, *it rains* is the adverbial clause, and *the flowers bloom* is the main clause.

The old lady is hungry because she has no money.

Here *the old lady is hungry* is the main clause, *because* is the subordinating clause, and *she has no money* is the adverbial clause.

Where there is smoke, there is fire.

Here *where* is the subordinating conjunction and *there is smoke* is the adverbial clause.

She danced very well, as if she is a good dancer.

Here *she danced well* is the main clause, *as if* is the subordinating conjunction, and *she is a good dancer* is the adverbial clause.

She is more beautiful than her sister.

Here *she is more beautiful* is the main clause, *than* is the subordinating clause, and *her sister* is the adverbial clause.

Prepositions

A preposition shows the relationship of a noun to the remaining part of the sentence. The word preposition means 'place before'. There are two kinds of prepositions—simple and complex. Simple prepositions comprise a single word, e.g., *in, at, as, of,* and *from*. Complex prepositions are the combination of two or more words, e.g., *ahead of* and *due to*. Observe the use of the following propositions.

At, in   *At* is used with small towns and villages:

He lives at Pilani.

In is used with cities and countries:

My sister works in Delhi.

By, with   *By* is used before the doer of an action:

The dog was killed by a lion.

With is used before the name of a tool:

The judge gave the decision with the pounding of his gavel.

Between, among   *Between* is used with two persons or things:

The father divided the property between the two children.

Among is used with more than two persons or things:

The father divided the property among all the children.

Red Crown got the preposition wrong.

Beside, Besides   *Beside* means 'by the side of':

I sat beside the vice chancellor.

Besides means 'in addition to':

Besides teaching, she also does English coaching.

In, within   *In* means at the end of a certain period:

I will finish my work in two days. (i.e., at the end of two days)

Within means before the end of a certain period:

I will finish the assignment within two days. (i.e., before the end of two days)

Right use of prepositions after verbs, nouns, and adjectives

Sometimes, errors occur due to the wrong use of prepositions after verbs, nouns, and adjectives. The Online Resource Centre contains tables which list tips for correct prepositional usage.

The following are examples of the use of prepositions:

Take the time *to* do it right.

Go *with* the faith of our mothers.

Walk *in* the footsteps of our fathers.

That is the question *for* you.

That is what it is all *about*.

Articles

Articles in English are used to clarify if a noun is specific or not specific. Basically, articles are adjectives. Like adjectives, articles also modify nouns and are used before nouns or adjectives. There are two types of articles:

Definite articles (*the*): These are used for specific nouns.

Indefinite articles (*a, an*): These are used for non-specific nouns.

The is used to refer to specific or particular nouns; *a/an* is used to modify non-specific or non-particular nouns. That is why, *the* is a definite article and *a/an* is a non-definite article. For example,

I am reading the book.

Here, but *the book I am reading* is specific. So *the* is used. Whereas, in

Let's read a book.

The book that I am going to read is not specific. It may be any book.

Indefinite articles

'a' and 'an' when used signifies that the noun modified is indefinite, referring to any member of a group. For example,

My son wants a dog on his birthday.

This refers to any dog. One cannot find out which dog as it is not found yet.

When I was in Mumbai, I bought an aquarium.

Here the non-specific thing is an aquarium. There are various types of aquarium but there is one that was bought.

1. Use of *a* or *an* depends on the sound that begins the next word:
   - a + singular noun beginning with a consonant: *a girl, a cycle, a ball, a chair, etc.*
   - an + singular noun beginning with a vowel: *an aeroplane, an egg, an apple, an idiot, an orphan, etc.*

- a + singular noun beginning with a vowel but having consonant sound: a user, a university, a unicycle, etc.
- an + singular noun beginning with consonant but having vowel sound: *an honest man, an honour, etc.*

2. If the noun is modified by an adjective, the use of *a* and *an* depends on the initial sound of the adjective that immediately follows the article. For example, a broken egg, an honored man, a university event.
3. Indefinite articles are also used to indicate membership in a group. For example, he is a teacher, Melissa is an American.
4. Indefinite articles are used for countable nouns. For example, I need a glass of water, he gave me a pen.

Definite article

The following are some uses of 'the'.

1. The definite article is used before singular and plural nouns when the noun is specific or particular. The signals that the noun is definite; that it refers to a particular member of a group. For example:
   The boy who stole my purse was caught by the police.

Here, the boy is a specific boy who stole the purse.

2. 'The' can be used with uncountable nouns. For example:
   'I love to sail over the water' (some specific body of water) or 'I love to sail over water' (any water).

   'He spilled the milk all over the floor' (some specific milk, perhaps the milk you bought earlier that day) or 'He spilled milk all over the floor' (any milk).

3. 'The' is used before
   - names of rivers, oceans, and seas: *the Nile, the Pacific*
   - points on the globe: *the Equator, the North Pole*
   - geographical areas: *the Middle East, the West*
   - deserts, forests, gulfs, and peninsulas: *the Sahara, the Persian Gulf, the Black Forest, the Iberian Peninsula*

4. 'The' is not used before
   - names of most countries/territories: *Italy, Mexico, Bolivia*; however, *the Netherlands, the Dominican Republic, the Philippines, the United States*
   - names of cities, towns, or states: *Seoul, Manitoba, Miami*
   - names of streets: *Washington Blvd., Main St*
   - names of lakes and bays: *Lake Titicaca, Lake Erie*, except with a group of lakes such as *the Great Lakes*
   - names of mountains: *Mount Everest, Mount Fuji*, except with ranges of mountains like *the Andes* or *the Rockies* or unusual names like *the Matterhorn*
   - names of continents: *Asia, Europe*
   - names of islands: Easter Island, Maui, Key West, except with island chains like *the Aleutians, the Hebrides,* or *the Canary Islands*

Omission of Articles

Articles are not used before few nouns like the following.

- names of languages and nationalities: *Chinese, English, Spanish, Russian*
- names of sports: *volleyball, hockey, baseball*
- names of academic subjects: *mathematics, biology, history, computer science*

SUMMARY

It is difficult to write for technical documents without having the basic knowledge of vocabulary and grammar.

Possessing rich vocabulary provides people with the confidence to face any communication situation boldly. Vocabulary can be enhanced significantly by knowing how to change words from one form to another, such as from nouns to adjectives. It can also be remarkably improved by being well-versed with prefixes and suffixes, synonyms, idioms, confusables, one-word substitutes, homonyms, homophones, eponyms, and phrasal verbs.

To write effectively, one must know the rules of grammar. Grammar begins with the parts of speech—noun, adjective, pronoun, verb, adverb, preposition, conjunction, and interjection.

Imperatives are used to give commands, instructions, etc. and modality has to do with the expressions of possibility and necessity. There are three types of tenses—present, past, and future. Nouns and noun phrases have significant roles in a sentence. After the knowledge of base verbs, infinitives, progressives, perfectives, passives, and modals, attention should be drawn to noun phrases.

A preposition shows the relationship of a noun with the remaining part of the sentence. A conjunction is a word that joins two words, phrase, or clauses. A phrase is a group of words without a subject and a predicate, however, a clause has a subject as well as a predicate. In a sentence, the subject and the verb should be in agreement to each other to make it correct.

EXERCISES

1. Why do we need rich vocabulary during communication?
2. English words have undergone a big transformation since the language was developed first. Give some examples to justify.
3. Can you substitute one word with another synonym without careful decision? Why or why not?
4. Which word in English has got the maximum synonyms? How many?
5. Explain the following terms:
   - Homonyms
   - Antonyms
   - Eponyms
   - Homophones
6. What are the ways to enrich the vocabulary?
7. From the alternatives given below select the one that fits the most:
   (a) To make (oneself or person) aware of _____
   - (i) acknowledge
   - (ii) acquaint
   - (iii) admire
   - (iv) affinity
   (b) Person who is unable to pay his/her debts _____
   - (i) barbaric
   - (ii) benign
   - (iii) bankrupt
   - (iv) benevolent
   (c) Spoken or written in two languages _____
   - (i) bias
   - (ii) bilingual
   - (iii) biography
   - (iv) bewilder
   (d) Top surface of room _____
   - (i) charm
   - (ii) deal
   - (iii) deceit
   - (iv) ceiling
   (e) Pertaining to ancient (Greek and Roman) literature and art _____
   - (i) deceit
   - (ii) depth
   - (iii) classic
   - (iv) deter
   (f) Exact statement of meaning _____
   - (i) determine
   - (ii) dictionary
   - (iii) decency
   - (iv) definition
   (g) Shaking and vibration at the surface of the earth _____
   - (i) efficiency
   - (ii) disaster
   - (iii) fiscal
   - (iv) earthquake
   (h) Great rising and overflowing of water _____
   - (i) jogger
   - (ii) knob
   - (iii) floor
   - (iv) flood
   (i) To play for money by betting _____
   - (i) hamper
   - (ii) idiom
   - (iii) knock
   - (iv) gamble
   (j) The place where two or more roads or railway lines meet _____
   - (i) justice
   - (ii) junction
   - (iii) junk
   - (iv) majesty

(k) To uncover, to expose to view _____
    (i) discover   (iii) disclose
    (ii) disapprove   (iv) disobey

(l) The state of having a serious mental illness
   _____
    (i) majority   (iii) loyalty
    (ii) madness   (iv) mandatory

(m) To take part in an activity _____
    (i) payment   (iii) participate
    (ii) pause   (iv) pacify

(n) A measurement of the speed at which something happens _____
    (i) rural   (iii) rate
    (ii) routine   (iv) raise

(n) Connected with the practical use of machinery _____
    (i) tentative   (iii) technical
    (ii) terminate   (iv) tiresome

(o) Part of body between hips and ribs
   _____
    (i) warn   (iii) whim
    (ii) weary   (iv) waist

(p) Unhappy or showing unhappiness
   _____
    (i) scale   (iii) sad
    (ii) sale   (iv) salad

(q) State of being vacant _____
    (i) vapour   (iii) vacancy
    (ii) verbal   (iv) vague

(r) The limited number or amount of people or thing _____
    (i) quarter   (iii) quest
    (ii) quota   (iv) quiet

(s) To force somebody to do something
   _____
    (i) observe   (iii) oblige
    (ii) object   (iv) obvious

(t) A join made by tying together two pieces or ends of string _____
    (i) kit   (iii) knee
    (ii) knot   (iv) knob

8. Give one word for the following:
  (a) Person who collects coins
  (b) A person who does not believe in religion
  (c) The act of killing one's wife
  (d) A post without remuneration
  (e) Mania of stealing articles
  (f) Words different in meaning but similar in sound
  (g) Walking in sleep
  (h) An unexpected stroke of good luck
  (i) One who totally abstains from alcohol
  (j) A speech by the orator at the end of the play
  (k) To break off proceedings of the meeting for a time
  (l) The place where bricks are baked
  (m) Official in charge of museum
  (n) Allowance due to a wife from her husband from her separation
  (o) One who cannot be corrected
  (p) Lack of enough blood

9. Give the meaning of the following words:
  reticent, euphony, sybaritic, alleviate, denouement, ramshackle, bucolic, serene, deluge, boisterous, meander, myriad, vituperative, sidereal, bigwig, soliloquy, gist, enervate, fulsome, winsome, meretricious, nonchalant, noisome, serendipity, usurp, clandestine

10. Give American English equivalent of the following words:
  (a) Aerial   (f) Pram
  (b) Bonnet (of car) (g) Postbox
  (c) Handbag   (h) Candy floss
  (d) Lavatory   (i) Dustbin
  (e) Lift   (j) Chemist

11. Make the nouns from the following words:
  (a) Derive   (e) Hospitalize
  (b) Arrive   (f) Examine
  (c) Swim   (g) Write
  (d) Speaking

12. Identify the error in the statements and rewrite them:
  (a) The management committee has requested me not to take any action unless and until I don't see all the documents.
  (b) The manager of the bank was busy; so he/she asked them to come and see him between two to three in the afternoon.
  (c) The article should not exceed more than five hundred words.
  (d) He/she signed upon the application form.
  (e) The economic policy of the government is not quite in variance with the idea of the socialist pattern of society.
  (f) This is one of two types of communication receivers that is available for general use.

(g) Rural women congregated the mud roofs when the colourful procession passed through the street.

(h) The ability of plan, organize and coordinate work is all fundamental to working with deadline.

13. Add prefixes to the following words to produce their antonyms:

(a) apposite
(b) happy
(c) action
(d) equate
(e) able
(f) adroit
(g) digestion
(h) symmetrical
(i) principled
(j) ethical
(k) moral
(l) abate
(m) professional
(n) equivocate
(o) competent
(p) advantage
(q) disputable
(r) decisive
(s) comfort
(t) flexible
(u) behave
(v) educated
(w) exceptional
(x) sensitive

14. Add suffixes to the following words in consonance with the meanings given against them:

(a) gene _____: the science of heredity
(b) diction _____: lexicon
(c) gene _____: family tree
(d) guileless_____: naïve
(e) listless_____: tiredness
(f) feck_____: weak
(g) micro_____: integrated computer circuit
(h) flaw_____: immaculate
(i) klepto_____: irresistible desire to steal
(j) sinus_____: cold problem
(k) bio_____: study of living beings
(l) impure_____: state of being impure
(m) prefer_____: the act of choosing

15. Write two synonyms for the following words:

(a) amalgamation
(b) conflagration
(c) receptacle
(d) tycoon
(e) letter
(f) cliché
(g) terse
(h) vociferous
(i) taciturn
(j) appropriate

16. Choose an idiom that could be used in reply to the following.

(a) Where's Mark?
(b) My interview starts in ten minutes.
(c) Rachel's getting married!
(d) My life's a complete disaster!
(e) Paul says he's met George W. Bush.

(f) I'd better not get home late again tonight. It would be the third time this week.

17. Identify the pairs of homophones from the following clues.

Example: One word means a place for keeping aircraft; the other word means a shaped piece of wood, metal, etc. on which you can hang clothes.

Answer: Hangar/hanger.

(a) One word means simple; the other means an aircraft.
(b) One word means expected; the other word means condensed vapour.
(c) One word is nautical; the other is central to the body.
(d) One word means connections; the other is an animal.
(e) One word means an occasion; the other is a herb.
(f) One word means to hit; the other is a vegetable.
(g) One word means permitted; the other means audible.
(h) One word means a singer; the other means a sum of money.
(i) One word means an animal; the other means an undercover fighter.
(j) One word means kind; the other means searched for.
(k) One means excluded; the other means a poet.
(l) One word means a day; the other means a sweet.
(m) One word means pursued; the other means pure.
(n) One word means a woolly South American animal; the other means a Buddhist monk in Tibet or Mongolia.
(o) These are the names of two particular people; one is a macho man; the other is a poet.

18. Read each sentence and fill in the blank, choosing the correct homophone.

(a) Please _____ our sincere apologies. (accept, except)
(b) You will _____ your deposit if you cancel the order. (lose, loose)
(c) _____ going to the fair tomorrow morning. (Their, They're)

(d) Your opinion will not _____ my decision. (affect, effect)

(e) He is far _____ young to go by himself. (to, too, two)

(f) She _____ the capitals of every state. (new, knew)

(g) He's a man _____ opinion I respect. (who's, whose)

(h) They set _____ at dawn. (forth, fourth)

(i) New roads will link the _____ cities of the area. (principal, principle)

(j) I would _____ extreme caution. (advice, advise)

19. From the given alternatives, choose the one that is closest in meaning to the underlined word.

(a) No player can <u>flout</u> the rules and remain in the team. (boast about, disobey, make a mockery of)

(b) Developing <u>indigenous</u> technology is important to lead the nation to self-sufficiency. (native, intelligent, capitalistic)

(c) He/she felt <u>mortified</u> at his son's failure to pass the exam. (depressed, humiliated, unhappy)

(d) His <u>impeccable</u> taste and sense of humour have helped his career. (remarkable; unbelievable, flawless)

20. Match the words in column A with the phrases that best match their meanings in column B:

A		B	
(a)	conflagration	(i)	cut the part
(b)	receptacle	(ii)	list of names
(c)	biotic	(iii)	very huge
(d)	pedagogue	(iv)	the study of animals
(e)	fusion	(v)	maladroit
(f)	zoology	(vi)	of living beings
(g)	amputate	(vii)	scientist
(h)	gawky	(viii)	dais
(i)	boffin	(ix)	remove the internal organ
(j)	roster	(x)	teacher
(k)	rostrum	(xi)	container
(l)	behemoth	(xii)	an uncontrolled fire
(m)	eviscerate	(xiii)	combination of atoms
(n)	agenda	(xiv)	alphabetical list of sources such as books and articles
(o)	bibliography	(xv)	long research report
(p)	thesis	(xvi)	brief written communication
(q)	memo	(xvii)	the list of individual items to be discussed in a meeting

21. Write two synonyms for the following words:

amalgamation	cliché
conflagration	terse
receptacle	vociferous
tycoon	taciturn
letter	appropriate

22. Use gerund instead of infinitive in the following sentences:

(a) To become an engineer has been his long-cherished dream.

(b) To apply for the position of a technician needs at least three years' experience.

(c) To become a good teacher requires lot of knowledge and compassion.

(d) To answer the question, you need a lot of confidence.

(e) To participate in the seminar, you must have good preparation.

(f) To die is not indicative of victory.

(g) It is difficult to understand the nuances of delivery techniques.

(h) His main concern was to maintain eye contact with the audience.

23. Fill in the blanks with suitable comparative adjectives:

(a) Platinum is _____ (costly) than diamond.

(b) Sodium is _____ (much) reactive than gold.

(c) Pluto is the _____ (far) planet in the solar system.

(d) A very few classes are _____ (interest) than mathematics class.

(e) Zinc is _____ (malleable) than diamond.

(f) Travel by air is the _____ (fast) mode of transportation.

(g) Cast iron is _____ (durable) than aluminium.

(h) The summer here is _____ (long) than the winter.

(i) A day on Mars is slightly _____ (long) than a day on Earth.

(j) Cars are _____ (safe) than motorbikes.

24. Rewrite the following sentences transforming them into impersonal passive forms:

(a) You must take care to attend the classes regularly.

(b) We do the overhauling of the machines at fixed periods of time.

(c) They have opted for new methods of innovation in training.

(d) We have taken a lot of care to generate alternate methods of energy.

(e) We experience difficulties in operating this new lathe machine.

(f) We pass electric current through electrode.

(g) Coal miners extract millions of tons of coal every week.

(h) A skilled operator can carry out many operations on a lathe.

25. Fill in the blanks with suitable prepositions:

(a) What is the time _____ your watch?

(b) Is the flight _____ time?

(c) She was born _____ a hamlet called Pilani.

(d) Please be comfortable and sit _____ the armchair.

(e) Looking _____ the blackboard, he emphasized the proper use of audiovisual aids.

(f) This is the worst earthquake-hit area _____ the country.

(g) Ramesh comes to work by car, but I prefer to come _____ foot.

(h) He held his breath _____ seven minutes.

(i) I'll see you _____ home when I get there.

(j) He usually travels to Chennai _____ train.

26. Check the following sentences for subject–verb agreement:

(a) The number of jobs are increasing in the technology sector.

(b) Neither of the printers are working.

(c) She is one of those people who keeps calm in an emergency.

(d) Each of the runners were going at top speed.

27. Fill in the blanks with appropriate forms of the verbs given in the brackets:

(a) The sun _____ (rise) at 6:03 this morning.

(b) Listening is as important as _____ (speak) in a group discussion.

(c) Everything _____ (change) with time.

(d) The equipment _____ (begin) to give trouble before the guarantee expired.

(e) He joined us 12 years ago and he still _____ (work) for us.

(f) Go down this road, _____ (turn) left, and the road leads to the industrial area.

(g) She will be excellent as a receptionist as she _____ (possess) good manners.

(h) Cotton _____ (catch) fire easily.

(i) I will phone you when she _____ (come) back.

(j) Look at those black clouds. It _____ (rain).

(k) Perhaps we _____ (visit) Kolkata next week.

(l) By 2030, robots _____ (take over) many of the jobs that people do today.

(m) I'm sure she _____ (pass) the exam.

(n) The train _____ (arrive) at 5.00 pm.

(o) Did you think you _____ (see) me somewhere before?

(p) The principal _____ (want) to speak to you.

(q) She _____ (is) unconscious since morning.

(r) I know about that film because I _____ (watch) it twice.

28. Complete the following conditional sentences:

(a) If a person is making a professional oral presentation, _____.

(b) If the speaker and listener belong to different cultures and different values, _____.

(c) If you had prepared well, you _____.

(d) _____, the car will not start.

29. Pick out the noun phrases from the following sentences:

(a) Have you ever tried working with a scientific calculator?
(b) Troglodyte prefers staying in the caves.
(c) He dislikes having to punish his subordinates.
(d) Travelling in scorching heat gives me no pleasure.
(e) Working with an intellectual is a refreshing delight.
(f) Your doing such a thing boggles me.
(g) Early to bed is a good adage.
(h) To win an award is my dream.
(i) I tried to get the problem right.
(j) Did you enjoy writing this book?

30. Supply the noun phrase:
(a) Do you dream _____?
(b) _____ surprised me.
(c) I don't intend to _____.
(d) _____ is not simple.
(e) _____ seems dishonest.
(f) I hope _____.
(g) Lions like _____.
(h) She enjoys _____.
(i) I want _____.
(j) His mother promised _____.

31. Use conjunctions to fill in the blanks:
(a) He ran away _____ he was afraid.
(b) She is more intelligent _____ him.
(c) _____ you want it, it shall be done.
(d) Make haste, _____ you will be late.
(e) _____ he was industrious, I encouraged him.
(f) Blessed are the merciful, _____ they shall obtain mercy.
(g) _____ you are mistaken, _____ I am.
(h) I will stay _____ you return.
(i) Bread _____ milk is wholesome food.
(j) I hear _____ your brother is in Switzerland.

32. Join the following pairs of sentences with suitable conjunctions. Make necessary changes wherever required.
(a) My mother is well.
My father is ill.
(b) He is not a knave.
He is not a fool.
(c) You must be quiet.
You must leave the room.
(d) The pied piper played.
The children danced.
(e) I lost the prize.
I worked assiduously.

(f) We went to the dance show.
We did not get the seat.
(g) Soldiers fought for the country.
Soldiers died for the country.
(h) He remains cheerful.
He has been wounded.
(i) I have a cricket bat.
I have a set of stumps.
(j) The captive fell down on his knees.
The captive pleaded for mercy.

33. Put the correct form of words in the given blanks:

Noun	Adjective	Verb	Adverb
_____	able	enable	ably
Attraction	_____	attract	attractively
Friend	friendly	_____	*
Education	educational	educate	_____
Beauty	beautiful	_____	beautifully
Speed	speedy	speed	_____
Politeness	_____	*	politely
_____	popular	*	popularly
Thought	_____	think	*

34. Expand the given compound nouns so as to bring out their full meaning:

Commander-in-chief	Electromagnetism
Black bird	Clinical medicine
Blowlamp	Bank balance
Stepping stone	Banner headline
Standby	Artificial respiration
Peacock blue	

35. Make plural of the following compound nouns:

Coat-of-mail	Passer-by
Son-in-law	Man-of-war
Daughter-in-law	Stepdaughter
Maidservant	

36. Use proper articles in the sentences given below:
(a) You are _____ best player.
(b) _____ Ganga is _____ sacred river.
(c) The bellwether knows _____ way.
(d) Let us participate in _____ group discussion effectively.
(e) Mumbai is _____ very costly place to live in.
(f) Maldives is _____ island.
(g) Today _____ European came to my office.
(h) He returned after _____ hour.
(i) Japanese is _____ easy language.
(j) March is _____ third month of the year.

37. Adverbs are used as intensifiers. Show three different functions of adverbial intensifiers with the appropriate examples.

38. Select the correct form of the verb given below and write it in the space.

(i) One of the boys _____ punished.
   (a) was
   (b) were

(ii) Neither the children nor their mother _____ admitted.
   (a) were
   (b) was

(iii) All that glitters _____ not gold.
   (a) is
   (b) are

(iv) Two and two _____ four.
   (a) make
   (b) makes

(v) Mohan as well as his friend _____ guilty.
   (a) is
   (b) are

(vi) A hundred kilometres _____ good distance.
   (a) is
   (b) are

(vii) The general with all his soldiers _____ killed.
   (a) were
   (b) was

(viii) The poet and philosopher _____ my friend.
   (a) are
   (b) is

(ix) Peas and cheese _____ an expensive dish.
   (a) are
   (b) is

(x) Many a flower _____ born to blush unseen.

(a) is
(b) are

(xi) Two weeks _____ a long time.
   (a) is
   (b) are

(xii) Two third of the crop _____ ruined.
   (a) is
   (b) are

(xiii) Either your eyesight or your brakes _____ bad.
   (a) is
   (b) are

(xiv) The teacher or the student _____ wrong.
   (a) is
   (b) are

(xv) Each of the two candidates _____ paid his subscription.
   (a) have
   (b) has

(xvi) Either of these ideas _____ acceptable.
   (a) are
   (b) is

(xvii) The data _____ now ready for analysis.
   (a) are
   (b) is

(xviii) None of the alternatives _____ acceptable.
   (a) is
   (b) are

(xix) More than one car _____ damaged.
   (a) were
   (b) was

(xx) The public _____ tired of everyday strikes.
   (a) is
   (b) are

Index

Active voice 460
Adjectives 465
  Interrogative 467
  of demonstration 467
  of number 466
  of quality 466
  of quantity 466
Adverbs 469
  Intensifiers 470
  Two forms 469
Antonyms 437
Argumentative skills 107
Articles 475
  Definite 476
  Indefinite 475

Barriers to communication 22
  Interpersonal 27
  Intrapersonal 24
  Organizational 31
Basic sounds of English 68
  Oral and nasal 69
  Phonetic transcription 73
  Problem sounds 75
  Rules of pronunciation 74
  Voiced and unvoiced 69
  Vowels and consonants 69
Brochure 198
Bulletin 200
Business letters 359
  Claim letters 368
  Collection letters 360
  Cover letters 378
  Credit letters 359
  Letters of enquiry 366
  Order placement letters 368
  Sales letters 372

Categories of reports 276
  Analytical 277
  Event 277
  Formal and informal 282

  Individual and group 282
  Informative 277
  Long and short 281
  Oral and written 281
  Periodic and special 277
Chronemics 41
Communication networks 11
  Formal network models 11
  Informal network models 12
Collaborative writing 256
Communication styles 6
  Aggressive 6
  Assertive 6
  Passive 6
  Passive-aggressive 6
Comparison of adjectives 467
  Comparative degree 467
  Positive degree 467
  Superlative degrees 467
Comprehension techniques 207
  Author's viewpoint (inference) 209
  Non-verbal signals 207
  PQRST 215
  Predicting the content 213
  Punctuation 209
  Questions 211
  Reader anticipation 210
  Sequencing 226
  Skimming and scanning 207
  Structure of paragraphs 209
  Structure of the text 209
  Study skills 218
  Summarizing 210
  Typical questions 211
  Understanding discourse coherence 224
  Understanding the gist 214
Conditional sentences 462
Conferences 194
  Planning and preparation 196
  Procedure 197
  Significance 195
Confusables 438

Conjuctions 471
Conjunction classes 472
  Coordinating conjunctions 472
  Subordinating conjunctions 472
Controlling nervousness and stage fright 132
  On-camera techniques 134
  Strategies for reducing stage fright 133
  Visualization strategies 133
Conversations 101
  Argumentative skills 107
  Strategies for effectiveness 102
  Types 101
Creating indexes 258
Critical reading 221
Critical thinking 222
Cross-cultural variations 43
  Language 43
  Politics and law 44
  Religion and beliefs 43
  Social organization 44
  Technology 44
  Values and attitudes 43

Dialogue writing 108
Dialogue types 109

Effective Googling 223
Effective writing 238
  Collaborative writing 256
  Creating indexes 258
  Journal writing 253
  Right words and phrases 238
  Sentences 244
Emails 393
  Advantages 393
  Effectiveness and security 400
  Etiquette 396
  Limitations 394
  Style, structure, and content 394
Eponyms 442
Event report 277
Expert technical lecture 136

Formats of reports 282
  Letter 284
  Manuscript 282
  Memo 282
Future tense 458
  Future continuous 459
  Future perfect 460

Future perfect continuous 460
Simple future 458

Gerunds 449
Group communication 166
  Forms 167
Group discussions 170
  Cooperative atmosphere 172
  Discussing problems and solutions 171
  Intervention 175
  Persuasive strategies 173
  Reaching a decision 175
  Speaking 171
  Turn-taking strategies 173

Haptics 41
Homonyms 442
Homophones 442
Humour 132

Idioms 437
Infinitives 450
Information design and development 255
Interviews 139
  Objectives 139
  Types 140
Intonation 80

Job interviews 142
  Categories 142
  Preparing for interviews 146
  Process 149
  Stages 143
Journal writing 253

Letter writing 342
  Business letters 359
  Layout 349
  Principles 353
  Purpose 343
  Seven Cs 343
  Significance 343
  Structure 344
Listening 48
  Barriers 60
  Importance 49
  Meaning and art 48
  Types 58
Listening modes 55
  Active versus passive listening 56
  Global versus local listening 58

Media interviews 161
Meetings 183
  Duration 185
  Minutes 191
  Notice and agenda 185
  Preparation 184
  Procedure 189
  Purposes 184
  Time 185
  Venue and set-up 188
Memos 382
  Classification 384
  Purpose 384
  Structure 386
  Style 390
Messages 341
  Bad news 341
  General message 341
  Negative message 342
  Positive news 341
Mother tongue influence (MTI) 87

Newsletter 200
Noise 23
  Channel 23
  Semantic 23
Non-verbal communication 35
  Chronemics 41
  Haptics 41
  Kinesics 36
  Proxemics 40
Nouns 446
  Collective 446
  Common 446
  Compound 446
  Countable 446
  Proper 446
  Uncountable 446

One-word substitutes 438
Organizational GD 175
  Brainstorming 176
  Delphi technique 176
  Nominal group technique 176

Paralinguistic features 89
Passive voice 460
Past tense 457
  Past continuous 457
  Past perfect 458

Past perfect continuous 458
Simple future 458
Simple past 457
Perspectives in communication 14
  Environment 15
  Ethics and etiquette 15
  Language 14
  Past experiences 14
  Prejudices and feelings 14
  Visual perception 14
Phrasal verbs 443
  Phrasal prepositional verbs 445
  Prepositional verbs 444
  Particle verbs 444
Planning a letter 358
  Audience awareness 358
  Clarity of purpose 358
  Deciding the content 358
  Understanding the need for response 359
Poor listening 49
PQRST technique 215
Précis writing 265
  Guidelines 265
  Samples 267
  Steps 265
Prefixes 435
Prepositions 474
Presentations 114
  Nuances of delivery 125
  Outlining and structuring 118
  Planning 115
  Visual aids 134
Present tense 455
  Present continuous 455
  Present perfect 456
  Present perfect continuous 457
  Simple present 455
Press conferences 162
  Preparation 163
  Process 163
Prewriting 284
  Audience 284
  Interpreting information 289
  Making an outline 289
  Organizing the material 289
  Purpose and scope 284
  Sources of information 288
Principles of letter writing 353
  Clarity and conciseness 354
  Correctness and completeness 355

Courtesy and consideration 357
Positive approach 356
'You' attitude 353
Proposals 313
    Characteristics 315
    Definition 313
    Evaluation 320
    Purpose 313
    Structure 316
    Style and appearance 319
    Types 314
Proxemics 40
Public speaking 94

Quizzes and interjection 132

Reading comprehension 206
Reports 273
    Categories 276
    Characteristics 275
    Formats 282
    Importance 274
    Objectives 275
    Structure 292
    Types 304
Research paper 409
    Characteristics 409
    Components 409
Résumés 158
    Design and structure 159

Seminars 197
Sentence stress 78
Sociolinguistic competence 88
Speaking 68
    Articulation 91
    Clarity 85
    Confidence 84
    Fluency 86
    Mother tongue influence (MTI) 87
    Negotiation skills 96
    Pace/rate 90
    Paralinguistic features 89
    Pauses 92
    Pitch 91
    Pronunciation 91
    Quality 90
    Sociolinguistic competence 88
    Voice modulation 92
    Volume 90

Speaking and negotiation skills 96
Structure of reports 292
    Main text 300
    Prefatory parts 293
    Supplementary parts 303
Style matrix 7
Subject–verb agreement 452
Suffixes 435
Symposia 197
Synonyms 436

Technical communication 2
    Flow 9
    Importance 3
    Levels 7
    Objectives and characteristics 4
    Process 4
    Styles 6
Technical description 423
    Guidelines for writing 423
Techniques for good comprehension 207
Technology in communication 15
Tenses 454
Traits of a good listener 53
    Further contributions 55
    Non-evaluative 53
    Paraphrasing 53
    Reflecting hidden feelings 54
    Reflecting implications 54
    Responding non-verbally 55
Types of reports 304
    Feasibility 304
    Incident 304
    Introductory 304
    Laboratory test 305
    Marketing 304
    Progress 304
    Project 305

Word stress 76
Writing for the web 254
    Audience 255
    Clarity 255
    Effective style 256
    Formatting 256
    Proofreading 256
Writing the report 305
    First draft 305
    Revising, editing, and proofreading 306

About the Authors

Meenakshi Raman was formerly Professor and Head of the Department of Humanities and Social Sciences at Birla Institute of Technology and Science (BITS) Pilani. She has over 25 years of teaching experience, authored many books on communication skills, guided several PhD students, and was actively involved in administration at the Pilani and Goa campuses of BITS.

Sangeeta Sharma is a Professor in the Department of Humanities and Social Sciences, BITS Pilani. Her contributions include classroom delivery, corporate connect and training, writing books and research papers, and presenting papers at national and international conferences. She has published more than 25 books and 56 research papers in the area of gender studies, advertising, cultural studies, and communication in reputed journals. She is a licensed trainer for Springboard Development Program (UK).

Related Titles

PERSONALITY DEVELOPMENT AND SOFT SKILLS 2E [9780199459742]

Barun Mitra, *Formerly Professor of English, Indian Institute of Technology Kharagpur*

The book aims to provide crucial insights into various facets of developing one's personality as well as to improve written, verbal, and non-verbal communication skills. Special attention has been paid to the specific needs of a job aspirant, such as writing of effective CVs, participation in group discussions, tackling job interviews, and to hone one's public speaking and speed-reading skills.

Key Features
- Provides inputs on avoiding common mistakes in speaking English
- Provides several case studies, examples, and illustrations to elucidate the concepts discussed
- Contains several classroom-based activities for students to develop their personalities and enhance their soft skills

COMMUNICATION SKILLS 2E [9780199457069]

Sanjay Kumar, *Consultant* &
Pushp Lata, *Associate Professor of English, Department of Humanities and Social Sciences, BITS, Pilani*

Communication Skills is a textbook designed for undergraduate students of engineering for a two-semester course in technical English to develop their linguistic and communicative competence. The aim of the book is to help students acquire the ability to speak and write English effectively in real-life situations.

Key Features
- Covers English grammar in detail with plenty of examples, practice tests, and exercises
- Contains numerous samples of business letters, reports, proposals, paragraphs, essays, and email correspondence
- Provides tips on editing and proofreading
- Includes interesting illustrations in the text and a Wisewell Quips series at the end of the chapters that emphasize the nuances of the English language

EFFECTIVE TECHNICAL COMMUNICATION: A GUIDE FOR SCIENTISTS AND ENGINEERS [9780195682915]

Barun Mitra, *Formerly Professor of English, Indian Institute of Technology Kharagpur*

Effective Technical Communication is designed to serve as a practical guide and useful resource for scientists, engineers, and researchers. It addresses the need of practitioners engaged in the exchange of technical information to effectively share their ideas with, and make an impact on their peers.

Key Features
- Acquaints readers with key communication techniques
- Includes all forms of communication including technical papers, reports, and proposals
- Provides illustrative examples of concepts and forms of communication
- Explores emerging trends in communication, including email and voicemail

Online English Learning & Assessment Solution

Oxford Achiever is a web-based English learning and assessment system developed with the aim of building students' language skills and enhancing employability along an individualized, self-paced learning path

Diagnostic Test

Practice Exercises

Level Promotion

Progress Charts

Learning Profile

36 Progressive Levels

6 Interconnected Skills

Listening, Speaking, Reading, Writing, Grammar, & Vocabulary

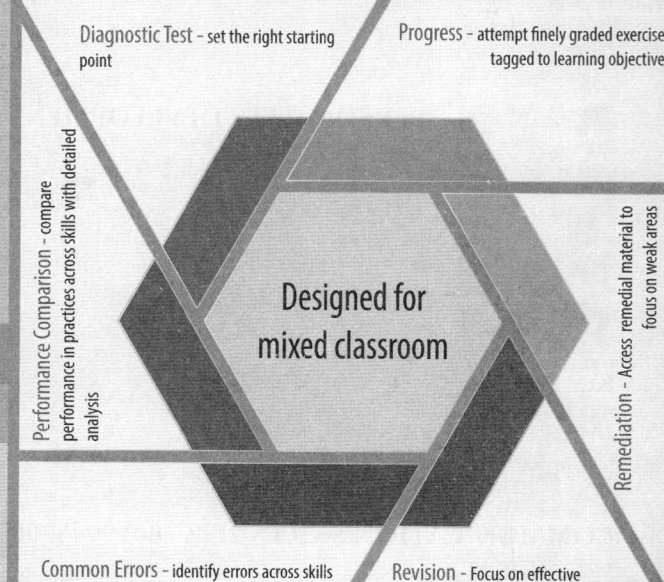

Diagnostic Test - set the right starting point

Progress - attempt finely graded exercise tagged to learning objective

Performance Comparison - compare performance in practices across skills with detailed analysis

Designed for mixed classroom

Remediation - Access remedial material to focus on weak areas

Common Errors - identify errors across skills

Revision - Focus on effective learning derived from previous levels

Students start with the Diagnostic Test, a digital assessment that assigns a level in each skill thereby creating a Personalised Learning Path for each student

Benefits

- Access - anytime, anywhere
- Excellence in transactional English
- Development of self-learning aptitude

STUDENT

- Continuous evaluation
- Automated reports
- Easy identification of problem areas

FACULTY

- Premium OUP offering
- Minimal resource requirements
- Measurable impact

INSTITUTE

For demonstration and queries, please contact

RVKS Raju
Senior Zonal Manager, Sales
☎ +91 9866191820 ✉ rvks.raju@oup.com

Ashes Saha
Senior Zonal Manager, Sales
☎ +91 9830708011 ✉ ashes.saha@oup.com